S0-BQM-075

Progress in Radio-Oncology II

Progress in Radio-Oncology II

Editors

K. H. Kärcher, M.D.
H. D. Kogelnik, M.D.
G. Reinartz, M.D.

University Clinic for Radiotherapy and Radiobiology of Vienna
Vienna, Austria

Raven Press ■ New York

Raven Press, 1140 Avenue of the Americas, New York, New York 10036

Papers or parts thereof have been used as camera-ready copy as submitted by the authors whenever possible; when retyped, they have been edited by the editorial staff only to the extent considered necessary for the assistance of an international readership. The views expressed and the general style adopted remain, however, the responsibility of the named authors. Great care has been taken to maintain the accuracy of the information contained in the volume. However, neither Raven Press nor the editors can be held responsible for errors or for any consequences arising from the use of information contained herein.

The use in this book of particular designations of countries or territories does not imply any judgement by the publisher or editors as to the legal status of such countries or territories, of their authorities or institutions or of the delimitation of their boundaries.

Some of the names of products referred to in this book may be registered trademarks or proprietary names, although specific reference to this fact may not be made: however, the use of a name without designation is not to be construed as a representation by the publisher or editors that it is in the public domain. In addition, the mention of specific companies or of their products or proprietary names does not imply any endorsement or recommendation on the part of the publisher or editors.

Authors were themselves responsible for obtaining the necessary permission to reproduce copyright material from other sources. With respect to the publisher's copyright, material appearing in this book prepared by individuals as part of their official duties as government employees is only covered by this copyright to the extent permitted by the appropriate national regulations.

Library of Congress Cataloging in Publication Data
Main entry under title:

Progress in radio-oncology II.

Papers presented at the Second International Meeting on Progress in Radio-Oncology, Baden near Vienna,
Austria, 1981, sponsored by the Austrian/International
Club for Radio-Oncologists.
Includes bibliographies and index.
1. Cancer—Radiotherapy—Congresses. I. Kärcher,
K.-H. (Karl-Heinz), 1923– . II. Kogelnik, H. D.
III. Reinartz, G. IV. International Meeting on Progress
in Radio-Oncology (2nd : 1981 : Baden, Austria)
V. Austrian/International Club for Radio-Oncologists.
[DNLM: 1. Neoplasms—Radiotherapy—Congresses. QZ 269
P964 1981]
RC271.R3P758 1982 616.99'40642 81-40545
ISBN 0-89004-783-9

Preface

The Austrian/International Club for Radio-Oncologists was founded *inter alia* to promote the exchange of information among the international community of radio-oncologists and radiobiologists, as well as to provide Austrian specialists in all medical disciplines a chance to exchange views with leading protagonists of clinical and experimental radiotherapy through the informal discussion sessions linked to the triennial international meetings. The chapters presented herein are derived from the Club's Second International Meeting on Progress in Radio-Oncology (Baden near Vienna, Austria, 1981), which was principally devoted to many detailed presentations of clinical, biological, and physics data relating to four main topics in radiotherapy—particle-beam therapy, radiosensitizers and other "radiomodifying" drugs, the result of altering fractionation schedules, and hyperthermia as an adjuvant therapy.

Evidence has continued to accumulate that neutrons show a therapeutic advantage for particular sites and types of cancer. Various therapy protocols, including the "shrinking field" technique, that have been devised to make use of and allow for the special radiobiological characteristics of these particles are examined. However, the presentation of conflicting evidence alerts us to the many unanswered questions surrounding the role of neutron therapy in cancer care. It also becomes clear that the intercomparison of results is currently very difficult due to the physical differences between the neutron beams at the approximately 20 centers around the world employing neutrons clinically. It is hoped that the newer neutron generators, with better beam definition and a lower gamma component, will provide data more amenable to analysis without interference from technical artifacts, making it possible to make meaningful comparisons with photon-beam therapies. Results with pion beams and heavy charged particles are also considered.

New data from studies with hypoxic cell sensitizers are presented, and extensive trials are yielding encouraging results. Consideration of the sensitizer mode of action has led to the formulation of principles for conducting an organized search for new compounds that may have improved action and/or reduced neurotoxicity. The role of sulphydryl-containing radioprotectors is also discussed.

Various, at present unconventional, fractionation regimes are examined, both from the point of view of clinical practice and that of radiobiology. Experimental evidence is presented that hyperfractionation (which in the limit approaches low-dose continuous irradiation) can improve the therapeutic ratio and shows better normal-tissue tolerance. The pilot study conducted by the Radiation Therapy Oncology Group (RTOG) represents the first major trial of this treatment modality and interesting results were presented.

The value of hyperthermia as a single or adjuvant therapy against cancer has been the subject of much argument for nearly 100 years. It is only recently, however, that hyperthermia has begun to be investigated under controlled conditions. The technical problems, especially that of temperature measurement, are highlighted, and the difficulty of interpreting the data is emphasized. However, the results achieved with patients with advanced cancers are very impressive, and work is now concentrating on standardizing equipment and techniques as a preliminary to starting large-scale trials.

Among the 55 chapters presented herein are two introductory chapters, one dealing with patterns and quality of cancer care, and the other with the association of less-than-radical surgery with less-than-radical radiotherapy. The advantages of surgically removing the mass of hypoxic tumor, leaving radiation to eradicate the subclinical disease, are examined in detail, revealing a treatment approach that has yet to be given the attention it deserves.

Some 100 participants attended the presentations, representing the work of some 165 clinicians, radiotherapists, and radiobiologists from 17 countries. This publication makes the full papers available to a wider audience, and the wealth of data and information will ensure its continuing value as a reference source.

In addition, Professor Luther Brady kindly provided a closing paper that summarizes the points that aroused a particular measure of interest during the presentations and the ensuing round-table discussions.

It is my hope that the optimum patient-sparing treatments may, for many types of cancer, derive from less-than-radical surgery and radiotherapy, each used fully as part of a multidisciplinary approach.

Karl-Heinz Kärcher

Acknowledgments

I should like to express my gratitude to all persons and institutions that supported the undertaking, especially acknowledging the financial assistance provided by the "Bundesministerium für Gesundheit und Umweltschutz," Austria; the Edinburgh E. A. R. Congress Educational Trust, Edinburgh; Gerot-Pharmazeutika GmbH, Vienna; and Hoffmann-La Roche S. A., Basel.

Further, my sincere thanks are due to the three Scientific Secretaries of the meeting, H. Dieter Kogelnik, Gisela Reinartz, and Tibor Szepesi of the University Clinic for Radiotherapy and Radiobiology, Vienna, to Ingrid Beck, the Administrative Editor for these Proceedings, and her team, as well as to E. Robert A. Beck, who was always on hand to advise with the diverse aspects of publication.

Karl-Heinz Kärcher

Contents

Radiosensitizers

Altered Fractionation

Contributors

G. E. Adams
Radiobiology Unit, Physics Department
Institute of Cancer Research
Sutton, Surrey, United Kingdom

R. G. Ahier
Medical Research Council Cyclotron Unit
Hammersmith Hospital
London W12 OHS, United Kingdom

A. S. M. Al-Abdulla
Department of Radiology
Division of Radiation Oncology
The University of Texas Medical Branch at Galveston
Galveston, Texas 77550

K. K. Ang
Department of Radiotherapy
Academic Hospital St. Rafaël
B-3000 Leuven, Belgium

T. Arai
Division of Clinical Research
National Institute of Radiological Sciences
Anagawa, Chiba-shi, Japan

G. Arcangeli
Istituto Medico e di Ricerca Scientifica
Clinica Villa Flamina
I-00191 Rome, Italy

S. J. Arnott
Department of Clinical Oncology
University of Edinburg
Western General Hospital
Edinburgh EH4 2XU, United Kingdom

R. Arriagada
Institut Gustave-Roussy
F-94800 Villejuif (Val-de-Marne), France

Reba Baker
Department of Radiation Oncology
Indiana University School of Medicine
Indianapolis, Indiana 46223

V. Bär
Department of Oncoradiology
Postgraduate Medical School
Ráth György-u. 7/
9H-1122 Budapest, Hungary and
National Oncological Institute
Budapest, Hungary

J. P. Bataïni
Section Medicale et Hospitalière
Institut Curie
26 Rue d'Ulm
F-75231 Paris Cedex 05, France

J. J. Battermann
Antoni van Leeuwenhoek Hospital
Plesmanlaan 121
NL-1066 CX Amsterdam, The Netherlands

J. Bernier
Section Medicale et Hospitalière
Institut Curie
26 Rue d'Ulm
F-75231 Paris Cedex, France

D. van Beuningen
Institut fur Medizinische Strahlenphysik und Strahlenbiologie
Universitätskinikum Essen
Hufelandstrasse 55
D-4300 Essen, Federal Republic of Germany

H. I. Bicher
Henry Ford Hospital
2799 West Grand Boulevard
Detroit, Michigan 48202

H. Blattmann
Swiss Institute for Nuclear Research
CH-5234 Villigen, Switzerland

N. M. Bleehen
University Department and Medical Research Council Unit of Clinical Oncology and Radiotherapeutics
The University of Cambridge School of Clinical Medicine
Addenbrooke's Hospital
Hills Road
Cambridge CB2 2QQ, United Kingdom

H. J. G. Bloom
Royal Marsden Hospital and
Institute of Cancer Research
Downs Road
Sutton, Surrey SM2 5PT, United Kingdom

H. Börner
Institut Laue-Langevin
Grenoble, France

L. W. Brady
Department of Radiation Therapy and Nuclear Medicine
Hahnemann Medical College and Hospital of Philadelphia
230 North Broad Street
Philadelphia, Pennsylvania 19102

F. Brassow
Radiotherapy Department
University Hospital Hamburg-Eppendorf
Martinistrasse 52
D-2000 Hamburg, Federal Republic of Germany

F. Brunin
Section Medical et Hospitalière
Institut Curie
26 Rue d'Ulm
F-75231 Paris Cedex, France

R. D. Bugden
Royal Marsden Hospital and Institute of Cancer Research
Downs Road
Sutton, Surrey SM2 5PT, United Kingdom

Janice Burgin
Medical Research Council Cyclotron Unit
Hammersmith Hospital
Ducane Road
London W12 OHS, United Kingdom

T. Burke
Roswell Park Memorial Institute
666 Elm Street
Buffalo, New York 14263

S. E. Bush
Cancer Research and Treatment Center and
Department of Radiology
School of Medicine
University of New Mexico
Albuquerque, New Mexico 87131

J. Carlsson
Department of Physical Biology
The Gustaf Werner Institute
Box 531
S-751 Uppsala, Sweden

J. R. Castro
Mount Zion Hospital and Medical Center
PO Box 7921
San Francisco, California 94120

Mary Catterall
Medical Research Council Cyclotron Unit
Hammersmith Hospital
Ducane Road
London W12 OHS, United Kingdom

G. T. Y. Chen
Radiotherapy Section of Donner Laboratory
University of California Lawrence Berkeley Laboratory
Berkeley, California and University of California San Francisco School of Medicine
San Francisco, California 94143

C. Chenal
Hôpital Pitié la Salpétrière
Paris, France

M. Collier
Radiotherapy Section of Donner Laboratory
University of California Lawrence Berkeley Laboratory
Berkeley, California and
University of California San Francisco School of Medicine
San Francisco, California 94143

P. J. Conroy
University of Rochester
School of Medicine and Dentistry
Cancer Center
Rochester, New York 14642

I. Cordt
Swiss Institute for Nuclear Research
CH-5234 Villigen, Switzerland

J. M. Cosset
Institut Gustave-Roussy
F-94800 Villejuif (Val-de-Marne), France

J. Crawford
Swiss Institute for Nuclear Research
CH-5234 Villigen, Switzerland

S. Darai
Institute of Nuclear Medicine
German Cancer Research Center
D-6900 Heidelberg 1, Federal Republic of Germany

F. Dietzel
Radiologisches Institut der Städtischen Krankenanstalten Bayreuth
D-8580 Bayreuth, Federal Republic of Germany

S. Dische
Marie Curie Research Wing for Oncology
Regional Radiotherapy Centre
Mount Vernon Hospital
Northwood, Middlesex HA6 2RN, United Kingdom

W. Duncan
Department of Clinical Oncology
University of Edinburgh
Western General Hospital
Edinburgh EH4 2XU, United Kingdom

J. Dutreix
Institut Gustave-Roussy
F-94800 Villejuif (Val-de-Marne), France

Margareta Edgren
Department of Tumor Biology II
Karolinska Institute Medical School
S-104 01 Stockholm, Sweden

M. M. Elkind
Division of Biological and Medical Research
Argonne National Laboratory
Argonne, Illinois

R. D. Errington
Medical Research Council Cyclotron Unit
Hammersmith Hospital
London W12 OHS, United Kingdom

F. Eschwege
Institut Gustave-Roussy
F-94800 Villejuif (Val-de-Marne), France

C. F. von Essen
Swiss Institute for Nuclear Research
CH-5234 Villigen, Switzerland

L. E. Feinendegen
Institute of Medicine
Kernforschungsanlage Jülich GmbH
PO Box 1913
D-5170 Jülich, Federal Republic of Germany

S. B. Field
Medical Research Council Cyclotron Unit
Hammersmith Hospital
Ducane Road
London W12 OHS, United Kingdom

G. H. Fletcher
Division of Radiotherapy
The University of Texas M.D. Anderson Hospital and Tumor Institute
Texas Medical Center
Houston, Texas 77030

J. Forsberg
Department of Surgery
Akademiska Sjukhuset
Uppsala, Sweden

A. Fourcy
CEA Centre d'Etudes Nucléaires de Grenoble
Grenoble, France

J. F. Fowler
Gray Laboratory of the Cancer Research Campaign
Mount Vernon Hospital
Northwood, Middlesex HA6 2RN, United Kingdom

H. D. Franke
Radiotherapy Department
University Hospital Hamburg-Eppendorf
Martinistrasse 52
D-2000 Hamburg, Germany

J. C. Goffin
Centre Hospitalier de Tivoli
La Louvière, Belgium

T. W. Griffin
Department of Radiation Oncology
University Hospital
University of Washington
Seattle, Washington 98195

B. Grundei
Radiologisches Institut der Stádtischen Krankenanstalten Bayreuth
D-8580 Bayreuth, Federal Republic of Germany

J. Gueulette
UCL-Cliniques Universitaires St. Luc
B-1200 Brussels, Belgium

P. H. Gutin
School of Medicine
University of California
San Francisco, California 94143

A. Hess
Radiotherapy Department
University Hospital Hamburg-Eppendorf
D-2000 Hamburg, Federal Republic of Germany

L. R. Holsti
Department of Radiotherapy and Oncology
University Central Hospital
Helsinki, Finland

J. A. G. Holt
Radiotherapy and Oncology Centre
190 Cambridge Street
Wembley, Western Australia 6014, Australia

J. C. Horiot
Centre G. F. Leclerc
Dijon, France

N. B. Hornback
Department of Radiation Oncology
Indiana University School of Medicine
Indianapolis, Indiana 46223

Y. Hosobuchi
School of Medicine
University of California
San Francisco, California 94143

K. H. Höver
Institute of Nuclear Medicine
German Cancer Research Center
D-6900 Heidelberg 1, Federal Republic of Germany

S. Hurst
School of Medicine
University of California
San Francisco, California 94143

D. H. Hussey
Division of Radiotherapy
The University of Texas M.D. Anderson Hospital and Tumor Institute
Texas Medical Center
Houston, Texas 77030

P. Jakobsson
Department of Oncology
University of Uppsala
Uppsala, Sweden

C. Jaulerry
Section Medicale et Hospitalière
Institut Curie
F-75231 Paris Cedex, France

R. H. Jesse
Department of Head and Neck Surgery
The University of Texas M.D. Anderson Hospital and Tumor Institute
Houston, Texas 77030

R. Johnson
Roswell Park Memorial Institute
666 Elm Street
Buffalo, New York 14263

K. H. Kärcher
University Clinic for Radiotherapy and Radiobiology of Vienna
A-1090 Vienna, Austria

A. B. M. F. Karim
Department of Radiotherapy
Academic Hospital
Free University
NL-1007 MB Amsterdam, The Netherlands

K. Kawashima
Division of Clinical Research
National Institute of Radiological Sciences
9-1, 4-chome
Anagawa, Chiba-shi, Japan

H. D. Kogelnik
University Clinic for Radiotherapy and Radiobiology
Vienna
Alser Strasse 4
A-1090 Vienna, Austria

S. Kramer
Department of Radiation Therapy and Nuclear Medicine
Jefferson Medical College
Thomas Jefferson University
Philadelphia, Pennsylvania 19107

R. Krishnamsetty
Roswell Park Memorial Institute
666 Elm Street
Buffalo, New York 14263

J. Kummermehr
Abteilung Strahlenbiologie
Gesellschaft fur Strahlen- und Umweltforschung mbH
Ingolstädter Landstrasse 1
D-8042 Neuherberg, Federal Republic of Germany

Sharon Lamb
School of Medicine
University of California
San Francisco, California 94143

G. E. Laramore
Department of Radiation Oncology
University Hospital
University of Washington
Seattle, Washington 98195

B. Larsson
Department of Physical Biology
The Gustaf Werner Institute
Box 531
S-751 Uppsala, Sweden

Margaret Laverick
Richard Dimbleby Department of Cancer Research
St. Thomas's Hospital Medical School
London SE1 7EH, United Kingdom

M. P. Law
Medical Research Council Cyclotron Unit
Hammersmith Hospital
Ducane Road
London W12 OHS, United Kingdom

W. Lierse
Institute of Anatomy
Hamburg University
Hamburg, Federal Republic of Germany

G. Linhart
Abteilung Strahlenbiologie und Strahlenschutz
Zentrum für Radiologie
Universität Giessen
Giessen, Federal Republic of Germany

B. Littbrand
Department of Oncology
University of Umeå
S-901 85 Umeå, Sweden

J. B. Little
Department of Radiation Therapy
Joint Center for Radiation Therapy
Harvard Medical School
Boston, Massachusetts 02114 and
Laboratory of Radiobiology
Department of Physiology
Harvard School of Public Health
Boston, Massachusetts

J. T. Lyman
Radiotherapy Section of Donner Laboratory
University of California Lawrence Berkeley Laboratory
Berkeley, California and
University of California San Francisco School of Medicine
San Francisco, California 94143

A. R. Mackay
School of Medicine
University of California
San Francisco, California 94143

H. Madoc-Jones
Department of Therapeutic Radiology
Tufts-New England Medical Center
171 Harrison Avenue
Boston, Massachusetts 02111

A. Malcolm
Department of Radiation Therapy
Joint Center for Radiation Therapy
Harvard Medical School
Boston, Massachusetts 02114 and
Laboratory of Radiobiology
Department of Physiology
Harvard School of Public Health
Boston, Massachusetts

M. H. Maor
Division of Radiotherapy
The University of Texas M.D. Anderson Hospital and Tumor Institute
Texas Medical Center
6723 Bertner Avenue
Houston, Texas 77030

Carol Marshall
Department of Radiation Oncology
Indiana University School of Medicine
1100 West Michigan Street
Indianapolis, Indiana 46223

A. Michalowski
Medical Research Council Cyclotron Unit
Hammersmith Hospital
Ducane Road
London W12 OHS, United Kingdom

M. Molls
Institut fur Medizinische Strahlenphysik und Strahlenbiologie
Universitätsklinikum Essen
Hufelandstrasse 55
D-4300 Essen, Federal Republic of Germany

Jane E. Morgan
University Department and
Medical Research Council Unit of Clinical Oncology and Radiotherapeutics
The University of Cambridge School of Clinical Medicine
Addenbrooke's Hospital
Hills Road
Cambridge CB2 2QQ, United Kingdom

K. Morita
Department of Radiation Therapy
Aichi Cancer Center
81-1159 Kanokoden
Tashiro-cho, Chikusa-ku
Nagoya 464, Japan

S. Morita
Division of Clinical Research
National Institute of Radiological Sciences
9-1, 4-chome
Anagawa, Chiba-shi, Japan

C. C. Morris
Medical Research Council Cyclotron Unit
Hammersmith Hospital
Ducane Road
London W12 OHS, United Kingdom

H. Mühlensiepen
Institute of Medicine
Kernforschungsanlage Jülich GmbH
PO Box 1913
D-5170 Jülich, Federal Republic of Germany

F. Mundinger
Abteilung Stereotaxie und Neuronuklearmedizin
Neurochirurgische Universitäts-klinick
Klinikum der Albert-Ludwigs-Universität
Hugstetter Strasse 55
D-7800 Freiburg, Federal Republic of Germany

Y. Nakamura
Division of Clinical Research
National Institute of Radiological Sciences
9-1, 4-chome
Anagawa, Chiba-shi, Japan

A. H. W. Nias
Richard Dimbleby Department of Cancer Research
St. Thomas's Hospital Medical School
London SE1 7EH, United Kingdom

K. H. Njo
Department of Radiotherapy
Academic Hospital
Free University
De Boelelaan 1117
Postbus 7057
NL-1007 MB Amsterdam, The Netherlands

G. Notter
Sahlgrenska Sjukhuset
Göteborg, Sweden

Berkeley, California and
University of California San Francisco School of Medicine
San Francisco, California 94143

Y. Obata
Department of Radiation Therapy
Aichi Cancer Center
81-1159 Kanokoden
Tashiro-cho, Chikusa-ku
Nagoya 464, Japan

J. Overgaard
Radiumstationen
The Institute of Cancer Research and Department of Oncology and Radiotherapy
Nörrebrogade 44
DK-8000 Aarhus C, Denmark

Alison Pannell
Cancer Research and Treatment Center
University of New Mexico
Albuquerque, New Mexico 87131

E. Pedroni
Swiss Institute for Nuclear Research
CH-5234 Villigen, Switzerland

H. van Peperzeel
Academic Hospital
Utrecht, The Netherlands

Ch. Perret
Swiss Institute for Nuclear Research
Ch-5234 Villigen, Switzerland

Lester J. Peters
Institute of Oncology and Radiotherapy
Prince of Wales Hospital
University of New South Wales
High and Avoca Streets
Randwick, N.S.W. 2031, Australia

T. L. Phillips
Radiotherapy Section of Donner Laboratory
University of California Lawrence Berkeley Laboratory
Berkeley, California and Department of Radiation Oncology
University of California San Francisco School of Medicine
San Francisco, California 94143

S. Pitluck
Radiotherapy Section of Donner Laboratory
University of California Lawrence Berkeley Laboratory

Z. Pollak
Department of Oncoradiology
Postgraduate Medical School
Ráth György-u.7/9
Budapest, Hungary and
National Oncological Institute
H-1122 Budapest, Hungary

R. Porschen
Institute of Medicine
Kernforschungsanlage Jülich GmbH
PO Box 1913
D-5170 Jülich, Federal Republic of Germany

W. Porschen
Institute of Medicine
Kernforschungsanlage Jülich GmbH
PO Box 1913
D-5170 Jülich, Federal Republic of Germany

J. M. Quivey
Radiotherapy Section of Donner Laboratory
University of California Lawrence Berkeley Laboratory
Berkeley, California and
University of California San Francisco School of Medicine
San Francisco, California 94117

J. Raps
Middelheim Ziekenhuis
Antwerp, Belgium

A. Resouly
St. Mary's General Hospital
Portsmouth PO3 6AD, United Kingdom

L. Révész
Department of Tumor Biology II
Karolinska Institute Medical School
S-104 01 Stockholm, Sweden

I. L. Rodé
Department of Oncoradiology
Postgraduate Medical School
Ráth György-u. 7/9
H-1122 Budapest, Hungary

M. Salmo
Department of Radiotherapy and Oncology
University Central Hospital
Helsinki, Finland

M. Salzmann
Swiss Institute for Nuclear Research
CH-5234 Villigen, Switzerland

J. J. Santoro
Department of Therapeutic Radiology
Tufts-New England Medical Center
Boston, Massachusetts 02111

M. I. Saunders
Marie Curie Research Wing for Oncology
Regional Radiotherapy Centre
Mount Vernon Hospital
Northwood, Middlesex HA6 2RN, United Kingdom

W. Saunders
Radiotherapy Section of Donner Laboratory
University of California
Lawrence Berkeley Laboratory
Berkeley, California and
University of California San Francisco School of Medicine
San Francisco, California 94117

J. P. Saxton
Division of Radiotherapy
The University of Texas M.D. Anderson Hospital and Tumor Institute
Texas Medical Center
Houston, Texas 77030

K. E. Scheer
Institute of Nuclear Medicine
German Cancer Research Center
D-6900 Heidelberg 1, Federal Republic of Germany

E. Scherer
Clinic of Radiotherapy
University of Essen
Essen, Federal Republic of Germany

G. Schmitt
Department of Radiation Oncology
Alfried Krupp von Bohlen und Halbach Hospital
D-4300 Essen 1, Federal Republic of Germany

K. Schnabel
Institute of Nuclear Medicine
German Cancer Research Center
D-6900 Heidelberg 1, Federal Republic of Germany

A. V. Schratter-Sehn
University Clinic for Radiotherapy and Radiobiology of Vienna
A-1090 Vienna, Austria

E. van der Schueren
Department of Radiotherapy
Academic Hospital St. Rafaël
B-3000 Leuven, Belgium

R. Scott
Roswell Park Memorial Institute
Buffalo, New York 14263

P. W. Sheldon
Radiobiology Unit, Physics Department
Institute of Cancer Research
Sutton, Surrey, United Kingdom

R. E. Shupe
Department of Radiation Oncology
Indiana University School of Medicine
Indianapolis, Indiana 46223

D. W. Siemann
University of Rochester
School of Medicine and Dentistry
Cancer Center
Rochester, New York 14642

A. R. Smith
Cancer Research and Treatment Center and
Department of Radiology
School of Medicine
University of New Mexico
Albuquerque, New Mexico 87131

Nancy Smith
Cancer Research and Treatment Center
University of New Mexico
Albuquerque, New Mexico 87131

P. M. Stafford
Cancer Research and Treatment Center
University of New Mexico
Albuquerque, New Mexico 87131

R. Stark
Cancer Research and Treatment Center
University of New Mexico
Albuquerque, New Mexico 87131

E. S. Sternick
Department of Therapeutic Radiology
Tufts-New England Medical Center
Boston, Massachusetts 02111

I. J. Stratford
Radiobiology Unit, Physics Department
Institute of Cancer Research
Sutton, Surrey, United Kingdom

C. Streffer
Institut für Medizinische Strahlenphysik und Strahlenbiologie
Universitätsklinikum Essen
D-4300 Essen, Federal Republic of Germany

J. Subjeck
Roswell Park Memorial Institute
Buffalo, New York 14263

R. M. Sutherland
University of Rochester
School of Medicine and Dentistry
Cancer Center
Rochester, New York 14642

V. H. J. Svoboda
St. Mary's General Hospital
Portsmouth PO3 6AD, United Kingdom

T. Szepesi
University Clinic for Radiotherapy and Radiobiology of Vienna
A-1090 Vienna, Austria

Irena Szumiel
Department of Radiobiology and Health Protection
Institute of Nuclear Research
Warsaw, Poland

Howard D. Thames, Jr.
Department of Biomathematics
University of Texas M.D. Anderson Hospital and Tumor Institute
Houston, Texas 77030

M. Thellier
Laboratorie de Nutrition Minérale
Mont-Saint-Aignan, France

C. Tobias
Radiotherapy Section of Donner Laboratory
University of California Lawrence Berkeley Laboratory
Berkeley, California and University of California San Francisco School of Medicine
San Francisco, California 94117

E. L. Travis
Radiation Oncology Branch of the National Cancer Institute
Bethesda, Maryland 20014

K.-R. Trott
Strahlenbiologisches Institut
Universität München,
Munich, Federal Republic of Germany

H. Tsunemoto
Division of Clinical Research
National Institute of Radiological Sciences
Anagawa, Chiba-shi, Japan

M. Tubiana
Institut Gustave-Roussy
Rue Camille Desmoulins
F-94800 Villejuif (Val-de-Marne), France

D. F. H. Wallach
Department of Therapeutic Radiology
Tufts-New England Medical Center
171 Harrison Avenue
Boston, Massachusetts 02111

A. Wambersie
Unité de Neutronthérapie Experimentale
Faculté de Médecine
ULC-Cliniques Universitaires St. Luc
Avenue Hippocrate 54
B-1200 Brussels, Belgium

W. M. Wara
School of Medicine
University of California
San Francisco, California 94143

T. H. Wasserman
Division of Radiation Oncology
Mallinckrodt Institute of Radiology
Washington University School of Medicine
St. Louis, Missouri 61110

K. Watai
Department of Radiation Therapy
Aichi Cancer Center
81-1159 Kanokoden
Tashiro-cho, Chikusa-ku
Nagoya 464, Japan

M. Watanabe
Department of Radiation Therapy
Aichi Cancer Center
81-1159 Kanokoden
Tashiro-cho, Chikusa-ku
Nagoya 464, Japan

K. A. Weaver
School of Medicine
University of California
San Francisco, California 94143

R. R. Weichselbaum
Department of Radiation Therapy
Joint Center for Radiation Therapy
Harvard Medical School
Boston, Massachusetts 02114

H. Rodney Withers
Department of Radiation Oncology
UCLA Center for Health Sciences
Los Angeles, California 90024

K. H. Woodruff
Radiotherapy Section of Donner Laboratory
University of California Lawrence Berkeley Laboratory
Berkeley, California and
University of California San Francisco School of Medicine
San Francisco, California 94143

Introductory Papers

Progress in Radio-Oncology II, edited by K. H. Kärcher et al., Raven Press, New York © 1982.

Assessment of the Patterns and Quality of Care in Radiation Therapy in the United States of America*

S. KRAMER
Department of Radiation Therapy and Nuclear Medicine,
Jefferson Medical College,
Thomas Jefferson University,
Philadelphia, Pennsylvania,
United States of America

Abstract

Radiation therapy in the USA has been developed into a modern medical specialty in the 25 years from 1955 to 1980. Two groups have been active in developing guidelines and quality assessment: (1) The Committee for Radiation Oncology Studies (CROS), established in 1969, has provided the interface between the discipline and the National Cancer Institute. The CROS has developed a rational research plan for radiation oncology, training programmes, multi-disciplinary conferences, and guidelines for the clinical practice of radiation therapy. Details of these guidelines will be described with reference to personnel needs and equipment utilization; (2) The practice of radiation oncology has been reviewed and a national 'profile' established by the Patterns of Care Study. By developing a Facilities Master List and a statistically valid sampling survey of all strata of radiation therapy practice in the USA, from both process and outcome aspects, rational bench-marks have been established for nine diseases where radiation therapy plays a major role in cure. Our data indicate that considerable variation exists, but that, in general, large training institutions show better compliance and achieve better results than all other facilities. Part-time radiation oncologists and small facilities (under 200 patients per year) fall consistently below the national average.

Radiation therapy in the United States has developed into a well organized separate medical specialty in the 25 years from 1955 to 1980. In 1955 some 80 full-time radiation therapists were active mostly in university hospitals, utilizing orthovoltage equipment. Most of the clinical work was performed by part-time radiologists. By 1980 there were 1400 full-time therapists representing two-thirds of all physicians practicing radiation therapy. Megavoltage had become standard and utilized low, intermediate and high energies. Two-thirds of all patients were now being treated in community hospitals.

The reasons for the rapid development of the field are:

1. Recognition of the value of radiation therapy in cancer management.
2. The establishment as an independent discipline with a separate administration and budget and a separate department both in the university and the hospital structures.
3. Separate training programs and separate specialty board examinations.
4. Improved professional and technical reimbursement policies.

---

* Supported in part by PHS Grant Number CA 15978, awarded by the National Cancer Institute, Department of Health and Human Services.

Guidelines and quality assessment procedures for radiation therapy have been largely the responsibility of two groups: The Committee for Radiation Oncology Studies (CROS), and the Patterns of Care Study (PCS).

The CROS, formerly the Committee for Radiation Therapy Studies, was initially a group of 12 and later 15 senior radiation therapists who were elected by their peers. The work of this group was supported by the National Cancer Institute but not controlled by it. The group was initially established in 1959 and expanded in 1976. Dr. Gilbert Fletcher was the original chairman for the first 12 years, he was followed by the author and presently Dr. William Powers is the chairman.

The group established an interface with the National Cancer Institute primarily in four areas: Research, training, guidelines for radiation therapy and multidisciplinary conferences. The initial guidelines were published in a "Blue Book" in 1968 in an effort to integrate radiation therapy into the regional medical program. An updated revision was published in 1972 and a second revision has just been completed and is about to be published.

Table I gives the guidelines for the staffing of radiation oncology centers for medical and paramedical personnel and for equipment utilization guidelines.[1] These guidelines have had very wide circulation in the United States in the past and have been widely accepted and utilized by the profession, hospital administrators and by government agencies.

A study entitled "Clinical and Research Radiation Therapy in Cancer Care" was initiated in 1974 (1) (2). Its purpose was to establish a national profile of radiation therapy practice in the United States, to define gap areas capable of improvement and to develop an ongoing quality assessment program. We have utilized Donabedian's model of quality of care assessment in terms of structure, process and outcome (3). Initially we identified all the facilities in this country which provided radiation therapy by means of megavoltage equipment. Very complete facilities master lists were established in 1973, 1974 and again in 1977. Table II shows a comparison between 1973 and 1977 of the number of patients treated, the number of facilities and the number of full-time and part-time therapists engaged in practice. Table III shows the type of practice, i.e. community hospital, university hospital or free standing facilities and the distribution of patients between these strata. This information formed the basis of a stratification system which, in turn, enabled us to develop a statistically valid sample of all facilities in the United States.

To evaluate the processes of radiation therapy we chose ten diseases in which curative radiation therapy plays a major role. We developed a "Consensus of Best Current Management" in these diseases. This consensus was arrived at by multiple discussions with experts in the appropriate area. We defined a series of questions referring to the nature and extent of the tumor and the suitability of the patient to undergo curative radiation therapy. These questions were linked in decision trees which then logically led to the appropriate radiation therapy. Figure 1 shows such an idealized tree for carcinoma of the cervix.

A set of questionnaires were developed from the decision trees and served as a basis of our surveys. These surveys were performed on site by specially trained senior radiation therapy residents in an appropriately sized sample of facilities in each of the strata developed. The basis of the

---

[1] Quoted by permission of Dr. Carlos Perez, Chairman of the Subcommittee on the "Blue Book" revision.

Table I. Guidelines For Optimal Radiation Therapy Staffing For Patient Care*

Personnel Category	Functional Representation
Radiation Oncologist-in-Chief	1 per Center
Radiation Oncologist	1 per 200-250 Patients**
Radiation Physicist	1 per 400 Patients**
Treatment Planning Staff (Junior Physicist, Dosimetrist, and/or Treatment Planning Technician)	1 per 300 Patients**
Radiation Therapy Technologists	
Supervisor	1 per Center
Staff	2 per Megavoltage Unit***
Brachytherapy	1 or more
Simulator	1-2 per Simulator****
Treatment Aid Technologist	1 or more
Nursing Staff	1 per 300 Patients**
Social Worker	1 or more per Center
Dietician	1 or more per Center

* The above recommendations are exclusively for radiation therapy related patient care. If the complete primary care of the cancer patient is undertaken by the radiation oncology service additional staff will be required. Additional staff will also be required for administrative education and research activities of the Center.

** New patients treated per year.

*** Or at least 1 per 150 new patients per year.

**** Several of the above functions may be performed by a single professional or may be shared among several institutions, depending on patient load and number of procedures performed.

Table II. PCS: Facilities Master List Data

	1973	1977
Pts. Rx Mega.	304 020	350 028
Rx facilities	1 013	1 098
Therapists total	2 194	2 177
Full-time	1 080	1 414
Part-time	1 114	763
Full-time increase		334
Part-time decrease		351

Table III. PCS:New Cancer Cases by Institution

	First FML 1973	Third FML 1977
Community	190 120	222 797
University	66 006	76 343
Free Standing	36 943	37 070
Federal	10 951	13 818
Total	304 020	350 028

surveys were the radiation therapy charts and hospital charts at these facilities, as well as personal interviews with the radiation therapists at the facilities. Ten charts were chosen randomly at each of five sites of patients treated in that facility in 1973 and a second survey was undertaken a year later of five other sites of patients treated in 1973 and 1974. A total of 8334 charts were reviewed in this manner (Table IV).

These data were computerized and were compared with our consensus of best current management. A weighted procedural score was developed as

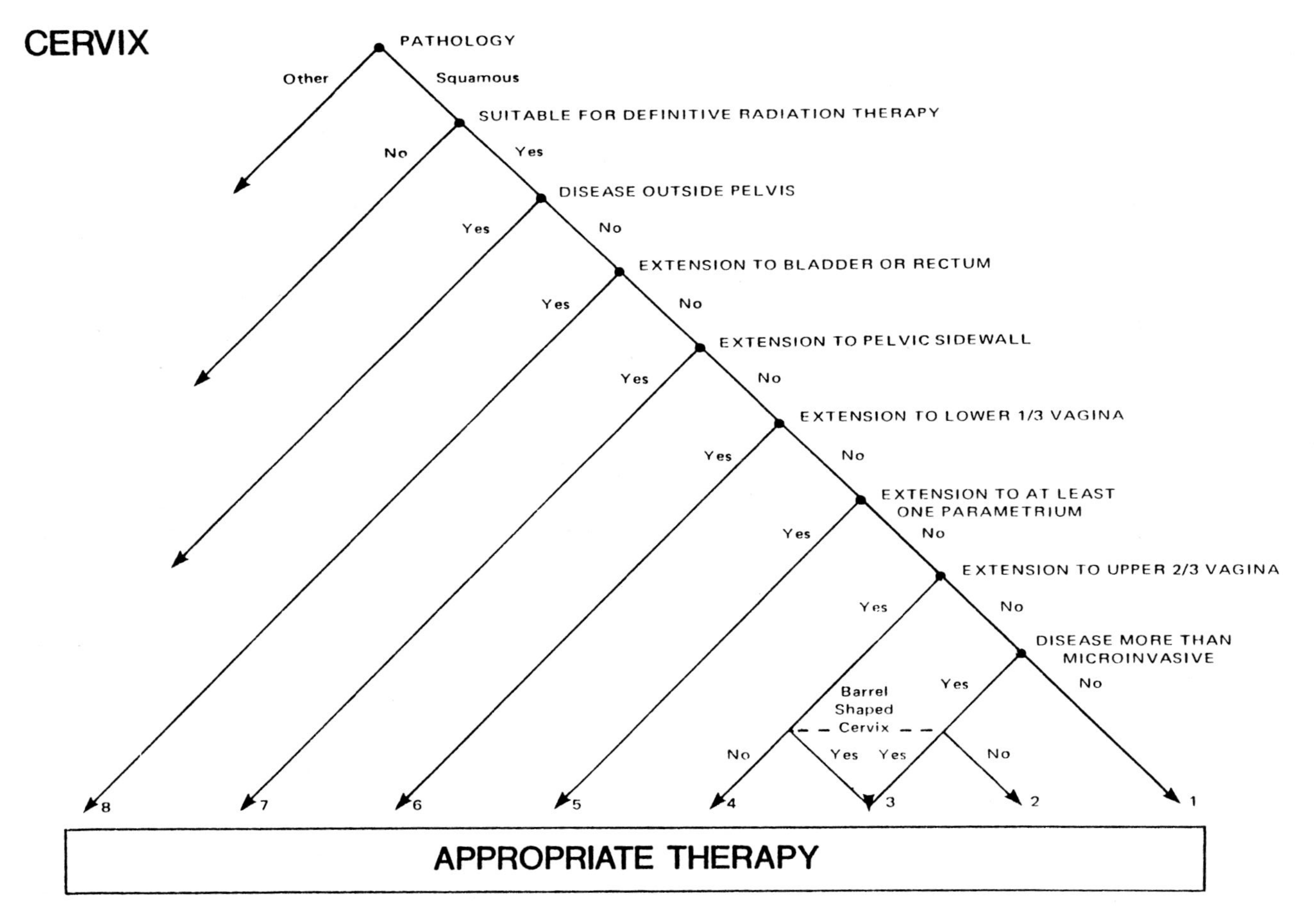

FIG.1. Idealized decision tree - cervix.

a numerical index of the compliance in the surveyed institutions with the consensus of best current management patient charts (4). Figure 2 gives the national data for carcinoma of the larynx and figure 3 shows the same information now broken down for those institutions providing training for residents, all other full-time departments and those headed by part-time radiation therapists. It becomes obvious that there is a very much higher degree of compliance in the training institutions, a somewhat lesser degree in the full-time and a very much worse degree of compliance in those headed by part-time therapists. However, it is noteworthy that in all types of strata there is a considerable spread between good compliance and less than good compliance. This pattern is seen quite consistently and figures 4, 5, 6 and 7 show some examples of process information obtained for carcinoma of the cervix and Hodgkin's disease. Those facilities with part-time heads, free-standing facilities and small facilities fall consistently below the national average, while those facilities who train residents universally exceed the national average.

Because of the variations in compliance with the process data and in order to try and establish the impact of process on the result of therapy, we undertook an outcome study three years after the original surveys. Three

Table IV. PCS: Charts Reviewed

Breast	1565	Bladder	1072
Testis	459	Prostate	901
Cervix	937	Larynx	1023
Hodgkin's	495	Ant. Tongue	556
Corpus	1018	Nasopharynx	308
	Total cases	8334	

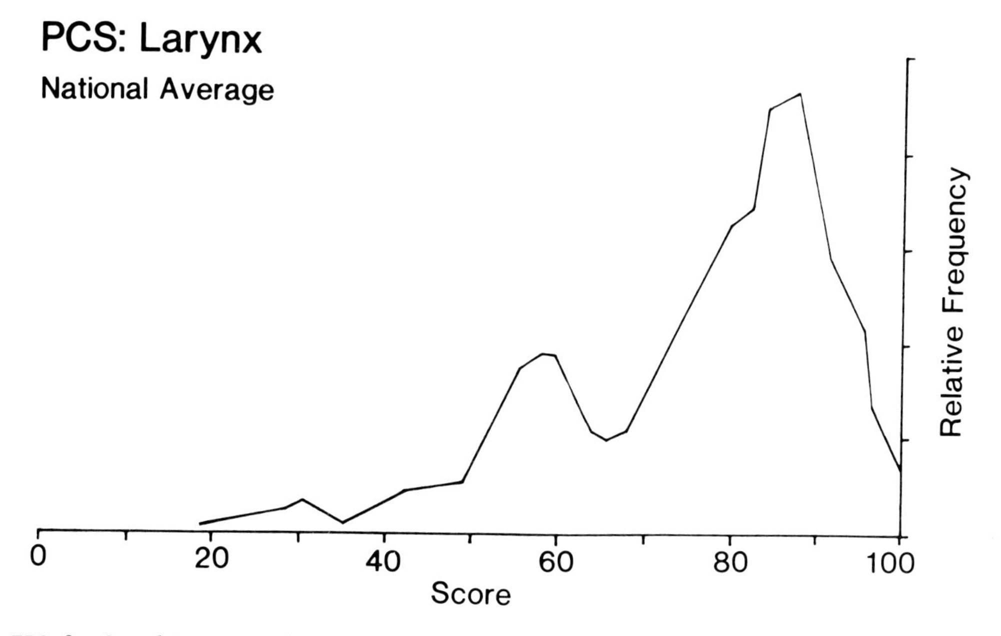

FIG.2. Compliance with process criteria.

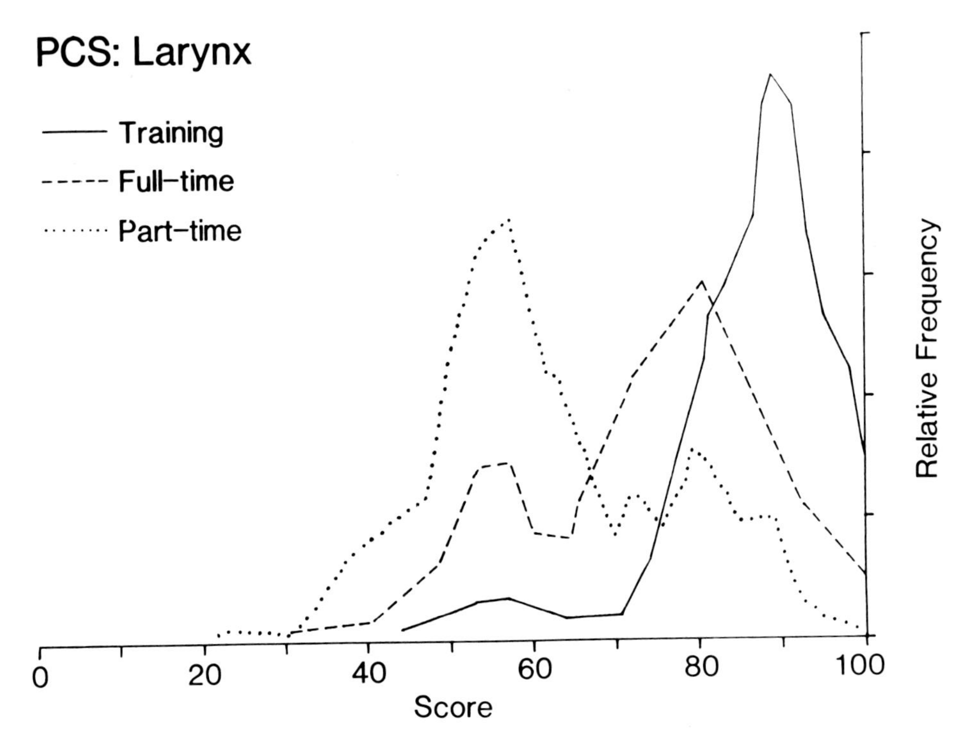

FIG.3. Compliance with process criteria by strata.

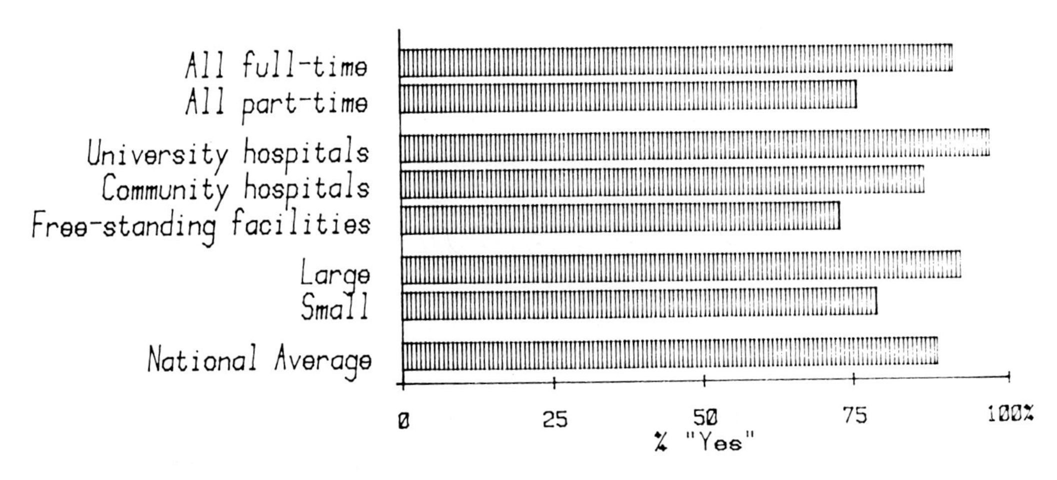

FIG.4. Process data in carcinoma of the cervix - data for radium applicators.

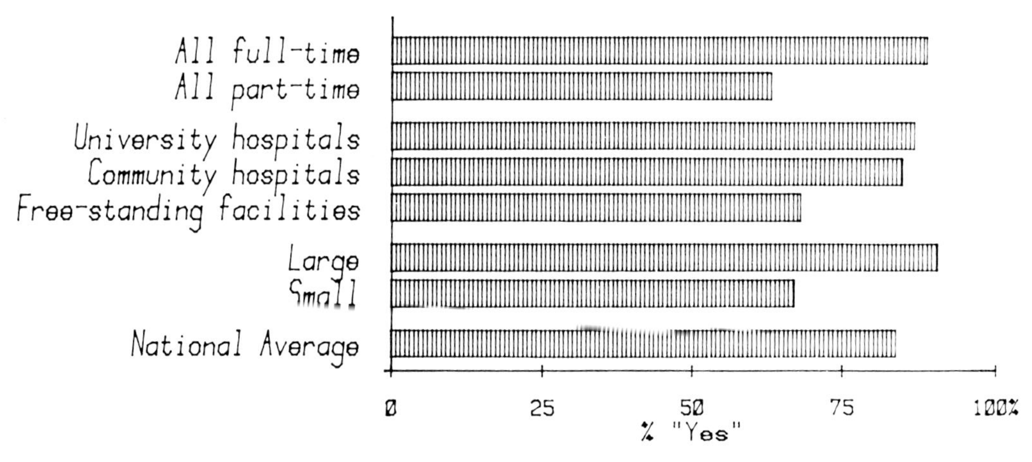

FIG.5. Process data in carcinoma of the cervix – data for external beam therapy.

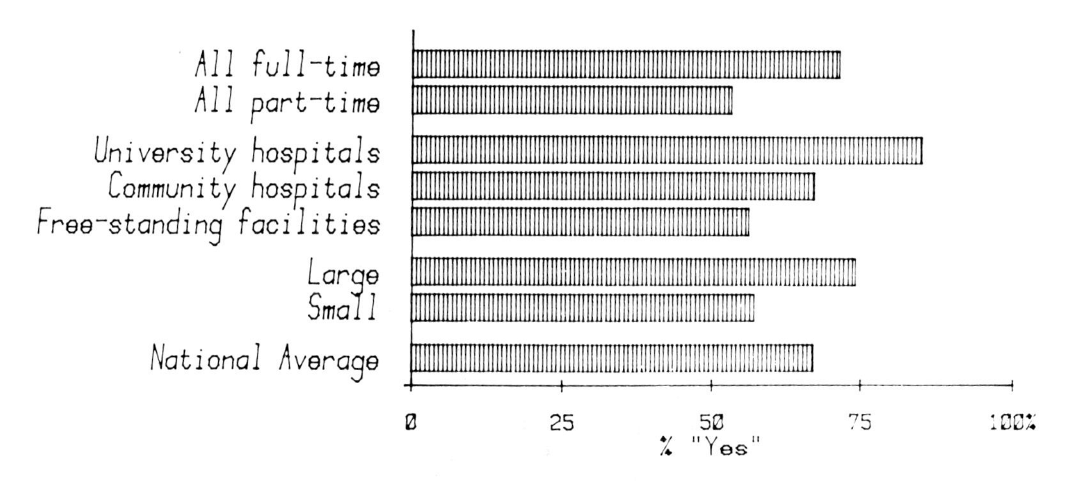

FIG.6. Process data in Hodgkin's disease – was a lymphangiogram taken?

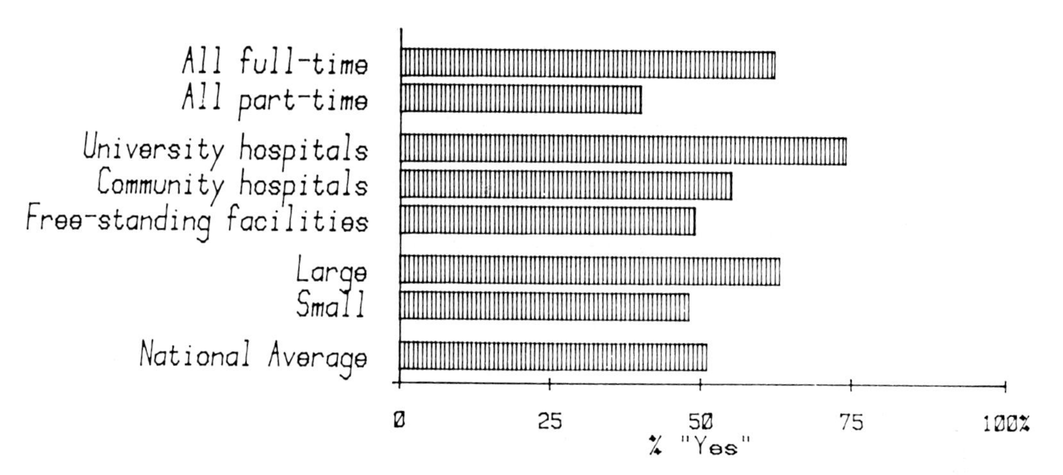

FIG.7. Process data in Hodgkin's disease - dosimetry calculation method.

types of outcome surveys were performed. In the first place, the same population that had been surveyed for process data was now examined to establish the outcome in these patients. Secondly, surveys were done in five institutions with the largest patient load in appropriate diseases and the charts of 50 patients randomly chosen were examined in each of these institutions. This type of survey was called "extended survey", and thirdly, special surveys were done in areas where our data base was small in order to enlarge our information in those areas.

Table V shows the national outcome benchmarks for carcinoma of the cervix in Stages I, II, & III. The recurrence rates for both pelvic and other recurrences are quite acceptable in Stages I and II, although somewhat high for Stage III. Noteworthy, too, is the fact that major complications, defined as complications requiring hospitalization of the patient for management, are unacceptably high in Stages I and II. Table VI compares the incidence of pelvic recurrences in the national average with the extended survey (the five institutions with large number of patients with carcinoma of the cervix). There is a marked difference in the pelvic recurrence rate between these data. The value of intracavitary irradiation of cervix cancer has long been stressed. Table VII shows that Stage IIIB intracavitary treatment is used more frequently in the five large institutions compared to the national average. Table VIII confirms the value of intracavitary therapy in local/regional control. Table IX demonstrates the effect of the general condition of the patient (as measured by the Karnofsky scale) on the incidence of pelvic recurrences and Table X confirms a higher incidence of complications in the younger patient group.

National outcome benchmarks for Hodgkin's disease are shown in Table XI. In field and out of field recurrences and major complications are acceptably low. Again, when comparing the results in institutions with large number of Hodgkin's disease patients with the national benchmark results, it can be

Table V. PCS: Carcinoma of the Cervix
US National Outcome Benchmarks
4 year Actuarial Results

	Stage		
	I	II	III
Pelvic Recurrence	7%	18%	59%
Any Recurrence	11%	28%	71%
Major Complications	11%	11%	14%
Survival	88%	82%	44%
Number of Patients	275	210	81

Table VI. PCS: Carcinoma of the Cervix
Pelvic Recurrences by Facility Type
4 Year Actuarial Results

	Facility Type		Number of Patients	
Stage	Regular Survey	Extended Survey	Regular Survey	Extended Survey
I	7%	1%	275	85
II	18%	15%	210	111
III	59%	22%	81	47

p < .05

Table VII. PCS: Carcinoma of the Cervix
STAGE IIIB ONLY
Use of Intracavitary Irradiation

Facility Type	Percent Using Intracavitary	Number of Patients
Regular Survey	65%	88
Extended Survey	88%	59

Table VIII. PCS: Carcinoma of the Cervix
STAGE IIIB ONLY
4 Year Actuarial Results

	Pelvic Recurrences		Number of Patients	
Treatment	Regular Survey	Extended Survey	Regular Survey	Extended Survey
No intracavitary	75%	57%	31	7
Use of intracavitary	43%	22%	57	52

Table IX. PCS: Carcinoma of the Cervix
4 Year Actuarial Results

Karnofsky	Pelvic Recurrences	Number of Patients
80 or less	37%	135
90 or 100	12%	439

p < .01

Table X. PCS: Carcinoma of the cervix
4 year actuarial results

Age	Major Complications	Number of Patients
Under 36	22%	49
36-50	13%	196
51-65	11%	241
Over 65	5%	147

$p < .01$

Table XI. PCS: Hodgkin's Disease
US NATIONAL OUTCOME BENCHMARKS
4 Year actuarial results

	Stages			
	IA	IIA	IIIA	All B
In field recurrences	3%	10%	18%	17%
Any recurrences	17%	26%	41%	50%
Major complications	10%	6%	7%	13%
Survival	87%	85%	77%	68%
Number of patients	110	148	74	66

Table XII. PCS: Hodgkin's Disease
4 year actuarial results

Facility type	Any Recurrence	Number of Patients
Regular survey	30%	399
Extended Survey	23%	246

$p < .05$

Table XIII. PCS: Hodgkin's Disease
STAGE IA & IIA
4 Year actuarial results

Technique	Any Recurrence	Number of Patients
Total nodal	17%	76
Extended field	20%	328
Involved field only	54%	22

$p < .01$

seen that the recurrence rate is significantly lower in the former (Table XII). The value of extended field irradiation, or total nodal irradiation, as compared to involved field only treatment in Stages IA and IIA disease is demonstrated in Table XIII, and Table XIV indicates that this higher incidence of recurrence is not due to in field recurrent disease. We have not been able to demonstrate a dose/response curve in Hodgkin's disease (Table XV).

Table XIV. PCS: Hodgkin's Disease
STAGE IA & IIA
4 year actuarial results

Technique	In field Recurrence	Number of Patients
Total nodal	9%	76
Extended field	8%	328
Involved field only	8%	22

Not sig.

Table XV. PCS: Hodgkin's Disease
4 Year actuarial results

Dose	In field recurrence	No. involved node groups treated
Below 3000 rad	8%	25
3000-3500	1%	78
3500-4000	4%	317
4000-4500	4%	402
Above 4500	6%	174

Not sig.

In reviewing 996 involved node groups treated by radiation therapy only, in whom the dose to each of these node groups had been calculated, the incidence of in field recurrences is essentially the same between 3000 rads and above 4500 rads. There is clearly need to confirm or refute these findings since an overall reduction of dose in young patients with Hodgkin's disease may well be of importance in terms of long term complications.

Finally, the data for the other sites examined in our study reflect essentially the same information. Particularly noteworthy is the fact that institutions who have training programs in radiation therapy consistently employ intracavitary irradiation more commonly in advanced carcinoma of the cervix, surgery combined with radiation therapy more often for advanced carcinoma of the larynx and for oral cancer and utilize interstitial irradiation in combination with external irradiation more commonly for early oral cancer.

CONCLUSIONS

The criteria for radiation oncology established by us have been very widely accepted. We have established a profile of the practice of radiation oncology in the United States and are now developing the methodology for an "on line" quality assessment program of radiation oncology. This should ensure a universally high level of quality of care.

Our success in obtaining the willing cooperation of our colleages in the field (97%) was due in large measure to the fact that we have worked through and within our professional organization - the American College of Radiology. This work has been done and has been perceived to have been done without government regulations on a voluntary basis by the profession and for the profession, for the benefit of our patients. We hope that other specialties in cancer therapy and perhaps other disciplines in medicine will follow our lead.

REFERENCES

[1] KRAMER, S.: The study of the patterns of cancer care in radiation therapy, Cancer 39 2 (1977) 780-787.
[2] KRAMER, S.; HERRING, D.: The Patterns of Care Study: A nationwide evaluation of the practice of radiation therapy in cancer management, Int. J. Radiat. Oncol., Biol. Phys. 1 (1976) 1231-1236.
[3] DONABEDIAN, A.: Evaluating the quality of medical care, Milbank Memori. Fund Q. 44 (1966) 166-206.
[4] McLEAN, C.J., DAVIS, L.W.: Discriminant analysis of radiation therapy procedures. The Patterns of Care process survey for carcinoma of the larynx, Cancer (1981, in press).

Progress in Radio-Oncology II, edited by K. H.
Kärcher et al., Raven Press, New York © 1982.

The Association of Irradiation with Less Than Radical Surgery in Various Types of Cancer*

G.H. FLETCHER
The University of Texas M.D. Anderson Hospital and
Tumor Institute,
Houston, Texas,
United States of America

Abstract

At mid-century, to cure cancer, either a radical surgical procedure was undertaken or, if irradiation was used, it had to be radical. In the ensuing 20 years, considerable progress had been made in radiobiology. This has shown that microscopic nests of cancer cells can be eradicated with less irradiation than gross masses. This has led to the use, pre- or post-operatively, of lesser doses of irradiation, which are well tolerated. The analysis of several series of patients treated with the combined treatment has shown that the overall local/regional controls were clearly superior to those obtained with either modality alone. It also became apparent that the surgical procedure does not need to be the conventional radical one. Examples of such combination treatment are given for head and neck squamous cell carcinomas, breast cancer, and soft-tissue sarcomas. It would be desirable that the basic principles of radiobiology be taught to all surgical oncologists so that, with an understanding of the underlying principles, the combined treatment modality could be used most effectively.

1. INTRODUCTION

At the turn of the century, radical surgical procedures were designed for various cancers, radical mastectomy, radical neck dissection, abdomino-perineal resection, radical hysterectomy, etc. These operations had to be radical to be curative and were thought to be the only hope for cure.

In 1936 Paterson structured the concept of radical X-ray treatment, an equivalent concept to radical surgical procedures [1]:

> "When radiation therapy is to be employed it really must be radical and treatment must be taken to the absolute limits of tolerance. It must be designed to include the whole potential tumour zone in the irradiated block of tissue with the dosage so balanced that the whole zone is as homogeneously irradiated as possible."

At the mid-century there were three rigid concepts used in the planning of treatment:

(1) The treatment modality must be either radiation or surgery;

* This investigation was supported in part by Grant CA06294 awarded by the National Cancer Institute, Department of Health and Human Services.

(2) There is an "all" or "none" cancerocidal dose linked with histology of the disease. Some tumours are radioresistant and irradiation has no place in their management;

(3) The surgical procedure has to be a radical procedure which was called "cancerwise". At that time a surgeon could not have conceived of performing a surgical procedure less than the conventional surgical procedure.

2. RATIONALE FOR COMBINING IRRADIATION AND SURGERY IN THE OPERATED AREA

Surgical failures result, despite the most radical procedure, because not all microscopic disease is removed in all patients. Failures of irradiation are experienced with gross masses, the frequency increasing as the volume of cancer increases [2]. Therefore, the two modalities of treatment are complementary, irradiation to eradicate the microscopic disease and the surgical procedure to remove the gross mass(es).

3. GUIDELINES FOR THE COMBINED TREATMENT, AND ADVANTAGES OF LIMITED SURGERY

Irradiation can be given either pre- or postoperatively. Clinical data from the University of Texas M.D. Anderson Hospital have shown that the local/regional control rates are identical with 5000 rad given preoperatively or 5000-6000 rad given postoperatively. The main criticism levelled at postoperative irradiation is that it does not diminish the potential to produce distant metastases through the surgical manipulations.

If postoperative irradiation is chosen, optimally the surgical procedure should be limited to the removal of the gross mass(es) (see Fig.1) [3] for the following reasons:

(a) With diminished surgical manipulations, there is less opportunity of throwing tumour cells into the blood stream;

(b) There is less scar tissue and therefore less possibility of hypoxia of the tumour cells left behind;

(c) Postoperative irradiation should be given 3 to 4 weeks following surgery. With very radical surgical procedures, oft-times using pedicles for reconstruction, it may take months before irradiation can be started. During that time there is repopulation of the tumour cells and a gross recurrence is commonly seen when the patient is ready for postoperative irradiation.

(d) With conservative surgery, there is a better quality of life.

3.1. Head and Neck

In squamous cell carcinomas of the head and neck, management of the metastases to the neck nodes has been traditionally a radical neck dissection which includes removal of the jugular vein and of the 11th nerve. This results, if it is done bilaterally, in marked oedema of the head with considerable discomfort until it is eased. The resection of the 11th nerve produces shoulder

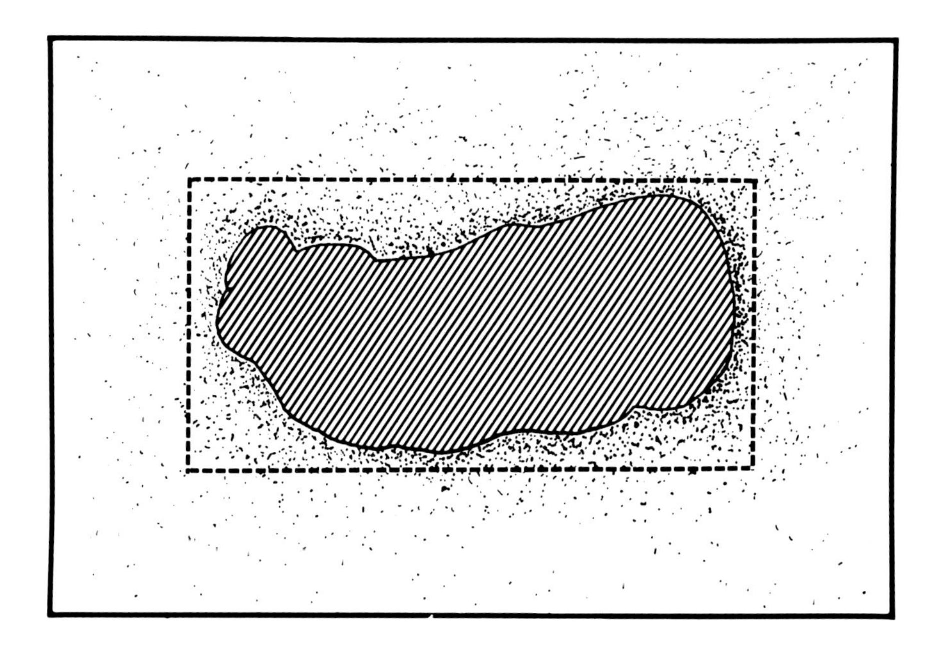

FIG.1. The dotted lines indicate a surgical excision that removes gross tumour with close margins. Within the larger rectangle, microscopic disease will be eradicated by irradiation. (Reproduced from Fletcher, G.H. [3], by kind permission of Alan R. Liss Inc., New York.)

dysfunction. With the use of pre- or postoperative irradiation, there is no need to remove all neck tissues since the irradiation will eradicate the microscopic nests of cells, so that modified neck dissections of various degrees can then be done. Shoulder function is never impaired with simple removal of the gross mass(es) to include the sternocleidomastoid muscle, and the jugular vein if a node(s) is attached to it (Fig.2).

3.2. Breast

From some series it has been strongly suggested that radical mastectomy for advanced breast cancers produces distant metastases [4]. In the 1950s, several studies were undertaken to find an anatomical reason. Through multiple blood samplings taken during a surgical procedure, it was shown that there is an increase in the tumour cell count in the circulating blood. In the radical mastectomy for breast cancer, the thinning of the flaps by peeling off the fat (there are no cancer cells in the fat) can be fraught with wound slough resulting in months of delay before irradiation of the chest wall can be started - since one should not start treatment until there is complete wound healing, or it may never heal. Thin flaps are devascularized and tumour cells which are

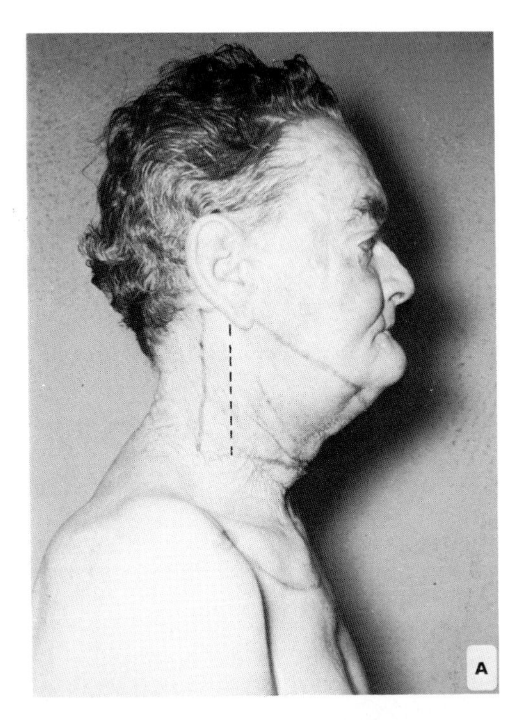

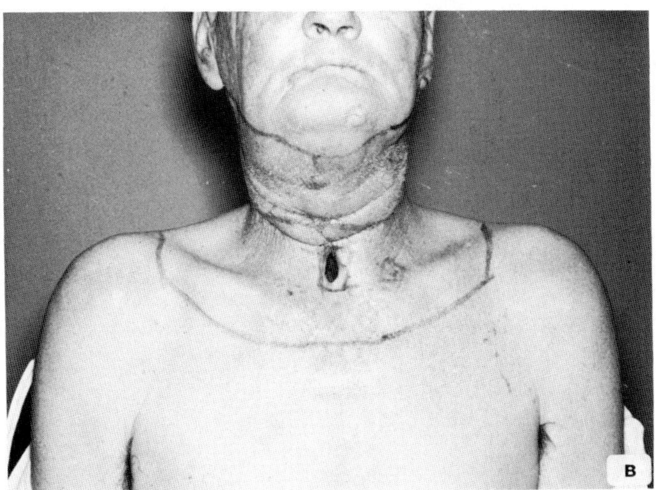

FIG.2. Clinical examination and lateral soft tissue radiograph showed in June 1970 an ulcerated lesion of the infrahyoid epiglottis with marked involvement of the pre-epiglottic space. A 1.3 cm node was palpable in the left subdigastric area and two 3.0 cm nodes were present in the right subdigastric area. Biopsy: squamous cell carcinoma. The plan of treatment was total laryngectomy with bilateral modified neck dissection and postoperative irradiation. The patient died 15 months later from distant metastases without evidence of disease above the clavicles. (Reproduced from Fletcher, G.H. [2], by kind permission of Lea and Febiger, Publishers, New York.)

in the lymphatics of the skin are less radiosensitive. A simple extended mastectomy (dissection of the lateral axilla) removes the mass in the breast and palpable axillary nodes without thinning of the flaps (Fig.3). The wound heals promptly and irradiation can always be started within 3 to 4 weeks.

Table I shows that in patients with advanced, sometimes very advanced, disease of the breast, similar in extent to that described in Haagensen's patients, there is 30% to 38% 10-year 'no evidence of disease' (NED) survival rate [5], whereas all the patients in Haagensen's series had died by 6 years, almost all of them prior to 5 years [4]. Furthermore, there is only a 15% incidence of local/regional failures in the University of Texas M.D. Anderson Hospital series compared with 47.7% in Haagensen's series [4]. For small volume breast cancer, a simple excision followed by irradiation yields cure rates similar to those obtained with classical radical mastectomy (Fig.4) [6], and the cosmesis is excellent, as may be seen in the patient of Fig.5 [7].

TABLE I. SURVIVAL AND LOCAL/REGIONAL FAILURE RATES IN PATIENTS TREATED WITH SIMPLE OR SIMPLE EXTENDED MASTECTOMY AND IRRADIATION (NO ADJUVANT CHEMOTHERAPY): 1955-1975 (ANALYSIS APRIL 1980)
(Adapted from Montague et al. [5])

	Stage III† (124 pts)	Stage IV† (280 pts)
Disease-free 10-year survival *	38%	30%
Sites of failure ‡		
Chest wall	15	34
Axilla §	5	13
Supraclavicular	6	11
Parasternal	-	-
Total patients with failure	17	48
(Percentage of failure)	(14%)	(17%)

* Berkson-Gage, not age adjusted.

† American Joint Committee Staging System.

‡ A patient may have more than one recurrence.

§ Of 18 patients with axillary failures, 14 had been treated with simple mastectomy and 4 with extended simple mastectomy.

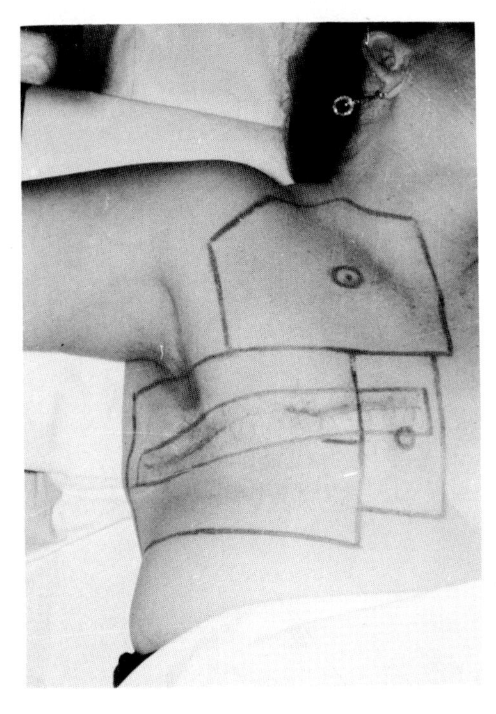

FIG.3. Patient, age 58, with a 6 cm x 6 cm central mass in the right breast with a 2 cm low axillary node. Extended simple mastectomy performed in June 1970 revealed 3 of 15 positive nodes. Postoperatively, 5000 rad was delivered to the chest wall and axilla, and 5000 rad given to the supraclavicular and internal mammary chain nodes; the 6 MeV electron beam was used to give a 1000 rad boost to a strip along the scar. In June 1973 there were metastases in both lungs and the patient expired 11 October 1974, with no evidence of disease on the chest wall or the regional lymphatics. (Reproduced from Fletcher, G.H. [2], by kind permission of Lea and Febiger, Publishers, New York.)

3.3. So-Called Radioresistant Tumours

Some examples of the combination of irradiation and surgery will be given in patients with tumours which were until recently considered to be radioresistant. This combination of two less than radical treatments can not only improve control rates but can also result in a considerable improvement of the quality of life. The conventional surgical procedure for malignant tumours of the parotid is a total parotidectomy with sacrifice of the facial nerve, leaving the patient with great disfigurement. With a less radical procedure which preserves the facial nerve, followed by irradiation, the control rates have been shown to be better than with the radical surgical procedure [8], and the cosmesis is immensely better, as shown by the photograph of a patient (Fig.6) [9].

In the 1930s the soft-tissue sarcomas were placed in the category of absolutely radioresistant tumours. With radical surgical procedures, including amputation and disarticulation when the tumour is close to a joint, the local recurrence rate is 29% [10].

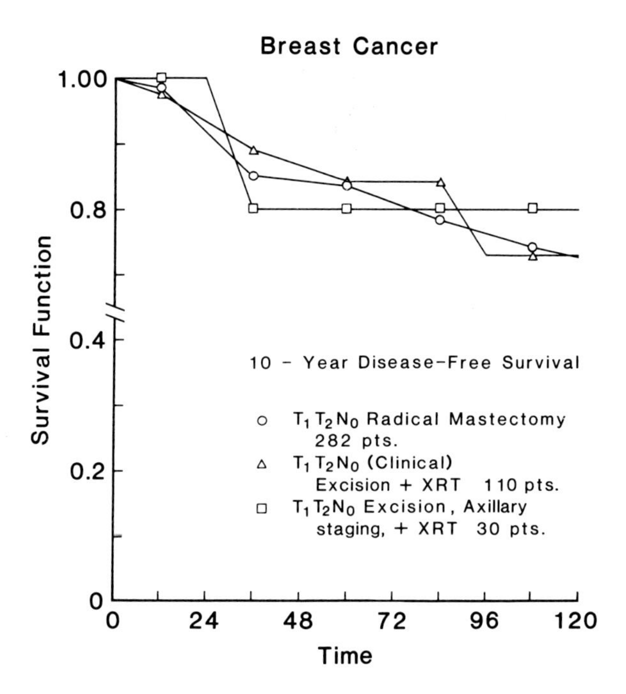

FIG.4. Ten-year disease-free survival rate of patients with histologically or clinically negative nodes. (Reproduced from Montague, E.D. et al. [6], by kind permission of Eleanor D. Montague, MD, and of Masson, Editeur, Paris.)

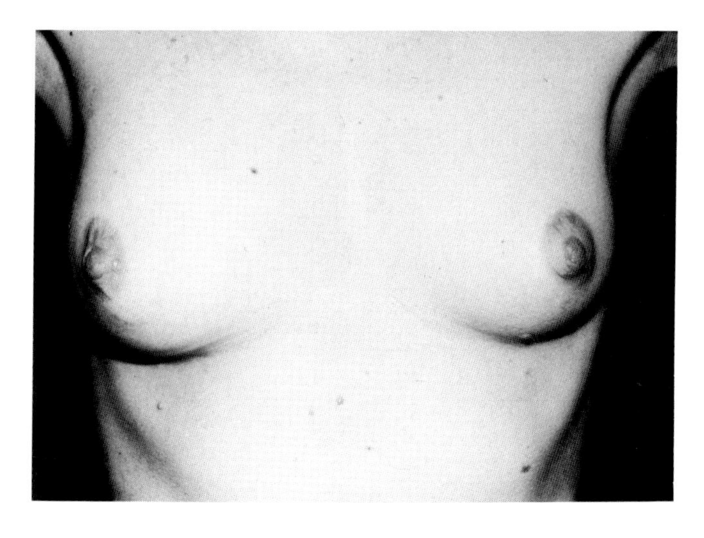

FIG.5. Thirty-five year old woman who presented on 26 July 1974 with a 3 cm cancer in the upper outer quadrant of the right breast. The tumour was excised and 5000 rad was given postoperatively to the breast and the regional lymphatics. She has remained well without evidence of disease 7 years after treatment. (Reproduced from Montague, E.D. et al. [7], by kind permission of the Editor of Cancer, Philadelphia.)

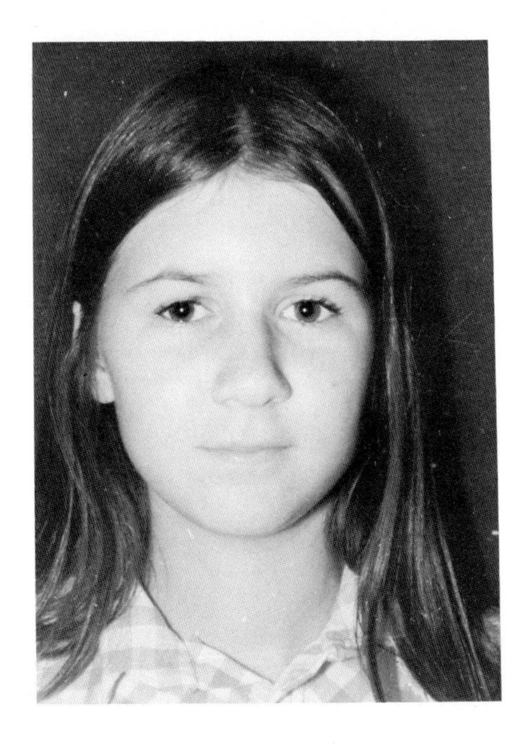

FIG.6. A 10-year old girl was admitted in June 1971 with a history of left parotidectomy performed at another hospital in April 1971 for a fixed pre-auricular mass that had grown rapidly. At operation, the main trunk of the facial nerve was sectioned to remove the tumour mass, and the nerve was then resutured. Histology: poorly differentiated adenocarcinoma. The patient received postoperative irradiation. The patient was without evidence of disease in August 1980. She has only minimal facial nerve weakness and the cosmetic result is excellent. (Reproduced from Tapley, N. duV. [9], by kind permission of the Ear, Nose and Throat journal, Insight Publishing Co., Inc., New York.)

It has been shown that the local control with conservative removal of the tumour mass followed by irradiation is effective (Table II) [11]. The patient shown in Fig.7 had a tumour which would have required a forequarter amputation that would have included removal of the shoulder and scapula, whereas simple excision followed by irradiation yielded excellent function [12,13].

4. SUMMARY

The fact that irradiation may not be the sole treatment in some tumours does not mean that it does not have a place in their management. This is well illustrated in the management of soft-tissue sarcomas. As a general rule, large tumours should not be treated with irradiation only if gross cancer can be resected simply. One must think of the quality of life available to the patient. In that context, conservative surgical removal with modest doses of irradiation may not only yield better results than either of the disciplines used alone, but also results in excellent function and cosmesis.

TABLE II. INCIDENCE OF LOCAL FAILURES (1963 THROUGH 1973): UNLIMITED FOLLOW-UP, MINIMUM 2 YEARS
(Reproduced from Lindberg, R.D. et al. [11], by kind permission of Robert D. Lindberg, MD, and of the Year Book Medical Publishers, Inc., Chicago, holders of the copyright © 1977.)

Trunk*		Upper Extremity		Lower Extremity	
Axilla and shoulder	1/6 (1)	Arm	5/15 (1)	Thigh	6/30 (1)
Buttocks	2/14	Elbow	0/3	Knee	3/16 (3)
Other sites	4/8 (1)	Forearm	2/13	Leg	6/13 (3)
		Wrist	0/4	Ankle	1/3
		Hand	0/8	Foot	1/3
Total	7/28(2)		7/43 (1)		17/65 (7)

* Intraabdominal primaries are excluded.
() Number in parentheses indicates number of recurrences outside the irradiated volume.

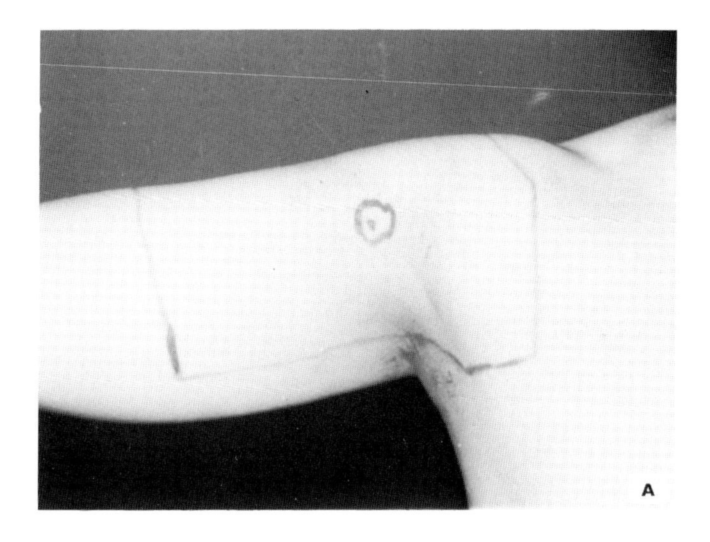

FIG.7. This 40-year old woman was admitted with pain on motion in the right arm. A 2 cm x 4 cm subcutaneous mass just anterior to the head of the humerus was excised on 17 November 1972 that proved to be a high grade fibrosarcoma. Using the shrinking field technique, a tumour dose of 6600 rad was delivered in 7 weeks. The patient is living free of disease at 8 years with excellent arm function. (Fig.7 A,C are reproduced from Lindberg, R.D. [12], by kind permission of the author and of Lea and Febiger, Publishers, New York; Fig.7 B from Tapley, N. duV. [13], by kind permission of John Wiley and Sons, Inc., New York.)

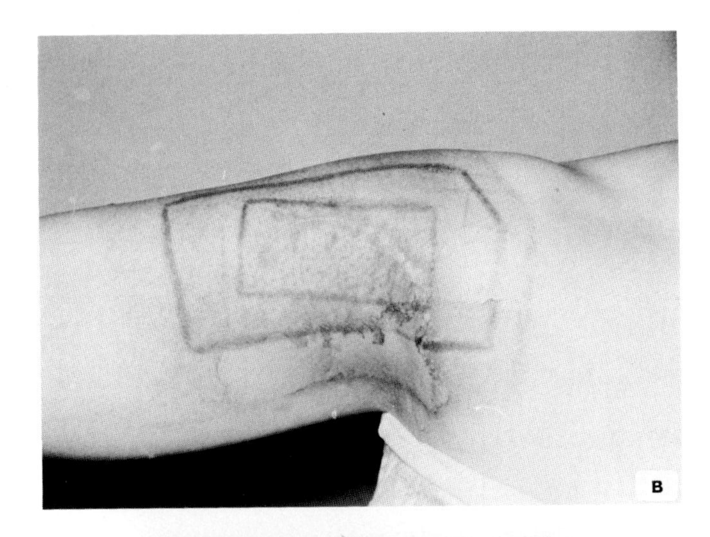

Fig. 7 (cont.)

To maximize the effectiveness of multimodality treatments, the co-operating specialists must have an understanding of the underlying principles. Not only radiotherapists, but also surgical oncologists should participate in a core programme with lectures in radiobiology and periods of clinical observation to understand why it is not necessary to use extraradical surgical procedures if the surgical procedure is combined with irradiation.

REFERENCES

[1] PATERSON, R.: The radical X-ray treatment of the carcinomata, Br. J. Radiol. 9 (1936) 671.
[2] FLETCHER, G.H.: Textbook of Radiotherapy, 3rd ed., Lea and Febiger, Philadelphia (1980).
[3] FLETCHER, G.H.: "Combination of irradiation and surgery", in International Advances in Surgical Oncology (MURPHY, G.P., Ed.) Vol.2, New York, Alan R. Liss, Inc. (1979) 55-98.
[4] HAAGENSEN, C.D., STOUT, A.P.: Carcinoma of the breast, II. Criteria of operability, Ann. Surg. 118 (1943) 859.
[5] MONTAGUE, E.D., SPANOS, W.J., Jr., FLETCHER, G.H.: "Evolution at M.D. Anderson Hospital of treatment policies for the primary management of nondisseminated breast cancer", in Proceedings of the International Congress on Senology (Hamburg, Fed. Rep. Germany, May 1980), Thieme Verlag, Stuttgart (1981, in press).
[6] MONTAGUE, E.D., SPANOS, W.J., Jr., AMES, F., et al.: "Conservation surgery and irradiation for the treatment of clinically favorable breast cancer", in Proc. 19th Nat. Conf. on Breast Cancer (San Diego, California, March 1981), Masson, Editeur, Paris (1981, in press).
[7] MONTAGUE, E.D., GUTIERREZ, A.E., BARKER, J.L., et al.: Conservative surgery and irradiation for the treatment of favorable breast cancer, Cancer 43 (1979) 1058.
[8] FLETCHER, G.H.: Indications for combination of irradiation and surgery, J. Radiol. Electrol. 57 (1976) 379.
[9] TAPLEY, N. duV.: Irradiation treatment of malignant tumors of the salivary glands, Ear, Nose Throat 56 (1977) 39.
[10] CANTIN, J., McNEER, G.P., CHU, F.C., et al.: The problem of local recurrences after treatment of soft tissue sarcoma, Ann. Surg. 168 (1968) 47.
[11] LINDBERG, R.D., MARTIN, R.G., ROMSDAHL, M.M., et al.: "Conservative surgery and radiation therapy for soft tissue sarcomas", in Management of Primary Bone and Soft Tissue Tumors, Year Book Medical Publ., Chicago (1977) 289-298.
[12] LINDBERG, R.D.: "Soft tissue sarcoma", in FLETCHER, G.H., Textbook of Radiotherapy, 3rd ed., Lea and Febiger, Philadelphia (1980) 922-942.
[13] TAPLEY, N. duV.: Clinical Applications of the Electron Beam, John Wiley and Sons, New York (1976) 239.

Particle Beam XRT

Progress in Radio-Oncology II, edited by K. H. Kärcher et al., Raven Press, New York © 1982.

Fast Neutron Beam Treatment of Advanced Squamous Cell Carcinomas of the Head and Neck Region*

T.W. GRIFFIN, G.E. LARAMORE
Department of Radiation Oncology,
University of Washington,
Seattle, Washington,
United States of America

Abstract

A final report is presented on the local control rate and length of survival for 100 patients with advanced squamous cell carcinomas of the head and neck region who received fast neutron teletherapy at the University of Washington during the period 1973 through 1977. Sixty-two patients were treated with neutrons alone and 38 were treated with a combination of neutrons and photons as part of a mixed beam fractionation scheme. The overall initial complete remission rate was 68% for the mixed beam group and 44% for the group treated with neutrons alone. Initial complete remission rates for the two groups of patients are given as a function of T-stage and N-stage, and actuarial curves are presented which show the time course of local control and survival for the two treatment groups. For T_3 and T_4 lesions the initial complete remission rate appears to be greater using the mixed beam form of treatment than using neutrons alone. Treatment to high dose levels using neutrons alone gave rise to significantly greater morbidity – both acute and late effects – than resulted from the mixed beam form of treatment. Local control rates and survival rates are compared with similar groups of patients treated with neutrons at other institutions.

In 1971 the National Cancer Institute initiated a study at the University of Washington to determine the efficacy of fast neutron radiotherapy for treating human malignancies. This was a phase I study which ran until 1977. The first two years were spent in adapting for medical use a cyclotron which was originally built for physics research purposes and in characterizing the radiobiological properties of the neutron beam. Clinical trials were initiated in September 1973 and ran until May 1977. Because this was a phase I investigation, in general only patients with advanced tumors having less than 10% 5-year survival were accepted into the study. The project was regional in extent with patients being referred from many institutions throughout the Pacific Northwest. Approximately 220 patients were treated, with many different tumor types and prior modes of therapy being represented. A total of 100 patients were treated for advanced tumors of the head and neck region with a curative intent. Sixty-two patients were treated with neutrons alone and 38 patients were treated with a mixed beam regime. This paper reports the survival data for this group of patients. Acute and chronic morbidity of treatment as well as complications associated with surgical salvage procedures for uncontrolled disease are also discussed.

---

* This investigation was supported by a grant from the National Institutes of Health, National Cancer Institute, number CA-12441-09.

METHODS AND MATERIALS

The patients in this study were treated with fast neutrons from a 60-inch fixed frequency cyclotron on the University of Washington campus. The beam characteristics of this machine have been extensively described elsewhere[1,2] and will only be briefly summarized here. The neutron beam is produced via a 22 MeV d $\rightarrow$ Be reaction which gives rise to a broad peak centered at 8 MeV. Only a single horizontal port is available for biomedical work and the patients in this study received their neutron radiation while sitting in a specially designed chair which could be accurately positioned in front of the port. Individual fields were shaped using iron blocks in front of collimators made from a water-containing plastic resin. Neutron-nuclei interactions produce a depth-dependent photon contaminate in the beam. At a depth of 10 cm this photon component contributes less than 10% of the beam energy[1] but is implicitly included in the term "$rad_{n\gamma}$". Patients were treated at a skin-source distance of 150 cm and at this distance the dose rate in air was 30-40 $rad_{n\gamma}$/min. For typical field sizes used in treating head and neck cancers, the 50% isodose line lies at a depth of $\sim$ 9 cm along the central axis. This roughly approximates the depth-dose characteristics of a ^{137}Cs unit.

Due to somewhat limited medical access to the cyclotron during the early years of the study various treatment fractionation schemes were investigated. To standardize the treatments to some extent, 300 $rad_{n\gamma}$ (approximately 900 rad equivalent) were given each week (dose rad equivalent = 3 x dose $rad_{n\gamma}$ + dose rad_{γ}). Three treatment patterns were investigated using neutrons alone: (1) 150 $rad_{n\gamma}$ on Mondays and Fridays, (2) 100 $rad_{n\gamma}$ on Mondays-Wednesdays-Fridays, and (3) 75 $rad_{n\gamma}$ on Mondays-Tuesdays-Thursdays-Fridays. In the early phase of the study, patients were arbitrarily assigned to either the 2- or 3-neutron-fractions-per-week pattern. In the later phase of the study, a 4-neutron-fractions-per-week pattern was adopted as standard by all neutron institutions in the United States. In this study no significant difference was noted among the three treatment patterns either in terms of tumor control or treatment morbidity.

Animal model studies at the University of Washington by Nelson-Rasey et al [3,4] subsequently suggested an enhanced therapeutic ratio for two neutron fractions plus three photon fractions per week. Based upon this work, a mixed beam (neutron/photon) treatment option was included in which 60 $rad_{n\gamma}$ was given on Mondays and Fridays and 180 rad_{γ} was given on Tuesdays-Wednesdays-Thursdays. In the remainder of this paper, we will consider separately the subgroups of patients treated with neutrons alone and with the mixed beam regime.

The treatment fields for all patients were set up using either a Northwest Medical Physics Cascade simulator or a General Electric orthovoltage X-ray unit, and were verified with beam films taken on the cyclotron and/or megavoltage unit. Patients receiving photon radiation were treated in the usual prone position while the neutron radiation was delivered with the patient sitting as noted above. The patients were treated using a "shrinking field" technique as described by Fletcher [5]. Most of the patients were treated using right and left parallel opposed weighted 1:1 at midplane. In the neutron-only group, the majority of patients received between 1900-2000 $rad_{n\gamma}$ to the primary site and any area of massive adenopathy with the total range of doses being between 1700-2200 $rad_{n\gamma}$. The fields were adjusted to limit the spinal cord dose to $\sim$ 1500 $rad_{n\gamma}$ which was then thought to be a safe limit. In the initial phase of the study, 21 patients did not receive prophylactic irradiation to the clinically uninvolved lower neck and supraclavicular areas while one patient received 1550 $rad_{n\gamma}$ to these areas. In the mixed beam group, the majority of patients received between 6500-7000 rad equivalent to the primary site and any area of massive adenopathy with the actual range of delivered doses being between 720-1050 $rad_{n\gamma}$ + 3780-4400 rad_{γ}. The spinal cord dose was limited to 4500 rad equivalent (typically 600 $rad_{n\gamma}$ + 2700 rad_{γ}). The entire group of mixed beam patients received prophylactic photon irradiation (4500-5040 rad_{γ}) to the clinically uninvolved lower neck and supraclavicular regions. For both groups of patients, the lower neck and supraclavicular regions were treated using a single anterior field with the delivered dose calculated at 3 cm depth. In general, a midline block was used in the anterior field. However, if this would have compromised the treatment, then no midline block was used and the field junction moved every two weeks.

A total of 100 patients with histologically-proven, advanced squamous cell carcinomas of the head and neck region were treated with curative intent using either neutrons alone or neutrons as part of a mixed beam regime. Only patients with previously untreated primaries are included. We exclude patients who presented with cervical adenopathy from an unknown primary, patients who failed in the neck or at the primary site after a "curative" surgical excision, patients who received a neutron boost after having received photon irradiation elsewhere, or patients who received an interstitial implant as part of their planned treatment. These patients are discussed in previous works dealing with local control rates for the cervical neck nodes [6,7] and for the various primary sites [7,8,9]. A total of 62 patients were treated using neutrons alone and 38 patients were treated using the mixed (neutron/photon) beam regime as defined above. Table I outlines the sites of origin of these tumors.

Combining all tumor sites together, a graphical display of the local control rates at the primary site for the two treatment groups is shown in Fig. (1).

TABLE I. LOCATION OF PRIMARY TUMOR SITE

	Mixed Beam	Neutrons Only
Location		
Nasopharynx	4	5
Oral Cavity	6	15 *
Oropharynx	21	28
Hypopharynx	6	10 *
Supraglottic Larynx	1	5
Total	38	63

* Includes one patient with two simultaneous primaries - T_2 oral cavity, T_3 hypopharynx.

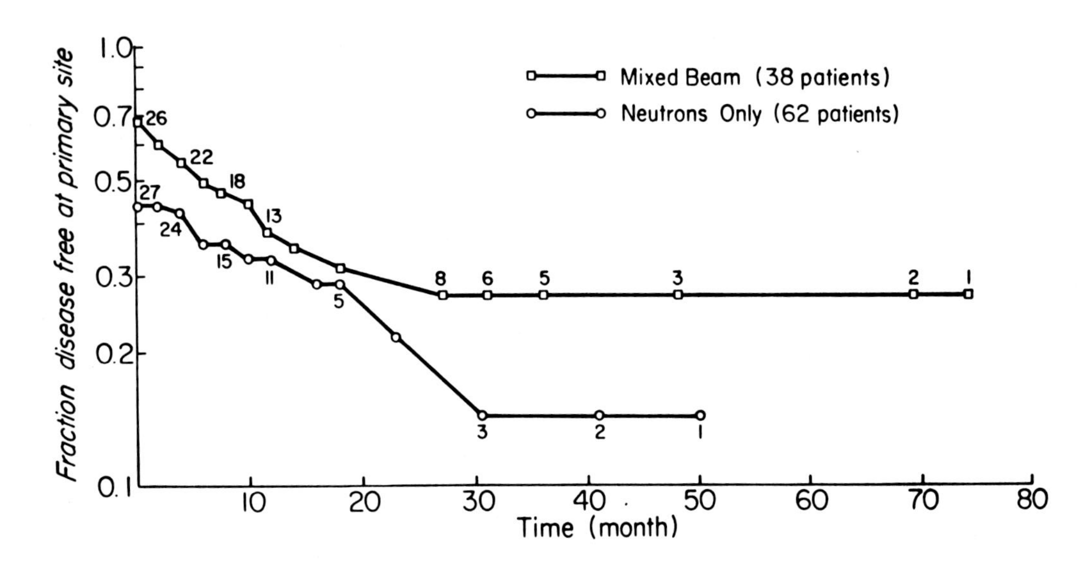

FIG.1. Fraction of patients with squamous cell carcinomas of the head and neck region who are disease free at the primary site as a function of time after completing therapy. (The curves were calculated using the actuarial methods [10] with the circular data points representing the subgroup of patients treated with neutrons alone and the square data points representing the subgroup of patients treated with a mixed beam regime. The ordinate is shown on a logarithmic scale and the numbers indicate the number of patients at risk for the indicated data points. For purposes of clarity only selected data points are shown.)

For the patients treated with the mixed beam regime, the initial complete remission rate was 68% compared with 44% for the patients treated with neutrons alone.

At two years from the time of completing treatment, the local control rates are 30% and 21% respectively for the mixed beam and neutron-only groups. However, the control rates do not flatten out until 30 months at which time the respective local control rates are 28% and 14%. Based upon the statistical Z-test [10], the difference in the initial complete remission rates is significant at approximately the 1% confidence level but the statistical significance between the two curves is lost for longer follow-up times. Because this was a non-randomized, retrospective study, it must be recognized that these two groups of patients are not equivalent and the somewhat larger number of T_4 lesions in the subgroup treated with neutrons alone may account in part for the difference between the two curves.

A total of 24 patients in the mixed beam group and 53 patients in the neutron-only group had clinically involved cervical neck nodes at the time of treatment initiation. The overall local control rate was 63% (15/24) for the mixed beam group and 53% (28/53) for the neutron-only group. The overall local control rates for the mixed beam and neutron-only groups were, respectively, 86% (6/7) and 82% (9/11) for stage N_1; 67% (2/3) and 38% (3/8) for stage N_2; and 50% (7/14) and 47% (16/34) for stage N_3. Thus there appears to be no significant difference in the nodal control rates on stage-by-stage basis. A geographical display of the time course of the cervical neck node control is given by Fig. 2.

Sequelae from the radiation therapy included the expected skin erythema, oral and pharyngeal mucositis reaction, dysphagia, increased skin pigmentation, decreased salivary function, and altered taste sensation in a large proportion of the patients. These sequelae occurred in both treatment groups and except for the increased degree of skin pigmentation (more pronounced for the neutron-only group) were about what would be expected from conventional photon irradiation to comparable dose levels.

In regards to acute effects, 8 patients in the neutron-only group developed a significant moist skin desquamation requiring treatment with Burrow's soaks and 5 of these patients needed breaks in treatment of 1 or 2 weeks. In contrast, only 3 patients in the mixed beam group developed a moist desquamation necessitating Burrow's soaks and none of these required a break in treatment.

In regards to late effects, in the neutron-only group two patients who had locally controlled tumors of the hypopharynx developed significant pharyngeal wall edema at respective times of 6 and 8 months after completing treatment. One of these patients died and the other required an emergency tracheostomy. Another patient with an uncontrolled hypopharyngeal tumor required an emergency tracheostomy at 4 months after completion of treatment but this could have been due to tumor progression as well as to treatment-related pharyngeal wall edema. In addition, one patient with an oral cavity tumor and 3 patients with oropharyngeal tumors who had a local control later developed a pronounced enough

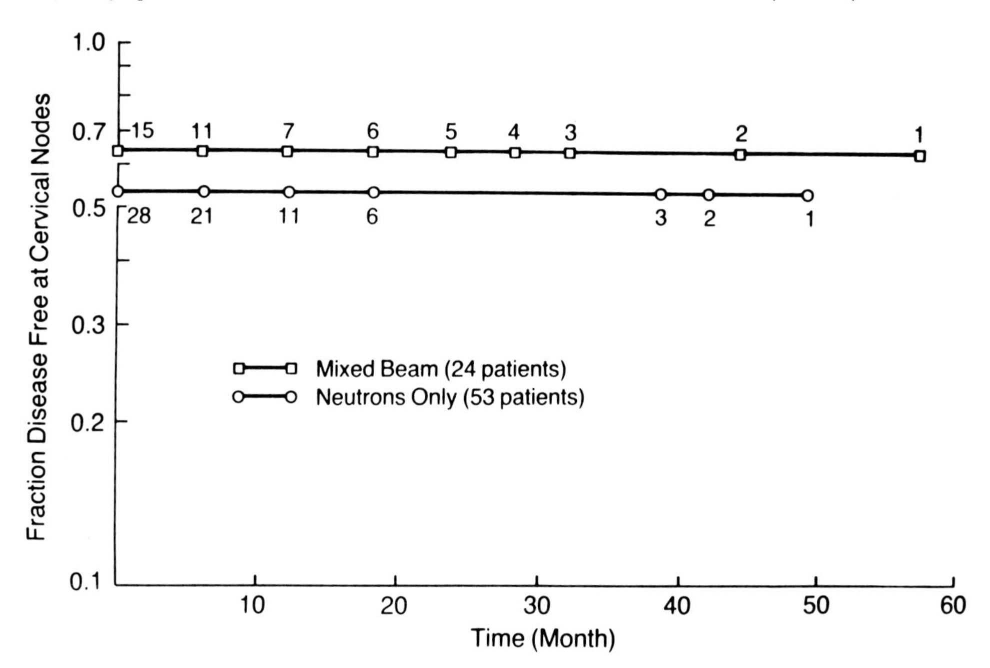

FIG.2. Fraction of patients with squamous cell carcinomas of the head and neck region who are disease free at the cervical nodes as a function of time after completing therapy. Only patients initially having positive nodes are included. (For details of curve calculation, etc., see caption to Fig.1.)

pharyngeal wall edema to cause a compromised nutritional status. This only occurred for one patient in the mixed beam group who had an oropharyngeal tumor.

Five patients with tumors of the oropharynx who received only neutron irradiation developed a cervical myelitis at times ranging between 7 and 39 months after completing treatment. This represents 50% of the patients with oropharyngeal tumors who were treated with neutrons alone who survived $\sim$ 10 months or longer. In each case, the patients were treated with lateral parallel opposed neutron fields and the dose to the spinal cord was limited to $\leq$ 1450 $rad_{n\gamma}$. The fields were then reduced and additional treatments given to bring the total dose to the primary tumor volume to between 1950 - 2050 $rad_{n\gamma}$. Review of the beam films verified that the spinal cord was excluded from the boost field irradiation. In all cases the lower neck and supraclavicular regions received prophylactic irradiation using a single anterior field. Four patients had this region treated with ^{60}Co radiation (between 4500-4950 rad_{γ}/ 5-6 weeks) and one patient received 1550 $rad_{n\gamma}$/5 weeks. In all cases a full midline block was used to exclude the possibility of any inadvertent overlap of the fields at the spinal cord and the radiation delivered to a depth of 3 cm. No such complication occurred for other tumor sites in spite of the same spinal cord dose being given. It may be that the posterior margin of the boost field was closer to the spinal cord for the oropharyngeal tumors and a somewhat greater amount of scattered neutron radiation at the spinal cord contributed to this.

The incidence of wound healing complications following surgical salvage procedures was also higher for the group of patients treated with neutrons alone. Ten patients in the neutron-only group underwent major surgery as a salvage procedure for persistent or recurrent tumor and 6 of these developed significant postoperative complications such as wound necrosis, fistula formation or carotid rupture. One other patient had an area biopsied that was suspicious for recurrent tumor and had severe healing difficulty (biopsy was negative). In contrast, 6 patients in the mixed beam group had major surgical salvage procedures and only 2 of these had significant postsurgical complications. The details of the surgical procedures and the nature of the complications are discussed more explicitly by Griffin et al[11] who also discuss the complications associated with planned combined pre- or postoperative neutron or mixed beam irradiation and surgery.

Overall survival curves for the two groups of patients are shown in Fig. 3. The difference between the two curves in part relates to the greater effectiveness of surgical salvage procedures for the mixed beam group but also reflects the somewhat more advanced tumors treated in the neutron-only group. For example, in the mixed group 11 patients survived more than two years from completion of treatment and 9 of these had N_0 disease. In neutron-only group, 6 patients survived more than two years after completing treatment and only 2 of these had N_0 disease.

DISCUSSION

High linear energy transfer (LET) radiation modalities such as neutron, pi-mesons, and heavy ions have different radiobiological properties compared to low LET modalities such as photons and electrons and some of these differences make them attractive for the treatment of certain tumor types[12]. Of these modalities, neutrons have thus far been the most extensively studied with

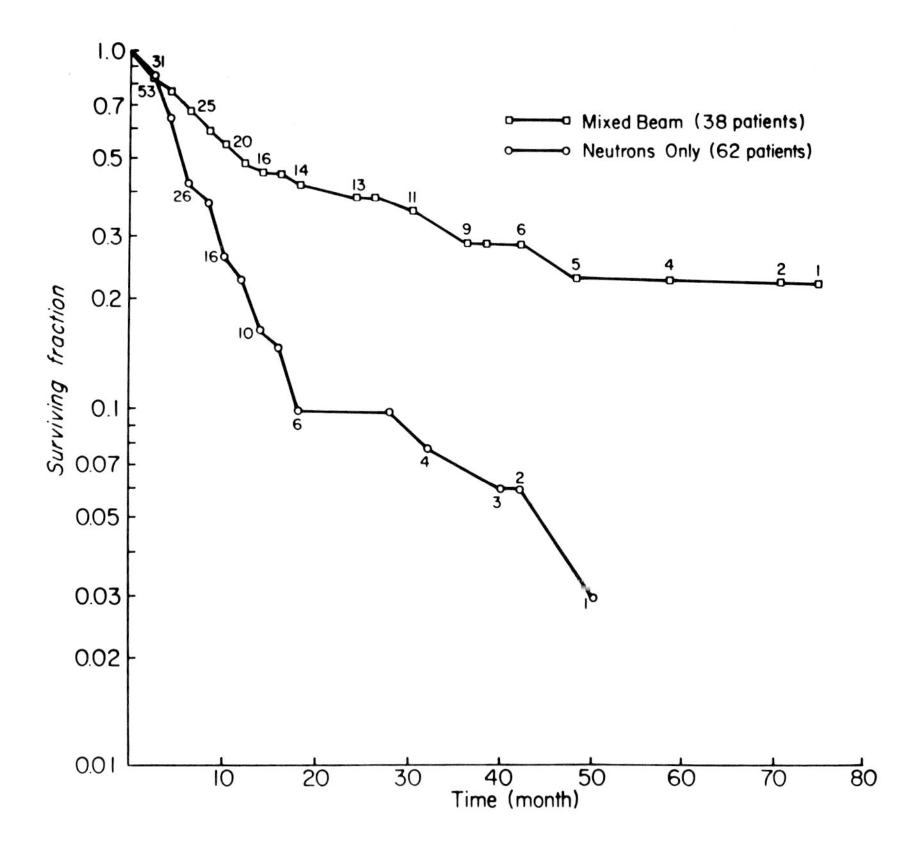

FIG.3. Surviving fraction of patients with squamous cell carcinomas of the head and neck region as a function of time after completing therapy. For the puposes of these curves, those patients whose treatments were stopped prior to receiving the planned total dose were assumed to have "completed" therapy at time t=0. (For details of curve calculation, etc., see caption to Fig.1.)

clinical work dating back to Stone and Larkin[13]. Although the radiobiological properties of fast neutrons are now better understood than in Stone's era, clinical trials are continuing to provide new information. This paper gives a final report on the results of a phase I pilot study conducted at the University of Washington on the efficacy of fast neutron radiation therapy for the treatment of advanced squamous cell carcinomas of the head and neck region. We studied both the use of neutrons alone and as part of a mixed beam (neutron/photon) treatment regime.

Table II compares our results for local control at the primary site and for survival with the results for other comparable groups of patients treated with fast neutrons at other institutions. The best local control rate is that of Catterall et al[14] of Hammersmith Hospital (London) who treated their patients with a total of 1560 rad neutron in 12 fractions over 4 weeks using the neutron beam produced by 16 MeV d $\rightarrow$ Be reaction. They analyze results for 70 patients and note a persistent local control in 53 of them. They had only one patient fail locally after having achieved an inital complete remission -- truly a remarkable result. However, their two-year survival rate of 28% is due in part

TABLE II. TWO-YEAR LOCAL CONTROL RATES AT THE PRIMARY SITE AND ACTUARIAL SURVIVAL RATES FOR PATIENTS WITH ADVANCED HEAD AND NECK CANCERS TREATED WITH FAST NEUTRON TELETHERAPY

	Local Control of Primary (2-year)	Actuarial Survival (2-year)
This Work		
Mixed Beam (38 patients)	30%	40%
Neutrons Only (62 patients)	20%	10%
Catterall _et al_ [14]		
Neutrons Only (70 patients)	76%	28%
*Maor _et al_ [15]		
Pilot Neutrons Only (49 patients)	43%	20%
Pilot Mixed Beam (25 patients)	35%	25%
Randomized Mixed Beam (41 patients)	60%	40%
**Battermann and Breuer [16]		
Neutrons Only (59 patients)	61%	17%

* Pilot neutron only includes 7 patients with salivary gland tumor; pilot mixed beam includes 2 patients with salivary gland tumors.

** Includes 11 patients with salivary gland tumors.

to 10 patients dying of treatment-related complications and 13 patients dying of distant metastases. Maor _et al_[15] report on three groups of patients treated at M.D. Anderson Hospital and Tumor Institute using the neutron beam produced by a 50 MeV → Be reaction. Their pilot study groups correspond most nearly to the patients whom we treated and their local control and survival rates are not too different from ours. However, it should be noted that their pilot groups contain patients with salivary gland tumors (noted in the table) and this type of tumor is currently felt to be one of the most responsive to fast neutron radiation. Later, they randomized patients to either mixed beam or photon irradiation because of a high complication rate associated with treatment by neutrons alone. The randomized mixed beam group was approximately comparable to the pilot mixed beam group as far as distribution of tumor stages and so the 60% local control rate may represent simply a reduced overall follow-up time. Battermann and Breur[16] report on a group of patients treated at the Antoni van Leeuvenhoekhus Hospital (Amsterdam) with a 14 MeV neutron beam produced by a d → T reaction. They show a high "persisting" local control rate but a 2-year survival not much different from ours. We also note that their treatment group includes 11 patients with salivary gland tumors.

In our study, the group of patients treated with neutrons alone had a somewhat more advanced disease state than the group treated with the mixed beam regime. If we compare the initial complete remission rates at the primary sites as a function of stage, we find, respectively, for the mixed beam and the neutron only groups 80% (8/10) versus 86% (6/7) for T_2 lesions, 70% (12/17) versus 48% (14/29) for T_3 lesions and 44% (4/9) versus 30% (8/26) for T_4 lesions. Although the distribution of tumor sites is not identical, these results suggest that for the larger tumors the local control rate with the mixed beam regime may be somewhat better than that achieved using neutrons alone. The local control rate for the cervical neck nodes noted by Griffin et al[6] also bears this out.

There were noticeable differences in the treatment-related complication rates -- both in terms of acute and late effects. Five patients in the neutron-only group developed a severe skin desquamation necessitating a break in treatment. This did not occur in the mixed beam group. Significant pharyngeal wall edema in patients with local tumor control occurred in 6 patients in the neutron-only group compared with one patient in the mixed beam group. Seven patients in the neutron-only group developed significant treatment-related neurological problems -- either cranial nerve damage or cervical myelitis -- and no such problems occured in the mixed beam group. Moreover, the complications of salvage surgery were also more severe in the neutron-only group (6/10) than in the mixed beam group (2/6). Other studies[11,14,15]also indicate a significant morbidity associated with high dose radiation therapy using neutrons alone. Hence, the current cooperative randomized study in the United Sates (RTOG 76-10) compares only the mixed beam form of treatment with conventional photon radiation. While this study addresses the question of comparing one type of neutron treatment with conventional photon irradiation, it does not address the questions of optimizing the given doses and overall time courses of neutron and photon irradiation in order to achieve maximum local control with a minimum of treatment-related morbidity. This latter question will be considered in future studies.

REFERENCES

[1] WEAVER, K., BICHSEL, H., EENMAA, J., WOOTTON, P.: Measurements of photon dose fraction in a neutron radiotherapy beam, Med. Phys. 4 (1977) 379-386.
[2] WOOTTON, P., ALVAR, K., BICHSEL, H., et al.: Fast neutron radiotherapy at the University of Washington, J. l'Assoc. Can. Radiol. (1975) 44-53.
[3] NELSON, J.S.R., CARPENTER, R.E., PARKER, R.G.: Response of mouse skin and C3HBA mammary carcinoma of the C3H mouse to X-rays and cyclotron neutrons: effect of mixed neutron-photon fractionation schemes, Eur. J. Cancer 11 (1975) 891-901.
[4] RASEY, J.S., CARPENTER, R.E., NELSON, N.J., PARKER, R.G.: Cure of EMT-6 tumors by X-rays or neutrons: effect of mixed fractionation schemes, Radiology 123 (1977) 207-212.
[5] FLETCHER, G.H.: Textbook of Radiotherapy, 2nd ed, Lea and Febiger, Philadelphia (1973).
[6] GRIFFIN, T.W., LARAMORE, G.E., PARKER, R.G., et al.: An evaluation of fast neutron beam teletherapy of metastatic cervical adenopathy from squamous cell carcinomas of the head and neck region, Cancer 42 (1978) 2517-2520.
[7] GRIFFIN,T., BLASKO, J., LARAMORE, G.: Results of fast neutron pilot studies at the University of Washington, High LET Radiation in Clinical Radiotherapy (BARENDSEN, G.W., BROERSE, J., BREUR, K., Eds.), Pergamon Press, Oxford (1979) 23-29.

[8] LARAMORE, G.E., BLASKO, J.C., GRIFFIN, T.W., GROUDINE, M.T., PARKER, R.G.: Fast neutron teletherapy for advanced carcinomas of the oropharynx, Int. J. Radiat. Oncol., Biol. Phys. 5 (1979) 1821-1827.

[9] LARAMORE, G.E., GRIFFIN, T.W., TONG, D., GROUDINE, M.T., BLASKO, J.C., KURTZ, J., RUSSELL, A.H., PARKER, R.G.: Fast neutron teletherapy for advanced carcinomas of the oral cavity and soft palate, Cancer 46 (1980) 1903-1909.

[10] AMERICAN JOINT COMMITTEE FOR CANCER STAGING AND END RESULT REPORTING: Manual for Staging of Cancer (1978).

[11] GRIFFIN, T.W., WEISBERGER, E.C., LARAMORE, G.E., TONG, D., BLASKO, J.C.: Complications of combined surgery and neutron radiation therapy in patients with advanced carcinoma of the head and neck, Radiology 132 (1979) 177-178.

[12] WITHERS, H.R.: Biological basis for high LET radiotherapy, Radiology 108 (1973) 131-137.

[13] STONE, R.S., LARKIN, J.C.: The treatment of cancer with fast neutrons, Radiology 39 (1942) 608-614.

[14] CATTERALL, M., BEWLEY, D.K., SUTHERLAND, I.: Second report on a randomized clinical trial of fast neutrons compared with X or gamma rays in treatment of advanced cancers of head and neck, Br. Med. J. 1 (1977) 1942.

[15] MAOR, M.H., HUSSEY, D.H., FLETCHER, G.H., JESSE, R.H.: Fast neutron therapy for locally advanced head and neck tumors, Int. J. Radiat. Oncol., Biol. Phys. (1981, in press).

[16] BATTERMANN, J.J., BREUR, K.: Results of fast neutron teletherapy for locally advanced head and neck tumors, Int. J. Radiat. Oncol., Biol. Phys. (1981, in press).

Progress in Radio-Oncology II, edited by K. H. Kärcher et al., Raven Press, New York © 1982.

High-LET Radiation Therapy at the National Institute of Radiological Sciences

H. TSUNEMOTO, S. MORITA, T. ARAI,
K. KAWASHIMA, Y. NAKAMURA
National Institute of Radiological Sciences,
Anagawa, Chiba-shi
Japan

Abstract

During the period from November 1975 to December 1980, 679 patients with locally advanced and radio-resistant cancers were treated with fast neutrons which were produced by bombarding a thick beryllium target with 30 MeV deuterons. In this clinical trial, emphasis was placed on the evaluation of local control and complications following the treatment. The effectiveness of fast neutron therapy was evaluated for carcinoma of the oesophagus and the larynx, Pancoast's tumour of the lung and osteosarcoma, on the basis of local tumour control. The survival rate of the patients with osteosarcoma or carcinoma of the oesophagus was found to be improved by fast neutron therapy. The cumulative survival rates of the patients who received either fast neutrons or photons are 63.3% or 17.6%, respectively. In the case of osteosarcoma, the therapeutic gain factor was greater when chemotherapy was combined with fast neutrons than when it was combined with photons. This was considered to be due to the better local control of the tumour by fast neutrons at a dose below the tolerance limit above which complications arise. It is concluded that the positive effects of high-LET radiations will be even more enhanced when radiations with appropriate dose distributions become available.

1. INTRODUCTION

A clinical trial with fast neutrons was initiated in November 1975 at the National Institute of Radiological Sciences, Japan. The aim of this trial was to assess the local control indications and to evaluate the late damage in normal tissues following fast neutron therapy. Up to December 1980, 679 patients with locally advanced or radioresistant cancers had been treated with 30 MeV (d-Be) neutrons produced by a medical cyclotron. The dose rate in the tissue-equivalent phantom was 42 rad (n,γ)/min for an 11.4 cm x 11.4 cm field at a TSD of 200 cm. Three types of treatment schedules were used: (i) neutrons only; (ii) a mixed schedule; and (iii) a neutron boost. To estimate the relationships between radiation doses and the effects of the treatment, the concept of a biologically equivalent 'time, dose and fractionation' factor (TDF) was applied [1].

2. RESULTS OF TREATMENT

Results of fast neutron therapy are discussed on the basis of experience with carcinoma of the oesophagus and osteosarcoma. Carcinoma of the oesophagus was selected as a suitable case for fast neutron therapy because its local control rates are low and recurrences appear within two years when conventional radiation therapy is used. Osteosarcoma was chosen as a typical radioresistant tumour which can hardly be controlled by photon beam therapy.

2.1. Carcinoma of the Oesophagus

Two series of clinical trials with fast neutrons were conducted, one comprising radiation therapy alone and the other preoperative irradiation. For radiation therapy alone, a dose equivalent to TDF 100-110 was applied to the target volume with a mixed schedule or a fast neutron boost, whereas for the preoperative irradiation, a dose of TDF 60 was delivered with fast neutrons only.

Figure 1 shows the cumulative survival rates of patients suffering from carcinoma of the oesophagus treated with either fast neutrons or photons. For the fast neutron series, the cumulative survival rate for a five-year follow-up period is 26%, whereas the rate decreases to 5.9% for the patients who received conventional radiation therapy. Since the prognosis depends strongly on the pathological grade of infiltration of the tumour cells into surrounding tissues, the relationship between the effects of radiation on the tumour (Ef) and the microscopic infiltration of the tumour cells through the oesophagial wall (a) was investigated. The results are shown in Table I. With the fast neutron beam, the destructive changes in the tumour cells seem to be rather marked at the level of the tissue beyond the oesophagus (a_2, a_3), and less marked at the superficial levels (a_0, a_1). The photon beam is less effective as compared with fast neutrons at all the levels. Accordingly, it is suggested that these pronounced pathological effects caused by the fast neutrons contributed to the excellent survival rate of 35.6% for the

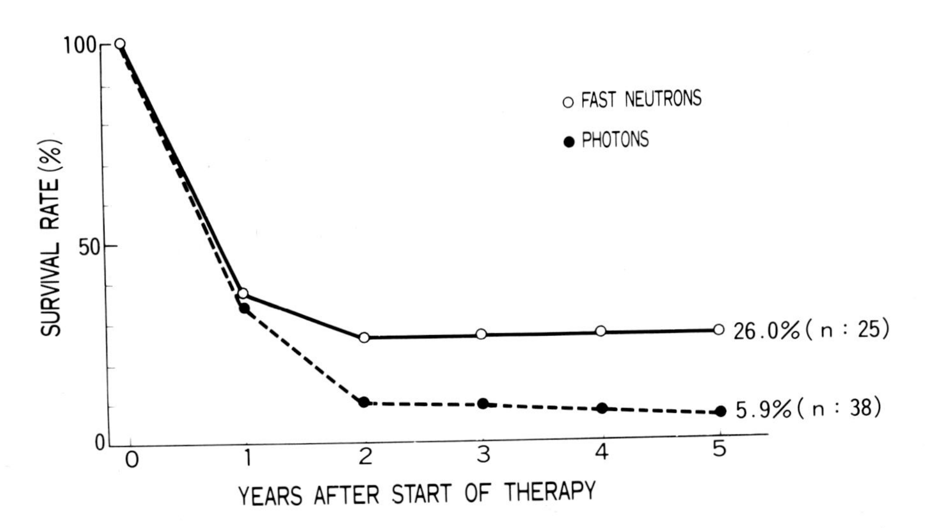

FIG.1. Cumulative survival rate of the patients suffering from carcinoma of the oesophagus treated with either fast neutrons or photons (data from T. Ishikawa, March 1981).

TABLE I. EFFECT OF FAST NEUTRONS AND PHOTONS ON CARCINOMA OF THE OESOPHAGUS: PATHOLOGICAL FINDINGS
(Data from T. Ishikawa, 1981)

	FAST NEUTRONS (N : 22)			PHOTONS (N : 70)		
	NO. PATIENTS	Ef_1	Ef_2, Ef_3	NO. PATIENTS	Ef_1	Ef_2, Ef_3
a_0, a_1	12	4/12 (33%)	8/12 (67%)	22	9/22 (41%)	13/22 (59%)
a_2, a_3	10	2/10 (20%)	8/10 (80%)	48	23/48 (48%)	25/48 (52%)

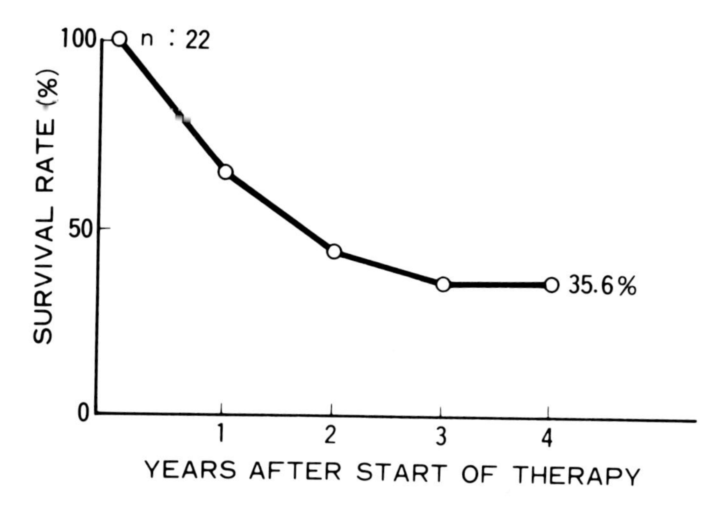

FIG.2. Cumulative survival rate of patients with carcinoma of the oesophagus (to March 1981): preoperative irradiation of fast neutrons.

patients who received preoperative irradiation (as shown in Fig.2).

2.2. Osteosarcoma

When a patient is diagnosed as having an osteosarcoma, physicians have to consider the possibility that the tumour cells have already invaded the subcutaneous soft tissues. Hence, chemotherapeutic drugs, such as adriamycin, were administered through the regional arteries prior to external irradiation with the fast neutron beam. After irradiation with a dose equivalent to TDF 120 to the target volume, a systemic chemotherapy was initiated to prevent dissemination of the tumour cells to the various organs. Therefore, chemotherapy was considered as an important therapy given to reduce local failure. When the pathological fracture became manifest, orthopaedic surgery was applied as befitted the condition of the lesion.

Figure 3 shows the cumulative survival rates of patients treated with either the fast neutron beam or a photon beam. The five-year

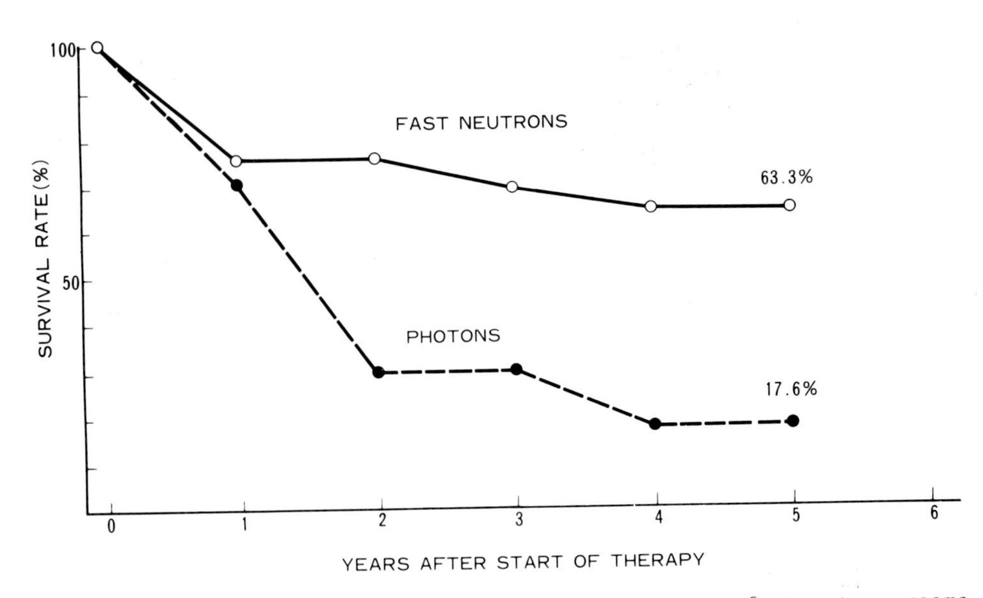

FIG.3. Cumulative survival rate of patients suffering from osteosarcoma treated with either fast neutrons or photons (data from N. Takada and M. Nakano, March 1981).

survival rate of patients who had received fast neutron therapy was 63.3%, whereas the rate was 17.6% when the photon beam was used. To evaluate the results more precisely, pathological studies were carried out on the specimens obtained when a biopsy or an amputation was done (Table II). Good response was seen in 66% of the specimens in the fast neutron therapy series, while it was only seen in 41% of the cases which received photon beam therapy. Skin reactions which developed after fast neutron therapy were no severer than those expected before the initiation of the clinical trial. The incidence of late skin reactions was almost identical for both series. On the other hand, the incidence was rather less in the fast neutron treatment than in the photon series when compared on the basis of the early reactions (Table III).

These results suggest that the improved therapeutic gain factor obtained when using the fast neutron beam is due to the better local control of the tumour without unexpected complications.

2.3. Glioblastoma Multiforme

The patients suffering from glioblastoma multiforme were first treated by a dose of 4000 rad of photons delivered over 4 weeks, and this was followed by a fast neutron boost with a shrinking field technique. As a standard treatment, a total dose equivalent to TDF 90-100 was given to the target volume. Crude survival rates were almost identical in the cases treated with fast neutrons or with photons (Fig.4). Since no severe complications have been observed in the fast neutron therapy patients, another clinical trial is under way using fast neutron therapy in combination with ACNU chemotherapy.

TABLE II. EFFECT OF FAST NEUTRONS AND PHOTONS ON OSTEOSARCOMA: PATHOLOGICAL FINDINGS
(Data from N. Takada, March 1980)

	NO. CASES	RESPONSE	
		POOR (1—3)*	GOOD (4—5)*
FAST NEUTRONS	35	12 (34%)	23 (66%)
PHOTONS	17	10 (59%)	7 (41%)

* 1: Non-destructive tumour nest.
2: Destructive tumour nest.
3: Non-viable tumour cells.
4: Atypical cells.
5: No tumour cells.

TABLE III. SKIN REACTION OF PATIENTS WITH OSTEOSARCOMA: FAST NEUTRON AND PHOTON THERAPY
(Data from N. Takada, March 1981)

	NO. CASES	COMPLICATION	
		EARLY REACTION	LATE REACTION*
FAST NEUTRONS	37	3 (8%)	15/28 (54%)
PHOTONS	17	5 (29%)	7/12 (58%)

* Patients receiving amputation were excluded.

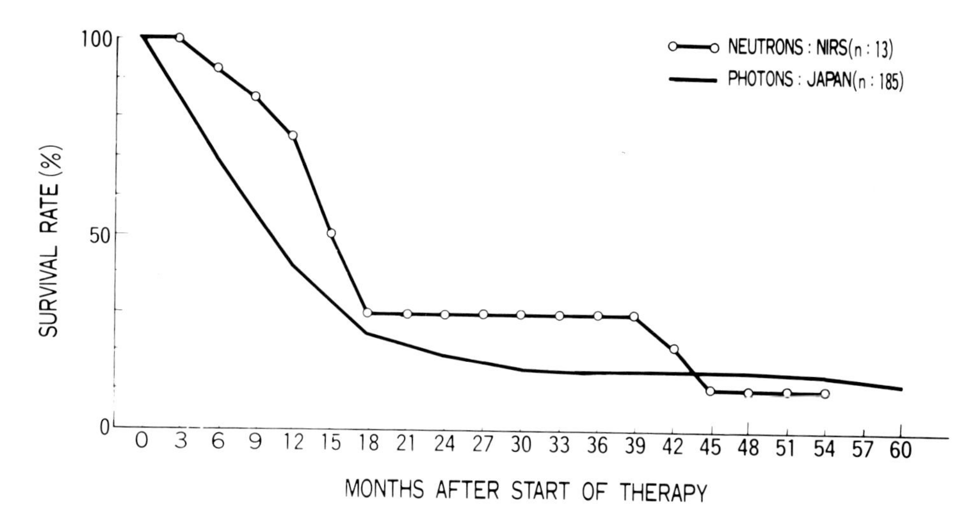

FIG.4. Relative survival rate of patients suffering from glioblastoma multiforme treated with either fast neutrons or photons (data from Y. Aoki, March 1981).

2.4. Skin Tolerance Dose

In view of the fact that complications in the normal tissues is the critical factor in high-LET radiation therapy, a study was made on the relationship between the tolerance dose and the area irradiated for various organs. Figure 5 shows the tolerance dose of the skin for neutrons only.

The relationship can be expressed by the following equations:

(1) Fast neutrons only: $TDF_{tol.} = 190 \cdot A^{-0.202}$

(2) Mixed schedule: $TDF_{tol.} = 128 \cdot A^{-0.052}$

where A is the area in square centimetres. As is apparent from the above equations, the slope for the mixed-schedule equation is less steep than that for fast neutrons only, suggesting that the complications arising from the mixed schedule might be slightly milder than those arising from a neutron-only schedule.

3. DISCUSSION

The results of clinical trials with 30 MeV (d-Be) neutrons suggest that high-LET radiations can contribute more effectively to improve the cure rate of certain cancers than can the conventional radiations. The results suggest also that normal tissues can tolerate fast neutrons well if the treatment schedules applied are appropriate. On the basis of results for local tumour control, carcinoma of the larynx and the oesophagus, Pancoast's tumour of the lung and osteosarcoma were considered suitable lesions for treatment with fast neutrons. (In our treatments of advanced carcinoma of the uterine cervix and the tongue, brachytherapy was necessarily applied, because the tumours were considered to be poorly controlled by the external radiation therapy alone, even when using fast neutrons.) Further studies are necessary to evaluate the effect of fast neutrons in cases of glioblastoma

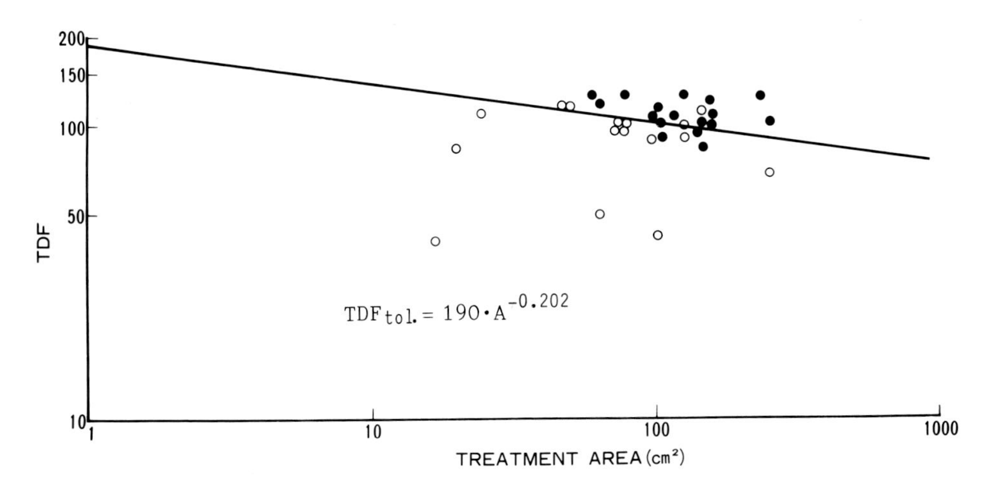

FIG.5. Relationship between tolerance dose of the skin ($TDF_{tol.}$) and the area of the field: fast neutrons only.

multiforme, especially with regard to the treatment policies and the complications. When the tumour cells have spread superficially into the skin, it is suggested that the malignant melanoma could be controlled by fast neutron therapy alone, while neutron therapy combined with surgery is indispensable to manage such a radioresistant tumour when the cells have infiltrated deeply into subcutaneous tissues. It is concluded that the usefulness of high-LET radiations in radiation therapy can be enhanced more definitely if such radiations become available with better depth-dose distributions.

REFERENCE

[1] NAKAMURA, Y.: Treatment planning method with the use of the TDF biological equivalent concept in fast neutron therapy, Nippon Acta Radiol. 38 (1978) 950.

Progress in Radio-Oncology II, edited by K. H. Kärcher et al., Raven Press, New York © 1982.

Results with a d + T Fast Neutron Machine in Amsterdam

J.J. BATTERMANN
Antoni van Leeuwenhoek Hospital,
Amsterdam,
The Netherlands

Abstract

The paper describes the five years' experience of neutron therapy with a 14 MeV (d+T) generator at Amsterdam. Interest was specially focussed on head and neck tumours, pelvic tumours, brain tumours and soft-tissue sarcomas. The preliminary results of some pilot studies and clinical trials are discussed, as are their implications for further therapy with heavy particles.

1. INTRODUCTION

Fast neutron teletherapy started in the Netherlands in 1975 after a period of extensive radiobiological investigation at the Radiobiological Institute TNO, Rijswijk. The Antoni van Leeuwenhoek Hospital was the second institute in Europe where fast neutrons were routinely applied for clinical purposes.

Clinical studies had started in London in 1966, at the Hammersmith Hospital, with a d(16 MeV)+Be cyclotron unit producing neutrons with a mean energy of 6.7 MeV. Already in 1975 the results of the first clinical trial on the application of fast neutrons in advanced tumours of the head and neck were published by Catterall et al. [1,2]. Head and neck tumours were chosen because of the relative ease of scoring tumour regression and normal tissue reactions and because of the poor beam characteristics of this cyclotron. The results of the trial favoured neutron therapy, even though criticism was voiced about the design of the trial.

Around 1975 the clinical investigation of fast neutron irradiation was also started in other cancer centres in Europe. In most centres relatively low-energy neutrons are used, derived from either a cyclotron or a d+T generator. In Amsterdam a d+T neutron generator, designed by Philips Medical Systems, was installed in 1975 to investigate the clinical application of fast neutrons. The beam characteristics of this machine are comparable with those of the Hammersmith cyclotron, but are far worse than those of modern megavoltage X-ray machines (Mijnheer et al. [3]). In spite of these poor beam characteristics, an attempt was also made to irradiate deep-seated tumours with fast neutrons.

In this paper the results of some pilot studies are described, and some preliminary data on clinical trials are given.

2. MATERIALS AND METHODS

Fast neutron therapy at Amsterdam is given with the Philips 14 MeV d+T neutron generator (Broerse et al. [4]). With a neutron output of approximately 10^{12} n/s, the dose rate at 10 cm depth in a water phantom is about 4 cGy/min at a source-to-skin distance of 80 cm. Thus the treatment time for 80 cGy per day will be around 20 minutes. If more fields have to be irradiated (such as six fields in the pelvis), or if wedges are used, the treatment time will be prolonged. As the 50% isodose curve is found at 9.5 cm, using a 9 x 11 cm^2 collimator, the depth-dose properties are slightly better than those from the Hammersmith cyclotron at a SSD of 120 cm.

The lifetime of the tube varies between 100 and 200 hours. Replacement takes only a few hours and does not interrupt the routine treatment. Separate neutron and gamma doses were measured in a water phantom with a small tissue-equivalent ionization chamber and a Geiger-Müller counter. The gamma contribution varies between 10% and 20%, depending on field size and depth. The dose is expressed as total dose (= neutron dose + gamma dose), with the gamma dose contribution given in brackets.

All patients are irradiated five times per week. The standard photon treatment policies were adapted for neutron therapy, but with an overall time of four weeks. The minimum tumour dose varied between 17 Gy and 24 Gy. This variation was caused by differences in field sizes and localization. In many cases the tumour was irradiated via two parallel opposed ports. When appropriate, wedge filters or a multi-field technique could be used. A computer program was used routinely to calculate the dose distribution.

In the pilot phase, interest was focussed on head and neck tumours, including tumours of the salivary glands, pelvic tumours, brain tumours and soft-tissue sarcomas. The results of these studies are discussed elsewhere (Battermann [5]), and will only be summarized here. Since 1978 controlled clinical trials have been running for advanced head and neck tumours, bladder tumours and rectal tumours.

3. RESULTS OF PILOT STUDIES, ACCORDING TO TUMOUR STAGE

3.1. Head and Neck Tumours

From November 1975 through May 1978, seventy patients with locally advanced tumours of the head and neck were treated. Eighteen were not evaluable for various reasons (treatment course not finished, 4 patients; combination of photon and neutron irradiation, 3 patients; previous photon irradiation, 3 patients; miscellaneous tumours, 8 patients).

The evaluable patients had a minimum follow-up of two years. In Fig.1 are shown the results obtained as survival and local disease-free curves for the whole group of head and neck tumours. Survival was poor because of the selection of patients. Since this was a pilot study, in general only patients with less than a 10% chance of a 5 year survival with conventional treatment modalities were accepted for the study. Hence, many patients had

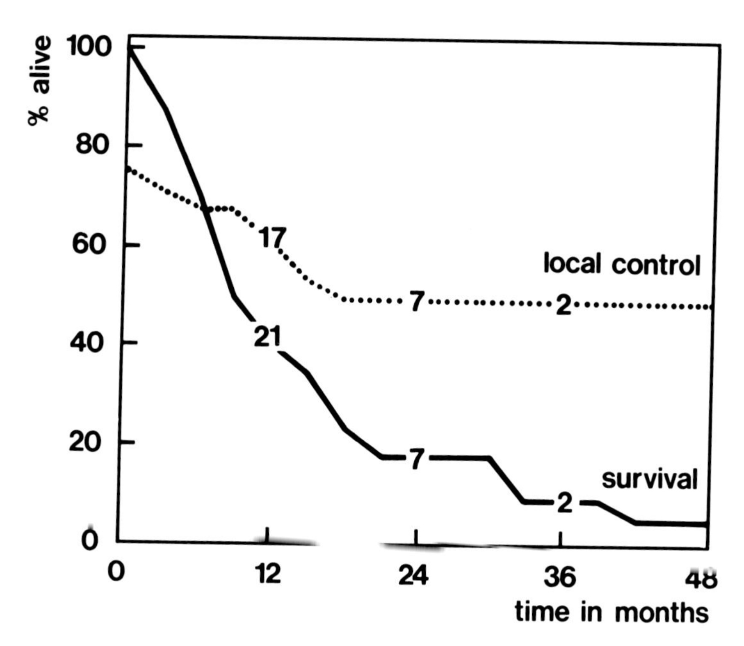

FIG.1. Actuarial survival and persisting local tumour control curves (Kaplan and Meier method) for fast neutron irradiation in 52 advanced head and neck tumours.

not only advanced primaries and cervical adenopathy, but also numerous medical problems concomitant with their advanced age.

Tumour staging was according to the recommendations of the UICC 1978. Table I gives the breakdown of the tumours according to stage. Patients with manifest distant metastases were not included in the study. Although in general only advanced lesions were accepted, 6 patients had relatively small lesions. Two of them had recurrent tumours after previous major surgery and died of tumour after 1½ and 13 months; 4 patients were judged unfit for surgery and were locally controlled but died of intercurrent disease after 5, 9, 13 and 17 months.

In most of the 46 patients with extended lesions, persisting local control was achieved; the exceptions were tumours in the anterior two thirds of the tongue (Table II). Persisting tumour in cervical lymph nodes was found in 7 out of 25 patients. Neck dissections performed in three of them showed fibrosis in only two cases. Wound healing was not prolonged after the neutron irradiation.

The good results obtained in 11 patients with tumours of the salivary glands are in agreement with the data described by Catterall [6]. The histology of the salivary gland tumours had no influence on the local cure rate. The histology of all other head and neck tumours was squamous cell carcinoma. No difference was found in prognosis or local cure rate with different histological grading of the tumours.

Apart from expected radiation sequelae such as mucositis, sore throat and dysphagia, major complications were found in 5 patients. One patient died of pneumonia, due to severe fibrosis

TABLE I. TNM STAGING OF HEAD AND NECK TUMOURS TREATED WITH FAST NEUTRONS IN THE PERIOD NOVEMBER 1975 THROUGH MAY 1978

Total	$T_2N_{0,1}$	$T_2N_{2,3}$	$T_3N_{0,1}$	$T_3N_{2,3}$	$T_4N_{0,1}$	$T_4N_{2,3}$
52	6	3	14	11	7	11

TABLE II. RESULTS OF FAST NEUTRON THERAPY IN THE HEAD AND NECK ACCORDING TO TUMOUR SITE

Site	total number	persisting control	one year survival	two years survival	severe complications
oral cavity and oropharynx	9	4	3	1	1 (severe fibrosis)
anterior 2/3 of tongue	11	2	7	-	-
base of tongue	6	5	-	-	-
larynx	8	6	5	1	2 {larynx necrosis, tracheostomy
hypopharynx	7	5	3	2	2 (cord damage
parotid gland	11	10	8	5	-
Total	52	32	26	9	5

in the floor of the mouth after a dose of 1950 cGy (230 cGy γ) given via two lateral ports of 16 x 14 cm^2. In general late fibrosis is more pronounced than after intensive photon therapy because of the 15% higher kerma for fatty tissue. Two out of 15 patients treated for primaries in the larynx or hypopharynx had focal necrosis of the larynx. In addition, two patients in this group showed severe nervous damage. The neurological symptoms appeared about 12 months after therapy and progressed to severe muscle weakness, but without complete paralysis. Because of the poor beam characteristics and the site of the tumour, part of the spinal cord received 16 Gy instead of 13 Gy.

3.2. Brain

Between November 1975 and November 1978, 22 patients with gliomas (grades III and IV) were treated with a neutron boost of 1160 cGy in 3 weeks, following whole-brain irradiation with photons at a dose of 30 Gy in 3 weeks (Battermann [7]). The results were disappointing, since 80% of the patients had clinical evidence of recurrent tumour and only 32% survived one year. As other investigations (Catterall [8], Parker [9], Duncan [10]) also came up with poor results, it was concluded at the Third Meeting on Fundamental and Practical Aspects of the Application of Fast Neutrons and Other High LET Particles in Clinical Radiotherapy that brain tumours are unsuited for neutron therapy (Dutreix and Tubiana [11]).

3.3. Pelvis

As advanced tumours of bladder and rectum have a poor prognosis with conventional photon therapy, these tumours were also considered suitable for a pilot study with fast neutrons. In Fig.2 the results are given as actuarial survival and local disease-free curves. The results after treatment with two opposed fields were more or less similar to the observations made by Stone almost 40 years earlier in that a good tumour regression was combined with severe late reactions (Table III). The complications were mainly caused by the poor penetration of our neutron beam, leading to overdoses of 20% to parts of the skin and intestine. The complication rate could be reduced significantly by the use of a six-field technique.

The complications seen after neutron irradiation with two opposed beams were compared with those seen after ^{60}Co irradiation for treatment of advanced bladder tumours during the years 1967 and 1968 (Battermann et al. [12]). A dose/effect curve was constructed, both for neutrons and for ^{60}Co gamma rays, for skin and intestinal complications, and for tumour control (Fig.3). As can

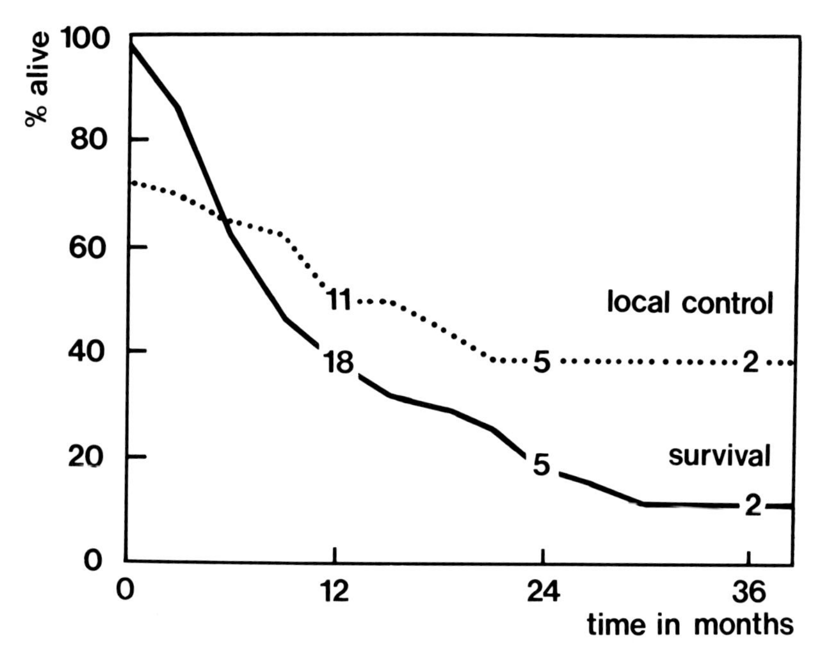

FIG.2. Actuarial survival and persisting local tumour control curves (Kaplan and Meier method) for fast neutron irradiation in 47 inoperable tumours from rectum and bladder.

TABLE III. RESULTS OF FAST NEUTRON THERAPY FOR BLADDER AND RECTAL TUMOURS, TREATED IN THE PERIOD MARCH 1976 THROUGH MAY 1978

	rectum	bladder
total number	25	22
persisting local control	14 [a]	11 [b]
> 50% initial tumour regression	4	3
< 50% initial tumour regression	7	8
complications	8	5
survival one year	11	7
two years	4	5
deceased of local tumour	4	7
metastases	9	5
complications	5	3
intercurrent diseases	3	4

[a] In 8 patients post mortem examination confirmed the clinical findings.
[b] In 5 patients post mortem examination confirmed the clinical findings.

be seen, a steep dose/response curve exists for tumour control and for radiation-induced normal tissue damage. The data were used in a logistic model (Cox [13]) to derive RBE values (Table IV). As the differences between the RBE for tumour control and serious normal tissue damage are small, the overall therapeutic gain for neutron irradiation of pelvic tumours will be small.

3.4. Soft Tissue

Only the results obtained in 22 patients with very extended sarcomas can be given. Eight of them were locally cured by neutron irradiation. In 6 of these 22 patients severe complications were noticed, especially in those patients treated for pelvic sarcomas. Three had intestinal damage, four had skin necrosis (one suffered from both skin and intestinal damage). As explained above, these complications were mainly caused by overdosing parts of the normal tissue when only two opposing fields were used.

These poor results in this group of patients are in contrast to observations made by others (e.g. Catterall and Bewley [6]; Tsunemoto et el. [14]; Salinas et al. [15]). In the series published by Catterall, local control was achieved in 21 out of 28 patients, with 30% complications. However, she treated most of these patients for minimal tumour residue after surgical excision, while in our series all patients had bulky tumours.

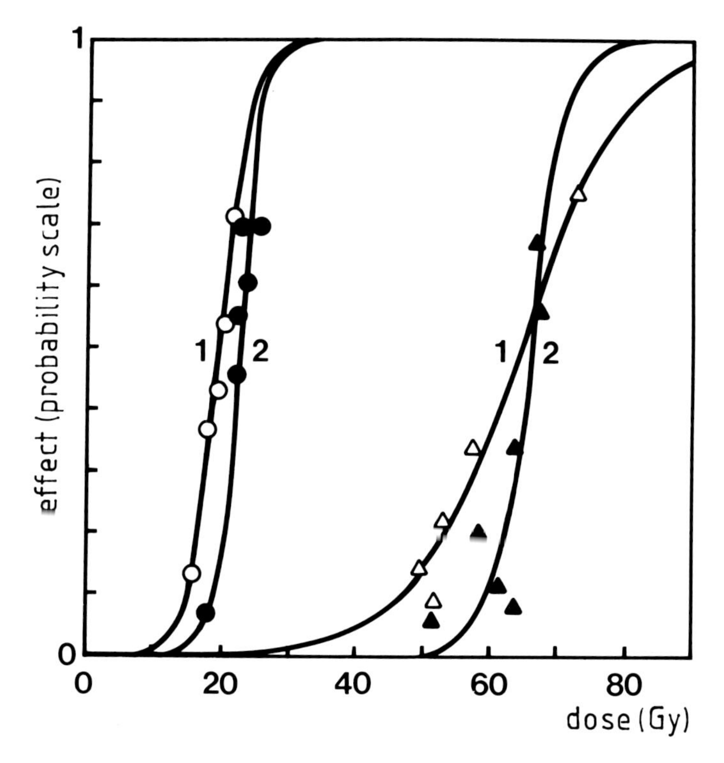

FIG.3. Dose/effect relationship for tumour control (1 - open symbols) and for normal tissue damage (2 - full symbols) after neutron and photon irradiation for T_{4B} bladder tumours. Round symbols represent results for neutrons, triangles for photons.

4. PRELIMINARY RESULTS OF CLINICAL TRIALS

Since May 1978 controlled clinical trials have been carried out at the Antoni van Leeuwenhoek Hospital. The head and neck protocol is in accordance with the combined Edinburgh, Essen, Amsterdam trial for advanced head and neck tumours (excluding salivary glands) as advocated by the High LET Therapy Group of the EORTC. Only some preliminary results will be summarized here. Table V gives the results of the head and neck trial. No significant differences were found in local tumour control or survival. However, it can be said that in the neutron arm the cases included were more advanced than in the photon arm. No serious complications were observed, except for one patient in the photon arm who died after dental treatment without antibiotic prophylaxis.

The results for the bladder and rectum studies are given in Table VI. In this study two dose levels in the neutron arm are introduced. The advantages of this policy are explained somewhere else (Breur and Battermann [16]). As all these pelvic tumours were randomized in one group, the numbers in each group are not equal. However, the results for both tumour sites are almost identical as far as local tumour control and survival are concerned. Up to May 1981, no fatal complications had been

TABLE IV. ESTIMATED RBE VALUES, USING THE COX [13] LOGISTIC MODEL
(from Battermann et al. [12])

	RBE	95% confidence interval
Serious skin damage	2.94	2.75 - 3.13
Serious intestinal damage	3.17	2.92 - 3.42
Bladder tumour control	3.19	2.73 - 3.65

TABLE V. PRELIMINARY RESULTS OF A CONTROLLED CLINICAL TRIAL FOR LOCALLY ADVANCED HEAD AND NECK TUMOURS
(MAY 1978 - MAY 1980)

	Photons	Neutrons
Total	13	20
Persisting local control	5	9
Survival ½ year	10	11
1 year	5	7
Salvage surgery	1 [a]	5 [b]
Complications	1	-
Deceased with local tumour	6	11
locally controlled	2	3

[a] neck dissection, one node negative.
[b] neck dissection, 2 nodes negative, 3 nodes positive.

noticed, which is the result of using a six-field technique. Only one patient needed a permanent colostomy. As in the head and neck trial, no significant differences are found in survival, but the tumour control seems to be better in the high-dose neutron arm as compared with the lower dose neutron arm or with the photon arm.

5. DISCUSSION

Results were given for some pilot studies, as well as preliminary results for two controlled clinical trials. A large number of patients are already being irradiated with fast neutrons in different cancer centres over the world; however, many more patients were needed to establish the more or less generally accepted treatment regimes currently used in photon therapy.

TABLE VI. PRELIMINARY RESULTS OF A CONTROLLED CLINICAL TRIAL FOR LOCALLY INOPERABLE RECTAL AND BLADDER TUMOUR (MAY 1978 - MAY 1980), USING TWO DOSE LEVELS IN THE NEUTRON SITE

	Photons	Neutrons 16 Gy	Neutrons 18 Gy
Bladder, total	5	9	11
Persisting local control	-	3	5
Survival ½ year	2	7	8
1 year	-	4	2
Complications	-	1	-
Deceased with local tumour	4	3	6
locally controlled	-	1	2
Rectum, total	19	15	10
Persisting local control	-	2	4
Survival ½ year	14	8	6
1 year	8	3	4
Complications	-	-	1
Deceased with local tumour	11	8	3
locally controlled	-	1	2

The improvement in treatment results by the application of fast neutrons will certainly not be as dramatic as at first expected by some clinicians. An extensive number of patients, treated in well designed clinical trials, will be needed to show the possible advantages of high-LET radiation. It was shown in our studies on human pulmonary metastases (Battermann et al. [17]) that only for a proportion of all studied tumours do fast neutrons have RBE values in excess of three. As a value of approximately three is generally accepted as the RBE for most normal tissues, a therapeutic gain with neutrons as compared with photons can be expected for only a limited number of patients.

Furthermore most neutron studies have been handicapped by suboptimal dose distribution and positioning. Improvement in dose distributions by the use of high-energy cyclotrons will be indispensable for further clinical investigations.

REFERENCES

[1] CATTERALL, M., SUTHERLAND, I., BEWLEY, D.K.: Br. Med. J. 2 (1975) 653.

[2] CATTERALL, M., BEWLEY, D.K., SUTHERLAND, I.: Br. Med. J. 1 (1977) 1642.

[3] MIJNHEER, B.J., ZOETELIEF, J., BROERSE, J.J.: Br. J. Radiol. 51 (1978) 122.

[4] BROERSE, J.J., GREENE, D., LAWSON, R.C., MIJNHEER, B.J.: Int. J. Radiat. Oncol., Biol. Phys. 3 (1977) 361.

[5] BATTERMANN, J.J.: Clinical Application of Fast Neutrons, The Amsterdam Experience, Thesis, University of Amsterdam, 1981.

[6] CATTERALL, M., BEWLEY, D.K.: Fast Neutrons in the Treatment of Cancer, Academic Press, London, and Grune and Stratton, New York (1979).

[7] BATTERMANN, J.J.: Int. J. Radiat. Oncol., Biol. Phys. 6 (1980) 333.
[8] CATTERALL, M., BLOOM, H.J.G., ASH, D.V., RICHARDSON, A., UTTLEY, D., GOWING, N.F.C., LEWIS, P., CHAUCER, B.: Int. J. Radiat. Oncol., Biol. Phys. 6 (1980) 261.
[9] PARKER, R.G., BERRY, H.C., GERDES, N.J., SOROSSON, M.C., SHAW, C.M.: Am. J. Roentgenol. 127 (1976) 331.
[10] DUNCAN, W., ARNOTT, S.J.: Eur. J. Cancer, Suppl. (1979) 31.
[11] DUTREIX, J., TUBIANA, M.: Eur. J. Cancer, Suppl. (1979) 243.
[12] BATTERMANN, J.J., HART, A.A.M., BREUR, K.: Br. J. Radiol. (1981, in press).
[13] COX, D.R.: Analysis of binary data, Methuen's Monographs, London (1970).
[14] TSUNEMOTO, H., MORITA, S., ARAI, T., KUTSUTANI, Y., UMEGAKI, Y.: in Treatment of Radioresistant Cancers (Proc. Int. Symp.), Elsevier/North-Holland Biomedical Press, Amsterdam, New York, Oxford (1979).
[15] SALINAS, R., HUSSEY, D.H., FLETCHER, G.H., LINDBERG, R.D., MARTIN, R.G., PETERS, L.J., SINKOVICS, J.G.: Int. J. Radiat. Oncol., Biol. Phys. 6 (1980) 267.
[16] BREUR, K., BATTERMANN, J.J.: Eur. J. Cancer, Suppl. (1979) 273.
[17] BATTERMANN, J.J., BREUR, K., HART, A.A.M., van PEPERZEEL, H.A.: Eur. J. Cancer 17 (1981) 539.

Progress in Radio-Oncology II, edited by K. H. Kärcher et al., Raven Press, New York © 1982.

An Interim Assessment of the Experience of Fast Neutron Therapy in Edinburgh

W. DUNCAN, S.J. ARNOTT
Department of Clinical Oncology,
University of Edinburgh,
Western General Hospital,
Edinburgh,
United Kingdom

Abstract

Clinical studies of fast neutron therapy were started in Edinburgh in March 1977. The treatment facility has an isocentric machine and a fixed horizontal beam. The neutron beam is produced by 15 MeV deuterons on a thick beryllium target. Six hundred and forty-one patients have been included in these studies in the four-year period to March 1981. Randomly controlled trials have been conducted since May 1977 and 376 patients have been recruited in this period. The local tumour response rates and morbidity observed is given for trials of patients with cerebral gliomas, squamous carcinoma of the head and neck region, transitional cell carcinoma of the bladder and adenocarcinoma of the rectum. This interim evaluation does not demonstrate any advantage of fast neutron radiotherapy as compared with photon therapy. However, greater numbers of patients require to be studied and followed-up for longer intervals. A definitive assessment of the randomly controlled clinical trials in Edinburgh should be possible at the end of 1982.

Clinical studies of fast neutron therapy began in Edinburgh in March 1977. The neutron beam is generated by the interaction of 15 MeV deuterons on a thick beryllium target. The deuterons are accelerated by a CS30 compact cyclotron.[1] Two treatment rooms are available, one providing a fixed horizontal beam and the other housing an isocentric machine, capable of 240° rotation. The beam collimators are similar in both rooms and consist of interchangeable wooden applicators. Both treatment heads have locations for beam bending wedges. Both have optical field definition and front and back pointers for accurate setting-up. Treatment planning is performed by a Varian computer system originally developed in Edinburgh.

The experience of the Edinburgh group in the four-year period until March 1981 is given in Table I. It will be seen that over half of the patients have been included in randomly controlled trials. Patients are still being recruited into so-called preliminary studies when they are ineligible for entry into a clinical trial or when no protocol exists for randomization. The small number of patients with brain tumours are being treated using a mixed modality schedule of two neutron fractions and three photon fractions each week.

[1] The Cyclotron Corporation.

TABLE I. EDINBURGH CYCLOTRON FACILITY: STUDY GROUPS (March 1977 to March 1981)

Study group	No. of patients
Preliminary neutrons	243
Preliminary, mixed schedule	22
Controlled trials, neutrons	333
Controlled trials, mixed schedule	12
Control trials, with surgery	31
Total	641

TABLE II. EDINBURGH CYCLOTRON FACILITY: TUMOUR SITES (March 1977 to March 1981)

Tumour site	No. of patients	Tumour site	No. of patients
Brain	84	Vertebral metastases	27
Head and neck	122	Secondary neck nodes	24
Bladder	101	Stomach	29
Oesophagus	57	Soft-tissue sarcoma	25
Rectum	50	Melanoma	25
		Salivary gland	19

The majority of patients are treated by 20 daily fractions over 4 weeks, all fields being treated daily. The central tumour dose varies between 1550 and 1685 rad total absorbed dose depending on site and field size. The gamma ray contribution is 6% to 11%. A number of patients are entered into studies in which each are given 3 fractions of neutrons per week over 4 weeks. At present the same total absorbed dose is prescribed to be given in 12 or 20 fractions in the same overall time. Patients with secondary neck nodes are being included in a randomly controlled trial comparing treatment given in either 3 or 5 fractions per week. It is hoped to evaluate both local tumour control rates and the morbidity of these regimes. The principal tumour sites that have been studied are given in Table II.

The first two randomly controlled studies started in Edinburgh were concerned with brain tumours and 'head and neck' cancer. The design of the brain trial was based on the experience of the Seattle group [1] and the trial conducted at the Hammersmith Hospital [2]. Patients with cerebral gliomas (Grades III and IV) were admitted to the trial following either biopsy or sub-total removal. Patients were stratified by the type of surgical procedure and by age (above or below 40 years). We would now advise that patients also be stratified by the degree of neurological deficit they exhibit (minor or major) as this has important prognostic implications. The treatment technique involved irradiating the whole brain. A parallel opposed pair of fields (usually 15 cm x 10 cm) was employed using 4 MV X-rays. To achieve a similar distribution of dose a third field was necessary when neutron therapy was given. The X-ray dose was 4750 rad in 20 fractions, and the 'neutron' total dose was 1380 rad (10% gamma).

A summary of the results of this trial is given in Table III. Seventeen deaths have occurred among the 18 patients treated with neutrons. All were assumed to have residual tumour in the brain. Fourteen of the 16 patients treated with photons have died, all presumed to have residual cerebral tumour. Two patients treated with neutrons developed unusual features of dementia and ataxia, which were considered to be signs of demyelinization. In one, evidence of white matter degeneration was found at autopsy, but not in the other. In one other patient treated with neutrons there was also evidence of abnormal white matter changes in the absence of any clinical features. The median survival was only seven months for neutron treated patients and ten months for those receiving photon irradiation. There are three survivors, one treated with neutrons, the others with photons. In the neutron treated group there was clinical or autopsy evidence of residual tumour in all patients and in three patients there were neurological changes, presumed to be a result of neutron irradiation.

It is now recognized that the RBE for white matter is surprisingly high [3], and it seems that it is much higher for normal brain than for high-grade gliomas. It would appear that further studies of fast-neutron irradiation alone for brain tumours cannot be justified, since a reasonable therapeutic ratio cannot be achieved.

One major investigation has been the trial of patients with previously untreated squamous cell carcinoma of the head and neck region. All sites were included except the naso-pharynx. All stages were included except stage T1 N0 in the oral cavity and oropharynx, and stage T2 N0 in the tongue when suitable for a single plane radio-nuclide implant. In the larynx T1 N0 cases were excluded and in the glottis also T2 N0 cases. The distribution of patients by site is given in Table IV and by stage in Table V. At the time of this interim analysis the distribution is similar in the neutron and photon treated groups. The neutron doses given depended on site and the volume irradiated. The total central absorbed dose varied between 1555 (7% gamma) and

TABLE III. EDINBURGH CYCLOTRON FACILITY: BRAIN TRIAL December 1980

	Neutrons	Photons
Total patients	18	16
Deaths	17	14
Tumour-related deaths	16	14
Autopsies	9	6
Abnormal white matter changes	2	0
Median survival (months)	7	10
Actuarial survival (at 18 months)	11.1%	16.2%

TABLE IV. EDINBURGH CYCLOTRON FACILITY: HEAD AND NECK TRIAL
Distribution by site, October 1980

Site	Neutrons	Photons
Oral cavity	15	21
Oropharynx	11	7
Larynx	13	11
Hypopharynx	5	6
Total	44	45

TABLE V. EDINBURGH CYCLOTRON FACILITY: HEAD AND NECK TRIAL
Distribution by staging, October 1980

Stage (UICC 1974)	Neutrons	Photons
T1 N3	0	1
T2 NO/1a	7	5
T2 N1/3	4	5
T3 NO/1a	14	15
T3 N1/3	14	14
T4 NO/1	5	5
Total	44	45

TABLE VI. EDINBURGH CYCLOTRON FACILITY: HEAD AND NECK TRIAL
Results by site, October 1980

Primary site	Local tumour control	
	Neutrons	Photons
Oral and oropharynx	48.0%	42.8%
Larynx and hypopharynx	50.0%	56.3%
Total	48.8%	47.7%

1685 (7% gamma) delivered in 20 daily fractions. The megavoltage X-ray dose lay between 5000 and 5500 rad using the same fractionation. When one examines the local tumour control rates there is no significant difference between the neutron and photon treated groups, either in the oral and oropharynx or in larynx and hypopharynx. No difference is seen in the response rates of the primary tumours alone or of the secondary nodes in the neck (Table VI and VII). Actuarial regression rates have been calculated for the whole group and the cumulative regression rate at six months after neutrons (71.3%) was similar to that after X-ray therapy (72.2%).

TABLE VII. EDINBURGH CYCLOTRON FACILITY: HEAD AND NECK TRIAL
Results by nodal status, October 1980

Node status	Local tumour control	
	Neutrons	Photons
N 1b	40.0%	50.0%
N 3	57.1%	45.5%
All nodes	52.6%	45.0%
< 3 cm	45.4%	42.9%
> 3 cm	62.5%	50.0%

TABLE VIII. EDINBURGH CYCLOTRON FACILITY: HEAD AND NECK TRIAL
Crude survival rates, October 1980

	Neutrons	Photons
3 months	95.0%	95.3%
12 months	40.1%	67.2%
24 months	36.1%	53.4%

It will be seen that the recurrence rate is also similar in both treatment groups. Salvage surgery was performed in almost equal numbers for persistent cancer or for radiation morbidity. The doses of neutron therapy used in the trial were selected after conducting pilot studies, which were used to indicate that the expected radiation morbidity would be comparable to that following megavoltage photons. The immediate reactions are similar, but we are seeing a greater number of serious late complications following neutron irradiation than after X-rays. This difference is not statistically significant and may represent a chance finding at the time of this interim review. The crude survival rates up to two years are given in Table VIII. There appears to be an advantage in photon therapy, but we expect the small difference recorded, which has no statistical significance, to diminish with increased numbers of patients and with a longer follow-up interval. Our evaluation at this time is that neutron irradiation is no better or no worse than megavoltage X-ray therapy for this group of tumours.

We have been interested also to compare the response of transitional cell carcinomas of the bladder to neutron irradiation with the response to megavoltage X-rays. Patients with all stages of the disease are included in this randomly controlled trial, provided they are suitable for radical treatment. Megavoltage X-ray therapy employs three fields (11 cm x 11 cm) and a central dose of 5500 rad in 20 daily fractions is given. A six field (11 cm x 11 cm) technique is required on the isocentric neutron therapy machine and a total dose of 1650 rad (10% gamma) is given in 20 daily fractions. The distribution of patients by stage and the local control rates for the neutron and photon treated groups are given in Table IX. It is clear that there is no difference in the tumour response assessed at six months after completion of treatment.

TABLE IX. EDINBURGH CYCLOTRON FACILITY: BLADDER TRIAL[a]
March 1981

Stage and tumour grade	Neutrons	Photons
T1 & T2 (all grades)	6	9
T3 (GI & II)	9	9
T3 (GIII)	11	10
T4 (all grades)	7	6
Total	33	34

[a] Local Control at 6 months 20 (60.6%) 22 (64.7%)

TABLE X. EDINBURGH CYCLOTRON FACILITY: SOFT-TISSUE SARCOMAS

Patients 18		
Complete response	Recurrence	Local control
10	3	7/18
Severe morbidity 6/13		

The morbidity of the small field beam directed technique at these dose levels is similar for both neutrons and photons. Serious problems with late radiation morbidity following neutron irradiation were encountered when a combined regional and small field technique was employed. In this technique the bladder received an additional total dose of 10 fractions of 35 rad using neutrons, compared with 10 fractions of 150 rad megavoltage X-rays, the remainder of the treatment being given by the regional technique. The RBE for such small dose fractions of 15 MeV(d+Be) neutrons is obviously much greater than the estimation of 5 which we had made originally.

A small group of patients with inoperable or recurrent adenocarcinoma of the rectum have also been studied and are now included in a randomly controlled trial. Numbers are very small indeed, but there is no evidence that the response rate after neutron therapy is different to that following X-ray therapy. Indeed our much larger experience of treating patients with cancer of the rectum in preliminary neutron studies gave a response rate similar to that which we would have expected to achieve with X-ray therapy. Another small group of patients with soft-tissue sarcomas of a variety of histological types have been treated electively with our fast neutron beam. Eighteen patients were suitable for critical evaluation, particularly in having measurable tumour for assessment. Table X gives the results, indicating that local control was obtained in seven of the 18 patients. The mean follow-up interval is short and so this figure may well be significantly reduced after a longer assessment period. The morbidity in this treated group was high and, in 13 patients who have been observed for more than twelve months after treatment, six had serious complications due to neutron therapy, which we feel is unacceptable. The morbidity was principally severe radiation changes in the skin and subcutaneous tissues, but two of these patients also had disabling impairment of joint mobility within the neutron treated volume.

Our overall experience has therefore been that neutron therapy may produce tumour regression rates similar to that achieved by good megavoltage photon therapy. The limitations of our neutron beam, especially its poor penetration, do increase the difficulties of field arrangements and of obtaining good dose distribution. These difficulties may largely be overcome. However it is apparent that there are real disadvantages in using fast neutrons. The increased absorption in fatty tissue will limit their application in cancer therapy. The fibrosis in subcutaneous tissues does produce a poor cosmetic end-result, and also secondary functional impairment. It also contributes to the development of lymphoedema. The lack of skin sparing is a much less serious disadvantage, but cannot be ignored when this feature of megavoltage X-rays is generally accepted as an important advantage.

A final reservation we have about neutron therapy concerns the relative disassociation of early and late effects on normal tissues [4]. Much more information is needed on the effects of neutron irradiation on normal tissues between two and five years after treatment, and also longer periods of time. It does seem that greater numbers of late sequelae may follow fast neutron irradiation than after X-ray therapy compared to similar levels of early reactions.

The efficacy of fast neutron radiation in cancer therapy must be judged primarily in relation to local tumour control and late normal tissue morbidity. Secondarily, there must be a worthwhile improvement in long-term survival rates. This does mean that the impact of any new form of local management, such as fast neutron therapy, will essentially be influenced by the natural history of the primary disease and its proclivity to metastasize in particular. Much more effort is required in collaborative randomly controlled trials to assess the role of fast neutron irradiation in cancer therapy, but its contribution is certain to be much less than supposed after the first enthusiastic assessment ten years ago.

REFERENCES

[1] LARAMORE, G.E., GRIFFIN, T.W., GERDES, A.J., PARKER, R.G., Fast neutron and mixed (neutron/photon) beam teletherapy for grades III & IV astrocytomas, Cancer 42 (1978) 96-103.

[2] CATTERALL, M., BLOOM, H.J.G., ASH, D.V., WALSH, L., RICHARDSON, A., UTTLEY, D., GOWING, N.F.C., LEWIS, P., CHAUCER, B., Fast neutrons compared with megavoltage X rays in the treatment of patients with supra-tentorial glioblastoma: a controlled pilot study, Int. J. Radiat. Oncol., Biol. Phys. 6 (1980) 261-266.

[3] HORNSEY, S., MORRIS, C.C., MYERS, R., WHITE, A., Relative biological effectiveness for damage to the central nervous system by neutrons, Int. J. Radiat. Oncol., Biol. Phys. 7 (1981) 185-189.
[4] WITHERS, H.R., FLOW, B.L., HUCHTON, J.I., HUSSEY, D.H., JARDINE, J.H., MASON, K.A., RAULSTON, G.L., SMATHERS, J.B., Effects of dose fractionation on early and late skin responses to γ-rays and neutrons, Int. J. Radiat. Oncol., Biol. Phys. 3 (1977) 227-233.

Progress in Radio-Oncology II, edited by K. H. Kärcher et al., Raven Press, New York © 1982.

The Progress of Neutron Therapy

Mary CATTERALL, R.D. ERRINGTON
Medical Research Council Cyclotron Unit,
Hammersmith Hospital,
London,
United Kingdom

Abstract

Attention is drawn to the increasing number of neutron therapy facilities. At present all these machines are so inferior to modern megavoltage apparatus that clinical trials which are truly controlled are impossible. Poor depth doses, wide penumbra, fixed field sizes and fixed positions of the beam unavoidably reduce the ability to deliver an adequate dose of neutrons to the tumour, and they increase the volume of normal tissues which are irradiated. These characteristics are especially important when compared with the deeply penetrating, sharply defined and flexible beams from linear-accelerator X-rays. There are encouraging signs that the new generation of machines will overcome these inadequacies and greatly improve dose distribution. In the meantime, trials at Hammersmith are mainly concerned with comparing the standard dose of 1560 rads given in 12 or in 9 fractions and the effects of an anti-inflammatory agent. The latter reduces the acute reaction but further data are required and are being collected. Good tumour control of melanoma is obtained when 1395 rads are given in 6 fractions over 12 days. The mechanically adjustable collimators designed at Hammersmith are a major advance and have improved treatments. They will be fitted to the isocentric head of the new high-energy machines in the UK and Seattle, USA.

The number of centres using fast neutrons in the treatment of solid tumours is increasing. From one at Hammersmith Hospital in 1970 there are now about 15 actually treating patients, or just about to start and 7 others to commence within the next three years. Four firms are at present actively producing cyclotrons for clinical use.

As new centres start treatments, the scrutiny of the clinical results attracts the attention of more and more onlookers. Computers are being used to enumerate complications and the demand increases for controlled clinical trials to compare neutron tumour control rates with those of photons. Such objectivity, although much more closely applied to neutrons than to electrons, chemotherapy or surgery, could produce a welcome and clear assessment of the value of neutrons in cancer treatment. However there are four important reservations to be made.

1. All the neutron machines now operating are grossly inferior to the modern megavoltage machines with which comparisons of treatments must be made. The most significant factors include poor penetration, wide penumbra and fixed positions of the neutron beams, fixed field sizes and poor output. Lack of reliability and availability are also major contributors to making clinical trials, which are truly controlled, impossible.

2. The margin between control of tumours with acceptable complications and failure to control with or without serious complications, is narrower with neutrons than with photons. This is not a detraction from the use of neutrons but it does underline the need to make neutron machines as good as photon ones. This equality especially applies to the penetration and penumbra of the beam, but also to the techniques of treatment used, because these will dictate the damage done to normal tissues and the delivery of a uniform and adequate dose to the tumours.

3. The users of neutrons must have available to them the necessary staff and hardware to plan treatments with CAT scans, simulation, neutrograms, activation analyses, shields, wedges and the facilities to alter plans during treatment to avoid exceeding tolerance doses to normal structures such as the cord, the eyes etc., while delivering sufficient dose to the tumour.

4. Computer programmes to assess tumour control and complications are at present misleadingly simple. They do not include highly significant factors such as the size of the volume treated, changes in the general condition of the patient, factors known to precipitate necrosis such as trauma and infection, and differences in treatment techniques. The quality of life is reduced to about 10 numerical grades and complications are translated into very broad groups.

These coarse assessments are necessary if large numbers of facts are fed into computers and some sort of report is to come out, but it is questionable whether they are appropriate to the subtleties of such a biologically complex group as patients with cancer. This situation is further complicated by the inferior characteristics of the neutron machines which are at present available for their treatment.

The clinical investigation at Hammersmith Hospital has been greatly improved and facilitated by the installation of mechanically adjustable field sizes. These were designed by the Physics and Engineering Sections of the Cyclotron Unit and will also be fitted to the isocentric head of the high energy cyclotron to be built in the Radiotherapy Department of Clatterbridge Hospital near Liverpool in the U.K. They have also been ordered by the group in Seattle for their new high energy cyclotron.

The "press-button" changing of the field size enables an infinite number of areas to be used, it reduces the work done by radiographers and reduces the time taken for treatment. Shields to protect normal structures are inserted into the collimator as required and they are kept in place by magnetic strips which adhere to the steel interior. Wedges are applied to the outside of beams. This system has been entirely reliable.

The cyclotron itself, which is 25 years old, required major maintenance during 1979 and 80 and patient numbers were reduced. It is now working well, although of course with the fixed horizontal and poorly penetrating beam and the wide penumbra of 7.5 MeV neutrons.

The results of the controlled clinical trial of advanced tumours of the head and neck have been published [1] and only a few comments are required here. Seventy per cent of the patients had nodes extensively involved; there were 76% complete regressions without recurrence on the neutron side compared with 19% on the photon side This trial has been criticised on the grounds that the control rate on the photon side is lower than in other trials and therefore indicates that these patients were undertreated. It has to be remembered however that the trial was started in 1971 and only patients with very advanced local disease were included. There was no age limit and many patients had co-existing diseases such as diabetes and hypertension; many would not have been fit enough to go into a hyperbaric oxygen tank. They were therefore a different population from those with similar tumours in other trials. It has also been stated that the improved control was at the expense of complications, but in many cases these were caused by the destruction of normal tissues by the tumour. The numbers of patients with controlled tumours and no serious complications were significantly greater in the neutron series. So also was the reduction in subsequent treatments necessary - salvage surgery, interstitial radiation, chemotherapy - after neutron therapy. This is of importance both economically and also in quality of life.

Tumours of the oropharynx, oral cavity, salivary glands and paranasal sinuses have responded well to the standard dose (1560 rads given in 12 fractions over 26 days) and special treatment techniques. The results from all stage III and IV tumours which have been treated are given in Tables I – IV.

These results are good but are not surprising. They reflect the known radiobiological and physical data that neutrons are more effective than photons against hypoxic cells and that neutrons are absorbed much less by bone than by soft tissues. The importance of hypoxia in tumours of the head and neck were demonstrated in the hyperbaric oxygen trial and the value of the differential absorption in bone and soft tissue is seen particularly in these sites, where the mandible or maxilla are in very close proximity to the tumours, which are of soft tissue.

The standard dose is now modified to 1395 rads in 6 fractions over 12 days for the treatment of malignant melanoma and results of tumour regression and lack of skin complications are excellent so far.

Trials are being conducted to compare the standard treatment with 1492 rads given in 9 fractions over 19 days, these patients are also randomised to receive, or not to receive, prednisolone by mouth during treatment and for 1 month afterwards. The steroid reduces the early reactions but its effect on later reactions and its possible effect on tumour control are not yet known.

Other trials are comparing neutrons with photons in the treatment of inoperable but superficially sited soft tissue sarcoma and in the prophylactic treatment of the operation site after excision of these tumours.

Fast neutron therapy in the U.K. has been given a great stimulus by the donation of £4 million by Liverpool (Lancashire) based charities. This money is for the purchase of a high energy cyclotron with an isocentric head. It was stipulated that the machine was to be placed in the Radiotherapy Department of Clatterbridge Hospital, about 10 miles from Liverpool. This department has 4 000 new patients each year. The Medical

TABLE I. TUMOURS OF OROPHARYNX TREATED BETWEEN JANUARY 1970 AND JANUARY 1980

Assessed February 1981, i.e. at least a one-year follow up

Stage	Nos. Treated	Complete Regression	Recurrence	Complication		Follow Up (Years)				
				Minor	Severe	1-2	2-3	3-4	4-5	5-7
III	6	6	O	O	4[x]	2	2 1 Alive			2 1 Alive
IV	9	9	O	1	4[o]	3 2 Alive	1 Alive	2	1	2 Alive
Total	15	15	O	1	8	5	3	2	1	4

Patients had locally advanced tumours but were in moderately good general condition. Five who were cachetic and unable to swallow before treatment are excluded.

x 2 due to spinal cord damage before cord tolerance to neutrons known.

o 2 precipitated by biopsy.

TABLE II. TUMOURS OF THE ORAL CAVITY TREATED BETWEEN JANUARY 1970 AND JANUARY 1980

Assessed February 1981, i.e. at least a one-year follow up

Stage	Nos. Treated	Complete Regression	Recurrence	Residual	Complication Minor	Complication Severe	Follow Up (Years) 1-2	2-3	3-4	4-5	5-6	6-7
III	7	6	1	1	1	1	1	3 2 Alive	1			2 2 Alive
IV	22	19	3	3	1	3	11	4	3 1 Alive		1 Alive	3 2 Alive
Total	29	25	4	4	2	4	12	7	4		1	5

Patients had locally advanced tumours but were in moderately good general condition.
Ten who were cachetic and unable to swallow before treatment are excluded.

TABLE III. SALIVARY GLAND TUMOURS TREATED BETWEEN JANUARY 1970 AND JANUARY 1980

Assessed February 1981, i.e. at least a one-year follow up

Histological Type	Numbers Treated	Complete Regression	Partial Regression	Recurrence after C.R.	Complications	Follow Up (Years)				
						1-2	2-4	4-6	6-8	8-10
Adenoid Cystic	18	18	0	4 at 13,30, 40 & 45 months after treatment	3 (All healed)	8 5 Alive	3 1 Alive	7 4 Alive		
Muco Epidermoid	7	7	0	0	2 Very mild		1	3 2 Alive	2 Alive	1 Alive
Adeno-carcinoma	6	6	0	2	0	1	1	4 2 Alive		
Anaplastic	5	4	1	?1	None	5 All died of mets				
Malignant Mixed	4	2	2 cyst remained	0	0	2		1	1 Alive	
Total	40*	37	3	6 ?7	None in long term	16 5 Alive	5 1 Alive	15 8 Alive	3 Alive	1 Alive

18 alive with no tumour, no symptoms from neutron treatment.

* 14 had tumours recurrent or residual after multiple "excisions" or RT or both.

TABLE IV. TUMOURS OF THE PARANASAL SINUSES TREATED BETWEEN JANUARY 1970 AND AUGUST 1980

Assessed February 1981, i.e. 6 month to 10 year follow up

Nos. Treated	Complete Regression	Partial Regression	Edge Extension or Recurrence	Complications		Follow Up (Years)						
				CNS	EYE	½-1	1-2	2-3	3-4	4-5	5-6	7-8
33	26	4	3	6	9	12 (4A	8 (3A)	4	2 A	3	3 (2A)	1 A

All tumours involved 2, and most 3 regions of the upper jaw.

Research Council has undertaken to manage the purchase and installation of the machine and associated buildings and there will be a Chair of Radiation Oncology. This cyclotron will be equal to modern megavoltage X-ray machines and, after an initial period of acquaintainceship with the new machine, truly controlled clinical trials will be undertaken comparing neutron effects with those from photons in tumours of the abdomen and pelvis.

REFERENCE

[1] CATTERALL, M., BEWLEY, D.K.: Fast Neutrons in the Treatment of Cancer, Academic Press, London, Grune and Stratton, New York (1979).

Progress in Radio-Oncology II, edited by K. H. Kärcher et al., Raven Press, New York © 1982.

Clinical Results of Pion Radiotherapy at LAMPF*

S.E. BUSH, A.R. SMITH
Cancer Research and Treatment Center, and
Department of Radiology, School of Medicine,
University of New Mexico,
Albuquerque, New Mexico

P.M. STAFFORD, Nancy SMITH, R. STARK, Alison PANNELL
Cancer Research and Treatment Center,
University of New Mexico,
Albuquerque, New Mexico

United States of America

Abstract

Clinical trials of the therapeutic use of negative pi-meson irradiation have been conducted by the Cancer Research and Treatment Center of the University of New Mexico and the Los Alamos National Laboratory since 1974. 173 patients have been treated for a variety of primary and metastatic neoplasms. 96 patients treated with curative intent have now been followed up for a minimum of one year. Tumour response and normal tissue reactions have been analysed with respect to tumour site and dose. Crude survival statistics range from 0 of 11 patients treated for unresectable adenocarcinoma of the pancreas to 13 of 15 with Stage T3 and T4 carcinoma of the prostate. Acute tolerance of most epithelial tissues is approximately 4500 π rads in 36 fractions over 7 weeks. Severe chronic reactions have been uncommon. Implications for future studies are discussed.

1. INTRODUCTION

The theoretical advantages of negative pi-meson (pion) radiotherapy were predicted by Fowler and Perkins in 1961 [1]. In 1974 the first cancer patients were treated with pions using the Biomedical Channel at the Los Alamos Meson Physics Facility (LAMPF) under the auspices of the Cancer Research and Treatment Center of the University of New Mexico. Since that time, 173 patients have been treated for local/regional or metastatic neoplasms of a variety of primary sites. Preliminary results of treatment of 96 of these patients have been reported previously [2,3]. Additional observations of tumour and normal tissue responses in this group of patients and in a subsequently treated group of 33 patients form the basis of this report.

* These investigations were supported by United States Public Health Service Grant No. 5 P01 CA 16127 from the National Cancer Institute and by the United States Department of Energy.

2. MATERIALS AND METHODS

The pion channel used for treatment of patients in this series provides a fixed, vertical, minimally divergent beam with a dose rate of approximately 5 rad/min for treatment volumes of approximately 2 litres. Characteristics of the channel, as well as immobilization, treatment planning, simulation and port check techniques have been reported previously [4-7].

One hundred and seventy-three patients with locally advanced neoplasms, with or without known metastatic disease, have been treated with pions between 1974 and 1980. All but six of these patients were treated under Phase I-II studies intended to elucidate tumour and normal tissue responses to this type of radiation in a variety of anatomic sites. Ninety-six of these patients (including one with simultaneous primary lesions of the prostate and urinary bladder) were treated with curative intent and followed for a minimum of one year. All but five patients, who received doses of 1250-1650 π rads to conedown volumes in the planned combined modality management of malignant gliomas using conventional megavoltage irradiation of the whole brain, received a minimum prescribed dose of 2700 π rads, the lowest dose at which complete regression of evaluable disease occurred. The minimum tumour dose was 80% of the prescribed dose designated in most cases. Only 24 patients in this group received doses of 4500 π rads or more, i.e. doses in the range currently accepted as tolerance for most clinically relevant volumes and sites. Table I shows the distribution of patients excluded from analysis and includes 33 patients with follow-up of less than 1 year, 9 patients treated for subcutaneous metastases and 15 patients with known distant metastases at the time of treatment. Twenty patients received doses less than 2700 π rads.

The patient population analysed for potential cure includes 36 patients treated for advanced neoplasms of the head and neck. This group includes 32 patients with T3 and T4 squamous carcinomas, three patients with minor salivary gland tumours and one with anaplastic carcinoma of the nasopharynx. Twenty-three patients had malignant gliomas. Astrocytoma, Grade III, was diagnosed in 8 cases and glioblastoma multiforme in 15. Fifteen patients had

TABLE I. PION PATIENTS 1974-1980

Total Cases	173
Exclusions:	
Follow-up < 1 year	33
Skin Metastases	9
Distant Metastases	15
Low Dose Pilot	7
Low Dose Beam	10
Low Dose Medical	3
Total Exclusions	77
Total Curative Cases	96

locally advanced adenocarcinoma of the prostate and 11 had locally unresectable adenocarcinoma of the pancreas. Twelve additional patients had locally advanced lesions of miscellaneous sites. Thirty-three patients including 13 with glioma, 6 with prostatic carcinoma, 5 with pancreatic carcinoma, 2 with carcinomas of the head and neck, and 7 with miscellaneous other tumours treated after June 1980 are analysed separately.

3. RESULTS

Tables II and III show statistics for crude survival and local control as related to dose range and site of disease with follow-up of 12 to 46 months. The survival rate was lowest for adenocarcinoma of the pancreas, with none of 11 patients surviving. All expired with clinical or autopsy evidence of local disease, although only 4 received a dose of 4000 π rads and only 1 received 4500 π rads. Distant metastases were documented pre- or post-mortem in nine of 11 patients with involvement of the liver as the sole site of distant disease in seven cases

TABLE II. SURVIVAL OF PION PATIENTS TREATED WITH CURATIVE INTENT BY SITE AND DOSE

Dose (rads)	<2000	<4000	>4000	Total
Head & Neck	-	6/12	8/24	14/36 (39%)
Brain	3/5	2/12	1/6	6/23 (26%)
Prostate	-	3/4	10/11	13/15 (87%)
Pancreas	-	0/6	0/5	0/11 (0%)
Other	-	4/8	0/4	4/12 (33%)

TABLE III. LOCAL CONTROL IN PION PATIENTS TREATED WITH CURATIVE INTENT BY SITE AND DOSE

Dose (rads)	<2000	<4000	>4000	Total
Head & Neck	-	6/12	11/24	17/36 (47%)
Brain	1/5	1/12	1/6	3/23 (13%)
Prostate	-	3/4	10/11	13/15 (87%)
Pancreas	-	0/6	0/5	0/11 (0%)
Other	-	5/8	2/4	7/12 (58%)

Fourteen of 36 patients with advanced neoplasms of the head and neck survive from 15 to 40 months following therapy. The 14 survivors are without evidence of disease. Three patients expired with intercurrent or metastatic disease without local/ regional disease.

Six of 23 patients treated for malignant gliomas survive from 13 to 28 months following therapy. The median survival for patients with Grade III gliomas treated more than 12 months ago is 16 months; that for patients with Grade IV lesions is 12 months. Three patients, all receiving treatment for Grade III tumours survive without evidence of disease. Criteria for local control include stability of neurological syndrome, absence of steroid dependence, and absence of contrast enhancement or mass effect on a CT scan.

Thirteen of 15 patients treated for T3 or T4 adenocarcinoma of the prostate survive between 15 and 44 months. One patient expired seven months following therapy with biopsy-proven hepatic metastases, while a second patient expired as a consequence of unrelated chronic obstructive pulmonary disease 20 months following 3800 π rads to the prostate. The latter patient had microscopic evidence of persistent carcinoma within the prostate. A third patient has had biopsy confirmation of locally persistent disease 13 months following 4300 π rads for a T4 prostatic carcinoma.

Twelve patients were treated for lesions of miscellaneous sites; they included three patients with advanced transitional cell carcinomas of the urinary bladder and two each with carcinomas of the oesophagus, lung and rectum. Also in this category, three patients were treated for carcinomas of the stomach, uterine cervix and skin. Two of three patients with bladder lesions had local control, although one expired of unrelated, intercurrent GI bleeding, while the third survives with local persistence. One patient with a Pancoast tumour survives without evidence of disease at three years, but a second patient with adenocarcinoma of the lung expired with local recurrence two years after therapy. Both patients with oesophageal carcinoma expired, although one was free of tumour at autopsy. Both patients with rectal carcinoma expired with metastatic disease although one of these was also free of local disease at autopsy. The patient with stomach cancer died of local recurrence, while a patient with a large basal cell carcinoma of the face, which had invaded the bone, remains free of disease at two years. The single patient with Stage IIIB carcinoma of the cervix expired with lung metastases and apparent small bowel necrosis one year after combined pion and interstitial template irradiation. Review of autopsy material in determining the role of irradiation in the small bowel injury is pending.

Crude survival rates are 87%, 39%, 26% and 0% for patients with neoplasms of the prostate, head and neck, brain, and pancreas respectively, with corresponding local control statistics of 87%, 47%, 13% and 0%. Table IV shows local control statistics for 67 patients (with 68 tumours) treated for cure with pions alone. Nine of 31 patients (29%) receiving less than 4000 π rads had no evidence of local/regional disease at last follow-up or death while 17 of 37 (46%) receiving greater than 4000 π rads had local control.

Table V shows crude survival and local control statistics for the group of 33 patients treated after June 1980 and followed for

TABLE IV. LOCAL CONTROL AFTER PION IRRADIATION ALONE BY SITE AND DOSE

Dose (rads)	<4000	>4000	Total
Brain	1/10	1/6	2/16
Head & Neck	1/4	4/12	5/16
Prostate	3/4	10/11	13/15
Pancreas	0/5	0/5	0/10
Other	4/8	2/3	6/11

TABLE V. PION PATIENTS TREATED JULY 1980 TO OCTOBER 1980

	Crude Survival	Local Control
Brain	8/13 (62%)	3/13 (23%)
Head & Neck	2/2 (100%)	2/2 (100%)
Pancreas	5/5 (100%)	0/5 (0%)
Prostate	6/6 (100%)	6/6 (100%)
Other	6/7 (86%)	3/7 (43%)

periods of 6 to 9 months. Eight of 13 patients with astrocytomas, of whom 3 have no evidence of disease, survive. Both patients treated for carcinomas of the head and neck and all 6 patients with prostatic carcinomas survive and have no evidence of active disease. All 5 patients with pancreatic carcinoma survive although none have had regression of mass disease by CT scan.

Acute reactions to pion radiotherapy have been systematically recorded according to a scale of 0-4 as follows: 0 – nil; 1 – skin erythema, mucosal injection, mild dysuria or diarrhoea ⩽ 4 stools per day, etc.; 2 – dry desquamation, patchy mucositis, moderate dysuria, diarrhoea with mucus (⩾ 5 stools per day), etc.; 3 – moist desquamation, confluent mucositis, severe dysuria with bladder spasms, diarrhoea with blood, etc.; and 4 – acute necrosis. Average sums of acute reactions were obtained by summing severities of all reactions for individual anatomic sites as follows: head and neck – mucosa, skin, salivary glands; pelvis – skin, rectum, bladder; thorax – skin, dysphagia; abdomen – nausea, diarrhoea; brain – skin. An analysis of acute reactions is shown in Table VI, demonstrating a trend to more severe reactions at all sites except pelvis and brain in the higher dose range, although these data ignore such potentially contributory factors as daily fraction size, treatment volume, and hyperfractionation.

TABLE VI. ACUTE INJURY RELATED TO PION IRRADIATION IN DOSES > 2700 π RADS BY SITE AND DOSE RANGE[a]

Dose Range	Site	Number of Patients	Average Sum of Acute Reactions
<4000 rad	Head & Neck	13	5.1
	Pelvis	8	4.6
	Thorax	3	2.3
	Abdomen	6	1.8
	Brain	6	1.2
>4000 rad	Head & Neck	24	5.8
	Pelvis	12	4.9
	Thorax	1	3.0
	Abdomen	6	2.2
	Brain	12	1.2

[a]See text for explanation of scoring system

The average sum of acute reactions for eleven patients (10 with carcinomas of the head and neck and one with carcinoma of the stomach) receiving greater than 5000 π rads was 6.2.

Chronic reactions following pion irradiation have been recorded according to the late effects scoring system of EORTC/RTOG in which severity of reaction ranges from 0 (i.e. no detectable reaction) to 5 (i.e. death ascribable to radiotherapy). Only two patients had severe chronic effects related to pion irradiation alone. A 70-year-old female had chronic, severe laryngeal oedema after 5000 π rads for a T4 squamous carcinoma of the larynx, and a second patient had severely symptomatic pulmonary fibrosis after 4000 π rads for a large adenocarcinoma of the lung. Thirteen patients treated for glioma have been autopsied, and, although all had gross or microscopic evidence of residual tumour, none had pathognomonic histological evidence of radiation injury to normal brain.

Table VII shows a comparison of average summed chronic reactions for various treatment sites and dose ranges of 2700-4000 π rads and greater than 4000 π rads. All patients received pions alone with curative intent. The average summed reactions for head and neck patients increased from 2.1 to 2.8 comparing low and high dose categories, with a corresponding increase from 1.4 to 2.8 for patients with pelvic primaries. Twenty-six of 31 patients in the "other" category were treated for glioma or pancreatic carcinomas and had minimal late effects usually consisting of Grade 1 skin reaction.

4. DISCUSSION

One hundred and seventy-three patients have completed pion therapy since 1977 and a subpopulation of 96 treated with curative intent and followed for a minimum of 1 year provides information of particular importance regarding survival, tumour control, and

TABLE VII. CHRONIC INJURY RELATED TO PION IRRADIATION ALONE IN DOSES > 2700 π RADS BY SITE AND DOSE [a]

Dose Range	Site	Number of Patients	Average Sum of Chronic Reactions
<4000 rad	Head & Neck	4	2.1
	Pelvis	7	1.4
	Other	19	0.9
>4000 rad	Head & Neck	12	2.8
	Pelvis	13	2.8
	Other	12	0.9

[a] See text for explanation of scoring system

normal tissue reactions. This group of patients has included a variety of disease sites, histological patterns, and combinations of conventional modalities of treatment with pions. Such heterogeneity of clinical material, as well as the relatively short duration of follow-up makes comparison with historical series impossible.

These data demonstrate that pion radiotherapy will effectively eradicate even advanced neoplasms of certain types and will provide local control of such lesions for periods up to 3½ years. This benefit may obtain with moderate but tolerable acute reactions and acceptable chronic reactions in most cases. The data regarding acute reactions suggest that most epithelial tissues will tolerate doses of approximately 4500 π rads in 125 π rad fractions over seven weeks to clinically relevant volumes, although severity of acute reactions begins to increase rapidly at doses above this range. The incidence of chronic injury related to pion irradiation appears to increase disproportionately as compared with acute injury in the group of patients receiving more than 4000 π rads. For example, average summed acute reactions increased from 5.1 to 5.8 for head and neck sites receiving less than or greater than 4000 π rads, a 14% increase, while average summed chronic reactions increased from 2.1 to 2.8 (33%) for the same patients. The average summed acute reactions for patients receiving treatment to the pelvis increased from 4.6 to 4.9 (7%) for that group receiving greater than 4000 π rads as compared with those receiving less than this dose, although the average summed chronic reactions increased from 1.4 to 2.8 (100%). These injuries have manifested primarily as subcutaneous fibrosis, chronic oedema, mucosal and cutaneous damage. Chronic injury to oral and pharyngeal mucosa is comparable to that of rectal mucosa for similar dose ranges. Chronic injury, particularly fibrosis, may not manifest itself for periods 12 months or more following high-dose pion therapy.

Early follow-up of this population of patients provides encouraging although anecdotal results for a variety of sites particularly including some lesions of the head and neck, gliomas and locally advanced prostate cancer. Non-randomized trials will continue for the malignant gliomas and advanced carcinomas of the uterine cervix. Phase III trials have been activated for Stage III and IV squamous carcinoma of the oral cavity and pharynx, and inoperable or locally recurrent adenocarcinoma of the rectum in order to compare conventional irradiation and pion radiotherapy in randomized, prospective trials. Continued accession of patients with locally advanced lesions in various sites as well as continued follow-up of previously treated patients may be anticipated to provide useful and necessary information for guidance of future clinical studies and assessment of the efficacy of pion radiotherapy.

REFERENCES

[1] FOWLER, P.H., PERKINS, D.H.: The possibility of therapeutic applications of beams of negative π mesons, Nature (London) 189 (1961) 524-528.

[2] KLIGERMAN, M., TSUJII, H., BAGSHAW, M., WILSON, S., BLACK, W., METTLER, F., HOGSTROM, K.: "Current observations of pion radiation therapy at LAMPF", in Treatment of Radioresistant Cancers (ABE, M., SAKAMOTO, K., PHILLIPS, T.L., Eds) Elsevier/North-Holland Biomedical Press, Amsterdam, (1979) 145-157.

[3] KLIGERMAN, M.M., BUSH, S.E., KONDO, M., WILSON, S., SMITH, A.: "Results of Phase I-II Trials of Pion Radiotherapy", Proc. 2nd Ann. Rome Int. Symp. Biological Bases and Clinical Implications of Tumor Radioresistance (in press).

[4] PACIOTTI, M., BRADBURY, J., HUTSON, R., KNAPP, E., RIVERA, O.: Tuning the beam shaping section at the LAMPF biomedical channel, IEEE Trans. Nucl. Sci. NS-24 (1977) 1058.

[5] KLIGERMAN, M.M., HOGSTROM, K.R., LANE, R.G., SOMERS, J.: Prior immobilization and positioning for more efficient radiotherapy, Int. J. Radiat. Oncol., Biol. Phys. 2 (1977) 1141-1144.

[6] HOGSTROM, K.R. SMITH, A.R., KELSEY, C.A., SIMON, S.L., SOMERS, J.W., LANE, R.G., ROSEN, I.I., VON ESSEN, C.F., KLIGERMAN, M.M., BERARDO, P.A., ZINK, S.M.: Static pion beam treatment planning of deep seated tumors using computerized tomographic scans at LAMPF, Int. J. Radiat. Oncol., Biol. Phys. 5 (1979) 875-886.

[7] TSUJII, H., BAGSHAW, M., SMITH, A., VON ESSEN, C., METTLER, F., KLIGERMAN, M.: Localization of structures for pion radiotherapy by computerized tomography and orthodiagraphic projection, Int. J. Radiat. Oncol., Biol. Phys. 6 (1980) 319-425.

Progress in Radio-Oncology II, edited by K. H. Kärcher et al., Raven Press, New York © 1982.

Clinical Radiotherapy with Heavy Charged Particles at Lawrence Berkeley Laboratory*

J.R. CASTRO, W. SAUNDERS, K.H. WOODRUFF, J.M. QUIVEY, T.L. PHILLIPS, G.T.Y. CHEN, J.T. LYMAN, M. COLLIER, S. PITLUCK, C. TOBIAS
Radiotherapy Section of Donner Laboratory,
University of California Lawrence Berkeley Laboratory,
Berkeley, California
and
University of California San Francisco School of Medicine,
San Francisco, California,
United States of America

Abstract

A clinical radiotherapeutic trial utilizing heavy charged particles in the treatment of human cancers has accrued over 300 patients since 1975. Heavy charged particle radiotherapy offers the potential advantages of improved dose localization and enhanced biologic effect. Target sites have included selected head and neck tumours, malignant glioma of the brain, carcinoma of the oesophagus, carcinoma of the stomach, carcinoma of the pancreas, soft tissue sarcoma, and other locally advanced, unresectable or recurrent tumours. Phase III prospective clinical trials have been started utilizing helium ion radiotherapy; Phase I-II studies are under way with heavier particles such as carbon, neon, silicon, or argon ions in order to prepare for prospective Phase III trials. These studies are supported by the United States Department of Energy and National Institutes of Health.

Since 1975, a heavy charged particle clinical radiotherapy trial has been conducted at Lawrence Berkeley Laboratory beginning with patient irradiations using helium ions produced at the 184-inch Synchrocyclotron and later with heavier particles such as carbon, neon, and argon nuclei produced at the Bevalac (the Bevalac is an acronym for a combination of the Super HiLac Linear Accelerator as injector into the Bevatron circular accelerator for final acceleration). With this combination, high intensities of heavier nuclei are possible permitting clinical radiotherapy applications.

The total patient accrual in the clinical trial has reached 301 patients irradiated with particles plus 22 patients treated as controls for a randomized study of irradiation of carcinoma of the pancreas (Table I). Of these, 244 patients have been treated with helium ions either solely or in combination with photon irradiation. Fifty-seven patients have received all or part of their irradiation with one of the heavier particles, either carbon, neon, or argon ions.

* Supported by National Institutes of Health, National Cancer Institute Grant 5 PO1 CA-19138, 2 R10 CA-21744, and United States Department of Energy Contract W7405-ENG-48.

TABLE I. HEAVY PARTICLE CLINICAL TRIAL PATIENTS 1975-07 to 1981-11-03

Anatomic region	Helium	Heavy particle
Head/neck	16	12
Intracranial	21	10
Eye	39	--
Thoracic	35	8
Abdomen/retroperitoneal	112	22
Pelvis	15	1
Other	6	4
Total	244	57
Low-LET photon patients (Randomized Pancreas Trial)	22	
Consultation only, not accepted for particle radiation therapy	92	
Total all patients referred	415	

The clinical trial can be divided into two general phases:

1. Evaluation of improved dose distribution without significant biologic advantage by use of helium ion irradiation.

2. Evaluation of improved dose distribution and enhanced biologic effect by irradiation with heavy charged particles such as carbon, neon, silicon, or argon ions.

The aim of the helium ion radiation therapy trial has been to test the potential clinical advantage of improved dose localization. In our opinion, no significant enhancement of biologic effect is present with helium ions, at least when a spread out Bragg peak is utilized as is appropriate for most clinical tumor target volumes. This has been borne out by pretherapeutic biologic studies which have shown that only minimal OER reduction is present in this configuration and by observation of tumor regression in a variety of clinical sites. While a significant advantage appears present clinically for selected target sites where a higher tumor dose may be given because of the physical properties of helium nuclei irradiation, there has not been an apparent improved regression of bulky, hypoxic, or necrotic tumors such as advanced lesions of the head and neck area.

With helium ions, we have now irradiated 69 patients with localized, unresectable carcinoma of the pancreas. All of these patients have received at least 5000 equivalent rads in 6 weeks with most patients receiving 6000 equivalent rads in 7½ weeks. After a pilot study was completed, 44 patients have been entered in a randomized trial in the treatment of carcinoma of the pancreas. This trial contrasts 6000 rads of helium ion irradiation plus 5-fluorouracil chemotherapy against 6000 rads low LET photon irradiation plus 5-fluorouracil chemotherapy. The trial is coordinated by the Northern California Oncology Group and the Radiation Therapy Oncology Group who provide the required statistical backup, quality control, and other logistical services (Fig. 1).

To date no significant difference in survival has been demonstrated between the study and control arms. Overall, including the pilot patients irradiated for carcinoma of the pancreas, about 15 percent of patients remain without evidence of disease from 6 months to 5 years post therapy. Although autopsy data is not present for all patients, there are a small number of patients who appear to have expired of distant metastases without recurrence in the irradiated area. The local and regional control rate within the irradiated volume therefore appears to be approximately 20 percent (Table II).

In our series, the incidence of developing liver metastases appears very high. This is true despite careful study of liver function, liver scans, and palpation and observation at time of laparotomy. Because of this high frequency of occult liver metastases, it seems important in the future to consider biopsy of the liver prior to acceptance into clinical trial and to devise a better technique of elective liver treatment than the current chemotherapy. This might include elective whole liver irradiation to doses of 2500 to 3000 rads or multi-drug chemotherapy.

Helium ions have also been used in a variety of other target sites where improved dose localization might be expected to show potential advantage. These have included:

1. Selected head and neck tumors around the base of skull and cervical spine as well as in the paranasal sinuses.

2. Localized soft tissue sarcomata.

3. Localized, unresectable carcinoma of the stomach.

4. Carcinoma of the esophagus.

5. A variety of locally advanced and/or recurrent neoplasms in the abdomen and pelvis.

Among these have been a group of 35 patients with localized ocular melanoma who have been irradiated with helium ions using a modified Bragg peak of 14 to 23 millimeters, delivering 7000 (19 patients) or 8000 (16 patients) equivalent rads in 5 fractions over 7 to 9 days. Extremely sharp lateral edges and the distal falloff of the beam have permitted precise dose localization with sparing of the critical structures of the eye in most patients. Thus, for nearly all patients vision has been preserved with control of the tumor in the eye. The direction of the gaze, angulation of the beam, and shape of the treatment aperture as well as the beam penetration depth and range modulation are established in a planning session using a computerized treatment planning program. The tumor localization is

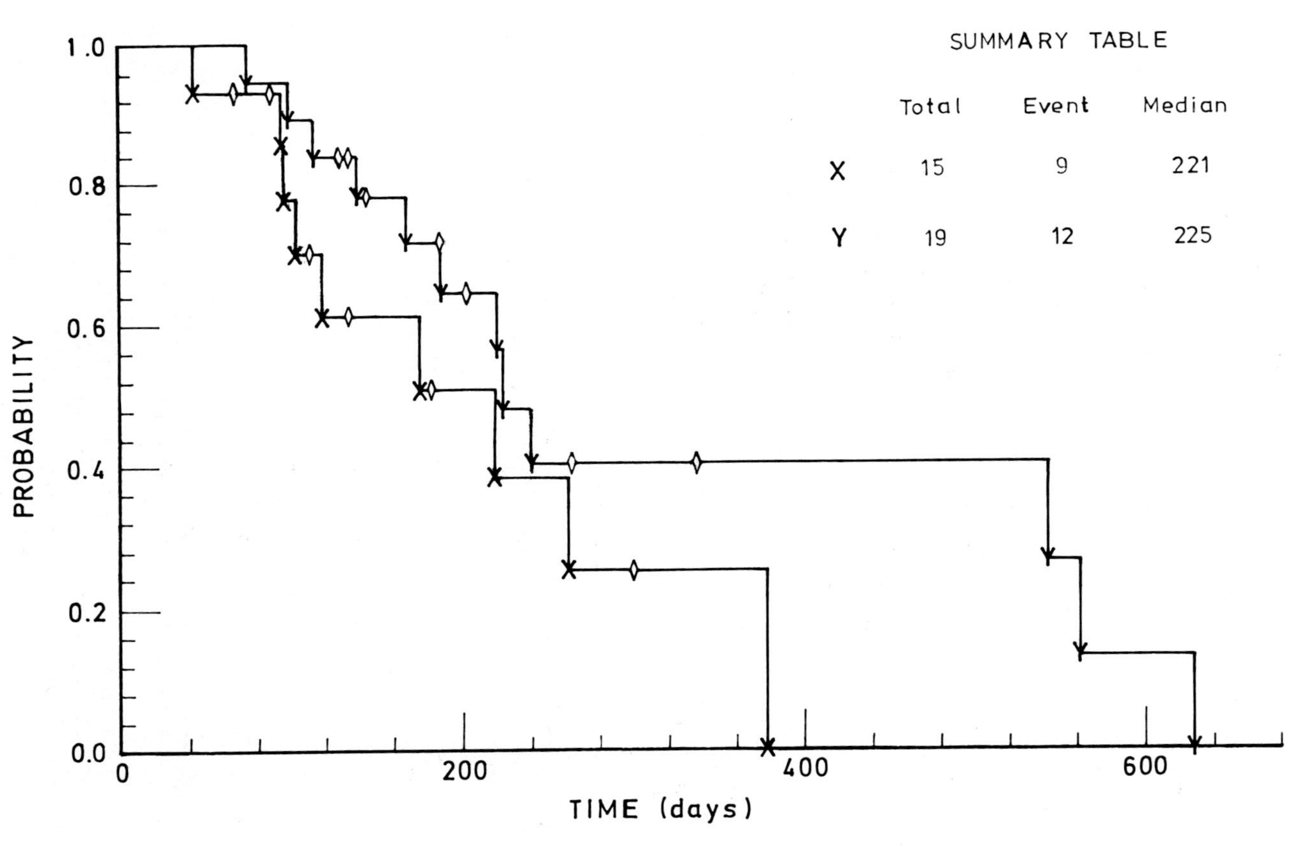

FIG.1. Carcinoma of the pancreas. Time to local failure only. Period assessed, 1978-08-31 to 1980-12-31. The significance of the difference between these Kaplan-Meier curves was tested by a Mantel-Cox test ($p = 0.07$).
(X – photons; Y – helium)

TABLE II. UNRESECTABLE CARCINOMA PANCREAS, PILOT STUDY
Data assessed 1981-04

Minimum 5000, maximum 6400 ^{60}Co rads equivalent	
35 patients { 26 treated helium only; 9 photon plus helium boost	
No evidence of disease (NED)	2 (32,62 months)
Died, intercurrent disease, NED	1 (15 months)
Died, complications, NED	1 (9 months)
Local control, distant metastases	3 (8,29,30 months)
Persistence	28
Control in XRT field	7/35 (20%)

facilitated by the operative placement of radio-opaque tantalum rings about the base of the tumor. Coordinates of the rings are obtained from orthogonal x-ray films. These data and measurements of the tumor height and the AP diameter of the globe as determined from an ultrasound study are used to locate the tumor relative to the structures of the eye. The dose and fractionation have been chosen following initial results reported from the Harvard Cyclotron with proton irradiated patients. Of the 35 treated patients (Table III), there are 4 with failure to control the tumor in the radiation field, 2 of which appear to be marginal misses due to technical problems. In one, the patient was re-irradiated and has subsequently had shrinkage of his tumor although with nearly complete loss of vision. This was because the tumor lay near the optic nerve and it unavoidably had to be included in the re-irradiation field in order to cover the tumor.

We have a strong clinical impression of improved clinical results in selected tumors in and around the base of the skull and cervical spine as well as some patients with localized soft tissue sarcomata. In these patients because of the improved dose localization with charged helium nuclei, we feel that a higher tumor dose has been able to be delivered relative to the dose to adjacent normal tissues. Thus, although not a randomized prospective trial, there appear to be selected target sites in which improved dose localization with charged particles such as helium ions offers a clinical advantage.

With helium ions, the dose fraction sizes have been 200 equivalent rads per fraction, 4 fractions per week to total doses of 6000 to 7000 equivalent rads. We have utilized this dose prescription in order to facilitate comparison to low LET photon irradiation. With clinical evaluation of skin, intestinal, and mucosal reactions, the RBE has appeared to range from 1.2 in the proximal portion of the spread out Bragg peak to 1.4 in the distal peak for fraction sizes of approximately 200 equivalent rads each.

We have not encountered any unexpected or any untoward toxicity with helium ions with the exception of gastric and duodenal hemorrhagic radiation reaction in some patients treated for carcinoma of the pancreas. This has appeared to have an incidence of serious radiation damage to the

TABLE III. OCULAR MELANOMA
1979-03-19 to 1980-12-31

IIIa. RESPONSE

Response	7000 rads		8000 rads[a]	
Tumor shrinkage	13/19	68%	5/11	46%
Stable	3/19	16%	6/11	55%
Failure	3/19[b]	16%	0/11	0%

[a] Five patients have been treated too recently to evaluate, although one appears to have a growing, probably progressive lesion.

[b] Two of these patients failured early in the program and probably represent marginal misses. Two of these patients were enucleated and one was re-irradiated.

IIIb. FOLLOW-UP TIME

	7000 rads (months)	8000 rads (months)
Range	3.7 - 34.8	0.2 - 12.8
Median	15	4.6

stomach and/or duodenum in approximately 10 to 15 percent of patients. It must be remembered however that few of these patients survive for long periods of time and therefore the true complication rate may be higher. Some of the long-term survivors have indeed had gastric outlet obstruction because of stenosis and/or fibrosis of the distal stomach or the proximal duodenum. This has required bypass surgery in some instances. This is not an unexpected effect since even with charged particles, a portion of the distal gastric mucosa as well as the proximal duodenum must unavoidably be included in the high dose target volume. However, with improved treatment planning using upright CT scans in treatment position, we hope to reduce the size of the target volume and thus perhaps lessen the incidence of the gastric and duodenal morbidity.

For 29 patients with localized unresectable carcinoma of the esophagus irradiated with helium ions, we have not been able to discern any clinical improvement over historical control patients treated with megavoltage irradiation. In our series at least 50 percent of the patients have had local failure with helium ion irradiation. A significant clinical advantage from the improved dose localization utilizing helium nuclei in the treatment of carcinoma of the esophagus does not appear to be present. We therefore propose to continue this prospective non-randomized clinical trial substituting a heavier particle such as neon or silicon nuclei for helium. It is hoped that a biological and physical advantage will accrue and offer therefore improved chances of local control in carcinoma of the esophagus (Table IV).

A small group of 12 patients with locally unresectable gastric carcinomata have been irradiated with helium ions. Several of these patients have had remissions of their disease, one of whom has survived almost 3 years without evidence of recurrence. We are intrigued by this rather unexpected result of some patients achieving local control of gastric tumors with helium irradiation. We propose to continue this series utilizing a heavier particle such as neon or silicon ions in order to study the effect of enhanced biologic effect as well as improved dose localization on unresectable gastric tumors.

We have had access to heavier particle beams such as carbon, neon, silicon, and argon ions at the Bevalac only since 1977 and with less frequency than at the 184-inch cyclotron. Accordingly the Phase I-Phase II heavy particle study has been under way for a relatively short time and only 66 patients have been irradiated to date. Most of these have been treated either with neon or carbon ions as limited access to the argon ion beam has been possible.

The goals of this study are:

1. Evaluation of acute and sub-acute response of normal tissue such as mucosa, skin, and intestine.

2. Initial evaluation of tumor response.

3. Development of effective treatment techniques using carbon, neon, silicon, and argon ions.

4. Clinical evaluation of physical and biological dose distributions available with these beams.

In particular we intend to confirm the RBE estimates for various tissues provided by pretherapeutic biologic studies. Thus, we have irradiated several patients with metastatic advanced tumors paying careful attention to skin, mucosal, and pulmonary reactions. In addition, several patients with advanced carcinoma of the pancreas not eligible for the randomized helium clinical trial have been irradiated with carbon and neon ions. Additionally, such lesions as primary or metastatic malignancies of the brain, localized unresectable abdominal or pelvic tumors such as recurrent colon cancer, or a few patients with carcinoma of the lung have been irradiated.

This Phase I-Phase II study of heavy particles is planned to accrue an additional 40 patients before July 1981. The data accumulated in this pilot study will be utilized to select the single best heavy particle for future prospective Phase III trials. We expect to begin such randomized trials for 1-3 tumor sites in March 1982.

A few patients have shown interesting clinical regression of tumors when irradiated with argon ions. Although the doses delivered were low (1500 to 2000 rads in 4 fractions), one patient with metastatic melanoma and another with metastatic fibrosarcoma have had regression of tumor with continued control of the irradiated area for periods ranging from 6 months to 18 months.

TABLE IV. SQUAMOUS CELL CARCINOMA OF THE ESOPHAGUS

Assessed at 1981-04-20

No. of patients	NED [a]	Local control	Local failure, with or without distant metastases	Distant metastases only	Died, intercurrent disease, NED	Unknown	Follow-up
29	5	9	18	2	2	2	2–24 months

Average survival	Patients with local control	12 months
Average time-dose fraction	Patients with local control	106
Average survival	Patients with local failure	9 months
Average time-dose fraction	Patients with local failure	107
Average survival	Patients alive and well	12 months
Average time-dose fraction	Patients alive and well	109

[a] No evidence of disease.

It is quite possible that an optimal ion for Phase III heavy particle irradiation will prove to be a heavy nucleus such as neon or silicon. Our clinical impression of patients treated with carbon ions does not show any significant difference from those treated with helium ions. On the other hand, the few patients irradiated with argon, the heaviest particle available to us, appear to show promising results although the data is scanty. Argon, however, has a less favorable depth dose distribution than neon or silicon.

This does corroborate the results in pretherapeutic testing in which the lowest OER across the spread out Bragg peak was present with the argon nuclei irradiations. Silicon which is intermediate between neon and argon looks promising clinically because of a low OER as well as a practical dose distribution superior to that of argon. We have therefore put a high priority on further pre-clinical radiobiologic studies of silicon with the possibility of utilizing this ion for future clinical trials.

With completion of Phase I-Phase II studies of heavy particles, we hope to begin prospective Phase III trials in 1982. Prior to that time we will continue pretherapeutic radiobiological characterization of heavy ion beams as an essential foundation to clinical trials. Extensive data has been utilized from these investigations in our past and ongoing clinical studies. For the future we particularly desire more information relative to the responsive normal tissue to heavy ion beams, particularly for CNS tissue as well as additional data relative to tumor response to heavy particle beams.

We continue to extend investigations into physical characterizations of beam quality as well as development of better methods of beam delivery for patient irradiation with charged particles. This includes development of beam scanning techniques which will provide an improved dose delivery system with optimal dose localization.

Radioactive beam imaging continues to be under study at Lawrence Berkeley Laboratory in order to assist in radiotherapy treatment planning by providing a means of localizing the high dose zone (the stopping region of charged particle beam) and establishing whether it conforms to the planned target volume.

We anticipate an additional 5-10 years of study in order to fully evaluate these heavy charged particle beams and characterize their role, if any, in clinical radiotherapeutic practice.

SELECTED BIBLIOGRAPHY

BLAKELY, E.A., TOBIAS, C.A., NGO, F.Q.H., YANG, T.C.H., SMITH, K.C., CHANG, P.Y., YEZZI, M.J.: Comparison of helium and heavy ion beams for therapy based on cellular radiobiological data, Int. J. Radiat. Oncol., Biol. Phys. 4 (1978) 93.

CASTRO, J.R., QUIVEY, J.M., LYMAN, J.T., CHEN, G.T.Y., PHILLIPS, T.L., TOBIAS, C.A.: Particle radiotherapy at Lawrence Berkeley Laboratory, Cancer 46 (1980) 633-641.

CASTRO, J.R., QUIVEY, J.M., LYMAN, J.T., CHEN, G.T.Y., TOBIAS, C.A., ALPEN, E.L.: "Clinical experience with helium and heavy ion radiotherapy", in Proc. 1st Int. Sem. Uses of Proton Beams in Radiotherapy, Moscow, 1977.

CASTRO, J.R., QUIVEY, J.M., PHILLIPS, T.L., LYMAN, J.T., CHEN, G.T.Y., TOBIAS, C.A.: Radiotherapy with heavy charged particles at Lawrence Berkeley Laboratory, J. l'Assoc. Can. Radiol. 31 (1980) 30-34.

CASTRO, J.R., QUIVEY, J.M., SAUNDERS, W.M., WOODRUFF, K.H., CHEN, G.T.Y., LYMAN, J.T., PITLUCK, S., TOBIAS, C.A., WALTON, R.E., PETERS, T.C.: Clinical results in heavy particle radiotherapy. Biological and medical research with accelerated heavy ions at the Bevalac 1977-1980, Lawrence Berkeley Laboratory Rep. 11220 (1980).

CASTRO, J.R., TOBIAS, C.A., QUIVEY, J.M., CHEN, G.T.Y., LYMAN, J.T., PHILLIPS, T.L., ALPEN, E.L., SINGH, R.P.: "Results of tumor treatments with alpha particles and heavy ions at Lawrence Berkeley Laboratory", in High LET Radiations in Clinical Radiotherapy (BARENSDEN, G.W., BROERSE, J.J., BREUR, K., Eds), Eur. J. Cancer, Suppl. (1979).

CHAPMAN, J.D., BLAKELY, E.A., SMITH, K.C., URTASUN, R.C., LYMAN, J.T., TOBIAS, C.A.: Radiation biophysical studies with mammalian cells and a modulated carbon ion beam, Radiat. Res. 74 (1978) 101-110.

CHAR, D.H., CASTRO, J.R., QUIVEY, J.M., CHEN, G.T.Y., STONE, R.D., CRAWFORD, J.B., IRVINE, A.R., HILTON, G.F., SCHWARTZ, A., LONN, L.I. BARRICKS, M.: Helium ion charged particle therapy for choroidal melanoma, Ophthalmology 87 (1979) 565-570.

CHEN, G.T.Y.: "Use of computers in charged particle radiotherapy", in Proc. 3rd World Conf. Medical Informatics (LINDBERG, D., KAIHARA, S., Eds), North Holland Publ. Co., New York (1980) 4-8.

CHEN, G.T.Y., CASTRO, J.R., QUIVEY, J.M.: Heavy charged particle radiotherapy, Ann. Rev. Biophys. Bioeng. 10 (1981) 499-529.

CHEN, G.T.Y., HOLLEY, W.R.: "Charged particles in medical diagnosis", in Advances in Radiation Protection and Dosimetry in Medicine (THOMAS, R., PEREZ-MENDEZ, V., Eds), Plenum Press, New York (1980) 289-316.

CHEN, G.T.Y., SINGH, R.P., CASTRO, J.R., LYMAN, J.T., QUIVEY, J.M.: Treatment planning for heavy ion radiotherapy, Int. J. Radiat. Oncol., Biol. Phys. 5 (1979) 1809-1819.

CHEN, G.T.Y., SINGH, R.P., LYMAN, J.T., QUIVEY, J.M., CASTRO, J.R.: Treatment planning for charged particle radiotherapy, Int. J, Radiat. Oncol., Biol. Phys. 5 10 (1979) 809-819.

LYMAN, J.T., HOWARD, J.: Dosimetry and instrumentation for helium and heavy ion beams, Int. J. Radiat. Oncol., Biol. Phys. 3 (1977) 81-85.

QUIVEY, J.M., CASTRO, J.R., CHEN, G.T.Y., MOSS, A., MARKS, W.M.: Computerized tomography in the quantitative assessment of tumor response, Br. J. Cancer 41 4 (1980) 30-33.

RAJU, M.R., BAIN, E., CARPENTER, S.G., HOWARD, J., LYMAN, J.T.: Cell survival measurements as a function of depth for a high energy argon ion beam, Radiat. Res. 84 (1980) 158-163.

TOBIAS, C.A., ALPEN, E.L., BLAKELY, E.A., CASTRO, J.R., CHATTERJEE, A., CHEN, G.T.Y., CURTIS, S.B., HOWARD, J., LYMAN, J.T., NGO, F.Q.H.: "Radiobiological basis for heavy ion therapy", Treatment of Radioresistant Cancers (ABE, M., SAKAMOTO, K., PHILLIPS, T.L., Eds), Elsevier/North Holland Biomedical Press, Amsterdam (1979) 159-182.

Progress in Radio-Oncology II, edited by K. H. Kärcher et al., Raven Press, New York © 1982.

Clinical Experience with Pi-Meson Therapy at SIN

C.F. von ESSEN, H. BLATTMANN, I. CORDT,
J. CRAWFORD, E. PEDRONI. Ch. PERRET, M. SALZMANN
Swiss Institute for Nuclear Research,
Villigen,
Switzerland

Abstract

The first treatments of patients with the SIN Piotron were initiated in November 1980. In order to develop a safe progression of treatment for this new and unique beam configuration, it was planned to treat in phases of increasing complexity. Phase Ia involved the use of a sector of 15 beams (of the 60 available) with the focus of the stopping pions on a circumscribed area of skin. Patients with multiple skin metastases received pion therapy and X-ray therapy in comparable geometric and dosimetric arrangements to separate skin lesions. A scale of acute skin reactions was used to score reactions and the reaction levels were compared. Only two out of six patients were able to complete the scheduled treatments because of technical beam failures; nevertheless it was possible to compare effects from 6 pion treated fields with 11 X-ray treated fields. RBE values of about 1.4 for 4 to 12 fractions corresponded to values reached in similar trials in Los Alamos and Vancouver. The next stage involved the treatment of single larger lesions with up to thirty beams and involved the complicated arrangement of patient immobilization in individual impression moulds in cylindrical treatment shells, the localization by an X-ray positioning device, and treatment inside a cylindrical water bolus. These technical developments will be discussed, together with the dosimetric and clinical results.

The Piotron is a large solid acceptance angle superconducting channel built for the use of negative pi-mesons in radiotherapy. It offers a new approach in that particles are directed into the target by radially converging beams. This approach requires dynamic scanning in order to achieve 3-dimensionally shaped treatment volumes. The potential advantages of negative pi-mesons include the Bragg-peak dose distribution and the high LET or "star" component in the peak. These advantages are enhanced by the unique radial beam array. This paper discusses the initial clinical experience with the Piotron.

The Piotron

The concept of a large acceptance angle pion channel for cancer therapy was developed at Stanford University by Boyd, Schwettman and Simpson (1). A prototype relying on an electron production beam was constructed but did not achieve clinical usefulness.

Benefiting from the Stanford experience, G. Vécsey and associates at SIN constructed a similar device (2) which has achieved most of the design specifications and is in clinical operation. The design (Fig. 1) allows efficient production of pions from a relatively small proton current (up to 20 microamperes) striking a target of molybdenum or beryllium. The pions are divided into 60 channels and deflected twice to enter the treatment volume radially. The momentum and momentum band for all 60 channels can be chosen and each beam can be switched on or off by individual slit shutters.

Dosimetry

Special equipment has been designed to measure the spatial distribution of dose from single and multiple beams. Cylindrical water tanks of different radii were used to measure the dose distributions by means of remotely controlled scanning with a tissue equivalent gas ionization chamber.

The beam spot of 60 beams stopping at the isocenter of a 30 cm radius water phantom has a full-width, half-maximum (FWHM) diameter, equivalent to the 50% isodose contour, of about 4.4 cm in all directions. The rate at the minimum for a proton current of 18 microamperes is 20-25 rads/minute for one liter of treatment volume.

Radiobiology

Pion dose response curves were determined for the same treatment geometry from colony formation or irradiated Chinese hamster ovary cells using a gel technique developed by Skarsgard and Palcic (3). These experiments were performed by Dr. J. Tremp of the Strahlenbiologisches Institut of the University of Zurich. The RBE value, using 140 kVp x-rays as reference radiation, was approximately 1.3 for a single exposure at the pion peak (4).

Other experiments have been performed for a geometry of 60 convergent beams using the mouse foot system for single and multiple fractions (E. Fröhlich and associates of the Strahlenbiologisches Institut of the University of Zurich)(5) as well as experiments planned or in progress involving mouse jejunal crypt survival, mouse lung, and rat spinal cord systems. RBE values to date are consistent with findings from other pion facilities with the expected increase of RBE at the pion peak with the number of fractions for various end-points.

First Clinical Trials

The goal of the medical project is to test how effective pi-mesons are in the treatment of cancer. However, the effects

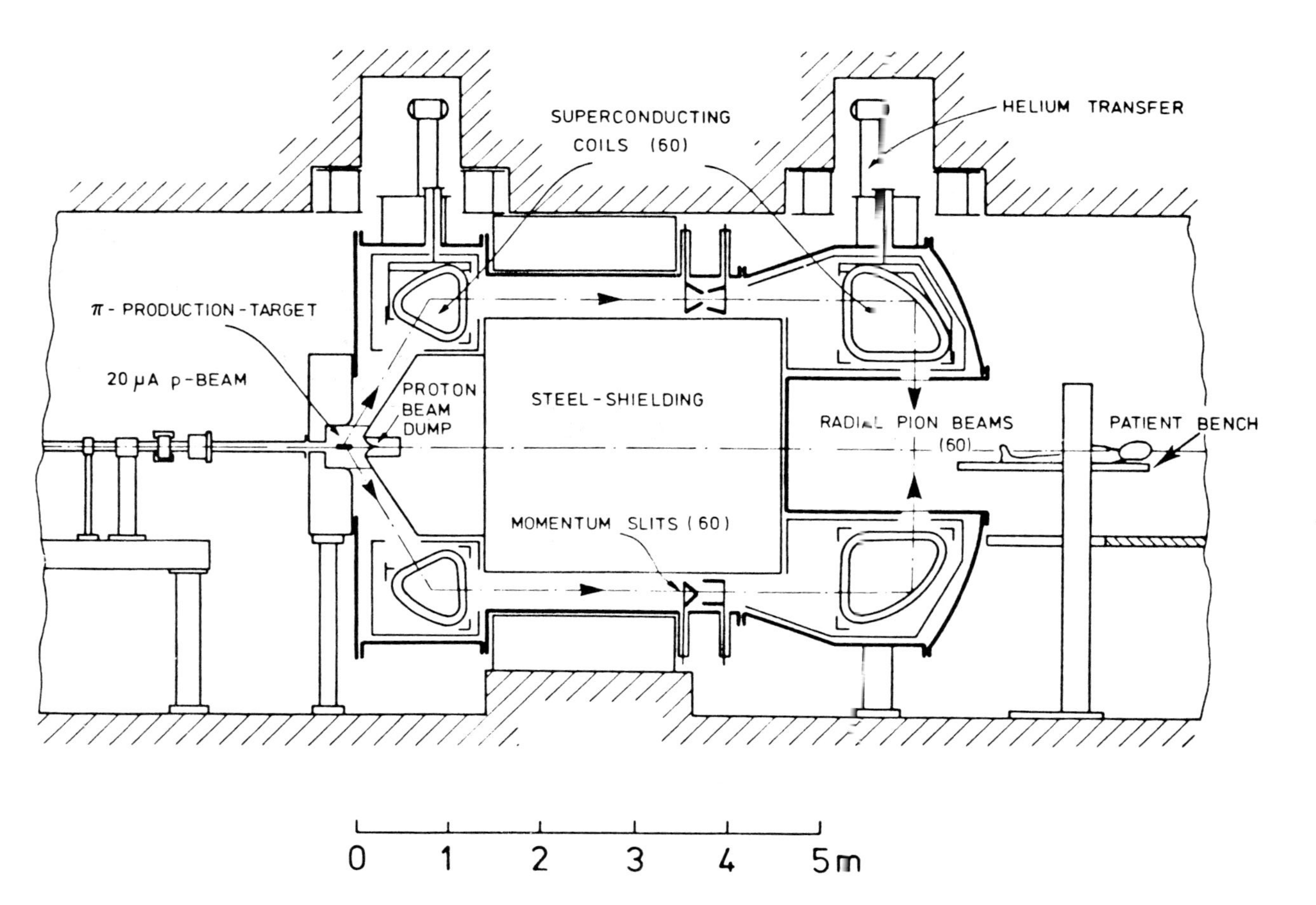

FIG.1. Cross-section of the Piotron.

of pi-mesons must be compared first with those of photons in well-studied normal tissues such as the skin, since quite usually the dosage (which relates to local effectiveness against cancer) is limited by damage inflicted to surrounding normal tissues. Later a therapeutic comparison must be made with the best available treatment methods for specific sites and tumor types. Accordingly the clinical program is divided into 3 classically defined phases.

Phase Ia is designed to determine the RBE (relative biological effectiveness when compared to conventional photons) of pions with small treatment volumes upon the skin of patients with metastatic skin nodules. The first clinical trial will be described below.

Phase Ia' is a newly conceived transitional step from the simple, small dose distribution achieved with the Phase Ia beam geometry to the more complex method of Phase Ib. It will be used to treat patients with somewhat larger metastatic or surgically recurrent nodules on or near the body surface. The intricate steps of creating an individually moulded couch, performing a localizing and planning CT-scan, "cylindrizing" the patient in the region of the treatment volume, and, finally treating through the water bolus ring, will all be necessary. However, the actual treatment would be relatively simple, employing a section of perhaps 30 beams, without movement of the patient.

Phase Ib carries this approach one step further by incorporating movement of the patient during treatment so that the isocenter of the multiple beams will scan the treatment volume (raster scan). Patients with larger metastatic or surgically recurrent tumors would be treated in this phase. At one stage, exploration of the ring scan (pion momentum changing during treatment) will be initiated.

Phase II finally will employ the technical development of the previous steps for the radical (with curative intent) treatment of locally and regionally advanced primary neoplasms. Sites that are under consideration, initially, include bladder, uterus, prostate, and rectum in the pelvic region; pancreas, biliary tract, and liver in the abdomen. Head and neck cancer and brain tumors will be considered later.

Phase III will consist of randomized clinical trials of pion therapy and the "best available" forms of radiotherapy for the curative treatment of tumors in sites where, from Phase II, pions appear promising.

Description of the Current Clinical Experience

Phase Ia: A dose distribution was selected in order to treat a circle of about 4 cm to a 1 cm depth of the skin. This was accomplished by selecting a sector of 15 pion beams and designating a semi-cylindrical bolus of about 18 cm thickness to allow the stopping region to reach the desired skin depth. The pion momentum was then carefully determined (180 MeV/c) and the dose distribution confirmed by 3-dimensional ion chamber dosimetry in a water phantom.

A collimator was designed to approximate the dose distribution of x-rays from a 300 kVp therapy unit with the dose distribution of pions in the special 15 beam configuration. The surfaces irradiated could not be precisely matched but approximated 4 cm in diameter. The fall-off was greater with pions, therefore a medium pion isodose of 90% was selected to compare to the 100% dose level of x-ray.

The reactions were read and scored according to a scale developed by L. Cohen (6) which involves 8 steps of visible acute skin reactions. Clinical observations were done by at least 2 clinicians supplemented by color photographs.

Doses were calculated to give an advanced dry desquamative reaction (level 5 in the Cohen scale) with x-rays and reduced for pions by a factor of $\frac{1}{1.5}$ or 0.67.

The patient 81-1 with the diagnosis of metastatic melanoma had over 41 distinct lesions scattered over the entire trunk. Because of logistic difficulties in carrying out the entire planned therapy due to patient illness the planned course of therapy including fractionation schemes of 1, 4 and 12 fractions could not be accomplished. The completed program is shown in Table I.

Treatments carried out in 10 fractions over 19 days and 3 fractions over 15 to 16 days could be compared. With serial observations over between 46 and 57 days from the beginning of treatments it is possible to make two RBE estimates as follows:

For 10 fractions in 19 days the pion dose of 2364 rads gives a reaction level between those of x-rays at doses of 3500 and 3330 rads. From this it can be calculated that the RBE lies between 1.48 and 1.41.

For 3 fractions in 15 or 16 days the pion dose of 1577 rads lies slightly below those of x-rays at doses of 2340 and 2115 rads. Thus the RBE appears to lie below 1.34.

The patient 81-2, also with multiple metastases from malignant melanoma, received pion therapy to one lesion and x-ray therapy to 2 lesions using the same experimental configuration. All lesions received 4 fractions of treatment. Acute reactions at 30 days of observation (the date of this writing) are not yet fully developed but appear not to be departing from an RBE level in the region of 1.2 to 1.4.

All melanoma nodules in pion treated sites as well as x-ray treated sites have demonstrated varying degrees of regression.

TABLE I. PROGRAMME OF TREATMENT OF PATIENT 81-1:
PHASE I - SKIN REACTIONS

	Skin Dose*	No. Fractions	No. Days	Maximum Acute Skin Reaction
PIONS	2612	10	19	7
	2480	10	19	6.5
	2364	10	19	5
	1577	3	16	4
	1509	3	11	3.5
X-RAYS	3500	10	19	6
	3330	10	19	4
	2944	8	12	4
	2944	8	12	4
	2800	8	12	3.5
	2664	8	12	3
	2820	4	15	5.5
	2700	4	15	5
	2340	3	15	5
	2115	3	15	4.5

*Pion factors: 180 MeV/c; dose rate 20 rad/min; field sizes – 2.3 cm mean diameter (95%), 3.8 cm mean diameter (85%), 5 cm mean diameter (75%).

X-ray factors: 250 kV; 12 mA; HVL 5 mm Pb; FSD 7.5 cm; dose rate 116 rad/min; field sizes – 4 cm X 4 cm (100%), 6 cm X 6 cm (with designed penumbra).

Discussion

Clinical data from similar phase I trials at LAMPF (Los Alamos Meson Facility) (7) and TRIUMF (Tri-University Meson Facility, Vancouver, B.C.) (8)as well as radiobiological studies on experimental animals (5) indicate that an RBE ranging from 1.4 to 1.6 could be expected for fractionated treatment with small peak volumes from 10 or more fractions.

It is not expected that the RBE values at SIN would vary widely from these studies and this has not been found to date. These studies relate to a limited field in a single tissue for acute reactions with fractionation schemes not yet applied in clinical practice such as in Los Alamos because of logistic reasons. It is interesting to note the possibly lower RBE of less than 1.4 for 3 fractions which has not previously been studied clinically. This, however, conforms to the experimental studies of high-LET radiation which demonstrate a positive correlation of RBE with fraction number.

Data on late effects and upon RBE values for different tissue, treatment volumes, and dosage protocols must be gradually accumulated with clinical experience and by careful progression of clinical studies hand in hand with experimental animal irradiation studies.

REFERENCES

[1] BOYD, D., SCHWETTMAN, H.A., SIMPSON, J.: A large acceptance pion channel for cancer therapy, Nucl. Instrum. Methods 111 (1973) 315-331.
[2] VĚCSEY, G., HORVATH, I., ZELLWEGER, J.: "Superconducting medical facility at SIN", in Proc. 6th Int. Conf. Magnet Technology (Bratislava), Slovak Academy of Science (1977) 361-368.
[3] SKARSGARD, L.D., HENKELMAN, R.M., LAM, G.K.Y., PALCIC, B., POON, M.N.: Pre-clinical studies of the negative pi-meson beam at TRIUMF, Radiat. Environ. Biophys. 16 (1979) 193-204.
[4] TREMP, J.: Bestrahlung von Zellkulturen, SIN Jahresbericht (1980).
[5] FRÖHLICH, E.: Fraktionierte Hautbestrahlungen bei der Albinomaus mit Peak-Pionen, SIN Jahresbericht (1980).
[6] COHEN, L.: personal communication.
[7] KLIGERMAN, M.M., et al.: Initial comparative response to peak pions and x-rays of normal skin and underlying tissue surrounding superficial metastatic nodules, Am. J. Roentgenol. 126 (Feb. 1976) 261-267.
[8] GOODMAN, G., JACKSON, S., DOUGLAS, B.: personal communication.

Progress in Radio-Oncology II, edited by K. H. Kärcher et al., Raven Press, New York © 1982.

Shape of the Initial Part of the Cell Survival Curve

Implications for the RBE/dose relationship for small doses per fraction

A. WAMBERSIE, J. GUEULETTE
UCL — Cliniques Universitaires St. Luc,
Brussels,
Belgium

J. DUTREIX
Institut Gustave-Roussy,
Villejuif,
France

Abstract

The initial part of the survival curve for intestinal crypt cells has been assessed, in mice, from LD_{50} determination after selective abdominal irradiation. Three approaches were used: (i) Variation of LD_{50} as a function of number of fractions (up to 20) given within same overall times; (ii) Variation of LD_{50} as a function of irradiation time (up to 14 h) for continuous low-dose-rate irradiation. The conclusion is that the crypt cell survival curve has a steep negative initial slope (${}_1D_0$ about 4.5 Gy) and coincides with its initial tangent for doses up to about 3 Gy; (iii) Determination of the RBE/dose relationship for 50 MeV (d-Be) neutrons. The RBE increases from 1.8 for a single fraction (γ dose 10 Gy) to 2.0 and 2.2 for two equal fractions separated by 24 and 3.5 h respectively. For 10 fractions separated by 3.5 and 7 h, the RBE reaches 2.76 and 2.58, the γ doses being 2.5 and 2.9 Gy respectively. These values are consistent with those derived from the shape of the initial part of the cell survival curve as obtained by the two first approaches; the RBE is not expected to increase further for smaller doses per fraction.

In previous studies, the survival curve for intestinal crypt cells in mice was derived from LD50 determinations ("intestinal death").For this aim, variation of LD50 after abdominal irradiation was assessed (I) when varying the fraction number N (1) , (II) when modifying the dose rate (2);more recently, as a third approach (III), the cell survival curve was derived from the variation of RBE for fast neutrons as a function of dose per fraction (3).The main assumption for deriving the cell survival curve from early intestinal tolerance (a "non-quantitative" gross tissue reaction) is that LD50, for the compared irradiation schemes and animal batches, corresponds to the same survival level of the stem cell population.The experimental procedures have been chosen in order to make this assumption as reasonable as possible.

I. Fractionated irradiation with γ-rays.

Influence of fraction number N on LD50 was systematically investigated using fractionated irradiation up to 20 fractions. More precisely, the increase of LD50 after abdominal irradiation was assessed when N equal fractions D_N were replaced by 2N equal fractions delivered over the same overall time. From these data an LD50 isoeffect curve $(N.D_N)$ as a function of N (or as a function of the dose per fraction D_N) could be derived, and from there the survival curve for intestinal crypt cells.

When deriving this cell survival curve from early intestinal tolerance, several assumptions are needed which were already discussed (1):

i. the cell survival curve is the same at each fraction. This implies a full repair of sublethal injury between fractions and no change in radiosensitivity and cellular repair capacity during the course of irradiation. This last condition is equivalent to assuming that any eventual change in the distribution of the cells in the cycle, including resting cells, and in the kinetics of their proliferation does not modify the cell survival curve.

ii. the biological endpoint which is considered corresponds to a definite survival level.

iii. repopulation rate at any time during the irradiation is the same for the two schedules which are compared.

The survival curve for intestinal crypt cells derived from these experiments with fractionated γ-irradiation could be represented by the following expression using the two component model:

$$S = \left[e^{-D/{}_1D_o} \right] \left[1 - (1 - e^{-D/{}_nD_o})^n \right] \qquad (1)$$

with ${}_1D_o = 4.5$ Gy, $n = 20$, ${}_nD_o = 2.25$ Gy and if $1/D_o = 1/{}_1D_o + 1/{}_nD_o$, $D_o = 1.5$ Gy. As can be seen on Fig.1, the survival curve for ^{60}Co has an important negative initial slope ($-1/{}_1D_o$ with ${}_1D_o = 4.5$ Gy) and it coincides with its initial tangent for doses up to about 3 Gy, for which cell lethality is essentially due to direct lethal events.

II. Low dose rate γ-irradiation

Another approach to the problem is the use of low dose rate γ irradiation. In practice, one has to determine the increase of LD50 for intestinal death when an acute γ irradiation is replaced by a low dose rate irradiation, the dose rate being low enough to avoid lethality by accumulation of sublethal damage.

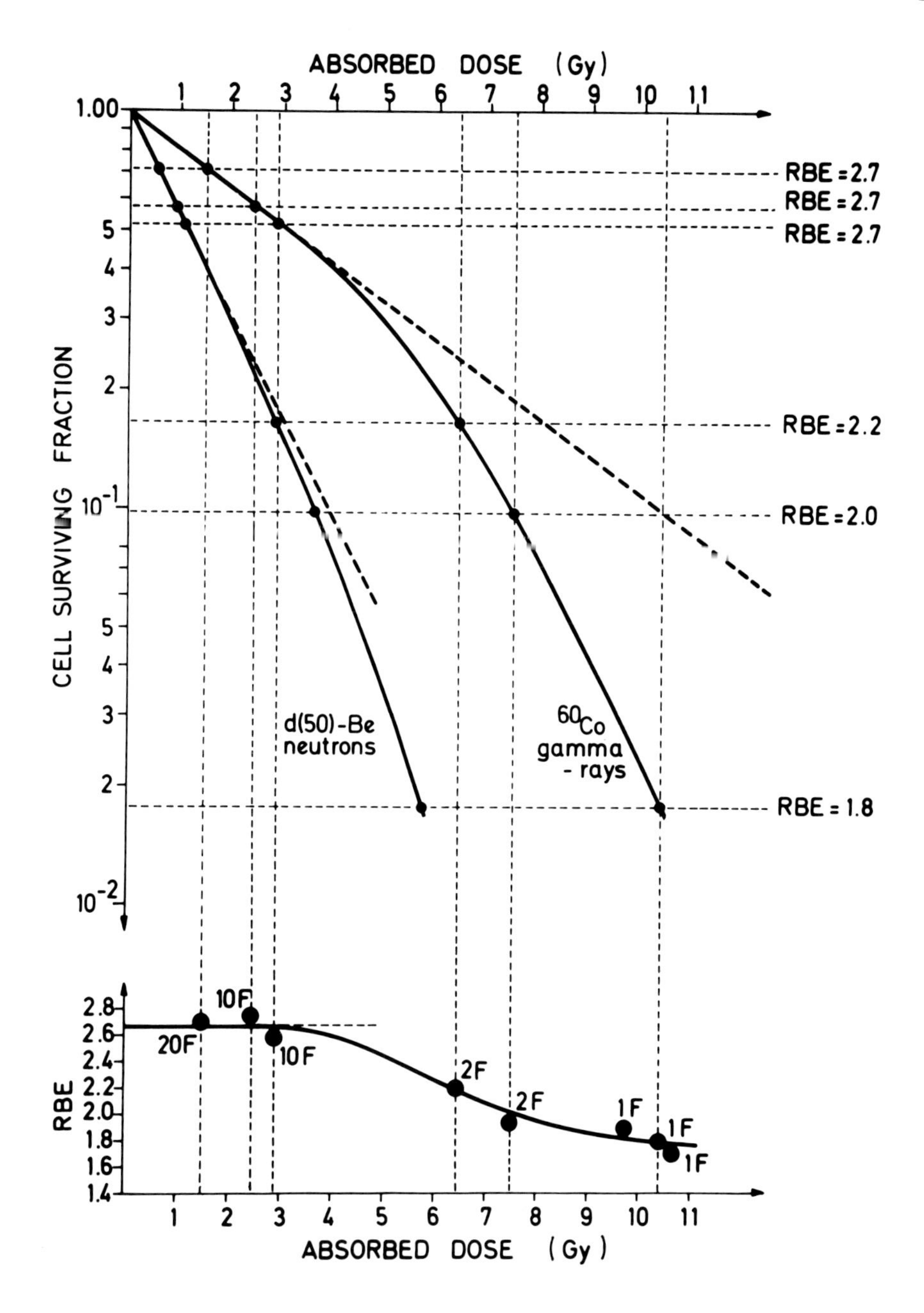

FIG.1. Survival curve for intestinal crypt cells derived from fractionated γ irradiation experiments; it can be represented by expression (1) in the text. The cell survival curve for d(50)-Be neutrons is derived in the same way from fractionated irradiation experiments, after normalization with the γ curve using an RBE value of 1.8 for a single dose. The lower part of the figure represents (solid line) the variation of RBE, as a function of the γ absorbed dose, deduced from the two upper curves. For comparison, the full circles are the RBE values experimentally derived (straightforward determination) using fractionated irradiations up to 20 fractions and listed in Table I.

At low dose rate, using the formulation and the values of the parameters of the former paragraph, survival can be expressed by:

$$S_1 = \exp(-D/{}_1D_o) \qquad (2)$$

with ${}_1D_o$ = 4.50 Gy.

As far as the experimental results are concerned, the following values were obtained. After ^{60}Co γ abdominal irradiation, LD50 at 6.5 days increases from 12.05 to 19.22 and 21.58 Gy when irradiation was delivered over 20 min, 6 and 14 hours respectively. After total-body irradiation, LD50 at 5.5 days increases from 9.92 to 15.20 and 16.83 Gy when irradiation was delivered over 20 min, 5 and 10 hours respectively (2). The observed additional doses for the 10 and 14 hour irradiations (9.53 and 6.91 Gy) correspond to the values expected from equations (1) and (2) (10.82 and 6.87 Gy) assuming that lethality was due mainly to direct lethal events. These experiments with low dose rate irradiation then support the model derived from the fractionated irradiation experiments and are consistent with the shape of the cell survival curve described by equation (1).

III. RBE of fast neutrons as a function of absorbed dose.

As a third approach to the determination of the shape of the initial part of the cell survival curve, the RBE of fast neutrons has been determined for decreasing doses per fraction. As the shoulder of the cell survival curve is far more important for γ-rays than for fast neutrons, RBE increases with decreasing dose. However the shape of the RBE/dose relationship at low doses per fraction depends on the shape of the initial part of the cell survival curve and mainly of the γ curve (we consider now fraction sizes of less than 3 Gy of γ-rays, for which the corresponding part of the neutron cell survival curve can be assumed to be exponential).

If for doses up to 3-4 Gy the cell survival curve for γ radiation is exponential and coincides with its initial tangent (as can be deduced from equation (1)), the neutron RBE will not increase anymore for doses per fraction lower than 3-4 Gy; it will in fact be equal to the initial slope ratio of the neutron and γ cell survival curves. On the contrary, if the initial slope were not so sharp and if the survival curve were separated from its initial tangent for smaller doses per fraction, the RBE would keep increasing.

The experiments performed with the d(50)-Be neutron beams produced at the cyclotron "CYCLONE" of Louvain-la-Neuve are in agreement with the model derived from the fractionated irradiation experiments (paragraph I). The results are presented in Table I: an RBE value of 2.7 ± 0.1 can be considered as an upper limit. As an example, Fig.2 presents the results of an experiment where neutron RBE is determined for 10 fractions separated by 3.5 hour intervals.

Table I. RBE of d(50)-Be neutrons, relative to ^{60}Co γ-rays, determined from LD50 after abdominal irradiation in mice.

Fraction number N	Interval between the fractions (hours)	Dose per fraction for γ-rays (Gy)	RBE
1 *	-	9.75	1.86
1 **	-	10.40	1.81
1 **	-	10.65	1.69
2	24	7.50	1.93
2	3.5	6.46	2.20
10	7	2.89	2.58
10	3.5	2.46	2.76
20	3.5	1.40	2.69

* LD50 at 5 days after total body irradiation.
** two independent determinations.

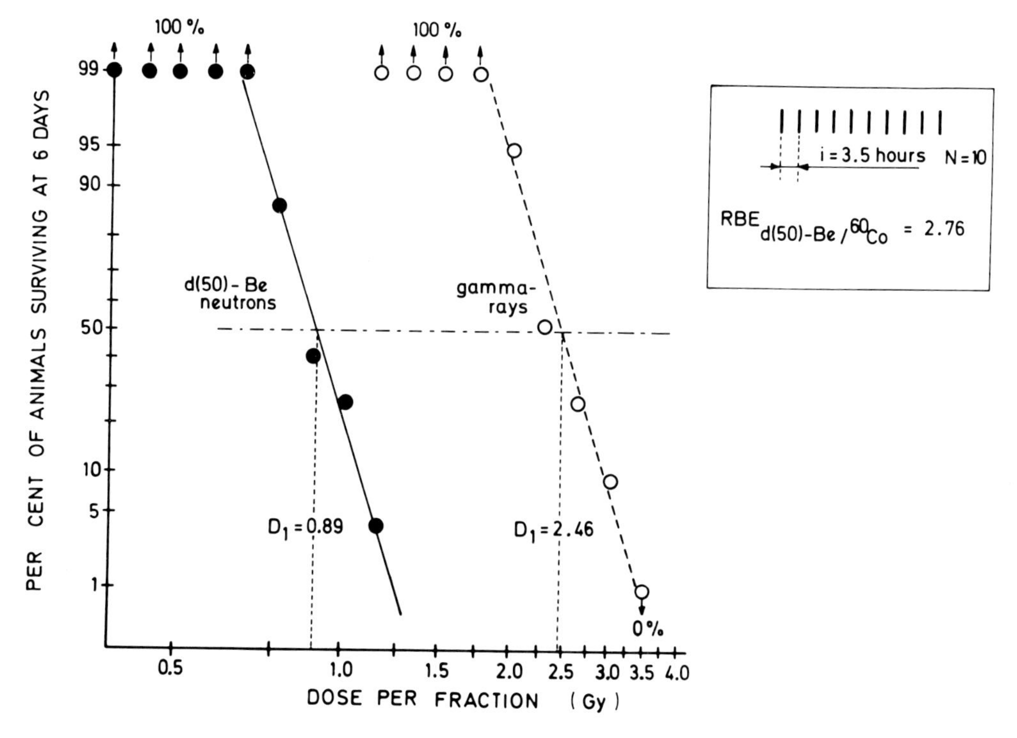

FIG.2. Survival rate at 6 days of the irradiated animals as a function of absorbed dose (in probit–log co-ordinates) for d(50)-Be neutrons (full circles) and ^{60}Co γ-rays (open circles). Irradiation is delivered in 10 equal fractions separated by an interval of 3.5 hours. The LD50 are (10 X 0.89) and (10 X 2.46) Gy respectively, which leads to an RBE of 2.76 for the d(50)-Be neutrons.

CONCLUSION.

Recent experiments with fast neutrons are in agreement with our previous conclusions derived from fractionated and low dose rate γ irradiation: the cell survival curve for intestinal crypt cells exhibits a marked negative initial slope ($-1/{}_1D_0 = 4.5$ Gy) and coincides with its initial tangent for doses up to about 3 Gy. For doses lower than 3 Gy, cell killing is mainly due to direct lethal events, which do not depend on fractionation or dose rate. The RBE for d(50)-Be neutrons increases up to 2.7 ± 0.1 for γ dose per fraction of about 3 Gy, but does not increase anymore for smaller doses per fraction.

As can be seen in Fig.1, there is a good agreement between RBE values directly determined for different fraction numbers and those deduced from the γ and neutron cell survival curves derived from fractionated irradiation experiments. The agreement between the three approaches for determining the initial part of the cell survival curve for intestinal crypt cell implies that the assumptions which were recalled in paragraph I were reasonable and valid for the intestinal crypt cell population. One should however stress that our conclusions derived from LD50 determinations (early intestinal tolerance) should not be extended without care to other tissues or effects.

REFERENCES

[1] WAMBERSIE, A., DUTREIX, J.: Eur. J. Cancer 10 (1974) 235.
[2] WAMBERSIE, A., STIENON-SMOES, M-R., OCTAVE-PRIGNOT, M., DUTREIX, J.: Br. J. Radiol. 52 (1979) 153.
[3] GUEULETTE, J., WAMBERSIE, A.: Int. J. Radiat. Biol. Relat. Stud. Phys., Chem. Med. 38 (1980) 111.

Progress in Radio-Oncology II, edited by K. H. Kärcher et al., Raven Press, New York © 1982.

Duodenal Ulcers as an Abscopal Effect of Thoracic Irradiation in Mice

A. MICHALOWSKI, Janice BURGIN
Medical Research Council Cyclotron Unit,
Hammersmith Hospital,
London,
United Kingdom

Abstract

Female CFLP mice irradiated to their thorax with either X-rays or fast neutrons developed peptic ulcers within 8 days of exposure. The steep X-ray dose/response curve for induction of duodenal ulcer gave an ED_{50} of approximately 14.5 Gy. As little as 6 Gy of fast neutrons was effective in some cases, but the neutron ED_{50} exceeded that for X-rays. The ulcers represented an abscopal effect of thoracic irradiation. Scattered radiation as simulated by whole-body X-ray treatment (1–5 Gy) caused a dose-dependent decrease in the frequency of duodenal lesions, possibly by decreasing gastric secretion. The greater amount of scattered radiation accompanying fast neutron exposure of the thorax was presumably responsible for the shallower dose/response curve of ulcer induction than that seen with X-rays.

Chance observations of peptic and intestinal ulcers in female CFLP mice irradiated to the thorax initiated the present study. In the first experiment a single dose of 14 Gy of cyclotron-generated fast neutrons ($\bar{E}$ = 7.5 MeV) was used while the animals were immobilized with Sagatal anaesthetic (pentobarbitone sodium, 0.06 mg/g body weight i.p.). Gross and microscopic examination following the spontaneous death of the mice uncovered a variety of gastro-intestinal lesions, the most common of which being the duodenal ulcer (30%). Occurring less frequently were erosions of the forestomach, small and large haemorrhagic erosions of the oxyntic mucosa and, in some cases, ulcers of the pyloric region. The haemorrhagic erosions of the stomach were usually accompanied by melaena, with both the small and large intestines filled with a tar-like material of haemoglobin derivatives. Ulcerations of the small and large intestines were also seen sporadically (Michalowski [1]).

The lesions would appear to be an abscopal effect since the dose of scattered radiation to the epigastric region was measured as 13% of that to the thorax, i.e. less than 2 Gy. Such a dose is unlikely to produce lesions of this severity. The large ulcers (fig. 1) were sharply delineated and surrounded by apparently normal mucosa totally uncharacteristic of direct radiation damage. In addition, in another series of thorax-irradiated mice the spleen weight of ulcer-bearing animals was nearly equal to that of controls, demonstrating that the dose of radiation absorbed in the epigastric region was negligible.

In contrast to the assertions of Phillips and Ross [2] that during the first 4 weeks following thoracic irradiation of mice, all deaths are

FIG.1. Duodenal ulcer on day 8 after thoracic irradiation.

secondary to oesophageal damage, the pattern of mortality of our mice was characterized by three peaks at days 5, 10 and 20, respectively, suggesting multiple causes of death that included the severe abdominal lesions. Likewise, weight loss of the animals (assumed by Kurohara and Casarett [3] as being solely attributable to severe oesophageal damage) proved to be an extremely unreliable predictor of oesophagitis. In particular, no difference in the degree of weight loss was seen between the animals dying before the onset of ulcerative oesophagitis and those showing its symptoms at death.

In a subsequent experiment the thorax of female CFLP mice was irradiated with a range of single doses of either X-rays or fast neutrons. The animals were sacrificed 8-10 days after irradiation and dose/response curves constructed. The X-ray curve for induction of duodenal ulcer (fig. 2) is seen to be steep, the ED50 being of the order of 14.5 Gy. However, with neutron irradiation the incidence of duodenal ulcer was much less clearly dose-dependent: as little as 6 Gy was effective in some cases, but the ED50 exceeded that for X-rays. There was excellent agreement between the results of the first and subsequent experiment (30% and 29% incidence of duodenal ulcers, respectively).

A working hypothesis was invoked to explain the above results. It was considered possible that thoracic irradiation induces duodenal ulcer, while scattered radiation absorbed by the remaining parts of the body acts so as to <u>prevent</u> ulcer formation. This would explain the findings in neutron-irradiated mice (fig. 2) in which the proportion of scattered radiation was greater than with X-rays. To test the hypothesis, 12-week

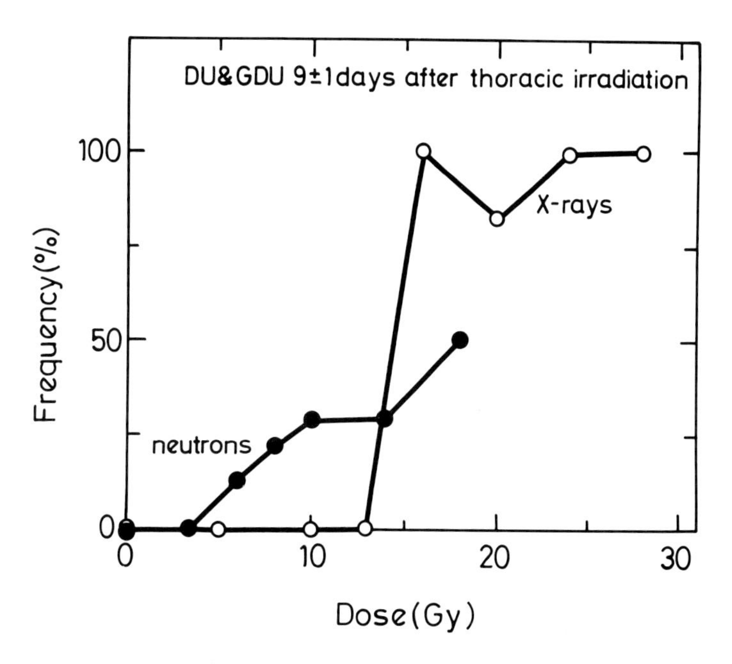

FIG.2. Dose/response relationship for duodenal/gastroduodenal ulcer induction by thoracic irradiation of mice with X-rays and fast neutrons. Five to nine animals were sacrificed and examined per dose-point.

old female CFLP mice were irradiated with X-rays in the following manner:

1) 60 animals received a dose of 16 Gy to the thorax only;
2) 59 mice were given a first dose to the thorax, followed within one minute by whole-body irradiation with doses ranging from 1 to 5 Gy, keeping the cumulative thoracic dose constant at 16 Gy;
3) 58 animals were treated with whole-body irradiation only from 1 to 5 Gy;
4) the remaining 10 mice served as unanaesthetized, unirradiated controls.

From groups 1 - 3 a total of 17 mice died on day 4 through 11 (Table I). These were not dissected and the causes of their death are not known. During the first week after exposure (days 3-7) a number of the thorax-irradiated mice (group 1) were sacrificed and their gastro-intestinal tract examined. Nine out of 30 animals (30%) exhibited duodenal ulcers (fig. 3), but no perforation or bleeding was evident. During this period the average weight of the gastric contents was significantly higher than that of the control group, with a gradual decrease to normal values by day 7. The remaining 29 mice irradiated to their thorax only were sacrificed on days 8 to 17, and of these 20 (69%) were found to carry duodenal ulcers.

During the same period all the remaining irradiated animals were sacrificed and dissected (Table I and fig. 3). Of those receiving whole-body irradiation only (group 3), none was duodenal ulcer-positive. Of the 45 mice receiving thoracic plus whole-body irradiation (group 2), 17 or 38% showed duodenal ulcers compared with 69% in group 1 during the same assay period. Thus the incidence was significantly lower than in mice irradiated to the thorax only. In addition it is seen in Table I and fig. 3 that the larger whole body doses produced a greater degree of protection from ulcer induced by irradiation of the thorax.

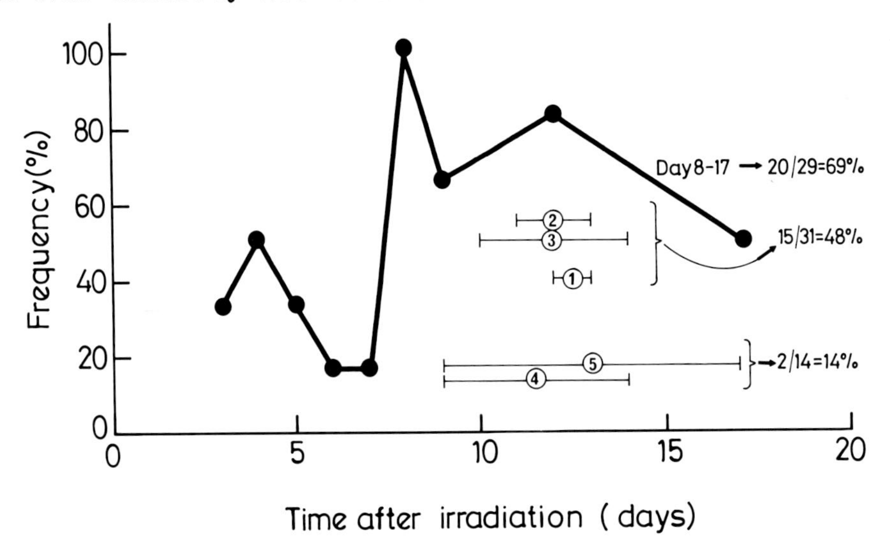

FIG.3. Incidence of duodenal ulcer versus time after exposure in mice irradiated to their thorax only (group 1, full points and bold line) and in those animals in which the same dose of 16 Gy of X-rays to the thorax was accompanied by simultaneous irradiation of the remaining parts of the body (group 2, thin horizontal bars indicating the extreme times of the assay with whole-body dose given in Gy in the circle).

TABLE I. TREATMENT, MORTALITY AND INCIDENCE OF DUODENAL LESIONS IN SACRIFICED MICE

Dose (Gy) Thorax	Dose (Gy) Remaining parts of the body	Spontaneous death on day n after irradiation, with number of animals dead in brackets	Day of sacrifice after irradiation	A = number of sacrificed mice with duodenal lesions; B = number of dissected mice (A/B)	(ΣA/ΣB [%])	
16	0	6(1)	3	2/6		
16	0		4	3/6		
16	0		5	2/6	9/30 [30%]	
16	0		6	1/6		
16	0		7	1/6		
16	0		8	5/5		
16	0		9	4/6	20/29 [69%]	
16	0		12	5/6		
16	0		17	6/12		
16	1	4(1)	12	3/5		
16	1		13	1/5		
16	2	6(1)	11	2/5	15/31 [48%]	
16	2		13	4/6		
16	3	5(2)	10	3/5		
16	3		14	2/5		17/45 [38%]
16	4	4(1) 6(1) 7(2)	9	1/3		
16	4		14	0/5	2/14 [14%]	
16	5	4(3) 5(2) 11(1)	9	1/2		
16	5		17	0/4		
1	1		12	0/12		
2	2		11	0/6		
2	2		13	0/6		
3	3		11	0/6		
3	3		13	0/6	0/56 [0%]	
4	4		10	0/6		
4	4		14	0/6		
5	5	6(1) 8(1)	10	0/5		
5	5		17	0/3		
0	0		3	0/4	0/10 [0%]	
0	0		6	0/6		

The inhibiting effect of scattered radiation, which is presumably responsible for the shallow dose/response relationship seen after thoracic neutron irradiation (fig. 2), may be due to either the prevention of ulcer formation or the acceleration of the healing process, presumably by decreasing the gastric secretion (Palmer [4]). However, the exact mechanism of ulcer formation distant from the irradiated area remains to be elucidated.

We conclude that:-

1) Irradiation of the thorax with either X-rays or fast neutrons results in peptic ulcer formation.
2) The radiation-induced duodenal lesions take up to 8 days to develop.
3) The lesions represent an abscopal effect of thoracic irradiation and scattered radiation accompanying thoracic exposure is <u>not</u> responsible for ulcer formation.
4) Scattered radiation as simulated by the whole-body treatment causes a dose-dependent <u>decrease</u> in the frequency of duodenal lesions.
5) The greater degree of scattered radiation accompanying fast neutron exposure of the thorax is presumably responsible for the shallower dose/response curve than that seen with X-rays.

Acknowledgements

Thanks are due to Ann Silvester for supplying an improved radiation jig and for radiation dosimetry, to Caroline Morris for irradiating mice, to Fred Paice for photography, to Michael Rogers for preparing microscopic slides, to Tom Wheldon for statistical evaluation, and to Doreen Wishart for typing the manuscript.

References

[1] MICHALOWSKI, A., Gastro-intestinal lesions in thorax-irradiated mice (abstract), Br. J. Radiol. (1981, in press).

[2] PHILLIPS, T.L., ROSS, G., Time-dose relationships in the mouse oesophagus, Radiology 113 (1974) 435.

[3] KUROHARA, S.S., CASARETT, G.W., Effects of single thoracic X-ray exposure in rats, Radiat. Res. 52 (1972) 263.

[4] PALMER, W.L. (ed.) Gastric Irradiation in Peptic Ulcer, The University of Chicago Press, Chicago and London (1974).

Progress in Radio-Oncology II, edited by K. H. Kärcher et al., Raven Press, New York © 1982.

Report of the M.D. Anderson Clinical Trial of Fast Neutron Therapy for Head and Neck Cancer*

D.H. HUSSEY, M.H. MAOR, J.P. SAXTON, G.H. FLETCHER
Division of Radiotherapy,
The University of Texas M.D. Anderson Hospital and Tumor Institute,
Houston, Texas

A.S.M. AL-ABDULLA
Department of Radiology,
Division of Radiation Oncology,
The University of Texas Medical Branch at Galveston,
Galveston, Texas

R.H. JESSE
Department of Head and Neck Surgery,
The University of Texas M.D. Anderson Hospital and Tumor Institute,
Houston, Texas

United States of America

Abstract

Between October 1972 and February 1980, 220 patients with locally advanced head and neck tumours were treated in the M.D. Anderson Hospital neutron therapy programme. Of these, 125 patients were treated in pilot studies and 95 in a randomized clinical trial. In the pilot studies, 49 patients were treated with neutrons alone, 31 with mixed-beam irradiation (2 neutron and 3 photon fractions per week), and 45 with conventional treatment (surgery, photons, or combined surgery and photons). There was no appreciable difference among the pilot studies with regard to local tumour control or survival. However, the patients in the conventional treatment pilot study had less-advanced disease than those in either of the other studies. The complication rates in the neutrons-only and conventional-treatment pilot studies were significantly greater than the complication rate in the mixed-beam study. In the randomized clinical trial, 54 patients were treated with mixed-beam irradiation and 41 with photon irradiation. The local control and survival rates for patients who were treated with mixed-beam irradiation were greater than those achieved with photons, although the difference in local control rates is not statistically significant.

Between October 1972 and February 1980, 220 patients with locally advanced head and neck tumors were treated with neutrons, a combination of

* This investigation was supported in part by grants CA-12542 and CA-6294, awarded by the National Cancer Institute, United States Department of Health, Education and Welfare.

neutrons and photons (mixed-beam), or conventional treatment in the U. T. M. D. Anderson Hospital neutron therapy program. [1] Of these, 125 patients were in pilot studies and 95 in a randomized clinical trial. This paper is a report of that experience. The principal objectives are: (1) to review the results of the neutrons-only, mixed-beam, and conventional-treatment pilot studies and (2) to compare the results of mixed-beam and photon irradiation in the randomized clinical trial. The data are analyzed in terms of local tumor control, frequency of complications and patient survival.

MATERIALS AND METHODS

Energy and Dosage Conventions

The neutron treatments were delivered with a neutron beam produced by 50-MeV deuterons incident on a thick beryllium target (50 $MeV_{d \rightarrow Be}$). The neutron beam doses are reported in terms of the physical dose expressed in rad including both the neutron and gamma components ($rad_{n\gamma}$). [2] The mixed-beam doses are reported in terms of: (1) the physical dose, listing the neutron beam and photon beam doses separately ($rad_{n\gamma}$ + rad), and (2) the total equivalent dose of the combined regimen (rad_{eq}). The equivalent doses (rad_{eq}) were determined by multiplying the physical dose delivered with neutrons by an RBE of 3.1 and adding this to the dose delivered with photons. [3]

Treatment Methods

The standard photon treatment policies at U. T. M. D. Anderson Hospital were adapted for the neutron therapy program (Table I). In general, the same treatment portals, total equivalent dose, and overall time were employed for neutron and mixed-beam therapy as would have been used with photon irradiation. [1] For most tumor sites, the aim was to deliver 6000-6500 rad_{eq} in 6-6½ weeks for moderately advanced tumors and 6500-7000 rad_{eq} in 6½-7 weeks for massive cancers.

The treatment methods listed in Table I refer to the management of the primary tumors. The regional lymphatics of the lower neck were usually treated with ^{60}Co gamma rays with a given dose of 4500 rad in 15 fractions in 5 weeks or 5000 rad in 25 fractions in 5 weeks. In all treatment groups, modified neck dissections were permitted for patients with bulky, but technically resectable lymph node metastases.

Patients in the neutrons-only pilot study were irradiated twice weekly, usually with fractions of 160 $rad_{n\gamma}$. A weekly dose of 320 $rad_{n\gamma}$ is equivalent to 1000 rad/5 fractions/week with ^{60}Co gamma rays, assuming an RBE of 3.1 relative to 200 rad ^{60}Co fractions. For the mixed-beam treatments, patients were irradiated with neutrons twice weekly with a fraction size of 65 $rad_{n\gamma}$ and with photons three times weekly with a fraction size of 200 rad.

The treatment for patients in the conventional-treatment group was determined by the site and extent of the cancer and the general condition of the patient. In the conventional-treatment pilot study, patients with oral cavity tumors were usually selected for treatment by surgery or combined surgery and irradiation, whereas those with oropharyngeal or hypopharyngeal lesions were treated by radiation therapy alone. The patients in the control arm of the randomized trial were treated by photon irradiation.

TABLE I. Treatment Methods and Average Tumor Dose

Study Treatment	Number Patients	Average Tumor Dose ± 1 S.D.*† Neutrons ($rad_{n\gamma}$)	Photons (rad)	Equiv Dose‡ (rad_{eq})
Pilot Studies				
Neutrons only	49	2128 ± 240		6593 ± 744
Mixed beam	31	808 ± 200	4299 ± 793	6803 ± 390
Conventional Tmt	45			
Surgery only	9			
Photons only	12		7060 ± 1367	7060 ± 1367
Surgery + photons	24		5692 ± 618	5692 ± 618
Randomized Trial				
Mixed beam	54	843 ± 168	4097 ± 652	6711 ± 477
Photons	41		7036 ± 316	7036 ± 316

* Including dose delivered with interstitial implants
† Abbreviations: S.D. = standard deviation
‡ Equivalent doses (rad_{eq}) were calculated by multiplying the neutron dose ($rad_{n\gamma}$) by 3.1 and adding this to the dose delivered with photons (rad).

PILOT STUDIES

Clinical Material

One hundred twenty-five patients were treated in the pilot studies--49 with neutrons alone, 31 with mixed-beam irradiation, and 45 with conventional treatment. With the exception of 7 patients in the neutrons-only study and 3 in the mixed-beam study who had adenocarcinomas or adenoid cystic carcinomas, all patients presented with squamous carcinomas.

There were more patients with oral cavity tumors in the conventional-treatment study (47%) than in either the neutrons-only (25%) or the mixed-beam studies (10%). On the other hand, a greater percentage of the patients receiving mixed-beam therapy had oropharyngeal tumors. Only 4% of the patients in the conventional-treatment study presented with recurrent tumors, as compared to 18% of those in the neutrons-only study and 29% of those in the mixed-beam study.

All of the patients presented with locally advanced carcinomas (stages $T_{3-4}N_{0-3}$ or $T_{0-2}N_{2-3}$ previously untreated tumors, or tumors that had recurred following surgical treatment). On the average, the patients in the mixed-beam study had less-advanced cancers than those in the neutrons-only study and more-advanced cancers than those in the conventional-treatment study (Table II).

TABLE II. Pilot Studies: Distribution of Clinical Material by Clinical Stage

	Neutrons Only	Mixed Beam	Conventional Treatment
Moderately advanced	13 (27%)	12 (39%)	25 (56%)
T_3N_{0-2}	10* (21%)	10 (32%)	24 (54%)
Recurrent tumors <6 cm in diameter	3 (6%)	2 (7%)	1 (2%)
Massive	36 (73%)	19 (61%)	20 (44%)
T_4N_{0-3}, or $T_{0-3}N_3$	30 (61%)	12 (39%)	19 (42%)
Recurrent tumors >6 cm in diameter	6 (12%)	7 (22%)	1 (2%)
TOTAL	49	31	45

* Includes 2 patients with stage T_XN_2 lesions

Results

The neutrons-only pilot study was analyzed in August 1977, and the mixed-beam and conventional-treatment pilot studies in March 1981. The average interval from the start of radiation therapy to the date of analysis was 46.6 months (range: 29-58 months) for the neutrons-only study, 51.1 months (range: 15-77 months) for the mixed-beam study, and 50.6 months (range: 20-67 months) for the conventional-treatment study.

There was no appreciable difference among the neutrons-only, mixed-beam, and conventional-treatment pilot studies with regard to local tumor control. The local control rates were 47% (23/49) for the neutrons-only study, 48% (15/31) for the mixed-beam study, and 44% (20/45) for the conventional-treatment study. Likewise, the survival rates for patients in the three pilot studies were about the same. At the time of analysis, 14% (7/49) of the neutrons-only patients, 29% (9/31) of the mixed-beam patients, and 24% (11/45) of the conventional-treatment patients were alive. However, four of the survivors in the mixed-beam study and three of those in the conventional-treatment study have been followed for periods of only 15-24 months.

There was a significantly greater incidence of major complications in the patients who received neutrons only (10/49, 20%) or conventional treatment (9/45, 20%) than in those who received mixed-beam irradiation (0/31,

0%). The majority of the complications in the conventional-treatment study were related to surgery. Three developed following surgery alone, five following surgery and postoperative radiotherapy, and one following photon irradiation alone. On the other hand, nine of the ten complications in the neutrons-only pilot study occurred following treatment with irradiation alone.

Actuarial local control, complication, and survival curves [4] are plotted in Figure 1. By this method of analysis, both the local control and survival curves are almost identical for the three pilot studies. However, the actuarial complication rates for the neutrons-only and conventional-treatment studies are significantly greater than the actuarial complication rate for the mixed-beam study (Wilcoxon two-tailed test: mixed beam vs neutrons only, $p = .02$; mixed beam vs conventional treatment, $p = .02$).

RANDOMIZED TRIAL

Clinical Material

Between January 1977 and February 1980, 95 patients with locally advanced squamous carcinomas of the head and neck were randomized to receive mixed-beam or photon irradiation. There were 54 patients in the mixed-beam group and 41 in the photon group. Only patients with stages $T_{3-4}N_{0-3}$, $T_{0-2}N_3$ cancers or recurrent tumors following surgical management were included.

The distribution of clinical material by tumor site was similar for both groups. Approximately two-thirds of the patients in each group presented with oropharyngeal cancers and only 10% had oral cavity tumors. The incidence of tonsillar fossa tumors was slightly greater in the photon group and the incidence of base of tongue tumors, nodal metastases from T_0 or T_X primary tumors, and recurrent tumors was slightly greater in the mixed-beam group.

The clinical stage distribution was similar for both groups (Table III). Approximately one-third of the patients in each group presented with moderately advanced cancers (T_3N_{0-2} or recurrent tumors <6 cm in diameter) and two-thirds with massive cancers (T_4N_{0-3}, $T_{0-2}N_3$ or recurrent tumors >6 cm in diameter).

Results

The data were analyzed in March 1981. The average interval from the start of radiation therapy to the date of analysis was 29.9 months (range: 12-49 months) for the mixed-beam group and 30.3 months (range: 13-49 months) for the photon group. The preliminary results are shown in Table IV and Figure 2.

The local control rates for the two groups were approximately the same--48% (26/54) for the mixed-beam group and 44% (18/41) for the photon group. However, the local control rates for base of tongue, pharyngeal wall, and recurrent tumors were slightly greater with mixed-beam irradiation than with photon irradiation (Table IV). These are tumors that are often difficult to control with conventional irradiation. There was no significant difference between the mixed-beam and photon control rates for tonsillar fossa and faucial arch tumors, cancers that often respond well to conventional radiotherapy. At the time of analysis, 35% (19/54) of the patients in the mixed-beam group were living compared to only 24% (10/41) of the patients in the

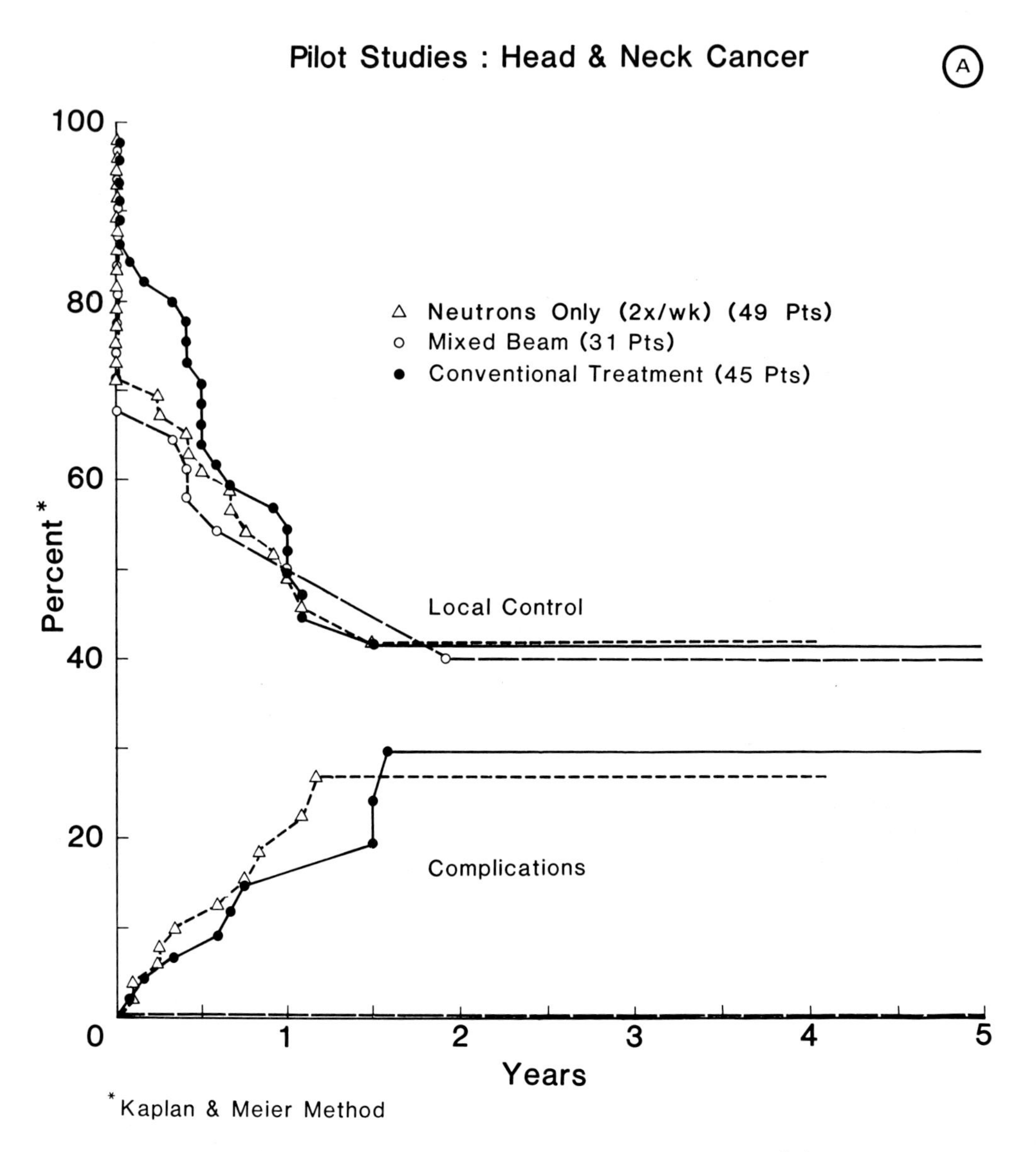

FIG. 1. Actuarial local control and complication curves (A) and survival curves (B) (Kaplan and Meier, [4]) for patients treated in pilot studies.

photon group. This difference is not statistically significant (χ^2 = 1.28; p = .26).

Three patients in each group developed major complications (6%). However, two of the three complications in the mixed-beam group developed following modified neck dissections (one wound dehiscence and one carotid rupture). The other (pharyngeal wall necrosis) occurred in a patient who was treated with radiotherapy alone to a dose of 6500 rad_{eq} for a T_3N_0 cancer of the base of tongue. In retrospect, this may have resulted from an over-

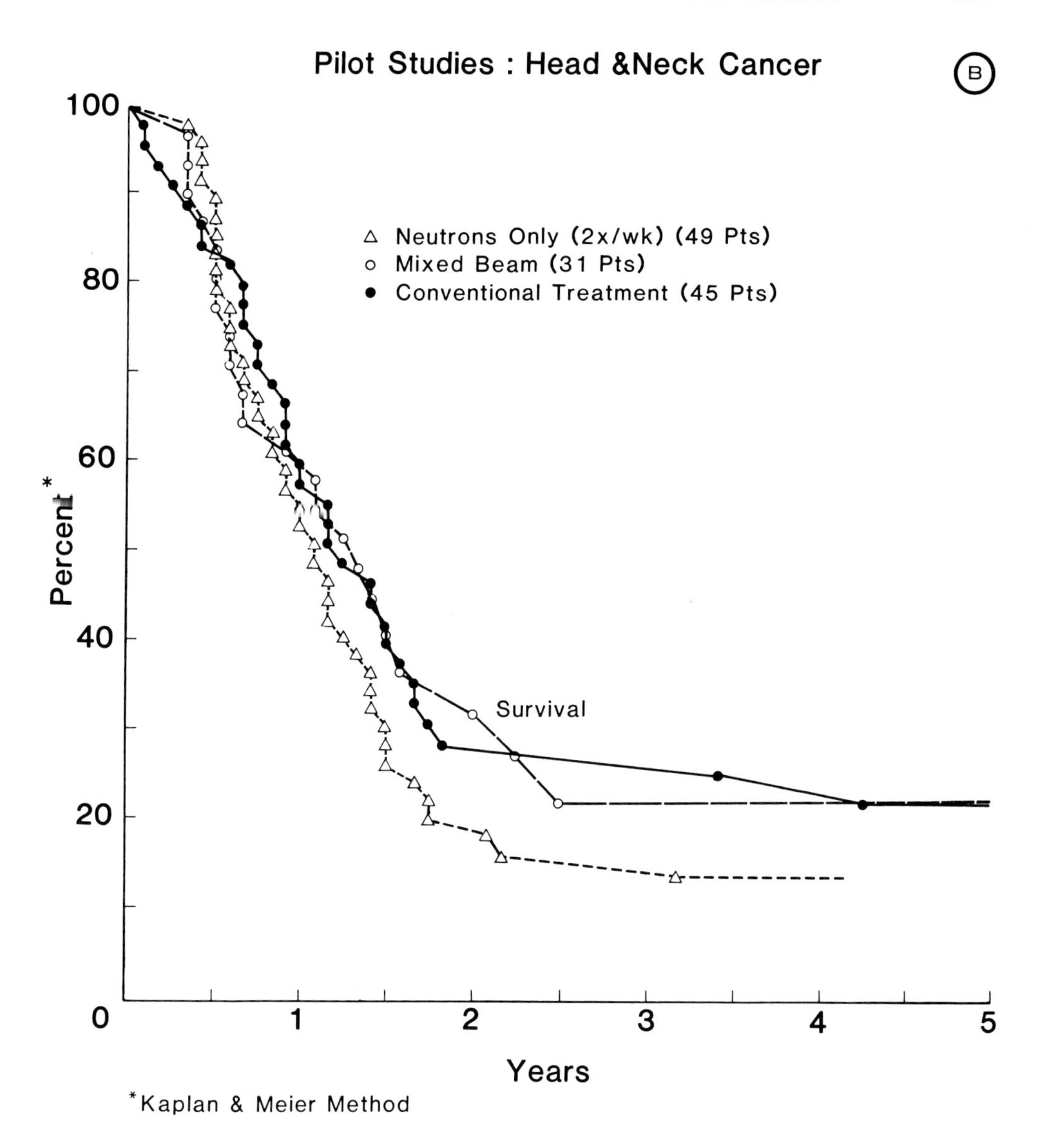

lap of the upper neck neutron portals with the lower neck photon portals. The three complications in the photon group occurred after treatment with radiotherapy alone. One patient developed pharyngeal wall necrosis (7000 rad), and one osteoradionecrosis (7000 rad), and one died of sepsis following a radium implant (5000 rad external irradiation plus 3000 rad with interstitial radium).

Actuarial local control, complication, and survival curves [4] were plotted because the follow-up interval was limited (Fig 2). The local control curve for the mixed-beam group is superior to that of the photon group, but the difference is not statistically significant (Wilcoxon two-tailed test: p = .21). The difference between the mixed-beam and photon survival curves is statistically significant (Wilcoxon two-tailed test: p = .03).

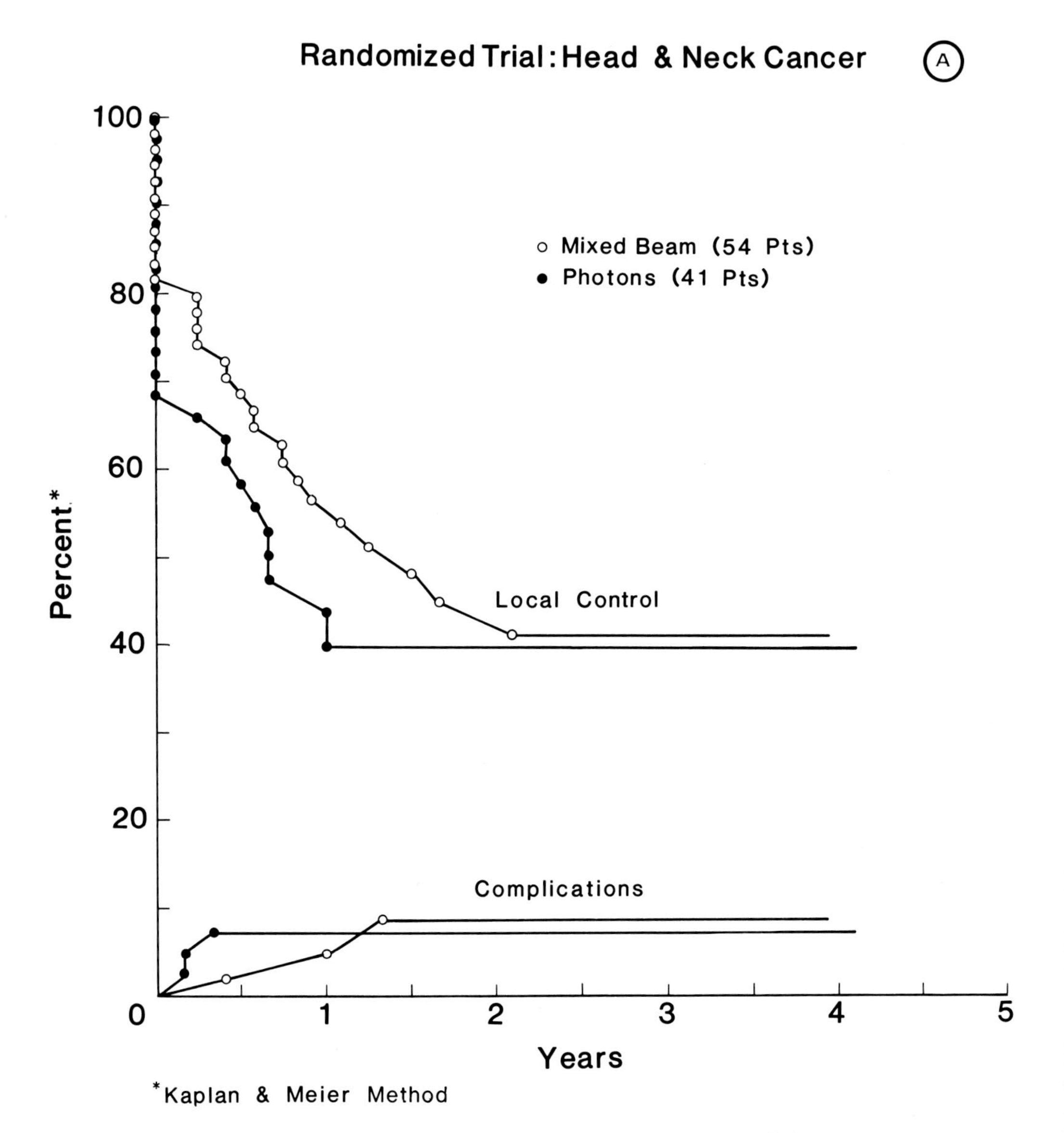

FIG. 2. Actuarial local control and complication curves (A) and (B) survival curves (Kaplan and Meier, [4]) for patients treated in the randomized trial.

DISCUSSION

The results of this analysis demonstrate the need for randomized clinical trials to evaluate the role of fast-neutron therapy in the treatment of head and neck cancers. Pilot studies are of value to establish fractionation schedules and treatment techniques, but are of limited usefulness for comparison purposes because of differences in clinical material and follow-up time.

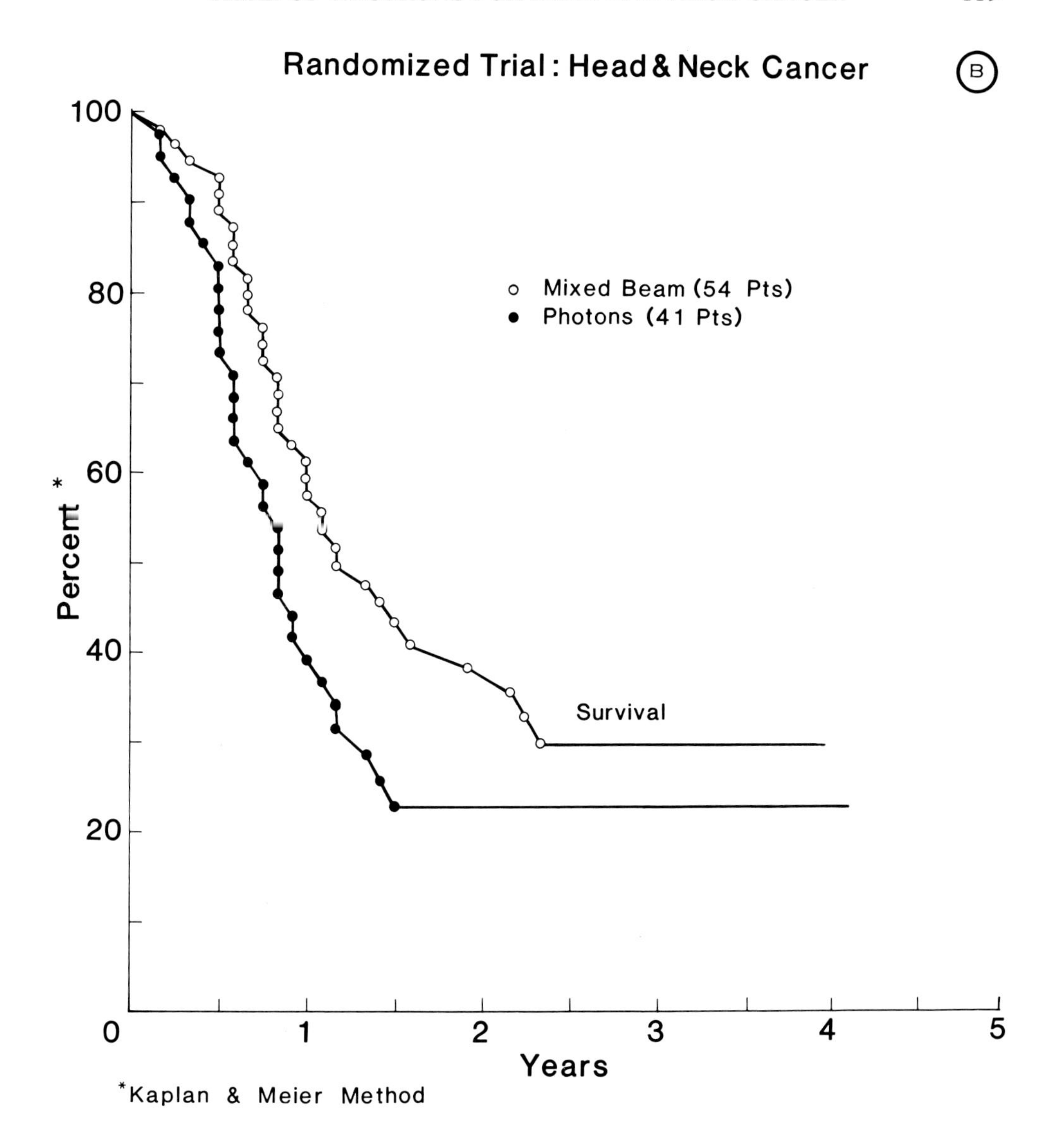

In this analysis, the local control and survival rates for the mixed-beam pilot study were the same as the local control and survival rates for the neutrons-only and conventional-treatment pilot studies (Fig 1). However, the complication rate with mixed-beam irradiation was significantly less than that observed in either of the other studies. The high complication rate in the conventional-treatment pilot study was mainly due to the high surgical morbidity for patients with massive tumors, whereas the high complication rate in the neutrons-only study resulted from treatment with irradiation alone.

In the randomized trial, the local control and survival rates for patients who were treated with mixed-beam irradiation were greater than those achieved with photons (Fig 2), although the difference in local control rates

TABLE III. Random Study: Distribution of Clinical Material by Clinical Stage (January 1977 - February 1980)

	Mixed Beam		Photons	
Moderately advanced	17 (32%)		14 (34%)	
T_3N_{0-2}		16 (30%)		14 (34%)
Recurrent tumor <6 cm in diameter		1 (2%)		0
Massive	37 (68%)		27 (66%)	
T_4N_{0-3}, or $T_{0-3}N_3$		33 (61%)		25 (61%)
Recurrent tumor >6 cm in diameter		4 (7%)		2 (5%)
TOTAL	54		41	

TABLE IV. Random Study: Local Control by Tumor Site (January 1977 - February 1980) Analysis - March 1981

	Mixed Beam		Photons	
Oral Cavity	1/6		1/6	
Floor of mouth		1/3		0/2
Oral tongue				0/3
Hard palate				
Buccal mucosa		0/2		
Gingiva		0/1		1/1
Oropharynx	17/36 (47%)		14/29 (48%)	
Faucial arch		3/7		2/5
Tonsillar fossa		4/9		7/11
Base of tongue		7/16		2/7
Pharyngeal wall		3/4		3/6
Larynx & Pyriform Sinus	0/2		2/3	
Nodal Metastasis (T_0 or T_X)	4/5		1/1	
Recurrent Tumor	4/5		0/2	
at primary site		2/2		0/1
in nodes		2/3		0/1
TOTAL	26/54 (48%)		18/41 (44%)	

is not statistically significant. The principal difference between the two groups was the result of improved control of base of tongue, pharyngeal wall, and recurrent tumors (Table IV).

REFERENCES

[1] Maor, M. H., Hussey, D. H., Fletcher, G. H., Jesse, R. H., Fast neutron therapy for advanced head and neck tumors, Int J Radiat Oncol Biol Phys 7 (1981) 155.

[2] Hussey, D. H., Fletcher, G. H., Clinical features of 16 and 50 $MeV_{d \rightarrow Be}$ neutrons, Eur J Cancer 10 (1974) 357.

[3] Hussey, D. H., Fletcher, G. H., Caderao, J. B.,"A preliminary report of the MDAH-TAMVEC neutron therapy pilot study", in Proceedings of 5th International Congress of Radiation Research, NYGAARD, O. F., ADLER, H. I., SINCLAIR, W. K., eds, New York, Academic Press (1975), 1106-1117.

[4] Kaplan, E. L., Meier, P., Nonparametric estimations from incomplete observation, Am Stat Assoc J 53 (1958) 457.

Progress in Radio-Oncology II, edited by K. H. Kärcher et al., Raven Press, New York © 1982.

Clinical Results After Irradiation of Intracranial Tumours, Soft-Tissue Sarcomas, and Thyroid Cancers with Fast Neutrons at Hamburg-Eppendorf*

H.D. FRANKE, A. HESS, F. BRASSOW
Radiotherapy Department,
University Hospital Hamburg-Eppendorf,
Hamburg

W. LIERSE
Institute of Anatomy,
Hamburg University,
Hamburg

Federal Republic of Germany

Abstract

Clinical experience with brain tumours at Hamburg-Eppendorf (1977–1979) has shown that better palliation or even cures are possible, in children as well as in adults, if the dose of fast neutrons is concentrated in the tumour and does not exceed the tolerance dose of the normal brain structures. The treatments have used either an isocentric multifield technique with fast neutrons only, or a mixed photon-neutron schedule. The dose limitations are similar for retroperitoneal soft-tissue sarcomas: large, deep-seated tumours of this type cannot be successfully irradiated homogeneously with fast neutrons. After homogeneous irradiation of soft-tissue sarcomas in other, more superficial regions of the body, with up to 16 Gy given over 4 weeks, the rate of local recurrences has been only 15%, without damage to the normal structures, 2–4 years after irradiation. For invasive stages of highly differentiated thyroid cancers with macroscopic residuals of tumour (pT3–4 N1–2 M0 R2), the rate of local recurrences in a pilot study involving fast neutrons (up to 16 Gy given over 4 weeks) was only 10%, without damage to the normal tissues, instead of a rate of 30%–60% to be expected after intense megavoltage therapy.

1. INTRODUCTION

The treatment of radioresistant tumours is a challenge to the radiotherapeutic oncologist. The radiosensitivity of the surrounding normal tissues limits the maximum dose that can be given and, therewith, the possibility of achieving a cure. Brain, spinal cord and intestine are especially sensitive to fast neutrons.

Already in 1969, seven years before beginning treatment of patients with our neutron generator[1], our experiments on brains

* This work is partly supported by the Bundesminister für Forschung und Technologie, Bonn, Federal Republic of Germany.

[1] DT neutron generator from Radiation Dynamics, Inc., USA, and facility designed in co-operation with AEG, Federal Republic of Germany, and the former company.

of guinea-pigs and rats with an uncollimated DT-neutron beam showed that after 60-360 rads there was an increasing accumulation of glycogen and acid mucopolysaccharides in glial cells, indicating disturbance of carbohydrate and protein metabolism; increasing the dose resulted in rising pycnosis in the non-dividing neurons and in the dividing glial cells, leading to demyelination [1]. Another disturbing sign was the very high RBE of fast neutrons, as compared with photons, in the central structures of the brain, such as in the thalamus and basal ganglia [2]. The morphometry showed that, with our collimated beam after a single dose of 360 rad total dose (DT,14 MeV neutrons), the rate of pycnotic cells was 2 to 3 times higher than for a dose of 115 rad (Table I).

2. CLINICAL EXPERIENCE

As a consequence of these experiments, we have, since 1976, irradiated only 23 patients with malignant glioblastoma and other intracerebral tumours with a mixed photon-neutron regime. The clinical experience [3,4] demonstrates the great sensitivity of the normal brain structures to irradiation with 13-20 Gy of neutrons after fractionated irradiation of the whole brain, as well as the neutron sensitivity of glioblastoma, which is resistant to photons. Hence it follows that with fast neutrons better palliation or possible cures can only be expected if we do not irradiate the whole brain, but concentrate the neutron dose in the tumour and do not exceed the estimated tolerance dose of 1100-1200 rads in the normal brain. This aim may be achieved using isocentric irradiation with more than two fields or with a mixed photon-neutron schedule (Fig.1).

The analysis of the 23 patients shows that the survival of the eight patients with glioblastoma multiforme (Table II) after treatment with the photon-neutron schedule is neither better nor inferior to photon therapy: 50% live longer than one year. One patient, treated with neutrons on the side of the tumour only, demonstrates the good effect of irradiation on the tumour itself; the interval before recurrence was 4 years.

TABLE I. EFFECTS OF 14 MeV (DT) NEUTRONS ON THE BRAINS OF GUINEA-PIGS

Area	Single dose (rad, n γ)	Foci of mucopolysaccharides (vol.%)	Rate of pycnotic glial cells in relation to normal glial cells (%)	Area of MPS foci 0.04 mm^2 (%)	0.08 mm^2 (%)
Brain	115	11.7	20	60	40
white matter	360	1.0	59	95	5
Cerebellum	115	2.9	25	95	5
white matter	360	0.9	54	95	5

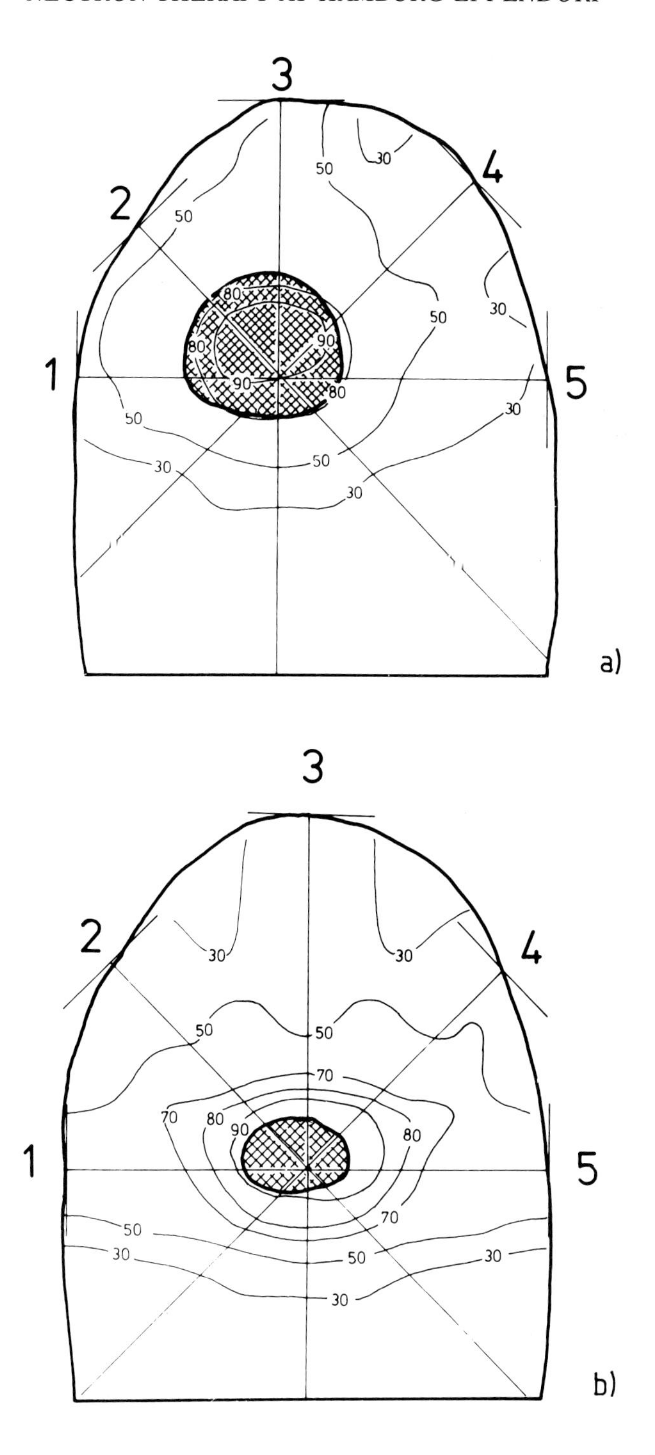

FIG.1. Computerized treatment planning with fast neutrons (DT, 14 MeV). (a) Isocentric five-field technique for a localized glioblastoma: Field size 4.2 cm x 10 cm; SSD 80 cm. (b) Isocentric five-field technique for a pituitary adenoma: Field size 4.2 cm x 4.2 cm; SSD 80 cm.

TABLE II. MALIGNANT GLIOBLASTOMA (INOPERABLE OR NOT RADICALLY OPERABLE), TREATED WITH A MIXED PHOTON-NEUTRON SCHEDULE

Age (years)	Tumour dose:		An effect seen:		Local recurrence	Treatment:		Living for (months)	Died after (months)
	neutrons (Gy, nγ)	photons (Gy)	clin. sympt.	CT scan		neutrons (Gy, nγ)	photons (Gy)		
31	4.00 over 2 weeks	40.00 over 4 weeks	+	+	+ after 4 years	–	42.5 over 4 weeks	52, good condition	–
46	4.00 over 2 weeks	40.00 over 4 weeks	+	(+)		7.20, over 2½ weeks	–		37
59	5.00 over 2 weeks	31.00 over 3 weeks	(+)	+	+ after 9 months	5.00, over 2 weeks	–		14
56	5.00 over 2 weeks	50.00 over 5 weeks	(+)	?	+				14
47	4.00 over 2 weeks	56.00 over 5 weeks	–	–	+				10
51	4.00 over 2 weeks	30.00 over 3 weeks	–	–	+				5
56	5.20 over 2 weeks	40.00 over 4 weeks	(+)	–	+				4
65	4.00 over 2 weeks	50.00 over 5 weeks	+	+	+				8

Four patients with chromophobic adenomas and one patient with acromegaly are living without signs of recurrence after periods of 19 to 41 months (Table III). All patients have shown a very good clinical and radiological reaction after irradiation [1]. Two of the four chromophobic adenomas are prolactinomas; the first one was treated without bromocriptine (Fig.2). The second patient was treated with bromocriptine, which was commenced soon after irradiation; the CT scan shows a complete shrinkage of the tumour; the patient lives now (May 1981) without bromocriptine and without a raised level of prolactine in the blood serum.

Two patients with inoperable recurrences after surgery of intracranial malignant hemangiopericytoma (Table IV) have shown complete shrinkage of the tumour and no local recurrence. One patient has presented in a very good condition for the last 3 years; the other is being treated for distant metastases in the lung and bones.

2.1. Children with brain tumours

Details of experience with brain tumours in children are shown in Table V. One child with a glioma of the optic nerve and chiasmal involvement is showing a slow but impressive shrinkage of the large tumour mass, with improved vision; the child is in a very good intellectual and general condition after a treatment with 9 Gy of neutrons and 30 Gy of photons (42 MeV X-rays) (Fig.3). The three others have spongioblastoma of the pons region and brain stem. Two of these have been living for the past 3 to 4 years in a relatively good condition; cerebral CT scans show tumour shrinkage and there have been no clinical or radiological signs of brain damage as a consequence of these doses (Fig.4).

2.2. Summary of our experience with mixed photon-neutron schedules in brain tumours

Our experience with the mixed photon-neutron schedule in our pilot study on intracranial tumours demonstrates surprising palliative effects, with some tendency to cures, yet without clinical or radiological signs of brain damage, even in children. We intend to reduce the photon fraction and to raise the neutron fraction, combined with an isocentric multifield technique, the treatment depending on the region of the brain and the individually acceptable dose distributions.

2.3. Soft-tissue sarcomas

Soft-tissue sarcomas are relatively resistant to photons. They need high doses, reaching the tolerance dose of the skin or even more. For this reason, we have, since 1976, treated 79 patients with different types of soft-tissue sarcomas with fast neutrons. For staging, we divided 64 of these patients into 3 groups (Table VI), which are summarized below:

A) Irradiated after radical surgery, homogeneous dose distribution, neutrons only.

TABLE III. ADENOMAS OF THE PITUITARY (POSTOPERATIVE RECURRENCES), TREATED WITH A MIXED PHOTON-NEUTRON SCHEDULE

Age (years)	Type of tumour	Locali-zation	Tumour dose: neutrons (Gy, nγ)	photons (Gy)	An effect seen: clin. sympt.	CT scan	Local recurrence	Living for (months)	Died after (months)	Comments
34	chromo-phobic (acido-phile)[a]	supra-sellar middle cran. fossa	3.2	50.0	(+)	+	–	41	–	without bromocriptine
30	chromo-phobic[a]	"	4.0	40.0	+	+	–	36	–	bromocriptine after irradiation
59	chromo-phobic	"	3.9	40.0	+	+	–	32	–	
61	chromo-phobic	"	4.8	44.0	+		–	32		
34	eosino-philic (acro-megaly)	"	13.0	–	+	+	–	19	–	acromegaly stopped

[a] Prolactinoma.

TABLE IV. MALIGNANT HEMANGIOPERICYTOMA IN THE BRAIN (POSTOPERATIVE RECURRENCES), TREATED WITH A MIXED PHOTON-NEUTRON SCHEDULE

Age (years)	Tumour dose: neutrons (Gy, nγ)	photons (Gy)	An effect seen: clin. sympt.	CT scan	Local recurrence	Metastases	Living for (months)	Died after (months)	Comments
47	5.6	40.0	+	+	–	–	36	–	in good condition
44	4.8	40.0	+	+	–	+ bones and lung	53	–	

TABLE V. BRAIN TUMOURS IN CHILDREN, TREATED WITH A MIXED PHOTON-NEUTRON SCHEDULE

Age (years)	Type of tumour	Operability	Tumour dose: neutrons (Gy, nγ)	photons (Gy)	An effect seen: clin. sympt.	CT scan	Local recurrence	Living for (months)	Died after (months)
5	glioma of the optic nerve and chiasma	–	9.1	30.0	+	+	–	53	–
12	pons tumour	–	4.8	44.0	(+)	(+)	–	39	–
6	spongioblastoma (pilocytotic astrocytoma) third ventricle	–	8.6	26.00	+	+	–	46	–
8	spongioblastoma (pilocytotic astrocytoma) pons	–	3.9	44.0	+	?	?	–	5

TABLE VI. SOFT-TISSUE SARCOMAS
Sixty-four patients treated with fast neutrons only (A+B) or with a mixed photon-neutron schedule (C).

Type of tumour	A) After radical surgery			B) Localized tumour, rest-tumour or recurrence				C) Inoperable tumour				Total No.
	No.	local recurr.	NED[a]	No.	regres-sion	local recurr.	NED[a]	No.	regres-sion	local recurr.	NED[a]	
Fibrosarcoma	3	–	2	11	10	2	10	9	6: (+) 3: –	6	2	23
Spindle cell sarcoma	2	1	1	–				2	2 (+)	1	1	4
Liposarcoma	4	–	4	5	5	1	4	6	4: (+) 2: –	4	2	15
Leiomyosarcoma	–							2	1: – 1: (+)	1	1	2
Rhabdomyo-sarcoma	2	–	2	4	4	–	4	5	3: (+) 2: –	5	–	11
Hemangioperi-cytoma	–			2	2	–	2	3	2: (+) 1: –	2	1	5
Malignant histiocytoma	1	–	1	1	–	1	–	2	1: (+) 1: –	2	1	4
	12	1	10	23	21	4	20	29	19:(+) 10: –	21	8	64
A+B:	35	5 (14%)	30				C:	29		21 (73%)	8	

[a] No evidence of disease.

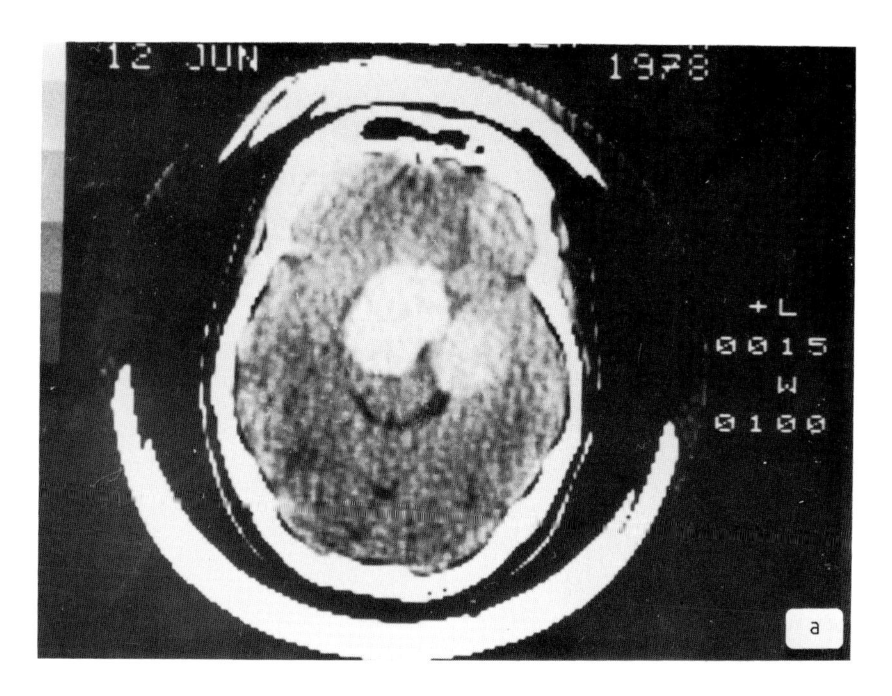

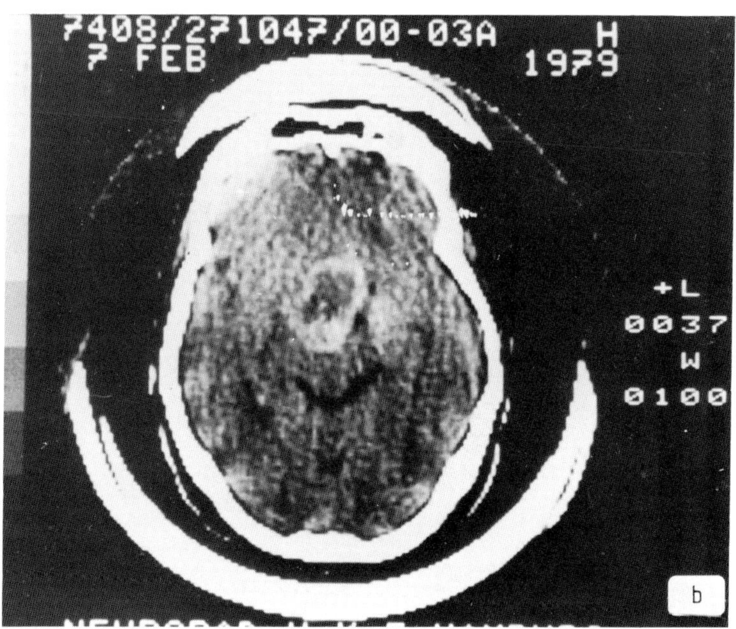

FIG.2. Treatment with a mixed photon–neutron schedule of an invasive chromophobic adenoma of the pituitary (prolactinoma), without bromocriptine therapy. (a) Tumour size before treatment (CT scan, June 1978). There is invasive tumour with large suprasellar extension, and blindness of the patient. (b) About 8 months after the beginning of treatment (CT scan, February 1979). There has been shrinkage of the tumour, with central necrosis (contrast medium enhancement reduced). Defect of visual field was partly resolved. Prolactin-secretion was detected one year later.

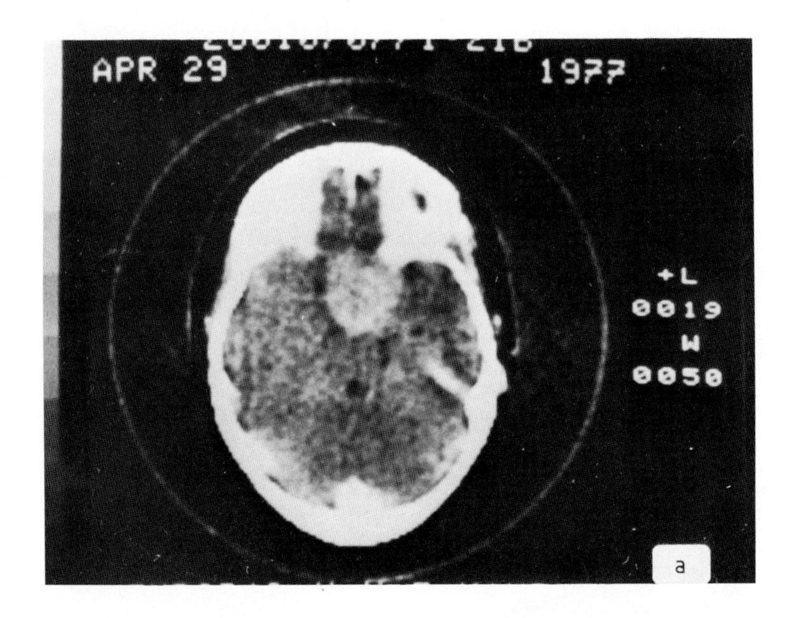

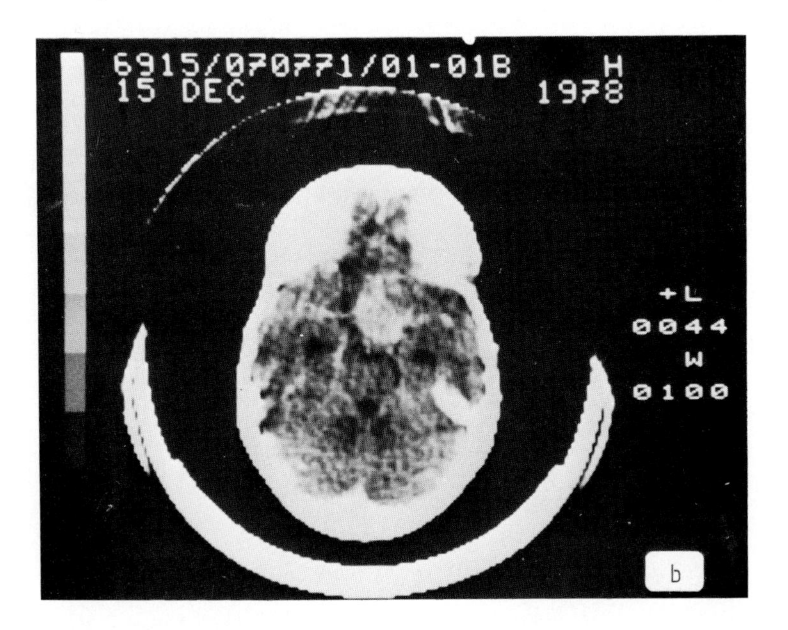

FIG.3. Treatment with a mixed photon–neutron schedule of a glioma of the optic nerve with chiasmal involvement in a child aged 5 years. (a) Tumour size at the end of treatment (CT scan, April 1977). (b) About 23 months after the beginning of treatment with 30 Gy of photons and 9.1 Gy of fast neutrons (CT scan, Decembe 1978). there is shrinkage of the tumour with some necrosis. (c) About 47 month after the beginning of treatment (CT scan, December 1980). There is continued shrinking of the tumour, and loss of contrast medium enhancement.

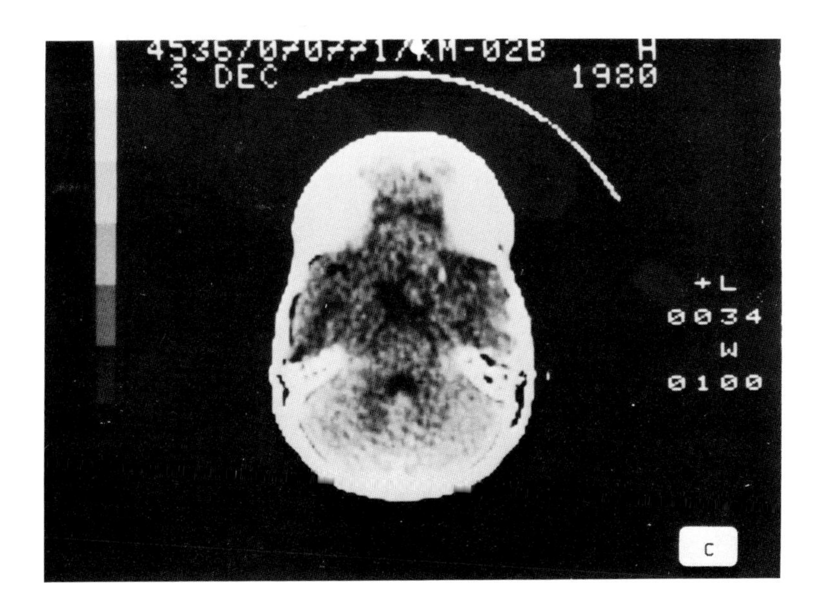

Fig. 3 (cont.)

B) Localized residual tumours or recurrences, homogeneous dose distribution, neutrons only.

C) Very large, inoperable tumours - mostly retroperitoneal, inhomogeneous dose distribution, mixed photon-neutron schedule.

The other 16 patients represent rare histological types, usually with only one patient in each histological group. For retroperitoneal soft-tissue sarcomas (Table VI,C) the dose limits are determined by the high sensitivity of the spinal cord and the intestine to fast neutrons; thus we cannot irradiate these large and deep-seated tumours homogeneously with fast neutrons (up to 16 Gy given over 4 weeks), like we do the more superficial tumours. The results in this group were mostly palliative, but we have had better temporary effects in these large, radioresistant tumours than are achieved with photons alone. After homogeneous irradiation of soft-tissue sarcomas in other regions of the body (Table VI, A,B), with up to 16 Gy given over 4 weeks, the rate of local recurrences is low, namely about 10% after 'radical' surgery, and about 20% after irradiation of residual tumours or recurrences after less-than-radical surgery. Two to four years after irradiation, we have seen no serious complication or damage in the normal tissues [1,2]. One important reason for this good result is the computerized treatment planning (Sidos-U program), which avoids 'hot spots' in normal tissues. Another reason is the small proportion of low-energy neutrons in our monochromatic energy spectrum of DT neutrons [2,5]. After irradiation, surgery is possible without disturbance of the healing process.

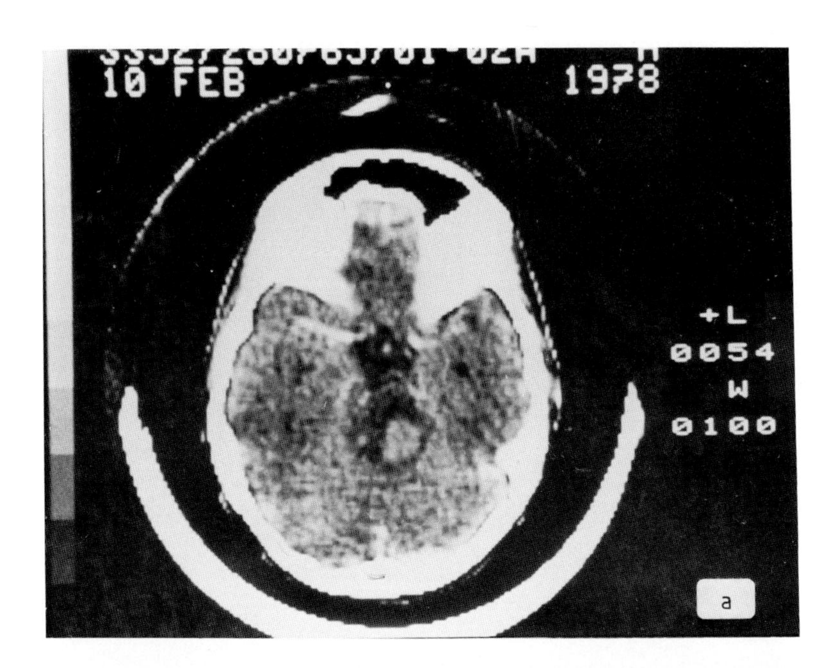

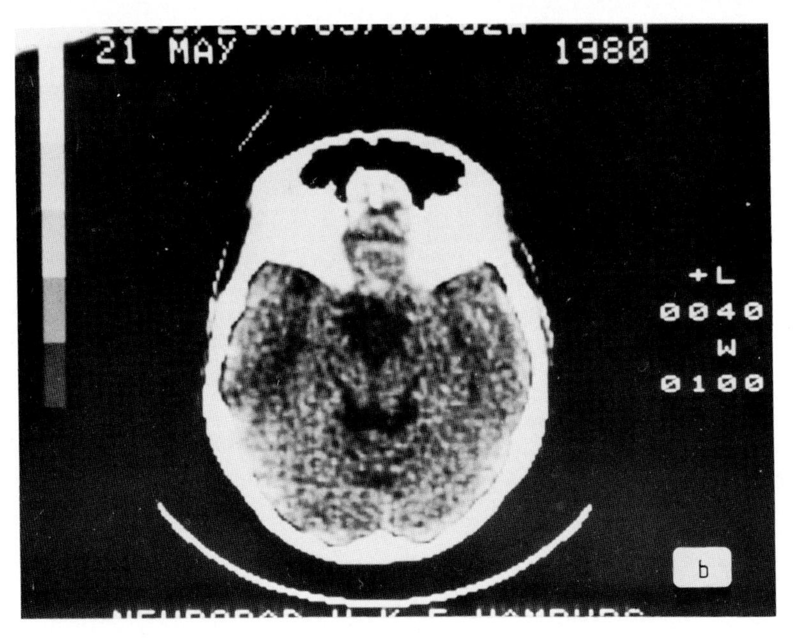

FIG.4. Treatment with a mixed photon-neutron schedule of a pons tumour in the bra
of a child aged 12 years. (a) Tumour size before treatment (CT scan, February 19
(b) 26 months after the beginning of treatment with 44 Gy of photons and 4.8 Gy
fast neutrons (CT scan, May 1980). There is hardly any demonstrable tumour, and
the cisterna laminae quadrigeminae is nearly normalized.

2.4. Thyroid cancers

The prognosis of invasive stages of highly differentiated thyroid cancers with macroscopic residual tumour after surgery and radioiodine treatment is bad [6]: the rate of local recurrences after additional external megavoltage therapy is high, about 30-60% in the first 3 to 5 years (Table VII). We irradiated 17 patients with invasive thyroid carcinoma and one patient with invasive parathyroid cancer with fast neutrons (Table VIII). The effect in 10 patients with highly differentiated (papillary and follicular) tumours ($pT_{3,4}$) is very impressive. Only one patient with a big residual tumour (more than 6 cm in diameter) showed only a partial regression of the tumour, with subsequent recurrence and metastases. The other 9 patients with invasive tumours have been free of tumour for 1.5 to 3.5 years, as two other patients have been who had invasive ($pT_{3,4}$) medullary carcinoma. The neutron dose to the spinal cord was very low, about 10% of the maximum dose (Fig.5); the regional lymph nodes outside the neutron field were irradiated with fast electrons (10 MeV), with nearly no dose to the spinal cord. The rate of oedema in the larynx has been very low, while the acute reactions have not been as intensive as after telecobalt therapy with 60 Gy given over 5 to 6 weeks. There was no fibrosis of or other late damage to the vocal cords.

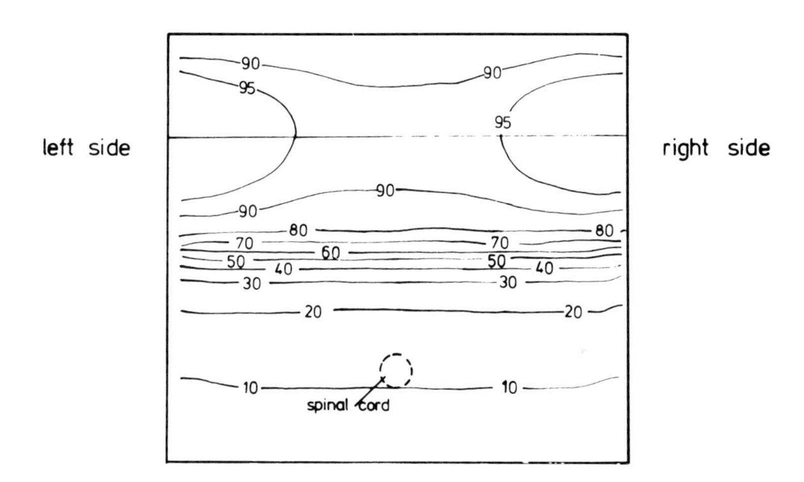

FIG.5. Computerized treatment planning (Sidos-U) for external irradiation with fast neutrons (DT, 14 MeV). Isodose distribution ($n\gamma$) for irradiation of thyroid cancer (wax mould with 0.3 mm lead at the inner surface of the mould). Diameter of the neck with moulds 14 cm; SSD 80 cm; field size 6 cm x 8 cm.

TABLE VII. DIFFERENTIATED THYROID CANCER DATA

Local/regional recurrences after treatment with megavoltage therapy

Author	Treatment: surgery	^{131}I	irrad.	Comments	Patients: No.	Age (years)	Recurrences: No.	(% of total)
Smithers 1969 [7]	+	+	–		21		9	43%
Tubiana et al. 1969 [8]	+	+	+	Surgery radical	48		5	11%
				Surgery not radical	19		4	21%
				200 kV- X-rays	14		10	71%
				Telecobalt	33		12	38%
Tubiana et al. 1975 [9]	+	+	+	Surgery not radical	17	< 40	3	18%
					18	> 40	5	28%
					35	all ages	8	23%
Glanzmann et al. 1979 [10]	+	+	+	follicular N_0	49		10	20%
				follicular N_{1-3}	21		14	67%
				papillary N_0	23		3	9%
				papillary N_{1-3}	24		8	33%

TABLE VIII. EXTERNAL IRRADIATION (1977-1979) WITH FAST NEUTRONS (DT, 14 MeV) OF INVASIVE GROWING TYPES OF THYROID CANCER AFTER INCOMPLETE SURGERY AND/OR RECURRENCES[a]

Histological classification	Number M	Number F	$pT_1N_1M_0$	$pT_2N_1M_0$	$pT_3N_2M_0$	$pT_4N_0M_0$	$pT_4N_1M_0$	$pT_4N_2M_0$	$pT_4N_3M_0$	N_{1-3}/Recurrence
Papillary	3				20 R_2	19 R_2		(†)19 R_2		
		5	42		18		40 R_2	17		
							†40 R_2			
Follicular	1					32 R_2				
		4		32		41 R_2	37			†5
Medullary	–									
		2				36			36 R_2	
Anaplastic	–									
		2						†6 R_2		†8
Number	4	13	1	1	2	4	3	3	1	2
Parathyroid carcinoma	1									48
Number	5	13	2		13					3

[a] p = pathohistological classification (UICC); R_2 = macroscopic residuals of the tumour (AJC, 1977); figures represent survival after irradiation in months; † = died with tumour; (†) = died without tumour; M = male; F = female.

REFERENCES

[1] FRANKE, H.D.: "Results of clinical application of fast neutrons at Hamburg-Eppendorf", in High LET Radiations in Clinical Radiotherapy, (BARENDSEN, G.W., BROERSE, J.J., BREUR, K., Eds), Pergamon Press, Oxford (1979) 51-59.

[2] FRANKE, H.D.: "Two years of experience with fast neutrons (DT, 14 MeV) in clinical tumor therapy at Hamburg-Eppendorf", in Progress in Radio-Oncology (Proc. Mtg. Baden, Austria; KÄRCHER, K.-H., KOGELNIK, H.D., MEYER, H.-J., Eds), G. Thieme, Stuttgart (1980) 8-27.

[3] PARKER, R.G., BERRY, H.C., GERDES, A.J., SORONEN, M.D., SHAW, C.M.: Fast neutron beam radiotherapy of glioblastoma multiforme, Am. J. Roentgenol. 127 (1976) 331-335.

[4] CATTERALL, M., BLOOM, H.J.G., ASH, D.V., WALSH, L., RICHARDSON, A., UTTLEY, D., GOWING, N.F.C., LEWIS, P., CHAUCER, B.: Fast neutrons compared with megavoltage X-rays in the treatment of patients with supratentorial glioblastoma: A controlled pilot study, Int. J. Rad. Oncol., Biol. Phys. 6 (1980) 261-266.

[5] SCHMIDT, R., MAGIERA, E.: Neutron and gamma spectroscopy for clinical dosimetry, Med. Phys. 7 (1980) 507-513.

[6] WOOLNER, L.B., BEAHRS, O.H., BLACK, B.M., McCONAHEY, W.M., KEATING, F.R.: Long-term survival rates of thyroid cancer patients, in Thyroid cancer (HEDINGER, Chr.E., Ed.), UICC Monograph Series Vol.12, Springer, New York (1969) 326-331.

[7] SMITHERS, D.W.: Thyroid carcinoma treated with radiodine, in Thyroid cancer (HEDINGER, Chr.E., Ed.), UICC Monograph Series Vol.12, Springer, New York (1969) 288-293.

[8] TUBIANA, M., LALANNE, C.M., BERGIRON, C., MONNIER, J.P., GERARD-MARCHANT, R.: Results obtained with radiotherapy in cases of thyroid cancer, in Thyroid cancer (HEDINGER, Chr.E., Ed.), UICC Monograph Series Vol.12, Springer, New York (1969) 279-288.

[9] TUBIANA, M., LACOUR, J., MONNIER, J.P., BERGIRON, C., GERARD-MARCHANT, R., ROUJEAU, J., BOK, B., PARMENTIER, C.: External radiotherapy and radiodine in the treatment of 359 thyroid cancers, Br. J. Radiol. 48 (1975) 894-907.

[10] GLANZMANN, Ch., HORST, W.: Behandlung und Prognose des follikulären und papillären Schilddrüsenkarzinoms. Erfahrungen mit kombinierter chirurgisch-radiologischer Therapie bei 200 Patienten aus dem Zeitraum 1963-1977, Strahlentherapie 155 (1979) 223-229.

[11] AMERICAN JOINT COMMITTEE FOR CANCER STAGING AND END-RESULTS REPORTING (AJC): Manual for staging of cancer, AJC, Chicago (1977).

Progress in Radio-Oncology II, edited by K. H.
Kärcher et al., Raven Press, New York © 1982.

Radiation Therapy of Bronchogenic Carcinomas with a Neutron Generator

K. SCHNABEL, S. DARAI, K.H. HÖVER, K.E. SCHEER
Institute of Nuclear Medicine,
German Cancer Research Center,
Heidelberg,
Federal Republic of Germany

Abstract

The relative biological effectiveness (RBE) of fast neutrons varies in different tumours and normal tissues. In comparison with low-LET radiation therapy, neutron therapy can only be of advantage if the RBE in the tumour is higher than in the surrounding normal tissue. The aim of all clinical studies is therefore to find and investigate such a tumour model. According to animal experiments undertaken by Shipley and Field, the RBE in Lewis lung cancer is much higher than in normal lung. Consequently bronchogenic carcinoma seems to be a good tumour model for evaluating neutron beam therapy. This was the reason why clinical efforts were concentrated on this type of tumour. Another advantage of bronchogenic carcinoma is that tumour regression, tumour growth and side effects (lung fibrosis) can be measured easily and precisely by radiography, scintigraphy or computerized tomography. The patients in the neutron group are treated with a total dose (N+γ) of 18 Gy in 20 fractions over 5 weeks. The total dose in the photon group (^{60}Co γ-rays) is 54 Gy, using the same fractionation scheme as for neutrons. By May 1981, more than 110 patients were treated. There are no relevant differences in the two groups up to now with regard to tumour regression, side effects and survival. The percentage of tumour progressions, however, seems to be lower in the neutron arm of the study.

1. INTRODUCTION

The relative biological effectiveness (RBE) of fast neutrons varies in different tumours and normal tissues. In comparison with low-LET radiation therapy, neutron therapy can only be advantageous if the RBE in the tumour is higher than in the surrounding normal tissue. The aim of clinical studies is therefore to find and investigate such a tumour model.

On the basis of animal experiments undertaken by Shipley [1] and Field [2], the RBE in Lewis lung carcinoma is much higher than in normal lung. However, the RBE for lung fibrosis is very low, much lower than the RBE for production of side effects in other organs (for example, see Broerse[3], Field [2], Fu [4], Geraci [5], Hornsey [6] and Phillips [7]). Another advantage of bronchogenic carcinomas is that tumour regression, tumour growth and side effects (lung fibrosis) can be measured easily and precisely by scintigraphy or computerized tomography. Therefore bronchogenic carcinoma seems to be a good tumour model for evaluating neutron beam therapy. These were the reasons that we concentrated our clinical efforts on this type of tumour.

We have been irradiating bronchogenic carcinomas with fast neutrons for three years. First, in the pilot phase, we treated only a few patients, and we observed tumour regression and side effects over a long period. Then we did a comparative study between fast neutrons and photons.

2. METHODS AND MATERIAL

The neutron source which is in use in Heidelberg is a generator which produces 14 MeV neutrons by the nuclear fusion of deuterium and tritium to give helium. The generator was manufactured by Haefely & Co in Switzerland. Its compactness allows it to be enclosed in a radiation shield which can be suspended on a gantry in the form of an isocentric radiation unit. The advantage of this neutron generator is its relatively high dose rate of 0.12 Gy/min at a distance of 1 m from the target. Photon treatment is carried out with a ^{60}Co unit (Gammatron S 80, Siemens). Its dose distribution is comparable to that of the neutron generator using identical treatment techniques and field sizes. The patients of the neutron group are irradiated with a total dose ($N+\gamma$) of 18 Gy in 20 fractions over a period of 5 weeks, while the photon group is treated with a total dose of 54 Gy, using the same fractionation scheme. This dose, according to the Ellis formula, corresponds to 60 Gy given in 30 fractions over 6 weeks. The technique we use for treating a centrally sited bronchogenic carcinoma comprises a ventral field and two oblique dorsal fields, in order to avoid radiation myelitis. If there is an additional mediastinal shift, one dorsal field is left out. Tumours in the periphery of the lung are, in most cases, treated with three fields.

For both neutrons and photons, radiation-treatment planning is done on the basis of computed-tomography (CT) data using special computer programs developed at our Institute. The advantage of these programs is that all inhomogeneities and their individual density variations can be taken into account in three dimensions. Tumour regression, renewed local tumour growth and side effects (lung fibrosis) are diagnosed by radiography, scintigraphy and computed tomography every three months. Scintigraphy of skeleton and liver and a CT scan of the brain are repeated every 6 months.

3. RESULTS

By 15 January 1981, we had irradiated 116 patients with bronchogenic carcinomas. Table I shows the different histologies and the number of patients. In the comparative study, we are concentrating our efforts on squamous cell carcinomas. We have treated 35 patients in the neutron group and 56 patients in the photon group. In only 30 and 46 cases, respectively, could the treatment be administered as prescribed, and only these cases can be evaluated as part of the study. The difference in the numbers of patients in the two groups is due to the greater down times of our neutron generator. Our comparative study is therefore not strictly randomized. Table II shows the distribution of the patients in both arms of the comparative study. On the basis of

TABLE I. STUDY OF BRONCHIAL CARCINOMAS

	Radiation	Histology	No. of patients
1. Pilot study	neutrons	squamous cell ca.	6
	neutrons	small cell ca.	1
2. Comparative study (neutrons or photons)	neutrons	squamous cell ca.	35
	photons	squamous cell ca.	56
	neutrons	large cell ca.	2
	photons	large cell ca.	6
	neutrons	adenocarcinoma	3
	photons	adenocarcinoma	3
3. Combined radiation study	neutrons + photons	squamous cell ca.	4
	recurrence after neutron therapy treated with photons	squamous cell ca.	2

TABLE II. COMPARATIVE STUDY OF BRONCHIAL SQUAMOUS CELL CARCINOMAS: DISTRIBUTION OF PATIENTS BETWEEN NEUTRON AND PHOTON ARMS

	Fast (14 MeV) neutrons	Cobalt-60 photons
Number of patients	35	56
suitable for evaluation	(30)	(46)
Age (years)	45–73	38–78
Stage	$T_2N_1M_0-T_3N_2M_1$	$T_1N_1M_0-T_3N_2M_1$
Previous treatment		
surgery	5	9
chemotherapy	5	6
radiotherapy	0	1
surgery + chemotherapy	3	1
Karnofsky index	30%–90%	40%–100%
Period of observation (months)	1.5–21	1.5–22

age, tumour stage, Karnofsky index and period of observation, our study is very well balanced. As far as previous treatment is concerned, no statistically relevant inconsistencies can be observed. Our clinical results are given in Table III. In most cases in both groups only partial tumour regression could be

TABLE III. COMPARATIVE STUDY OF BRONCHIAL SQUAMOUS CELL CARCINOMAS: CLINICAL RESULTS

	Fast (14 MeV) neutrons	Cobalt-60 photons
Number of patients	30	46
Local tumour control:		
none	2	1
partial	22	28
total	3	10
inconclusive	3	7
Lung fibrosis:		
none	20	27
+	4	7
++	3	9
+++	3	3
Osteoradionecrosis	2	0
Oesophagitis	0	6
Pericarditis	0	1
Myelitis	0	0
Subcutaneous fibrosis	4	0
Death	14	23
Renewed local tumour growth	4 (1 tot.)	13 (5 tot.)
Distant metastases after radiotherapy	6	10
Distant metastases without local recurrence	4	4

scored. Even using radiography, computed tomography, scintigraphy and bronchoscopy, it was difficult in many cases to distinguish precisely between a total or a partial tumour regression. If there was any doubt, the patient was assumed to have only partial tumour regression. This may explain the small number of patients estimated as having total tumour regression.

No statistically significant differences in the effect of neutrons or photons on the tumours have been noticed up to the present time. Even the dynamics of tumour regression do not differ in the two groups. The same results were observed for the production of lung fibrosis. The rate and degree of lung fibrosis and its time of appearance are similar in both groups. Osteoradionecrosis of the ribs occurred 9 months after neutron therapy in two cases. Oesophagitis has proved to be no real problem in neutron therapy, whereas it was observed to be severe in 6 cases after photon therapy. We have no explanation for this difference. Pericarditis occurred in one patient after photon therapy and severe subcutaneous fibrosis was noticed in 4 cases in the neutron group. Survival is identical in both groups. After one year 37% of our patients were still doing well; this percentage corresponds with the results obtained in many other institutions. Local tumour progression was seen in 4 out of 30 patients in the neutron group and in 13 out of 46 patients of the photon group. Since, as mentioned above, there are sometimes difficulties in distinguishing between partial and total tumour regression, the number of tumour progressions has been related to both. The difference in renewed local tumour growth is as yet not statistically significant. It is being observed continuously over a long period, so that evidence of any trend can be taken into account. Finally there is no difference in appearance of distant metastases between the two groups.

4. DISCUSSION

We have not observed any statistically significant differences between fast neutrons and photons in our clinical work on bronchogenic carcinoma; our preliminary results do, however, indicate that the percentage of tumour progressions after neutron therapy is lower than after photon therapy.

This corresponds very well with the results reported by Eichhorn et al. in 1977 and 1978 [8,9]. They treated bronchogenic carcinomas with a combined neutron and photon therapy regime which they compared with results of photon therapy alone. The histological examinations of the irradiated tumours in autopsy preparations showed a higher rate of tumour destruction in the therapy that included neutron irradiation, but the survival rate in the neutron group was markedly lower than in the photon group. Therefore an effect equivalent to an overdosage in the neutron and telecobalt therapy group cannot be excluded.

In our comparative study, survival is identical in both arms of the study. On the basis of a comparison of side effects in healthy tissues, a distinguishable overdosage in the neutron group can be excluded.

5. CONCLUSION

Bronchogenic carcinoma is an excellent tumour model for testing fast neutrons, because the RBE of neutrons in lung is very low in comparison with other tissues. Our preliminary results indicate that the local recurrence rate after neutron therapy is lower than after photon therapy, but the differences are not statistically significant at present. There is, however, a high probability that the therapeutic gain of fast neutrons was lower than expected. In a few months we will be starting a strictly randomized study and we hope to get a final answer to the question of whether neutrons are superior to photons in the treatment of bronchogenic carcinoma or not.

REFERENCES

[1] SHIPLEY, W.U., STANLEY, J.A., COURTENAY, V.D., FIELD, S.B.: Repair of radiation damage in Lewis lung carcinoma cells following in situ treatment with fast neutrons and gamma rays, Cancer Res. 35 (1975) 932.

[2] FIELD, S.B.: An historical survey of radiobiology and radiotherapy with fast neutrons, Curr. Top. Radiat. Res. Q. 11 (1976) 1.

[3] BROERSE, J.J., BARENDSEN, G.W.: Relative biological effectiveness of fast neutrons for effects on normal tissues, Curr. Top. Radiat. Res. Q. 8 (1973) 305.
[4] FU, K.K., PHILLIPS, T.L., ROWE, J.R.: The RBE of neutrons in vivo, Cancer 34 (1974) 48.
[5] GERACI, J.P., JACKSON, K.L., CHRISTENSEN, G.M., PARKER, R.G., THROWER, P.D., FOX, M.: Single dose fast neutron RBE for pulmonary and esophageal damage in mice, Radiology 120 (1976) 701.
[6] HORNSEY, S., KUTSUTANI, Y., FIELD, S.B.: Damage to mouse lung with fractionated neutrons and X-rays, Radiology 116 (1975) 171.
[7] PHILLIPS, T.L., FU, K.K.: Biological effects of 15 MeV neutrons, Int. J. Radiat. Oncol., Biol. Phys. 1 (1976) 1139.
[8] EICHHORN, H.J., LESSEL, A.: Four years' experience with combined neutron-telecobalt therapy. Investigations on tumour reaction of lung cancer, Int. J. Radiat. Oncol., Biol. Phys. 3 (1977) 277.
[9] EICHHORN, H.J., LESSEL, A., DALLÜGE, K.H.: "Five years of clinical experience of radiotherapy using a combination of neutrons and photons", in Progress in Radio-Oncology (Proc. Mtg. Baden near Vienna, 1978: KÄRCHER, K.H., KOGELNIK, H.D., MEYER, H.J., Eds), Georg Thieme, Stuttgart, New York (1980) 28.

Progress in Radio-Oncology II, edited by K. H. Kärcher et al., Raven Press, New York © 1982.

Results of a Pilot Study on Neutron Irradiation of Soft-Tissue Sarcomas

G. SCHMITT*, E. SCHERER
Clinic of Radiotherapy,
University of Essen,
Essen,
Federal Republic of Germany

Abstract

An analysis of a pilot study of neutron therapy of 70 patients with soft-tissue sarcomas is presented. In a group of 50 patients who had undergone macroscopically or microscopically subtotal surgery, 39 (78%) were free of recurrence after a mean follow-up period of 16.3 months (range: 7–33 months). In another group of 20 patients who had had partial tumour resection or only biopsies with gross tumour residual, local tumour control has been observed in 7 cases (35%). Despite these favourable results, there is as yet no evidence that neutron irradiation is superior to photon or electron irradiation in the postoperative treatment of soft-tissue sarcomas. To clarify this question, an EORTC/RTOG trial is being prepared.

1. INTRODUCTION

Despite some very encouraging results with preoperative radio and chemotherapy [1], surgery continues to be the primary step in the management of soft-tissue sarcomas.

Following extensive excision or amputation, the recurrence rate is still 25% [2] or 18% [3], respectively. For all types of radical surgery a recurrence rate of about 30% has been established. Up to 80% appear within the first two years after primary treatment.

These results were improved by postoperative irradiation. Control rates of 78%-100% were reported for tumours of the extremities after follow-up periods of 10 years (maximum), depending on the location, size and grade of the tumour [4,5].

Favourable results were achieved with neutron therapy of inoperable soft-tissue sarcomas at Hammersmith Hospital, London [6], and gave cause to inaugurate this treatment modality in other centres. First experiences have recently been published [7,8]. Particular interest is focused on postoperative neutron therapy, since the biological characteristics of the neutron beam suggest a higher effectiveness in terms of local tumour control as compared with photons or electrons. The following report gives a short analysis of our preliminary results.

---

* Present address: Department of Radiation Oncology, Alfried Krupp von Bohlen und Halbach Hospital, Alfried Krupp Strasse 21, D–4300 Essen 1, Federal Republic of Germany.

2. PATIENTS AND METHODS

Between January 1978 and March 1981, 89 patients with soft-tissue sarcomas received neutron irradiation using the Compact Cyclotron (CV 28, TCC, Berkeley, 14 $MeV_{d \rightarrow Be}$) of the University Hospital, Essen. The mean neutron energy was 5.8 MeV. The dose and fractionation regimen followed the criteria of the EORTC High LET Therapy Group, i.e. a total effective dose[1] of 16 Gy was administered in 16 fractions over four weeks, corresponding to a mean neutron dose of 15.6 Gy. The treatment portals overlapped the palpable tumour or the scar by at least two centimetres.

The results deriving from 70 patients treated between January 1978 and July 1980 have been analysed. The histological diagnoses are listed in Table I. 12 patients suffered from fibrosarcomas, 11 from synovial sarcomas and 10 from liposarcomas. There was no evidence of metastases in any patient at the beginning of radiotherapy.[2] Well differentiated tumours were seen in 40 cases, moderately differentiated tumours in 18 cases and poorly differentiated tumours in 12 cases.

Thirty-eight patients had stage $T_3N_0M_0$ tumours with a maximum diameter of 25 cm, 23 had stage $T_2N_0M_0$ tumours with a maximum diameter of 10 cm and 9 had stage $T_1N_0M_0$ tumours with a maximum diameter of 5 cm (UICC Classification, Geneva, 1978).

Table II shows the tumour localizations: 29 patients had tumours of the lower extremities, 22 had tumours of the trunk and shoulder girdle, 16 had tumours of the upper extremities and 3 had tumours of the head and neck. 50 patients had undergone macroscopically or microscopically subtotal surgery, in 15 others partial resection had been performed with gross tumour residual, while 5 others had only had biopsies.

TABLE I. HISTOLOGICAL DIAGNOSES OF 70 PATIENTS WITH SOFT-TISSUE SARCOMAS

Diagnosis	n	Diagnosis	n
Fibrosarcoma	12	Malignant haemangiopericytoma	2
Synovial sarcoma	11	Myxosarcoma	2
Liposarcoma	10	Pleomorphic rhabdomyosarcoma	2
Spindle cell sarcoma	7	Extraskeletal osteosarcoma	2
Chondrosarcoma	5	Angioliposarcoma	1
Pleomorphic sarcoma	4	Angiofibrosarcoma	1
Leiomyosarcoma	3	Dermatofibrosarcoma protuberans	1
Neurofibrosarcoma	3	Malignant histiocytoma	1
Malignant haemangioendothelioma	2	Malignant mesenchymoma	1

[1] TED = total effective dose = neutron dose + gamma dose / 3.

[2] The pre-treatment assessment included: blood count, ESR, chest radiograph, liver scan, liver function tests.

TABLE II. TUMOUR SITES OF 70 PATIENTS WITH SOFT-TISSUE SARCOMAS

Site	Patients	Site	Patients
Lower extremity	29	Upper extremity	16
Trunk and shoulder girdle	22	Head and neck	3

TABLE III. RESULTS OF NEUTRON THERAPY OF 70 PATIENTS WITH SOFT-TISSUE SARCOMAS

Mean follow-up period 16.3 months, range 7–33 months; overall control rate 46/70 (66%)

	Macroscopically or microscopically subtotal resection (50 patients)		Partial resection or no surgery (gross tumour residual) (20 patients)	
Local control	39	78%	7	35%
Partial control:				
regression[a]	1		2	
recurrence	4		5	
edge recurrence	6		1	
failure	0		5	
Patients alive with metastases	4		2	
Patients died from metastases	6		2	
Patients died from uncontrolled primary	1		2	

[a] Tumour volume $\leqslant$ 50% of the original extent.

3. RESULTS

Forty-six patients (66%) are free of recurrence after a mean follow-up period of 16.3 months (range:7-33 months; Table III). among the patients who had macroscopically or microscopically subtotal tumour resection, 78% (39/50) are free of recurrence. In the group of patients who had partial tumour resection or biopsies, this proportion is only 35% (7/20).

Serious complications were seen in 8.6% (6/70). In two cases severe fibrosis developed without impaired function of the adjacent joints, and in four cases ulcerations were observed. In three of these cases there was no evidence of recurrence. In one case an exulceration of the tumour occurred; the size of the portal was 10 cm x 15 cm in this case, positioned over the sacrum.

In four cases with recurrent tumour, amputations were required; in all other cases salvage surgery was performed.

The relatively short follow-up period does not yet show a correlation between histopathological tumour type and grade and survival time.

4. DISCUSSION

Our preliminary results of postoperative neutron irradiation of macroscopically or microscopically subtotally resected tumours demonstrate a relatively high control rate of 78%. In this group of patients, the 6 cases of edge recurrence might have been avoided by using larger treatment portals, so that a control rate of 90% might have been achieved.

This result strongly indicates that a generous margin of more than 3 cm is required to encompass the target volume, and that shrinking field techniques may be necessary to avoid untoward fibrosis.

The group of patients where gross tumour was left behind showed a considerably lower control rate of 35%. This figure corroborates tha data of Battermann and Breur [9], Ornitz and co-workers [7] and Schnabel [10], but is inferior to the results reported by Catterall (75%) [6] and Salinas and co-workers (69%) [8]. This group, however, used different dose and fractionation regimens as well as mixed beam schedules. One reason for the better results presented by Catterall [6] might be a more homogeneous dose distribution in large target volumes.

Our rate of serious side reactions is lower than in the above mentioned studies. This may be ascribed to the fact that the reference dose of 16 Gy TED in the target volume was rarely exceeded by more than 10%.

In spite of a rather short follow-up period of 16.3 months, the pilot data presented in this report seem favourable, since 38 patients had T_3 and 23 patients had T_2 tumours. However, there is as yet no evidence that neutron therapy is superior to photon or electron irradiation in the postoperative treatment of soft-tissue sarcomas. To answer this question, an EORTC and RTOG trial is being designed.

REFERENCES

[1] EILBER, F.R., MIRRA, J.J., GRANT, T.T., WEISENBURGER, T., MORTON, D.L: Is amputation necessary for sarcomas? Ann. Surg. 192 (1980) 431.

[2] CASTRO, E.B., HAJDU, S.I., FORTNER, J.G.: Surgical therapy of fibrosarcoma of extremities, Arch. Surg. 107 (1973) 285.

[3] CANTIN, J., McNEER, G.P., CHU, F.C., BOOKER, R.J.: The problem of local recurrence after treatment of soft tissue sarcoma, Ann. Surg. 168 (1968) 47.

[4] LINDBERG, R.D., MARTIN, R.G., ROMSDAHL, M.M., McMUTREY, M.J.: "Conservative surgery and radiation therapy for soft tissue sarcomas", in Management of Primary Bone and Soft Tissue Tumours (A Collection of Papers presented at the 21st Clinical Conference on Cancer, Houston, Texas, 1976), Year Book Medical Publ., Chicago, London (1977) 289-298.

[5] SUIT, H.D., RUSSELL, W.O., MARTIN, R.G.: Sarcoma of soft tissue: clinical and histopathological parameters and response to treatment, Cancer 35 (1975) 1478.

[6] CATTERALL, M., BEWLEY, D.K.: Fast Neutrons in the Treatment of Cancer, Academic Press, London, Grune & Stratton, New York (1979) 278-294.

[7] ORNITZ, R.D., HERSKOVIC, A., SCHELL, M., FENDER, S., ROGERS, C.C.: Treatment experience: locally advanced sarcomas with 15 MeV neutrons, Cancer 45 (1980) 2712.

[8] SALINAS, R., HUSSEY, D.H., FLETCHER, G.H., LINDBERG, R.D., MARTIN, R.G., PETERS, L.J., SINKOVICS, J.G.: Experience with fast neutron therapy for locally advanced sarcomas, Int. J. Radiat. Oncol., Biol. Phys. 6 (1980) 267.

[9] BATTERMANN, J.J., BREUR, K.: Fast neutron therapy for locally advanced sarcomas, Int. J. Radiat. Oncol., Biol. Phys., to be published.

[10] SCHNABEL, K.: DFVLR-Bericht: Tumortherapie mit schnellen Neutronen, Zeitraum 1.1.1978-31.12.1980 unpublished report.

Progress in Radio-Oncology II, edited by K. H. Kärcher et al., Raven Press, New York © 1982.

Biological Studies with Cold Neutrons*

An experimental approach to the LET problem in radiotherapy

B. LARSSON, J. CARLSSON
Department of Physical Biology,
The Gustaf Werner Institute,
Uppsala,
Sweden

H. BÖRNER
Institut Laue-Langevin,
Grenoble,
France

J. FORSBERG
Department of Surgery,
Akademiska Sjukhuset,
Uppsala,
Sweden

A. FOURCY
CEA Centre d'Etudes Nucléaires
de Grenoble,
Grenoble,
France

M. THELLIER
Laboratoire de Nutrition Minérale,
Mont-Saint-Aignan,
France

Abstract

To facilitate the study of high-LET radiation in tumour models and mammalian organs, an experimental situation based on the use of cold neutrons available at the end of a neutron beam guide at the high-flux reactor of the Institut Laue-Langevin, Grenoble, has been established and evaluated. The exposure conditions were designed so as to emphasize cellular responses caused by short-range nuclear fragments at LET-values around 100 keV/μm. Energy absorption in biological targets takes place with minimal production of sparsely distributed free radicals, the absorbed dose being ascribed near-exclusively to low-energy ions (less than 10 MeV/u) appearing in nuclear reactions following neutron capture in naturally occurring ^{1}H or ^{14}N, or added ^{10}B. The results of the initial experiments performed on dispersed cells and intestinal epithelium are summarized, and implications for radiotherapeutic research discussed.

1. INTRODUCTION

In radiotherapy, the critical damage to irradiated cells is mainly mediated through chemical action of ion radicals and short-lived free radicals of water and organic molecules. Such reactive species are formed along the tracks of all fast-travelling charged particles that interact with the

* The study was supported by the French and Swedish Natural Science Research Councils and the Swedish Cancer Society.

irradiated tissues [1]. Any alteration in character of the particle tracks is therefore of clinical significance, and has to be meticulously evaluated. A most significant parameter is the "Linear Energy Transfer" (LET, usually given in keV/μm). Qualitative as well as quantitative changes in biological effects as a function of LET have been investigated and the knowledge gained forms part of the basis for current considerations of the relative merits of various radiations in cancer treatment [2,3].

The average LET for tracks in therapeutic photon, electron or proton fields is below 10 keV/μm ("low LET"). The therapeutic fast neutron and pion fields feature wide LET spectra that give such radiations the character of "intermediary (or mixed) LET" radiations. Pure "high-LET" characteristics would be expected only when the LET in the tracks is uniformly of the order of 100 keV/μm.

Unfortunately well-defined high-LET radiations are difficult to obtain, at least when the aim is uniform treatment of multicellular systems. For such purposes, big accelerators permitting acceleration of heavy ions to high energies (100 MeV/u or more) have been the only tool available [4]. In this paper an alternative approach is presented, i.e. the use of well-defined beams of slow neutrons. Such neutrons give raise to heavy charged particles (< 10 MeV/u) that are created in reactions with nuclides with large cross-sections for neutron capture (cf. FIG. 1).

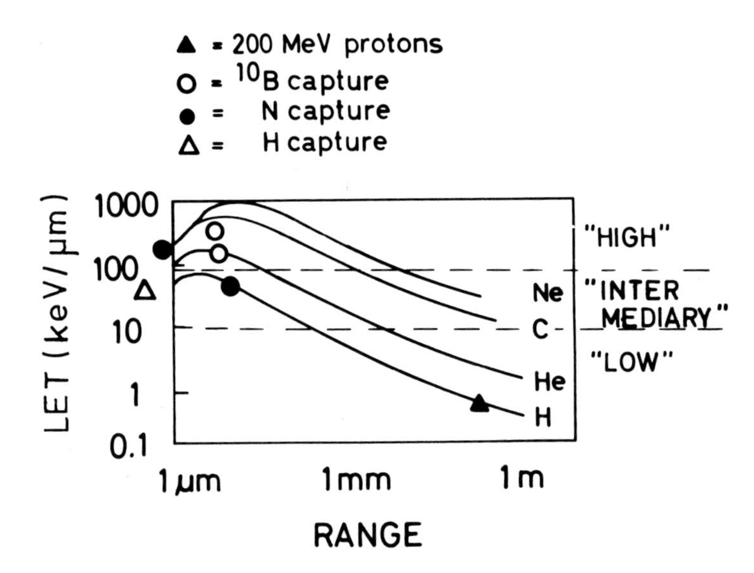

FIG.1. LET versus range for various accelerated ions [6]. Specially indicated are values for the high-LET fragments that appear at nuclear neutron capture, as emphasized in this paper, and 200 MeV protons, presenting low-LET particles.

To study radiobiological aspects of neutron capture reactions we have irradiated cultivated or organized mammalian cells with a beam of cold neutrons from the high-flux reactor at Institut Laue-Langevin, Grenoble, France. The experimental conditions were designed so as to emphasize cellular response caused by short-range nuclear fragments. In particular, products of the reactions $^{1}H(n,\gamma)^{2}H$ and $^{14}N(n,p)^{14}C$ were, normally, most important. By introduction of boron preparations in the system under study, additional effects of He and Li ions produced in the reaction $^{10}B(n,\alpha)^{7}Li$ were similarly demonstrated. The primary results and experiences are here presented and discussed.

2. MATERIALS AND METHODS

2.1 Cultivated cells

The V 79 hamster cell line was used to compare "reproductive survival", after irradiation with cold neutrons and ^{60}Co gamma rays, respectively. Areated or hypoxic suspensions of 400-40 000 cells in 0.5 ml medium were prepared, after trypsination and dilution of stock cultures in exponential growth, and kept for irradiation in air or nitrogen in cylindrical Lucite chambers (h = 4 mm, d = 12 mm). After irradiation, the colony-forming ability was tested by Petri-dish cultivation for ca. 10 days, followed by counting of colonies containing 50 cells or more.

2.2 Intestinal epithelium

For comparison of radiation response, without and with introduction of ^{10}B, the small intestine of Sprague-Dawley rats was used as an experimental model. Before irradiation, with the vascular supply intact, an intestinal segment was exteriorized and put into position for irradiation in a thin polycarbonate pocket. The irradiated section, ca. 2 cm long, was marked with sutures in the mesentery. Non-enriched boric acid at various concentrations, given in physiological saline, was used as a ^{10}B carrier. It was introduced (a) by intravenous injection, (b) by injection in the intestinal lumen, or (c) by intraarterial injection in combination with 40 μm degradable starch particles [5]. The rats were killed 4 days after irradiation. Semi-quantitative evaluation of damage to the intestinal epithelium was made by use of histological indicia, in a "blind test" carried out by two independent workers.

2.3 Irradiation procedure

Cold ($\bar{\lambda}$ = 11 A, E $\leq$ 10 meV) neutrons (FIG. 2) were available at position H 17 in the reactor hall. They emanated through an aluminium window of the small guide branch and presented themselves in a near-parallel beam with its axis ca. 60 mm above the permanent beam guide. Points of reference for use in beam monitoring and dosimetry were related to the geometry of the beam system and a simple optical bench was installed to facilitate alignment of filter, apertures and targets. Much effort was made to minimize contaminating radiations from the reactor core, beam guide, aluminium window and supporting structures. Main elements in the system were a 10 cm thick bismuth filter and a ^{6}LiF aperture, 20 mm in diameter, with 5 mm thick walls immediately in front of the target. The position, homogeneity and fluence rate of the beam were checked by neutron radiography and gold foil activation. Typically the "capture fluence rate equivalent" was 10^9 thermal neutrons/cm^2·s at the depth of the target structures (FIG. 3). The figure was varying by a factor of 2, however, and it had to be checked daily. The homogeneity of the field was

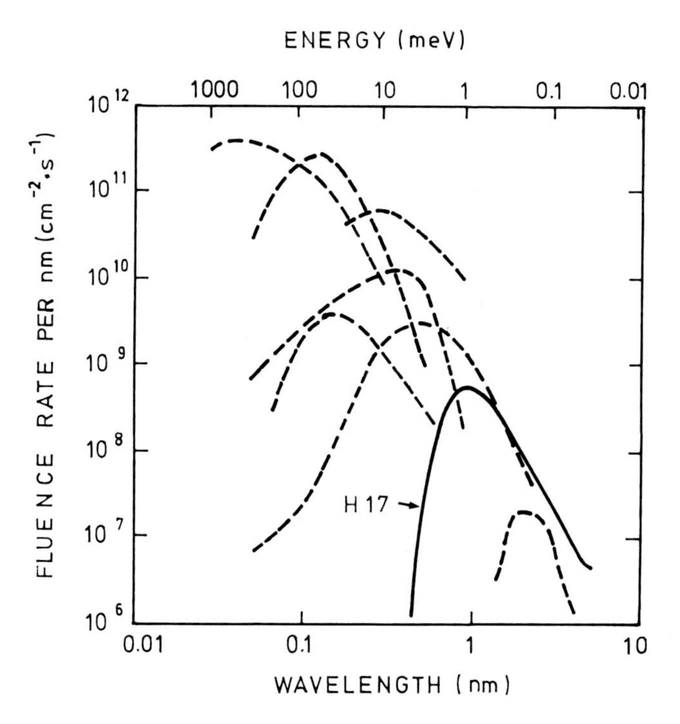

FIG.2. Examples of neutron spectra measured in various beams available at the high-flux reactor at Institut Laue-Langevin, Grenoble. Experiments here reported were made with the "cold" beam H 17. (Modified from Ref. [7]).

also varying within a factor of 2 but this defect was compensated for, in experiments with cultivated cells, by rotation of the target.

3. EXPERIMENTAL RESULTS

Doses are given as equivalent capture fluence for thermal neutrons, as the absorbed dose is dependent on the poorly known distribution of capture reactions. The accompanying gamma doses were low as estimated by thermoluminescence dosimetry, using ^{7}LiF, with and without ^{10}B impregnation.

3.1 Dose/response relationships for hamster cells

The LD_{50} for colony formation was 5×10^{12} n_t/cm^2 (ca. 2 Gy), the recorded survival, S, being an exponential function of dose for $100\,\% > 3 > 10\,\%$. In preliminary tests no effects of hypoxia on the survival parameters were seen while, for photons, an oxygen enhancement ratio of 2-3 was observed (FIG. 4).

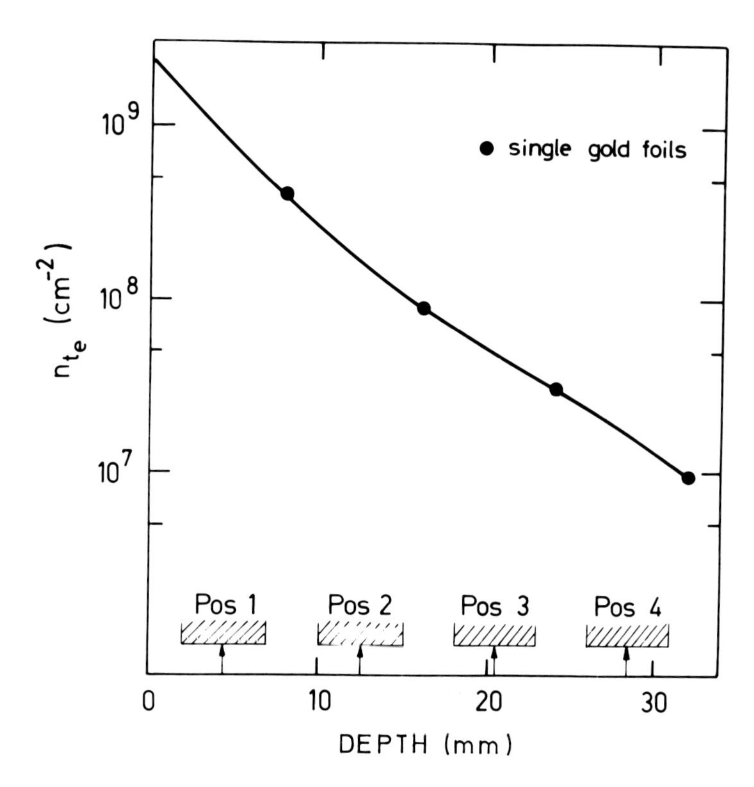

FIG.3. Fluence versus range measured by gold foil activation in the target arrangement employed. Four positions corresponding to different depths in Lucite are indicated. They were used for irradiation of dispersed cultivated cells at varying dose rate. Period of irradiation, 1 minute.

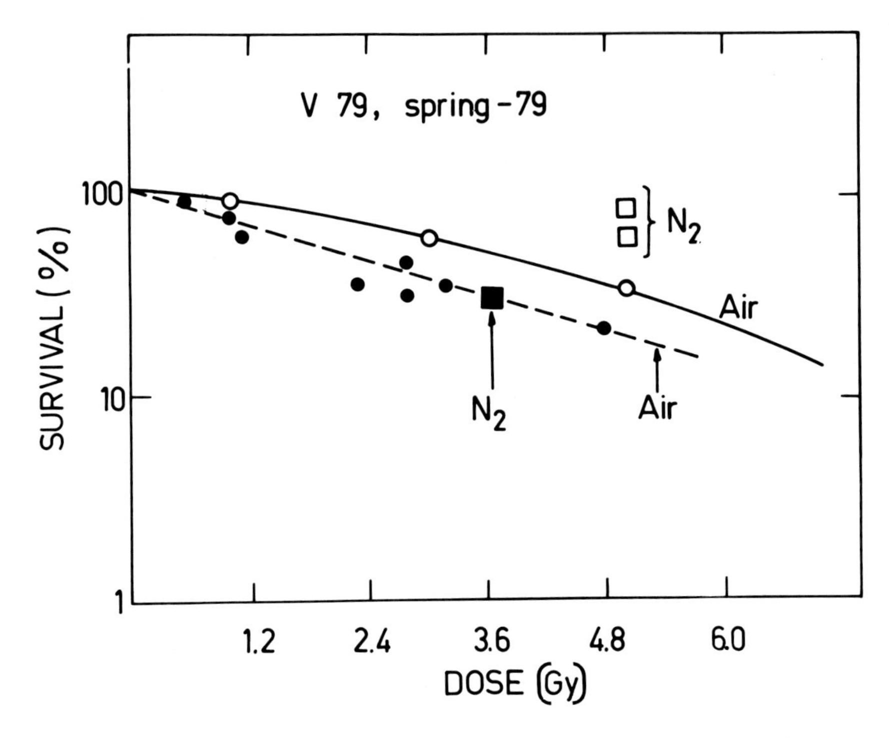

FIG.4. Survival curves based on colony forming ability of V 79 hamster cells irradiated under air (circles) or nitrogen (squares) with cold neutrons (black) or ^{60}Co photons (white).

These results support the view that capture reactions may cause a major component of "all or none" damage at the macromolecular level (e.g. double-strand breakage of DNA).

3.2 Intestinal epithelium

The highest doses of boric acid (31.5 mg I.A. in combination with starch microspheres; 270 mg I.V.; or 90 mg I.L.) invariably lead to a theoretically expected dramatic decrease of the fluence of neutrons necessary to induce a given histological damage (e.g. from 5×10^{12} n_t/cm^2 to less than 10^{12} n_g/cm^2) (TABLE I).

4. DISCUSSION

The arrangement thus investigated seems well suited for the continued study of neutron reactions, in natural as well as boron-loaded biological specimens. Indeed, such reactions have been employed at several centers [3],

TABLE I. OUTLINE OF RESULTS OBTAINED FOR V 79 CELLS AND RAT INTESTINE

Material	Condition	5×10^{12} n_t/cm^2 (ca. 2 Gy)	1×10^{12} n_t/cm^2 (ca. 0.4 Gy)
V 70 cells	normal	50% cells killed	10% cells killed
Intestine	normal	no obvious damage	no damage
	10 mg ^{10}B as H_3BO_3 I.V.	–	denudation
	3 mg ^{10}B as H_3BO_3 I.L.	–	denudation
	1 mg ^{10}B as H_3BO_3 I.A.	–	denudation
	∿1 mg ^{10}B as B-dextrane I.V.[a]	no obvious damage	no damage
	∿1 mg ^{10}B as B-dextrane I.V.[a]	–	no obvious damage
	dextrane without boron	–	no damage

[a] Similar experiments with slightly different outcomes.

in studies aimed at understanding the potentialities of boron-neutron therapy. Therefore, restrictions on contaminating low-LET tracks have been less stringent. The present situation appears favourable for experimental studies at near-pure high LET, even for comparison with beams of accelerated ions [5], where sparsely ionizing secondaries, including delta rays, frequently occur.

REFERENCES

[1] TOBIAS, C.A., BLAKELEY, E.A., NGO, F.Q.H., and YANG, T.C.H., in Radiation Biology in Cancer Research (MEYN, R.E. and WITHERS, H.R., Eds), Raven Press, New York (1980) 195-230.

[2] RAJU, M.R., in RAJU, M.R., Heavy Particle Therapy, Academic Press, New York (1980) 39-77.

[3] FOWLER, J.F., Nuclear Particles in Cancer Treatment, Hilger, Bristol (1981).

[4] TOBIAS, C.A., LYMAN, J.T., and LAWRENCE, J.H., in Progress in Atomic Medicine: Recent Advances in Nuclear Medicine, Vol. 3 (LAWRENCE, J.H., Ed.), Grune and Stratton, New York (1971).

[5] FORSBERG, J.O., JUNG, B., and LARSSON, B., Acta radiol. Oncology 17 (1978) 485.

[6] TOBIAS, C.A., Lawrence Berkeley Laboratory, Berkeley, University of California, private communication 1977.

[7] In Neutron Research Facilities at the ILL High Flux Reactor (2nd ed., Maier, B., Ed.), Grenoble (1980).

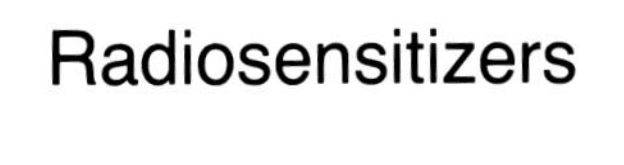
Radiosensitizers

Progress in Radio-Oncology II, edited by K. H. Kärcher et al., Raven Press, New York © 1982.

Radiation Sensitizers and Radiation Protectors Combined with Radiation Therapy in Cancer Management

L.W. BRADY
Department of Radiation Therapy and Nuclear Medicine,
Hahnemann Medical College,
Philadelphia, Pennsylvania

T.L. PHILLIPS
Department of Radiation Oncology,
University of California San Francisco School of Medicine,
San Francisco, California

T.H. WASSERMAN
Division of Radiation Oncology,
Mallinckrodt Institute of Radiology,
Washington University School of Medicine,
St. Louis, Missouri

United States of America

Abstract

The identification of the basic mechanisms involved in the radiation killing of cells through an understanding of radiation chemistry and radiation biology has permitted the elucidation of a number of modifying mechanisms. The repair of radiation-produced free radicals by reduction and their fixation by oxidation provides a potential for improving the results of clinical radiation therapy. A wide range of compounds has now been identified that exhibit electron affinity; the compounds mimic oxygen in their ability to sensitize the hypoxic cells, and show differential sensitization in tumours. Using sulphydryl-containing compounds to repair free-radical lesions in the DNA of aerated normal tissues has also proved possible. There are differential effects upon normal tissue versus tumour tissue through increased uptake of certain compounds in normal tissues and the decreased protective efficiency of these compounds in hypoxic cells. It has also been demonstrated that sulphydryl-containing radioprotective compounds are also active in protecting cells against the toxic effects of alkylating agent chemotherapy. The near future holds extensive promise for a combination of radiation therapy, cytotoxic chemotherapy, hypoxic cell sensitizers and radiation protectors. All of these when used in the appropriate circumstances may yield significant improvements in the therapeutic ratio and in the long-term control of tumours.

RADIATION SENSITIZERS

Adams et al. [1], in the early 1960s, proposed that not only oxygen but other electron affinic compounds could sensitize cells to radiation damage through fixation of the free radicals in DNA by oxidation. These presumptions have been confirmed by elegant radiochemical studies demonstrating the similarity between electron affinic compounds and oxygen.

It was almost immediately obvious that successful hypoxic cell sensitizers would need to meet basic criteria:

(a) A high electron affinity;
(b) Absence of toxicity to aerated cells at concentrations showing high hypoxic cell radiation sensitization;
(c) Ability to sensitize mammalian cells under conditions of hypoxia in vitro;
(d) Ability to reach pharmacologic concentrations with a sufficiently long half-life to allow tumour penetration;
(e) Sufficiently slow metabolism to allow full tumour penetration;
(f) Absence of whole-animal toxicity at concentrations sufficient to cause significant tumour sensitization.

New compounds that meet most of these requirements are the nitrofurans and the nitroimidazole derivatives (Fig.1). In vitro and in vivo studies with these agents have shown them to be effective hypoxic cell sensitizers [2-4]. The more electron affinic the molecule is, the more effectively it sensitizes hypoxic cells to radiation [5,6]. Experimental data suggest that the nitroimidazole derivatives actually react with DNA that has been previously activated by radiation [7]. The nitro group is the active site and either traps an electron or binds to the DNA molecule [6,8]. In both instances it will replace oxygen and cause fixation leading to irreversibility of radiation damage [9, 10]. The 5-nitroimidazole compounds appear to be significantly less potent, with approximately ten-fold higher concentrations required for the same degree of sensitization as that for 2-nitroimidazole compounds. The nitroimidazoles and nitrofurans do not increase the radiation effect in well-oxygenated cells. In addition, there is evidence to suggest that the nitrofurans and nitroimidazoles have a selective toxicity for hypoxic cells in the absence of radiation. Presumably, this toxicity relates to the metabolic reduction of the drugs to produce reactive intermediates which are toxic only under anaerobic conditions. The basic advantages of these drugs over hyperbaric oxygen are as follows:

(a) A slower rate of metabolism, allowing better diffusion into the tumour;
(b) Relative ease of administration and technical ease of radiation therapy;
(c) No problems with vasoconstrictive reflexes altering the blood flow and oxygen saturation in the area of the tumour.

The electron affinic compounds affect cellular radiosensitization predominantly by the oxidation of neutral target radicals. It has been shown that misonidazole completely inhibits lactate dehydrogenase, alpha-hydroxybutyric hydrogenase, aspartate amino-transferase, alanine amino-transferase, and creatine kinase.

The data from animal studies confirm that these compounds sensitize essentially all animal tumours known to contain hypoxic cells [11-15]. There was no evidence obtained to indicate a differential effect due to tumour histologic type or site, and there was no effect of cell cycle if the tumour were hypoxic.

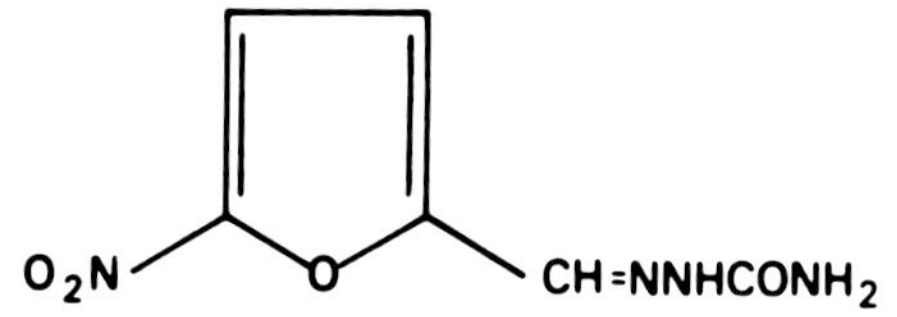

O_2N — O — $CH{=}NNHCONH_2$

a. Nitrofurazone

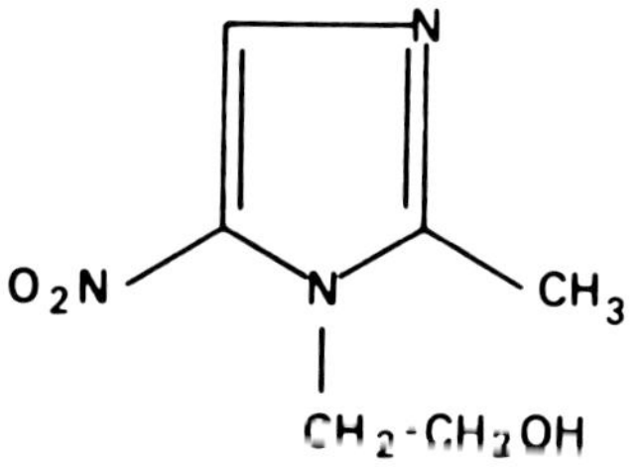

O_2N — N — CH_3, CH_2-CH_2OH

b. Metronidazole

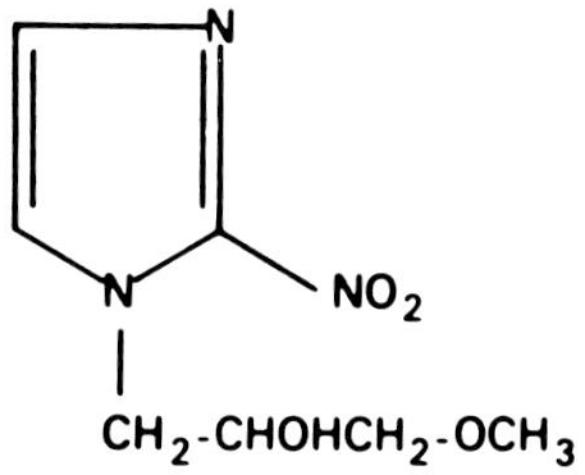

N — NO_2, CH_2-$CHOHCH_2$-OCH_3

c. Misonidazole
(Ro - 07 - 0582)

FIG.1. Hypoxic cell radiation sensitizers.

The accumulation of these data allowed for the rapid introduction of these compounds into clinical trials in Canada and the United Kingdom. Urtasun et al. [16,17] began the study of metronidazole, first determining the maximum tolerance and pharmacology and then utilizing it in randomized clinical trials in the treatment of malignant gliomas. He demonstrated that patients receiving metronidazole with radiation therapy had a significantly prolonged median survival. Although this study used an unconventional radiation fractionation scheme, it clearly indicated that under certain conditions metronidazole had a significant effect on survival of patients treated by the combination of radiation therapy and the drug. Dische et al. [18-21] subsequently explored the toxicity of misonidazole, first in human volunteers and then in volunteer patients to evaluate the pharmacology, toxicology, and the potential for sensitization of artificially hypoxic skin.

The factors allowing for differential sensitization with electron affinic compounds relate to the fact that these drugs sensitize only hypoxic cells with the usually used clinical radiation dosages, that the metabolic products are toxic to hypoxic cells and that 1-30% of tumour cells within a particular tumour are hypoxic, and only 0.1% of normal cells are hypoxic.

The data indicate that there is an enhancement ratio in animal tumours of about 1.82 ± 0.07. These data were collected by Sheldon et al. [22,23].

The ideal radiation sensitizer would be relatively non-toxic, penetrating into the deep tumour centre and deliverable in 20 to 40 daily doses yielding high blood levels that persist throughout the radiation therapy programme.

With that goal in mind, the United States National Cancer Institute, under the direction of Radiation Therapy Oncology Group, initiated phase I studies in July 1977 to elucidate the pharmacology of the compounds and to establish that adequate levels in tumours could be achieved for sensitization.

The second phase of testing involved the establishment of the efficacy of specific combinations of hypoxic cell sensitizers and radiation therapy. From the phase II trials currently completed with misonidazole, it is evident that each disease site imposes some limitation on the patient's tolerance of the drug and of a specific radiation fractionation pattern. Phase II studies are needed to establish the tolerance of both drug and radiation programmes. Phase III studies are built upon successful phase II studies with derived radiation/sensitizer combination contrasted at the same tumour site and stage with conventional radiation therapy programmes.

The objectives of the Radiation Sensitizer Program being carried out under the aegis of the Radiation Therapy Oncology Group are directed toward the determination of the effectiveness in increasing local control and survival, the establishment of tumour sites and histology where these combinations are most effective, the establishment of the optimum scheduling for radiation therapy and sensitizers, the development of optimum sensitizers, and the integration of clinical trial data into clinical practice.

The Radiation Therapy Oncology Group has as its short-term goal the completion of the Phase I Misonidazole studies, the initiation of the Phase I Desmethylmisonidazole studies, the initiation of Phase II Desmethylmisonidazole studies using tumours of the oesophagus, bladder and stomach, completion of the Phase III Misonidazole studies, and initiation of Phase I SR-2508 studies.

The phase I clinical trials establish a safe initial starting drug dose on several drug schedules. They have already established that high-pressure liquid chromatography and ultraviolet pharmacologic data can relate the qualitative and quantitative aspects of the dose-limiting factors to peripheral neurotoxicity. These data clearly show that the major risk factor was total misonidazole dose as a function of body surface area. From these experiments other major organ toxicities were identified, such as renal, bone marrow and hepatic.

The majority of the neurotoxicities occurred with total dosages greater than 12 g/m^2, and the majority of them were grade I or grade II type. The incidence of central neurotoxicity was 10%. The incidence of ototoxicity was 9%. It was observed that patients receiving dexamethasone or dilantin as well had a lower incidence of neurotoxicity than those receiving sensitizers only.

The Phase II studies under the aegis of the Radiation Therapy Oncology Group began in April 1978, and 15 studies have been opened as of January 1981, of which 5 studies have completed patient accrual. Over 467 patients have been entered into the various protocols, with a total of 372 patients evaluable. Table I gives the information relative to the five closed studies.

The phase II clinical trial for misonidazole with radiation therapy in gliomas has been completed [24]. The conclusion indicates that survival was equivalent to previous experience with radiation therapy only and that a randomized study should be undertaken in this disease. This study is currently under way, with toxicity from combined radiation therapy and misonidazole at a minimum and acceptability of the regimen.

The phase II study in head and neck cancer has been completed using misonidazole and a unique schedule of radiation therapy in advanced primary tumours of the head and neck (T3 and T4). Fazekas et al. [25] reported good tumour clearance at the end of therapy, with moderate toxicity. Of those patients completing the radiation therapy and misonidazole programme, 56% demonstrated a complete response, and 34% a partial response. A phase III randomized control study has been recommended based on these data.

A phase II clinical trial in patients with brain metastases has been completed [26]. This study utilized a twice-weekly fractionation scheme of radiation therapy in conjunction with twice-weekly administrations of misonidazole. Most patients had substantial improvement in neurologic function and there were patients with serial CT scans showing reduction of mass on CT scan. It was obvious that this regimen was equally tolerable to the previous best regimen identified by the Radiation Therapy Oncology Group (3000 rads in ten fractions in two weeks). A comparative phase III clinical trial has been initiated.

The study of misonidazole with radiation therapy for hepatic metastases in the phase II effort has also been completed [27]. This treatment revealed decreased symptoms in 51% of patients, increased Karnofsky scores in 64% of patients, decreased liver size in 42% of patients, decreased liver tests (return to near normal levels), and a median survival of 22 weeks in 41% of patients. This has been accomplished with no incidences of radiation hepatitis or nephritis. A phase III randomized trial has been initiated for liver metastases.

A phase II clinical trial of misonidazole with radiation therapy for advanced, Stage III, non-oat cell-cancers of the lung has also been completed [28]. Complete and partial responses of 17% and 54%, respectively, were observed on serial examinations of chest roentgenograms. The median survival was seven months, with 32% of the patients being alive at one year. This regimen proved to be tolerable and has been put into a phase III randomized study in locally advanced lung cancer.

Other phase II studies and other types and stages of tumours are continuing under the aegis of the Radiation Therapy Oncology Group (Table II). Several studies have been closed before completion of patient accrual and they are shown in Table III with the reasons for incomplete accrual.

The phase II clinical studies confirmed the dose-limiting neurotoxicity and established acceptable incidences of mild peripheral neuropathy with total dose limitations of between 10.5

TABLE I. COMPLETED RTOG PHASE II MISONIDAZOLE STUDIES

RTOG No.	Cancer type and stage	Date Opened	Date Closed	No. patients entered	No. patients evaluable	Radiation schedule	Misonidazole dose schedule
78-01	Malignant gliomas, grade III, IV	1978-04	1979-11	59	54	Whole brain: (400 rads, Mon + 150 rads, Tue, Thu, Fri) X 6 weeks + 180 rads X 5 boost Total dose = 6000 rads	2.5 g/m^2 every week X 6 weeks = 15 g/m^2
78-02	Head and neck cancers, T3, T4	1978-04	1979-06	50	45	250 rads (4 h after Miso) 210 rads (8 h after Miso) } every week X 5 weeks 180 rads every day X 3 d per week X 5 weeks, then 180 rads every day X 5 d per week X 2–3 weeks Total dose = 6600–7200 rads	2.5 g/m^2 every week X 6 weeks = 15 g/m^2; modified to 2.0 g/m^2 every week X 6 weeks = 12 g/m^2
78-12	Brain metastases (patients with other metastases also)	1978-07	1979-05	40	34	600 rads twice weekly X 3 weeks = 3600 rads	2.0 g/m^2 twice weekly X 3 weeks = 12 g/m^2
78-30	Liver metastases	1979-01	1979-11	50	44	300 rads daily X 7 d = 2100 rads	1.5 g/m^2 daily X 7 d = 10.5 g/m^2
78-14	Lung cancer (Stage III)	1978-08	1980-01	52	49	600 rads twice weekly X 3 weeks = 3600 rads	2.0 (reduced to 1.75) g/m^2 twice weekly X 3 weeks = 12 (10.5) g/m^2
		Total		251	226		

TABLE II. CONTINUING RTOG PHASE II MISONIDAZOLE STUDIES

RTOG No.	Cancer type and stage	Date opened	Radiation schedule	Misonidazole dose schedule
78-13	Widespread metastases	1978-07	600 rad, hemibody	5 (reduced to 4) g/m^2 one dose
78-15	Advanced or recurrent sarcomas or melanomas	1978-07 (modified 1979-07)	700 rad every week X 6 weeks = 4200 rad (600 rad twice weekly X 3 weeks = 3600 rad)	2.5 g/m^2 every week X 6 weeks = 15 g/m^2 (1.75 g/m^2 twice weekly X 3 weeks = 10.5 g/m^2)
78-23	Pediatric CNS tumours, recurrent	1978-12	300 rad X 10/2 weeks = 3000 rad	1.5 g/m^2 thrice weekly X 2 weeks = 9.0 g/m^2
78-32	Oesophagus, locally advanced	1979-01	400 rad X 12 fractions/ 4½ weeks = 4800 rad	1.0 g/m^2 X 12 = 12 g/m^2
79-03	Gliomas, neutrons	1979-07	300 rad (neutrons) X 4 + boost of 300 rad (neutrons) X 2	2.5 g/m^2 X 6 = 15 g/m^2
79-04	Head and neck, Stage III, IV	1979-06	{400 rad X 12 fractions/ 2½ weeks = 4800 rad Same + Misonidazole	 1.5 g/m^2 thrice weekly X 7 doses = 10.5 g/m^2
79-05	Pelvic, advanced local cancer	1979-10	1000 rad in 1 fraction every month X 3 months = 3000 rad	4.0 g/m^2 every month X 3 months = 12 g/m^2

TABLE III. CLOSED RTOG PHASE II MISONIDAZOLE STUDIES

RTOG No.	Cancer type and stage	Date Opened	Date Closed	No. patients entered	No. patients evaluable	Radiation schedule	Misonidazole dose schedule	Comments
78-21	Bladder, locally advanced	1978-10	1980-02	11	9	(400 rads every Mon + 200 rads every Wed, Fri) X 6 weeks = 4800 rads Bladder boost = 2000 rads/2 weeks	1.5 g/m^2 every week X 8 weeks = 12 g/m^2	Good local tumour control, 7 complete regressions Little Miso toxicity, 4 GR I Neuro study closed due to 3 small bowel obstructions
78-24	Head and neck, Stage III, IV	1978-12	1979-08	11	10	400 rads twice weekly, then thrice weekly = 4800-5600 rads	1.25 g/m^2 (reduced to 1.0 g/m^2) twice weekly, then thrice weekly X 5 weeks = 12-15 g/m^2	Study closed due to completing study
79-06	Cervix, IIIB, IVA	1979-07	1980-07	10	7	400 rads twice weekly X 5 weeks = 4000 rads + intracavitary boost	1.25 (reduced to 1.2) g/m^2 twice weekly X 5 weeks = 12.0-12.5 g/m^2 (1.25 g/m^2 X 2 during boost = 15 g/m^2	5/7 grade I neurotoxic. 1/7 grade II poor accrual due to high dose fractions

and 12 g/m^2. The acceptable dose schedules are [29]:

1. 3-4 g/m^2 as a single dose;
2. 1.25-1.5 g/m^2 daily for 5-7 days (total doses of 6.25-10.5 g/m^2);
3. 10.5 g/m^2 divided in equal doses over three weeks;
4. 12.0 g/m^2 given in divided doses over six weeks.

There was a tendency to find toxicity in patients who were elderly or who became dehydrated secondary to their primary cancer. As has been demonstrated in the phase I studies, most of the serious toxicities, central and peripheral, were preceded by mild peripheral neurologic signs or symptoms. Thus close monitoring of the patient during administration of the drugs is mandatory. This careful control of the patient will prevent any subsequent incidence of more serious neurotoxicities. Also to be investigated is the influence of dilantin or dexamethasone on the lowered incidence of neurotoxicity, as was observed in the glioma study. The phase II clinical studies have also established the fact that ultraviolet pharmacologic assay is a reliable monitoring technique.

As a consequence of these pilot studies, eight phase III studies have now been initiated under the aegis of the Radiation Therapy Oncology Group. These are listed in Table IV. These are disease specific in design to look at fixed dose schedules of radiation and misonidazole, generally randomized or stratified comparisons of radiation therapy and misonidazole against the best standard radiation therapy programme. The end results of the studies are tumour clearance, short-term and long-term toxicity, tumour-free interval, relapse rate and survival. Continued pharmacologic monitoring must be done to achieve the most appropriate relationships in drug administration in the radiation therapy programme. Thus far accrual of patients is proceeding smoothly, with over 400 patients being entered and over 200 patients receiving combined treatment.

The clinical trials of misonidazole in the United States of America have resulted in progressively increasing accrual of patients, with over 1000 patients having been brought into the programme. The dose formulation and pharmacologic monitoring are both effective. The toxicities, which are primarily moderate nausea and vomiting and mild to moderate peripheral neuropathy, are all within the acceptable standards in the management of the patient with cancer.

Continued efforts have been directed toward reduction of neurotoxicity by decreasing lipophilicity and increasing hydrophilicity. Efforts are being made to develop the use of intravenous preparations to improve the drug distribution within the tumour. The data from desmethylmisonidazole indicate an endogenously formed analogue, equal in vitro and in vivo radiosensitization but with a lower lipophilicity. Phase I studies have been initiated using desmethylmisonidazole to look at qualitative and quantitative toxicity, the pharmacology of the drug, and the dose-limiting toxicity for its administration.

TABLE IV. CONTINUING RTOG PHASE III MISONIDAZOLE STUDIES

RTOG No.	Cancer type and stage	Date opened	Randomization arms
79-15	Head and neck cancers, Stage II, III, IV	1979-09	1. XRT: 6600-7380 rads in 33-41 fractions at 180-200 rad/fraction, 5 fractions/week 2. XRT + Miso. XRT: 2 fractions every Mon, 1 fraction every Tue, Thu, Fri; 6600-7380 rads/6½-7 weeks Miso.: 2.0 g/m^2 every week X 6 weeks = 12 g/m^2
79-16	Brain metastases	1979-09	1. XRT: 300 rads every day X 5 d X 2 weeks = 3000 rads 2. XRT: 500 rads twice weekly X 3 weeks = 3000 rads 3. XRT (as in 1) + Miso.: 1 g/m^2 every day X 5 d X 2 weeks = 10 g/m^2 4. XRT (as in 2) + Miso.: 2 g/m^2 twice weekly X 3 weeks = 12 g/m^2
79-17	Lung cancer (Stage III-T_1-T_2, T_3, N_2)	1980-04	1. XRT: 200 rads every day X 5 d X 5 weeks = 5000 rads + 500-1000 rads boost 2. XRT (as in 1) + Miso.: 400 mg/m^2 every day X 5 d X 6 weeks = 12 g/m^2
79-18	Malignant gliomas, Grade III, IV	1979-11	1. XRT + BCNU XRT: 200 rads every day X 5 d X 6 weeks = 6000 rads BCNU: 80 mg/m^2 every day X 3 d X 8 weeks 2. XRT + BCNU + Miso. XRT: (400 rads every Mon + 150 rads every Tue, Thu, Fri) X 6 weeks + 900 rads boost = 6000 rads BCNU (as in 1) Miso.: 2.5 g/m^2 every week X 6 weeks = 15 g/m^2
79-25	Lung cancer, T_4 or N_3	1980-01	1. XRT: 600 rads twice weekly X 3 weeks = 3600 rads 2. XRT + Miso. XRT (as in 1) Miso.: 1.75 g/m^2 twice weekly X 3 weeks = 10.5 g/m^2

TABLE IV. (Continued)

RTOG No.	Cancer type and stage	Date opened	Randomization arms
80-03	Liver metastases	1980-05	1. XRT: 2100 rads in 7 fractions/1½ weeks 2. XRT (as in 1) + Miso.: 1.5 g/m^2 every day X 7 doses = 10.5 g/m^2
80-05	Cervix, III B, IV A	1980-08	1. XRT: 4600 rads pelvis + 1000 rads parametrial boost + intra-cavitary or external boost 2. XRT (as in 1) + Miso.: 0.4 g/m^2 every day X 30 d = 12 g/m^2
80-xx	Bladder B_2, C	—	1. XRT: 2000 rads in 5 fractions, 1 week pre-op. 2. XRT (as in 1) + Miso.: 1 g/m^2 every day X 5 d = 5 g/m^2 3. XRT pre-op. (500 rads in 1 fraction) + XRT post-op. (4500 rads/5 weeks)

RADIATION PROTECTORS

As with hypoxic cell sensitizers, it is in the initial radiochemical events that potential application of radioprotectors can be found. The radicals formed in the target molecule may be repaired by reduction rather than fixed by oxidation. Almost all cells contain significant sulphydryl compounds in the non-protein fractions, and these participate in such protective activities. Depletion of such sulphydryl compounds can be shown to cause significant cellular sensitization. The addition of exogenous sulphydryl compounds will significantly protect aerated cells but have only a small degree of effect on hypoxic cells. Thus it would appear that the addition of sulphydryl-containing compounds to the intracellular milieu might protect oxic cells to a much greater degree than hypoxic cells, much as electron affinic compounds sensitize hypoxic cells but not aerated cells.

The identification of this protection against ionizing radiation by sulphydryl-containing compounds was described by Patt et al. [30] in the late 1930s. This mechanism was recognized almost immediately as a potential method to protect troops against radiation exposure in atomic warfare. A large number of synthesis programmes were undertaken to find the most effective orally administered compounds. From these programmes came a number of promising radioprotective compounds, many of which were in the thiophosphate class [31]. WR-2721 [S-2 (3-amino-propropyl-amino ethyl phosphorothioic acid hydrate)] appears to be the most efficacious in terms of its therapeutic ratio. It shows significant protection at dosages far below its maximum tolerated levels. The relationship between the dose and the dose reduction factor is not linear, with much steeper increases in protection occurring at low concentrations rather than at high concentrations.

The use of protective compounds in clinical radiation therapy can only be predicted upon the presence of a differential effect on normal tissue vis-a-vis the tumour. If normal tissue could be protected against the radiation event, then total radiation dosages could be increased - yielding enhanced tumour cell kill with no increased price in normal tissue toxicities [32].

Yuhas and Storer [33] observed that WR-2721 appeared to cause differential protection of normal tissues in terms of skin reaction compared with tumour response. Yuhas later pointed out that WR-2721 appeared actively concentrated by a process that may be facilitated by diffusion in normal tissues, whereas it reaches slowly increasing concentrations over several hours in the tumour [31]. Thus, for the short time periods of 30-60 minutes after intravenous or intraperitoneal injection, concentrations would be far higher in most normal tissues than in the tumour, leading to significant differential radioprotection.

The basic factors causing differential radioprotection have to do with blood flow, the lipid content, binding after absorption, dephosphorylation, inactivation, and hypoxic cells not protected or poorly protected.

The animal data indicated that WR-2721 showed significant protection in bone marrow, skin, mucous membranes, salivary glands, and in the lining of the small intestine [34]. Protection of the liver and to a lesser degree the kidney was also noted [35,36].

This compound appeared to have major potential usefulness in the treatment of head and neck cancers, the treatment of abdominal malignancies, as well as with hemi-body or whole-body radiation therapy, where significant bone marrow protection would be an important component of the treatment programme.

The early phase I human testing programmes were conducted with oral administration of the compound showing only limited results. More recent explorations have been carried out using intravenous administration of the compounds. These initial phase I studies in the USA with the administration intravenously were carried out at the University of New Mexico by Kligerman et al. [37]. The major side effects have included nausea and vomiting, as well as hypotension secondary to the ganglionic blocking activity of WR-2721.

The Radiation Therapy Oncology Group has built its phase I evaluation upon the experience of the University of New Mexico trials. The objectives of the radiation protection programme are directed toward the determination of the sulphydryl-containing radioprotectors causing differential protection in human normal tissues and tumour, the establishment of differential properties leading to decreased morbidity with increased tumour control, the development of optimum dose schedules, the development of new protectors and the integration of these clinical trial data into clinical practice.

The phase II studies with radioprotectors are far different from those with the cytotoxic drugs or sensitizers. These studies are directed toward the determination of the maximum tolerated drug dose for a given radiation regimen. It is then necessary to establish the maximum tolerated radiation dosages. Tissues must be collected, and concentrations of the drug determined from the pattern that has been established in the animal studies. The phase II clinical trials will be directed primarily to head and neck cancers, abdominal malignancies including liver metastases, ovarian cancers and gastrointestinal cancers. Data from these studies will be used to design the phase III trials, where the efficacy of conventional radiation therapy will be compared with the efficacy and toxicity of radiation therapy using augmented doses in conjunction with the radioprotector, WR-2721.

REFERENCES

[1] ADAMS, G.E., FOWLER, J.F., WARDMAN, P., Eds: Hypoxic cell sensitizers in radiobiology and radiotherapy (Proc. 8th L.H. Gray Conf., Cambridge, UK, 1977), Br. J. Cancer 37, Suppl. III (1978).

[2] DENEKAMP, J., MICHAEL, B.D., HARRIS, S.R.: Hypoxic cell radiosensitizers: comparative tests of some electron affinic compounds using epidermal cell survival in vivo, Radiat. Res. 60 (1974) 119-132.

[3] ASQUITH, J.C., WATTS, M.E., PATEL, K., et al.: Electron affinic sensitization: V. Radiosensitization of hypoxic bacteria and mammalian cells in vitro by some nitroimidazoles and nitropyrazoles, Radiat. Res. 60 (1974) 108-118.

[4] CHAPMAN, J.D., REUVERS, A.P., BORSA, J., et al.: Nitrofurans as radiosensitizers of hypoxic mammalian cells, Cancer Res. 32 (1972) 2630-2632.

[5] TALLENTIRE, A., JONES, A.B., JACOBS, G.P.: The radiosensitizing actions of ketonic agents and oxygen in bacterial spores suspended in aqueous and non-aqueous milieux, Isr. J. Chem. 10 (1972) 1185-1197.
[6] GREENSTOCK, C.L., RUDDOCK, G.W., NETA, P.: Pulse radiolysis and ESR studies of the electron-affinic properties of nitroheterocyclic radiosensitizers, Radiat. Res. 66 (1976) 472-484.
[7] GRASLUND, A., EHRENBERG, A., RUPPRECHT, A.: Free-radical formation in γ-irradiated oriented DNA containing electron-affinic radiosensitizers, Int. J. Radiat. Biol. Relat. Stud. Phys., Chem. Med. 31 (1977) 145-152.
[8] WILSON, R.L., CRAMP, W.A., INGS, R.M.J.: Metronidazole (Flagyl): Mechanisms of radiosensitization, Int. J. Radiat. Biol. Relat. Stud. Phys., Chem. Med. 26 (1974) 557-569.
[9] CHAPMAN, J.D., REUVERS, A.P., BORSA, J., et al.: Nitroheterocyclic drugs as selective radiosensitizers of hypoxic mammalian cells, Cancer Chemother. Rep. 58 (1974) 559-570.
[10] CHAPMAN, J.D., GREENSTOCK, C.L., REUVERS, A.P., et al.: Radiation chemical studies with nitrofurazone as related to its mechanism of radiosensitization, Radiat. Res. 53 (1973) 190-203.
[11] COMMITTEE FOR RADIATION ONCOLOGY STUDIES RESEARCH PLAN FOR RADIATION ONCOLOGY, Radiation Sensitizers, Cancer 37 (1976) 2062-2070.
[12] DENEKAMP, J.: Testing of hypoxic cell radiosensitizers in vivo, Cancer Clin. Trials 3 (1980) 139-148.
[13] DENEKAMP, J., STEWART, F.A.: Sensitization of mouse tumours using fractionated x-irradiation, Br. J. Cancer 37 (1978) 259-263.
[14] KLIGERMAN, M., YUHAS, J.M.: personal communication.
[15] KOLSTAD, P.: The development of the vascular bed in tumours as seen in squamous cell carcinoma of the cervix uteri, Br. J. Radiol. 38 (1965) 216.
[16] URTASUN, R.C., BAND, P.R., CHAPMAN, J.D., et al.: Metronidazole as a radiosensitizer, New Engl. J. Med. 295 (1976) 901.
[17] URTASUN, R.C., STURMWIND, J., RABIN, H., et al.: High dose metronidazole; a preliminary pharmacological study prior to its investigational use in clinical radiotherapy trials, Br. J. Radiol. 47 (1974) 297.
[18] DISCHE, S., GRAY, A.J., ZANELLI, G.D.: Clinical testing of the radiosensitizer Ro-07-0582: II. Radiosensitization of normal and hypoxic skin, Clin. Radiol. 27 (1976) 159.
[19] DISCHE, S., SAUNDERS, M.I., LEE, M.E., et al.: Clinical testing of the radiosensitizer Ro-07-0582: experience with multiple doses, Br. J. Cancer 35 (1977) 567-579.
[20] GRAY, A.J., DISCHE, S., ADAMS, G.E., et al.: Clinical testing of the radiosensitizer Ro-07-0582: I. Dose tolerance, serum and tumour concentration, Clin. Radiol. 27 (1976) 151.
[21] THOMLINSON, R.H., DISCHE, S., GRAY, A.J., et al.: Clinical testing of the radiosensitizer Ro-07-0582: III. Response of tumours, Clin. Radiol. 27 (1976) 167.
[22] SHELDON, P.W., FOSTER, J.L., FOWLER, J.F.: Radiosensitization of C3H mouse mammary tumours by a 2-nitromidazole drug, Br. J. Cancer 30 (1975) 560-565.
[23] SHELDON, P.W., FOSTER, J.L., FOWLER, J.F.: Radiosensitization of C3H mouse mammary tumours using fractionated doses of X-rays with the drug Ro-07-0582, Br. J. Radiol. 49 (1975) 76-80.
[24] CARABELL, S.C., BRUNO, L.A., WEINSTEIN, A.S., et al.: Misonidazole and radiotherapy in the treatment of malignant glioma: A phase II trial of the Radiation Therapy Oncology Group, Int. J. Radiat. Oncol., Biol. Phys. (1980, in press).
[25] FAZEKAS, J.T.: Therapy as an adjuvant or as the primary therapy in head and neck cancer, Proc. Am. Soc. Therapeutic Radiologists (1980).

[26] PHILLIPS, T.L., NEWALL, J., ORDER, S.E., et al.: A phase II evaluation of misonidazole in patients with brain metastases, Int. J. Radiat. Oncol., Biol. Phys. 6 (1980) 1391-1392.
[27] LEIBEL, S.A.: Palliation of liver metastases with combined hepatic irradiation and Misonidazole - report of a Phase I-II study of the Radiation Therapy Oncology Group, Proc. Am. Society of Therapeutic Radiologists (1980).
[28] SIMPSON, J.R., PEREZ, C.A., PHILLIPS, T.L., et al.: Large fraction radiotherapy plus misonidazole in the treatment of advanced lung cancer: Report of phase I/II trial, Int. J. Radiat. Oncol., Biol. Phys. 6 (1980) 1391.
[29] WASSERMAN, T.H., STETZ, J., PHILLIPS, T.L.: Radiation Therapy Oncology Group clinical trials with Misonidazole, Cancer (1980, in press).
[30] PATT, H.M., TYREE, E.B., STRAUBE, R.L., et al.: Cysteine protection against x-irradiation, Science 110 (1949) 213-214.
[31] YUHAS, J.M.: Active vs. passive absorption kinetics as the basis for selective protectors of normal tissues against alkylating agents by WR 2721, Cancer Res. (1980, in press).
[32] HARRIS, J.W., PHILLIPS, T.L.: Radiobiological and biochemical studies of thiophosphate radioprotective compounds related to cysteamine, Radiat. Res. 46 (1971) 362-379.
[33] YUHAS, J.M., STORER, J.B.: Differential chemoprotection of normal and malignant tissues, J. Natl. Cancer Inst. 42 (1969) 331-335.
[34] WASHBURN, L.C., RAFTER, J.J., HAYES, R.L.: Prediction of the effective radioprotective dose of WR-2721 in humans through an interspecies tissue distribution study, Radiat. Res. 66 (1976) 100-105.
[35] PHILLIPS, T.L.: Rationale for initial clinical trials and future development of radioprotectors, Cancer Clin. Trials 3 (1980) 165-173.
[36] PHILLIPS, T.L., KANE, L.J., UTLEY, J.F.: Radioprotection of tumour and normal tissues by thiophosphate compounds, Cancer 32 (1973) 528-535.
[37] KLIGERMAN, M., YUHAS, J.M.: personal communication.

Progress in Radio-Oncology II, edited by K. H.
Kärcher et al., Raven Press, New York © 1982.

The Present Status of Clinical Studies with Misonidazole

S. DISCHE, M.I. SAUNDERS
Marie Curie Research Wing for Oncology,
Regional Radiotherapy Centre,
Mount Vernon Hospital,
Northwood, Middlesex,
United Kingdom

Abstract

Clinical experience with misonidazole began in 1974. After the administration of single large doses satisfactory serum and tumour concentrations were obtained. Radiosensitization of hypoxic skin was demonstrated and in some cases a greater response was seen in tumour nodules irradiated after the administration of the drug, when compared with radiotherapy used alone. Unfortunately, when multiple doses were given, the drug proved to be neurotoxic. A dose limit of 12 g per square metre of surface area given over a period of at least 18 days is now generally employed. Many clinical trials are under way in all parts of the world and many thousands of patients have already received the drug. The interim reports available suggest that benefit may be occurring in some studies but not in others, and we must await their completion. Research is being pursued for drugs with which higher radiosensitizing concentrations may be achieved in human tumours. Already several hundred such drugs have been tested in the laboratory and most are more potent sensitizers when used at the same dosage.

In 1973 misonidazole, then known as the Roche experimental drug Ro 07-0582, was found to be a highly effective radiosensitizer in laboratory trial. There was evidence to suggest that it might give a several-fold improvement in radiosensitization when compared with metronidazole which had been demonstrated to be a sensitizer the year previously. By 1974 the dramatic improvement shown with its use in the radiotherapy of animal tumours containing hypoxic cells led to its first administration to humans. Good plasma concentrations were reached and radiosensitization of artificially hypoxic skin demonstrated in man.[1,2] An increase in response was seen in some tumour nodules given radiotherapy combined with the drug, compared with radiotherapy alone in the same patient.[3] In 1974 misonidazole was given to patients in multiple doses. Although good, repeatable serum concentrations were achieved and good concentrations obtained in tumours, the drug was found to be neurotoxic, causing convulsions and brain damage with high doses and peripheral neuropathy when these were reduced.[4] Recent work has confirmed that the incidence of neuropathy can be correlated with the tissue exposure as indicated by the area under the curve plotting serum concentration against time.[4,5] By regulating total dose and monitoring serum concentration with further appropriate adjustments in dose the incidence of peripheral neuropathy can be minimised.[6]

In 1977, based on clinical experience to determine a safe scheme of administration, a maximum dose of 12 g per square metre of surface area was proposed for use over a period of not less than 18 d. With this regime a third of the patients may develop symptoms of peripheral neuropathy, but this is mild and transitory in most of the cases. In 1977 randomized controlled clinical trials were commenced and the number underway rapidly grew in 1979 and 1980. Clinical experience around the world has now extended to many thousands of patients and hundreds of radiotherapy centres are contributing cases to multi-centre trials.

Because of the limitation as to the total dose of misonidazole which can be given, much ingenuity has gone into the distribution of dose so as to give the maximum chance of benefit. There has been much discussion among radio-biologists as to the regime to be favoured on theoretical grounds, with opinion now moving towards administration with every treatment in a course of radiotherapy.[7] Unusual fractionation dose regimes have been employed and this is illustrated in Fig. 1.

We, and others, have seen remarkable responses when radiotherapy is given in combination with the drug in individual cases. However, we know only too well from experience how fallacious is such an assessment when it is entirely based on consecutive case experience. We must await results of the randomized controlled clinical studies underway. We have some interim reports concerning these clinical trials which are summarized in Table I.[8]

It can be seen that at this time there is some variation in the pattern; some trials have yielded no benefit, while others are showing advantage to those patients given the drug. When we add a consideration of reports of non-randomized trials, benefit seems most consistently reported in the

---

FIG. 1. Regimes for administration of misonidazole. The diagram illustrates the way in which a dose of 20 g can be distributed over a 4-week course of radiotherapy. The dose will be suitable for a patient with a surface area of 1.65 m^2, this being calculated from the advised maximum dose of 12 g per square metre of surface area.

A: One gram is given daily combined with radiotherapy on every occasion;

B: Radiotherapy and drug are given just twice weekly, leading to high drug and radiation doses on each occasion;

C: One dose of the drug is given weekly. Usually in these schemes a higher dose of radiation is combined with it on that day, smaller doses being given without it on subsequent days. An alternative is to give several large doses combined with radiotherapy before and/or after a conventional course of radiotherapy;

D: With this scheme, two radiation treatments are given with misonidazole on the day on which the large dose of drug is given. A third treatment is given as early as possible on the subsequent day when there is still some concentration of misonidazole in the tumour. Two further treatments are given without the drug later in the week;

E: Multiple treatments are given on each of three days of the week when misonidazole is administered.

radiotherapy of epithelial tumours in the head and neck region. Techniques which lead to the highest sensitizing levels being present with at least the majority of treatments seem most successful. We must wait until the large number of trials underway are reported so as to give secure data on which to base our evaluations of the drug. [5]

We must remember that the limitation of dose required with this drug considerably lowers the serum and tumour concentrations which can be

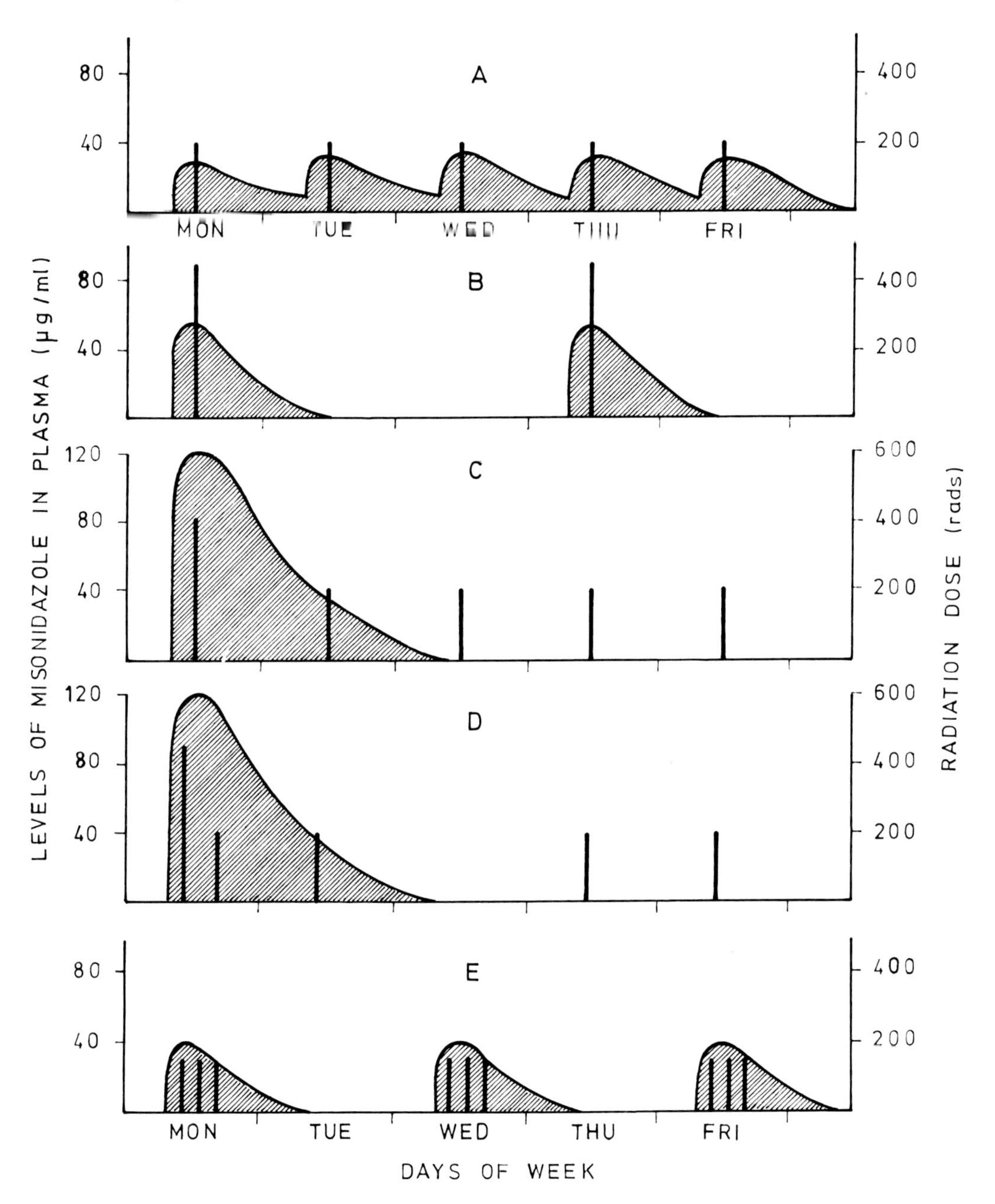

TABLE I Current state of clinical trials with misonidazole

Site	Senior Worker	Place	Scheme	Benefit
Oral Cancer	Sealy	Cape Town	B	Early margin of local control - subsequently lost
	Bataini	Paris	D	Significant improvement in survival
	Nervi	Rome	E	Significant improvement in local control
Brain - Glioblastoma	Bleehen	Cambridge	C	None
	Urtasun	Edmonton, Alberta	B	None
	Kogelnik	Vienna	C	Significant improvement in survival
Bronchus	Saunders	Mount Vernon	B	None

achieved. In a calculation based upon 742 observations of serum concentration of misonidazole,we have calculated that a dose of 1g per square metre of surface area will give 40 μg/ml in the serum at the usual time of treatment, i.e. $3\frac{1}{2}$ to 4h after administration.(5) Other workers have recorded very similar levels.(9) We know that in the majority of cases tumour concentrations, however, lie between 60 to 100% of that in the serum at the time the sample is taken. It is probable that within a tumour sample the concentration in hypoxic cells will tend to be lower than in the tissue as a whole. It seems, therefore, more realistic to consider that the hypoxic cells contain a concentration approximately 50% of that in the serum at the time of radiotherapy.

Knowing the serum concentrations which are reached we can calculate the concentration likely to be achieved in hypoxic cells and the enhancement factor for hypoxic cells using different fractionation regimes and the total dose limit of 12g per square metre of surface area (Table II).(5) Only the results of the randomized controlled clinical trials underway will tell us whether the concentrations achieved will give sufficient enhancement of response so as to lead to benefit. Also we must recognize the wide variety of the types of human malignant disease. Even in one site of origin we may see many different patterns of clinical course and pathological appearance. Although hypoxic cells must exist in considerable concentrations in all tumours presenting clinically,we do not know the extent to which they are the cause of failure after a conventional fractionated course of radiotherapy. Especially with the dose limitation which we must impose with misonidazole, this drug may only give significant benefit where hypoxia dominates as the cause of radiation failure.

TABLE II

Plateau plasma and levels in hypoxic tumour cells if 50% of the plasma concentration is achieved when misonidazole is given in different fractionation regimes using the dose limit of 12 g per square metre of surface area. Also shown are the enhancement ratios for hypoxic cells relating to the concentrations achieved

Fractionation	Dose (g/m^2)	Plateau Plasma Concentration		Probable levels in hypoxic tumour cells	
		(μg /ml)	Enhancement	(μg /ml)	Enhancement
6	2	80	1.65	40	1.50
10	1.2	50	1.55	25	1.40
20	0.6	24	1.40	12	1.20
24	0.5	20	1.35	10	1.20
30	0.4	16	1.25	8	1.15

While continuing the clinical testing of misonidazole,it is important that we continue to search for new radiosensitizing compounds which may prove to be superior in clinical practice. Much attention has been given recently to the nitroimidazoles which have a lower lipophilicity than misonidazole, for these combine, in animal studies, a shorter half-life in serum with a lower uptake in nervous tissue.[10] Our recent clinical experience with desmethyl-misonidazole, one of these compounds, has certainly shown a shorter half-life in serum compared with misonidazole, equally good tumour concentrations and lower levels in the cerebro-spinal fluid.[11,12] Unfortunately when the drug was administered in multiple doses to 23 patients to the same total dose as used with misonidazole, i.e. 12g per square metre of surface area, the incidence of peripheral neuropathy was similar to that with misonidazole.[13] Tumour responses were unusually good and work continues with this compound. There are, however, several hundred new radiosensitizers and the majority are more potent than misonidazole. The problem of predicting toxicity is still to be solved, but no doubt with an intensive effort methods will be evolved. At the moment the only reliable method is to cautiously administer any new drug to man.

Clinical experience so far recorded would suggest that we may show real benefit in some situations with misonidazole. The work of Ross Sealy in combining misonidazole with hyperbaric oxygen in the radiotherapy of advanced oral cancer is yielding most promising results[14], which may be those which we will achieve with a new drug with which higher radio-sensitizing levels may be achieved in hypoxic cells. An elevation of the radiosensitizing concentration by a factor of 2 or 3 may greatly improve upon the results currently emerging with misonidazole.

REFERENCES

[1] GRAY, A.J., DISCHE, S., ADAMS, G.E., FLOCKHART, I.R., FOSTER, J.L.: Clinical testing of the radiosensitizer Ro 07-0582. I. Dose tolerance, serum and tumour concentration, Clin. Radiol. 27 (1976) 151.

[2] DISCHE, S., ZANELLI, G.D.: Skin reaction – a quantitative system for measurement of radiosensitization in man, Clin. Radiol. 27 (1976) 145.

[3] THOMLINSON, R.H., DISCHE, S., GRAY, A.J., ERRINGTON, L.M.: Clinical testing of the radiosensitizer Ro 07-0582. III. Response of tumours, Clin. Radiol. 27 (1976) 167.

[4] DISCHE, S., SAUNDERS, M.I., LEE, M.E., ADAMS, G.E., FLOCKHART, I.R.: Clinical testing of the radiosensitizer Ro 07-0582: Experience with multiple doses, Br. J. Cancer 35 (1977) 567.

[5] DISCHE, S.: "Hypoxic cell sensitizers", Proc. 2nd Int. Symp. Biological Bases and Clinical Implications of Tumour Resistance (Rome, 1980), Masson Publishing USA Inc. (1981, awaiting publication).

[6] DISCHE, S., SAUNDERS, M.I., FLOCKHART, I.R., LEE, M.E., ANDERSON, P.: Misonidazole. A drug for trial in radiotherapy and oncology, Int. J. Radiat. Oncol., Biol. Phys. 5 (1979) 851.

[7] DENEKAMP, J., McNALLY, N.J., FOWLER, J.F., JOINER, M.C.: Misonidazole in fractionated radiotherapy: are many small fractions best? Br. J. Radiol. 53 (1980) 981.

[8] DISCHE, S.: Sensitizers in radiotherapy, Br. J. Hosp. Med., to be published.

[9] WASSERMAN, T.H., PHILLIPS, T.L., JOHNSON, R.J., GOMER, M.S., LAWRENCE, G.A., SADEE, W., MARQUES, R.A., LEVIN, V.A., RAALTE, Van G.: Initial clinical and pharmacologic evaluation of misonidazole (Ro 07-0582) an hypoxic cell radiosensitizer, Int. J. Radiat. Oncol., Biol. Phys. 5 (1979) 775.

[10] BROWN, J.M., WORKMAN, P.: Partition coefficient as a guide to the development of radiosensitizers which are less toxic than misonidazole, Radiat. Res. 82 (1980) 171.

[11] DISCHE, S., FOWLER, J.F., SAUNDERS, M.I., STRATFORD, M., ANDERSON, P., MINCHINTON, A., LEE, M.: A drug for improved radiosensitization in radiotherapy, Br. J. Cancer 42 (1980) 153.

[12] DISCHE, S., SAUNDERS, M.I., RILEY, P.J., HAUCK, J., BENNETT, M.H., STRATFORD, M.R.L., MINCHINTON, A.I.: The concentration of desmethyl-misonidazole in human tumours and in cerebro-spinal fluid, Br. J. Cancer, to be published.

[13] DISCHE, S., SAUNDERS, M.I., STRATFORD, M.R.L.: Neutotoxicity with desmethylmisonidazole, Br. J. Radiol. 54 (1981) 156.

[14] SEALY, R., WILLIAMS, A., LEVIN, W., BLAIR, R., FLOCKHART, I., STRATFORD, M., MINCHINTON, A., CRIDLAND, S.; "Progress in investigations in the use of misonidazole in head and neck cancer as a hypoxic cell radiation sensitizer and in combination with hyperbaric oxygen or hyperthermia", in Radiation sensitizers. Their Use in the Clinical Management of Cancer (Proc. Conf. on Combined Modality Cancer Treatment: Radiation Sensitizers and Protectors, Key Biscayne, Florida, 1979), Masson Publishing USA, Inc. (1980) 361-365.

Progress in Radio-Oncology II, edited by K. H. Kärcher et al., Raven Press, New York © 1982.

Clinical Experience with the Hypoxic Cell Radiosensitizer Misonidazole in Japan

K. MORITA, M. WATANABE, Y. OBATA, K. WATAI
Department of Radiation Therapy,
Aichi Cancer Centre,
Nagoya,
Japan

Abstract

Since January 1979, Phase I of a clinical study of combined radiotherapy and misonidazole treatment has been carried out in Japan. Of the 22 cases, for which the total dosage of misonidazole lay between 10.0 and 11.9 g/m^2 of surface area, 3 cases (13.6%) suffered neurotoxic reactions. When 12.0 g/m^2 or more of misonidazole was administered, toxic effects were seen in more than half of the cases. From these results, it is recommended that 10.0 g/m^2 total dosage should be used in Japan. In keeping within limits of safety, it is therefore recommended that 1.6 g/m^2 be administered once a week, 1.1 g/m^2 twice a week or 0.6 g/m^2 five times a week. In our Phase I trial, a remarkably good response to combined treatment has, on occasion, been obtained. Because of the encouraging results of this study, controlled clinical Phase II trials were started in October 1980. They involve more than 20 clinics. Patients suffering from glioblastoma, pulmonary cancer, oesophageal cancer, cervical cancer and malignant epithelial tumours of the head and neck are treated, using daily administration of small doses of misonidazole and radiotherapy.

1. PHASE I STUDY OF MISONIDAZOLE IN JAPAN

Following various kinds of fundamental experiments with the radiosensitizer misonidazole, a clinical Phase I study was started in January 1979 with the co-operation of six radiotherapy clinics in Japan [1]. In this report, we are presenting the results of our own Phase I experience in combination with the results of a Phase I study which has been carried out by Onoyama and Abe.

The trial comprised 101 patients with advanced carcinomas at various locations (lung, 33; oesophagus, 24; head and neck, 22; brain, 15; uterine cervix, 4; other sites, 3) and with limited survival times. In 46 cases 0.7 to 1.5 g misonidazole per square metre of body surface area were given once or twice a week with irradiation. In the remaining 55 cases, we employed daily treatment using 0.3 to 0.5 g/m^2 of the drug, administered on five days of each week.

1.1. Adverse Reactions of Misonidazole

1.1.1. GASTROINTESTINAL DISTURBANCES

Misonidazole was almost always given on an empty stomach. Anorexia and nausea were noted in 11 of our cases (10.9%). As these complications were usually not severe, misonidazole was, with one exception, given without interruption.

1.1.2. SKIN RASHES

Skin rashes due to misonidazole were recorded in two of our patients. A maculo-papular rash appeared on the arms, legs and trunk. In these cases the rashes were not especially troublesome, but did lead to the premature cessation of misonidazole administration. In four cases rashes appeared during the course of radiotherapy with misonidazole, but the misonidazole was not considered to have been responsible.

1.1.3. NEUROTOXIC REACTIONS

As is generally known, neurotoxicity limits the amount of misonidazole which can be given. The relationship between the total dose and dose regimes of misonidazole and the frequency of the neurotoxic effect is shown in Table I. In the four patients in whom misonidazole was given at a dose of 1.5 g/m^2 of surface area twice a week to a total dose of more than 9.0 g/m^2, all suffered a neurotoxic effect. In 84 cases receiving a total dose of less than 10.0 g/m^2, no neurotoxic effect was observed other than transient difficulties in hearing in two cases. In 22 cases receiving a total dose between 10.0 g/m^2 and 12.0 g/m^2, 3 cases (13.6%) suffered neurotoxic reactions. When the total dose of misonidazole exceeded 12.0 g/m^2, toxic neuropathy was seen in more than half of the cases.

Except in one case of transitory hallucination and an abnormal excited condition after administration of a total dose of misonidazole of 12.0 g/m^2 (1.5 g/m^2 per fraction, given twice a week),

TABLE I. INCIDENCE OF NEUROPATHY AND DOSE OF MISONIDAZOLE

Fraction size (g/m^2)	Number of fractions per week	Weekly dosage (g/m^2)	Tumour dose (g/m^2) Less than 8.0	8.0 to 9.9	10.0 to 11.9	12.0 to 13.9	More than 13.9	Total
1.5	1-2	1.5-3.0	0/ 2	2/ 4	-	2/ 2	-	4/ 8
1.0	1-2	1.0-2.0	0/ 9	0/ 7	1/ 9	0/ 1	-	1/ 26
0.7	1-2	0.7-1.4	0/12	-	-	-	-	0/ 12
0.5	5	2.5	0/12	0/ 5	2/10	-	-	2/ 27
0.3	5	1.5	0/21	0/ 2	0/ 3	0/ 1	1/ 1	1/ 28
Totals			0/56	2/18	3/22	2/ 4	1/ 1	8/101
Peripheral neuropathy			0	0	3	2	1	6
CNS disorders			0	0	0	1	0	1
Ototoxicity			0	2	0	0	0	2

no important central nervous system abnormalities were revealed. Six cases suffered a peripheral neuropathy which took the form of numbness and paraesthesia in the hands and feet. The symptoms gradually subsided with a cessation of misonidazole, but in two cases they persisted for more than six months. We have not experienced any motor loss in our patients.

We have seen a transient numbness and paraesthesia of hands and feet in three patients following one or two administrations of misonidazole in the dose range 1.0 to 1.5 g/m^2, but these effects disappeared after about 24 hours. The symptoms did not occur with the reduction of the given dose, but in one case the administration of misonidazole was interrupted.

In two cases given 1.5 g/m^2 of misonidazole twice a week, eighth nerve disturbances appeared. These symptoms were apparently transient and no permanent effects resulted.

From these results, it is concluded that in Japan a total dose of misonidazole of 10.0 g/m^2 can be used without causing severe toxic neuropathy. It is also recommended that the dosages of misonidazole to be used are either 1.6 g/m^2 once a week, 1.1 g/m^2 twice a week or 0.6 g/m^2 five times a week.

2. DETERMINATION OF A 'RADIORESISTANT' TYPE OF OESOPHAGEAL CANCER

2.1. Clinical Experience Obtained from Pre-operatively Irradiated Cases

In order to plan Phase II of this clinical trial, it was important to pinpoint which characteristics of a malignant lesion greatly influence the rate of local control of the malignant disease in question. For example, in oesophageal cancer, our experience has shown that ulcer-formation was one of the most important signs of a radioresistant lesion.

Since 1971, 95 cases of carcinoma of the oesophagus have been irradiated preoperatively in our clinic [2]. 40 Gy of 6 MV X-rays were given over 4 weeks, and surgical resection was performed 2 to 3 weeks after the irradiation. The effect of irradiation was investigated by histological examination of the resected specimen. In 32 cases (33.7%) no viable tumour cell was seen. The relationship between indications on X-ray radiographs before and after radiotherapy and the local control rate (i.e. the disappearance rate of viable tumour cells) is shown in Tables II and III. It has been our experience that, when a large and/or deep ulcer in the oesophageal tumour shadow is observed through X-ray examination before radiotherapy, and if the ulcer does not disappear in spite of irradiation to 40 Gy, it is usually difficult to obtain local control of the lesion by radiation therapy alone. It is then recommended that radiation therapy is combined with misonidazole therapy in such radioresistant cases, i.e. cases where radiotherapy alone offers poor prognosis for local control of neoplastic diseases.

2.2. Clinical Experience Obtained from Our Phase I Study of Oesophageal Cancer

In our Phase I study, 8 out of 15 patients with advanced carcinoma of the oesophagus experienced between 50 and 100%

TABLE II. RELATIONSHIP BETWEEN THE RESULTS OF X-RAY EXAMINATION IN OESOPHAGEAL CANCER BEFORE TREATMENT AND HISTOLOGICAL FINDINGS FOR THE SURGICAL SPECIMEN AFTER RESECTION

X-ray radiograph indications before treatment	Number of cases	Cases without viable tumour cells after radiotherapy
Superficial or tumorous	11	7 (64%)
Serrated or spiral without ulcer formation	25	10 (40%)
Serrated or spiral with ulcer formation	48	13 (27%)
Funnelled	11	2 (18%)
Total	95	32 (34%)

TABLE III. RELATIONSHIP BETWEEN THE RESULTS OF X-RAY EXAMINATION IN OESOPHAGEAL CANCER AFTER 40 Gy OF PREOPERATIVE IRRADIATION AND HISTOLOGICAL FINDINGS FOR THE SURGICAL SPECIMEN AFTER RESECTION

X-ray signs after 40 Gy of radiotherapy	Number of cases	Cases without viable tumour cells after radiotherapy
Disappearance of tumour mass without residual ulcer	21	16 (76%)
Residue of shallow ulcer Residue of tumour mass without ulcer formation	29	9 (31%)
Residue of deep and/or large ulcer with or without tumour mass Stricture	45	7 (16%)
Total	95	32 (34%)

objective regression of the tumour mass, although radiotherapy alone would probably have offered a poor regression rate in each case. The most striking result was observed with a 73-year-old man with a deep penetrating ulcer caused by oesophageal cancer (for which, from our experience it would have been impossible to achieve precise results with the usual radiation therapy). 60 Gy of 6 MV X-ray therapy combined with 10.0 g/m^2 of misonidazole in 20 fractions were given to this patient. After combined treatment, the filling-defect with the deep ulcer disappeared, though the irregularity of the oesophageal wall and the narrowing of the lumen due to scar formation remained. So far, he has survived without local recurrence for more than 1 year. His only complaint has been difficulty in swallowing due to narrowing of the oesophageal lumen.

Encouraged by this clinical experience, we are now pursuing a randomized controlled clinical trial.

3. CONTROLLED CLINICAL PHASE II TRIALS IN JAPAN

Controlled clinical Phase II studies involving more than 20 radiotherapy clinics throughout Japan were begun in October, last year. The tumours under study include glioblastoma, metastatic brain tumour, carcinoma of the uterine cervix (stage III and IVa), inoperable stage III lung cancer excluding small anaplastic cell carcinoma, advanced oesophageal cancer and inoperable epithelial tumour of the head and neck region (T_4 and N_3 group).

It is well known that the maximum enhancement effect of misonidazole is always seen with a large dose given in smaller fractions. But in recent years, it has been suggested that giving small daily doses of misonidazole with conventional daily irradiation may also be helpful [3-6], for the following reasons: (a) the relationship between the enhancement ratio of misonidazole and dosage is not linear. When daily doses of 0.5 g/m^2 misonidazole are employed in combination with 20 radiotherapy treatments (total misonidazole dose of 10.0 g/m^2 of surface area), an enhancement ratio of the order of 1.3 can be expected for hypoxic

TABLE IV. PLANNING OF THE CONTROLLED, RANDOMIZED PHASE II TRIAL IN JAPAN

IV-A. DAILY ADMINISTRATION OF MISONIDAZOLE

Regions: brain (glioblastoma and metastatic tumour), oesophagus, lung, uterine cervix, head and neck

Treatment	First course	Second course
Irradiation	2 Gy/fraction, 5 fractions/week	2 Gy/fraction, 5 fractions/week
Total	40 Gy over 4 weeks	20-25 Gy over 2 weeks
Misonidazole	0.5 g/m^2 per fraction, 5 fractions/week	None
Total	10.0 g/m^2 over 4 weeks	

IV-B. TWICE-A-WEEK ADMINISTRATION OF MISONIDAZOLE

Regions: oesophagus, lung

Treatment	First course	Second course
Irradiation	4 Gy/fraction, 2 fractions/week	4 Cy/fraction, 2 fractions/week
Total	40 Gy over 5 weeks	16-20 Gy over 2 weeks
Misonidazole	1.0 g/m^2 per fraction, 2 fractions/week	None
Total	10.0 g/m^2 over 5 weeks	

cells; (b) in order to avoid the possible disadvantages of large X-ray fractions to the healthy tissues, daily use of misonidazole with conventional daily radiotherapy is advantageous and is easily acceptable in many radiotherapy clinics; (c) if the independent cytotoxic effect of misonidazole is to be effective in tumour control, daily administration of the drug may lead to the greatest cell kill, because of the importance of 'duration of exposure' [6].

For these reasons, in the Japanese Phase II trials, misonidazole is usually given in small doses with each normal daily fraction in a fractionated course of radiotherapy lasting four weeks (20 fractions). However, in pulmonary and oesophageal cancer, misonidazole is given twice a week in larger doses combined with a high dose of radiation. The scheme of the Phase II trials is summarized in Table IV. The Phase II trials are to be continued for about two years in order to collect 200 cases in each group.

REFERENCES

[1] ONOYAMA, Y., ABE, M.: personal communication: an article 'Phase I study of hypoxic radiosensitizer Misonidazole in Japan' is to be published.

[2] MORITA, K., OBATA, Y., WATANABE, M., KARASAWA, K., TAKAGI, I.: Relationship between radiation effect for carcinoma of the oesophagus and x-ray findings before and after radiotherapy, Jpn. J. Clin. Oncol. 26 (1980) 6.

[3] DISCHE, S.: "Clinical experience with misonidazole", in Radiosensitizers of Hypoxic Cells (BRECCIA, A., et al., Eds), Elsevier, Amsterdam (1979) 221.

[4] DENEKAMP, J., McNALLY, N.J., FOWLER, J.F., JOINER, M.C.: Misonidazole in fractionated radiotherapy: are many small fractions best? Br. J. Radiol. 53 (1980) 981.

[5] KOGELNIK, H.D., REINARTZ, G., SZEPESI, T., SEITZ, W., WURST, F., MAMOLI, B., WESSELY, P., STARK, H.: Klinische Erfahrungen bei täglicher Gabe von Misonidazol, Strahlentherapie 156 (1980) 759.

[6] ADAMS, G.E.: "Hypoxic cell radiosensitizers in the future development of radiotherapy", in Radiosensitizers of Hypoxic Cells (BRECCIA, A., et al., Eds), Elsevier, Amsterdam (1979) 245.

Progress in Radio-Oncology II, edited by K. H. Kärcher et al., Raven Press, New York © 1982.

High-Dose Irradiation and Misonidazole in the Treatment of Malignant Gliomas

A preliminary report

H.D. KOGELNIK, K.H. KÄRCHER, T. SZEPESI,
A.V. SCHRATTER-SEHN
University Clinic for Radiotherapy and Radiobiology,
Vienna,
Austria

Abstract

Thirty-three patients with grade III and IV supratentorial astrocytomas who were entered into a randomized two-arm study to evaluate the effect of misonidazole in conjunction with postoperative irradiation are available for analysis. All patients received the same radiation dose: 6650 rads (43 MeV photons) in 31 fractions over $7\frac{1}{2}$ weeks. In the first, second and eighth week, 400 rads tumour dose were given on Mondays and Thursdays. From the third until the seventh week, daily tumour doses of 170 rads were administered (5 working days). Misonidazole was given orally 4–5 hours before irradiation on those treatment days when fractional radiation doses of 400 rads were used. The cumulative dose of misonidazole was 27 g, corresponding to 12.6–16.4 g/m^2. With a minimum follow-up time of one year, median survival for patients receiving postoperative irradiation alone is 35.1 weeks. Patients who were given misonidazole in combination with postoperative irradiation had a median survival of 57.1 weeks. The difference in the distribution of survival times between both groups is statistically significant ($p < 0.02$). There were no significant differences in the distribution of histological grades between the groups. From an analysis of variance for age, it cannot be ruled out that the survival difference between the two groups is at least partly due to differences in ages. Side effects of misonidazole were minimal, with only 2 patients experiencing mild paresthesias and 4 noting a transient tinnitus.

1. INTRODUCTION

The value of postoperative irradiation in the treatment of grade III and IV supratentorial astocytomas has become firmly established over the past two decades. Median survival following surgical resection and best conventional care ranges between 14 and 22 weeks [1,2]. A recently completed randomized study by the Brain Tumor Study Group (BTSG) indicated that patients receiving postoperative whole-brain irradiation with 5000 to 6000 rads had a median survival increased by 150% [1]. In the same study, the addition of BCNU failed to improve median survival; however, the number of survivors at 18 months was doubled. In a subsequent prospective, randomized study by the BTSG, the relative benefits of radiotherapy and nitrosoureas were further investigated [3]. The modest benefit in 18-month survival produced by the combination of BCNU plus radiotherapy as compared with radiotherapy alone was not significant at the 0.05 level. It was suggested in this study: "that it is best to use radiotherapy in the postsurgical treatment of malignant glioma and to continue the search for an effective chemotherapeutic regimen to use in addition to radiotherapy." On the other hand, a study by the EORTC Brain Tumor Group suggested no prolongation of the free interval with the use of CCNU; however, a prolongation of the total survival time was found when the drug was administered after relapse [4].

Although considerable progress has been made in all currently available treatment modalities, long-term survival for patients with high-grade astrocytomas is still poor, with less than 5% of the patients surviving 3 years. The failure to cure a significant proportion of patients with malignant gliomas by irradiation may originate in the radioresistance of the hypoxic tumour cell population. Therefore, on completion of our pharmacokinetic and toxicologic studies [5,6], a randomized trial was initiated at the University Clinic for Radiotherapy and Radiobiology of Vienna, exploring the radiosensitizing properties of misonidazole, currently the most efficient electron-affinic hypoxic cell sensitizer under investigation.

2. METHODS AND MATERIALS

Up to May 1981, 33 patients with a minimum follow-up time of one year are available for analysis. All patients had histologically confirmed grade III and IV supratentorial astrocytomas. Treatment was begun within 4 weeks of surgery. No prior radiotherapy to the head and neck or chemotherapy was allowed. Patients referred after surgery for a recurrence were also excluded. The lower age limit was 18 years. A reasonable functional ability was required, and patients with a modified Karnofsky performance status of less than 20 were not eligible for the study. Patients with adrenal steroid incompatibility and significant co-existing diseases were excluded.

All patients were hospitalized for most of the treatment time. Routine neurologic and ophthalmologic, and occasional otologic examinations were performed. Pre- and postoperative brain scans and electroencephalograms were obtained, in addition to preoperative computed tomography scans. Repeated examinations of the CBC, sedimentation rate, urinalysis, electrolytes and hepatic and renal parameters were performed. Serum concentrations of misonidazole were determined in the early phase of the study.

On the basis of random numbers, patients were randomized to receive one of two treatment schedules: irradiation alone or irradiation and misonidazole. Radiotherapy was identical for both treatment groups. A tumour dose (TD) of 6650 rads was given in 31 fractions over 7½ weeks using 43 MeV photons. In the first, second and eighth week 400 rads TD were given on Mondays and Thursdays. From the third until the seventh week, daily tumour doses (5 working-day week) of 170 rads were administered. The whole of the brain was treated until 4500 rads TD had been given (parallel opposing portals, field size ~ 16 cm x 12 cm). Then the field size was reduced (~ 10 cm x 10 cm) to cover the tumour area with a small margin of safety.

Misonidazole was given orally 4-5 hours before irradiation on those six treatment days when fractional radiation doses of

400 rads TD were used. The cumulative dose of misonidazole was 27 g (4.5 g per treatment day), corresponding to 12.6-16.4 g/m^2 (mean 14.2 g/m^2; 2.1-2.7 g/m^2 per treatment day). Chlorpromazine and thiethylperazine were given before the administration of misonidazole. Steroids and diuretics were used in conjunction with the high fractional doses of 400 rads during the first, second and eighth week; for the remaining treatment time these medications were usually omitted. Only patients at risk of developing epileptic seizures received antiepileptic medications.

3. RESULTS

Figure 1 shows the survival curves for both treatment groups. The superiority of the misonidazole group survival is significant (Mantel-Cox test, $p < 0.02$). Median survival is increased from 35.1 weeks for patients receiving irradiation alone to 57.1 weeks for patients treated by misonidazole and irradiation, a net increase in median survival of 22 weeks. There were 4 patients with grade IV tumours in the misonidazole group as compared to 3 patients in the group receiving irradiation alone. The mean age is 50.3 years for patients in the sensitizer group and 59.9 years for patients who only received radiation. Included

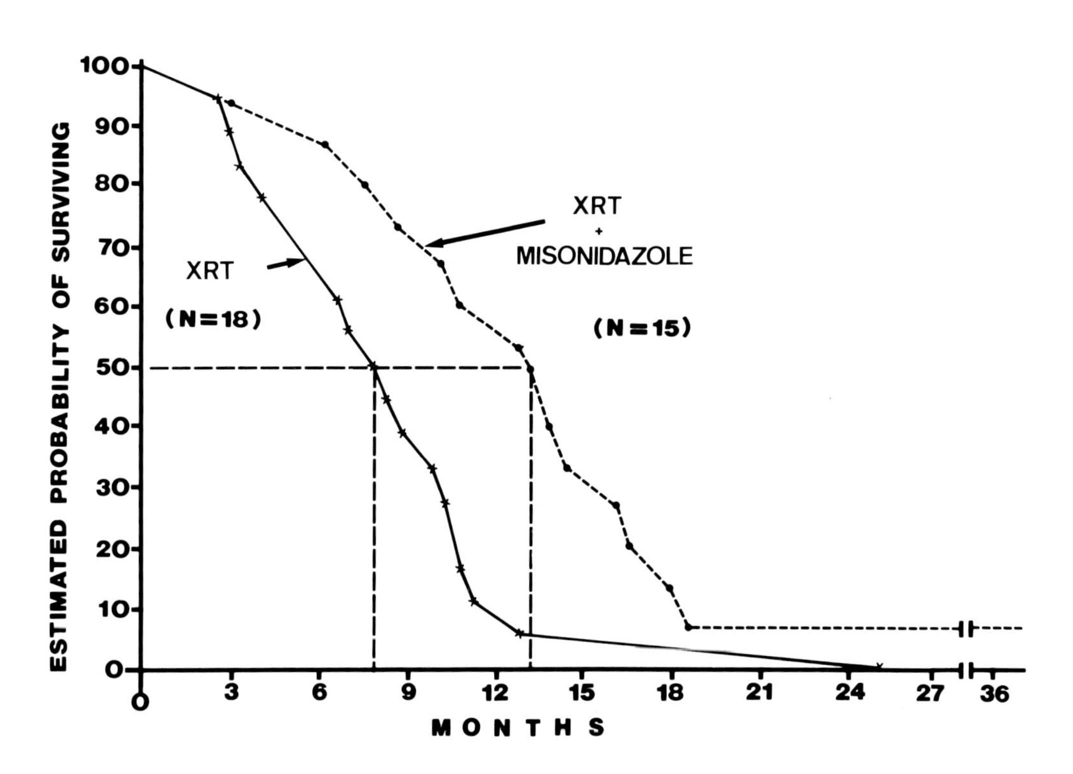

FIG.1. Kaplan-Meier survival plots for patients receiving irradiation alone (XRT) or irradiation plus misonidazole. Survival is statistically significantly better in the misonidazole group ($p < 0.02$), with an increase in median survival of 22 weeks.

among the 15 patients in the misonidazole group are 2 patients for whom treatment had to be interrupted for 4 and 5 weeks, respectively, and one patient for whom treatment had to be terminated at 5000 rads TD because of tumour progression.

In Table I, median survival times are shown separately according to the type of surgery. There is an indication of shorter survival times in patients who had subtotal removal of the tumour as compared with patients with "complete" surgery (analysis of variance, $p < 0.1$). This finding has been reported previously and must be weighed against a relatively high operative mortality with more extensive surgical procedures [7]. Only two patients with censored survival times were in the study, both in the misonidazole group with complete surgery. Hence the analysis of variance can be expected to reveal conservative results. Patients with subtotal and complete surgery seem to respond equally well to misonidazole (low F-value for the interaction 'surgical procedure X treatment'). Our relatively large proportion of patients with complete surgery stands in contradiction to the reports of others [7]. It is likely that some tumour was, in fact, left behind in the majority of our patients.

Mean serum levels 4 hours after the administration of 4.5 g (2.1-2.7 g/m^2) misonidazole were usually in the range of 90-160 µg/ml [6,8]. With these values significant sensitizer enhancement ratios may be expected [9]. There were no adverse effects of misonidazole on renal and hepatic parameters; all other haematologic findings also remained essentially within normal limits.

With the use of antinauseants 30 minutes before misonidazole administration, the drug was tolerated well. Although 12.6-16.4 g/m^2 (mean 14.2 g/m^2) of misonidazole were given, only 2 out of the 15 patients experienced mild paresthesias of the lower extremities, lasting a few weeks. In 4 of the 15 patients, a transient tinnitus occurred at the end of the second week of treatment, which subsided within a few days. No other significant side effects of misonidazole were found and no enhanced normal-tissue reactions could be observed in the interaction with irradiation.

4. DISCUSSION

Several factors influence survival of patients with grade III and IV supratentorial astrocytomas. Among the most important

TABLE I. MALIGNANT GLIOMA STUDY: MEDIAN SURVIVAL (WEEKS)
MINIMUM FOLLOW-UP TIME, 1 YEAR

TREATMENT	SUBTOTAL SURGERY (N = 15)	COMPLETE SURGERY (N = 18)	ALL PATIENTS (N = 33)
X-rays + misonidazole	50.1	60.4	57.1
X-rays	29.4	36.1	35.1

ones are the type of surgery performed, parameters of radiotherapy (mainly tumour dose and treatment volume), pathologic grade, the effect of age and the initial performance status.

Median survival is significantly better for patients with grade III tumours compared with those having grade IV lesions [10]. In our small series there was an even distribution of pathological grades and no further conclusions can be drawn. The initial performance status appeared to be equal between our treatment groups.

The importance of ensuring adequate coverage of the tumours by too small radiation portals was discussed by Kramer [11]. Patients irradiated with larger treatment volumes have a significantly better survival than those treated with limited fields [7]. Also, a clear-cut dose/effect relationship exists: patients receiving larger radiation doses have a significantly higher survival [7,10,12]. The dose used in our study corresponds to an NSD of 1890 rets, which equals about to 7000 rads in 7 weeks. A shrinking field technique, which is now generally recommended, was applied.

Patients with malignant gliomas who are under 45 years of age have significantly better survival times than patients who are older at diagnosis [10]. Figure 2 shows the survival times of our patients as a function of age. Obviously, there is a negative correlation between age and survival: the older the patients, the shorter the survival. This is valid for all 4 subgroups shown. When survival curves for patients 55 years of age or younger are compared with those for older patients,

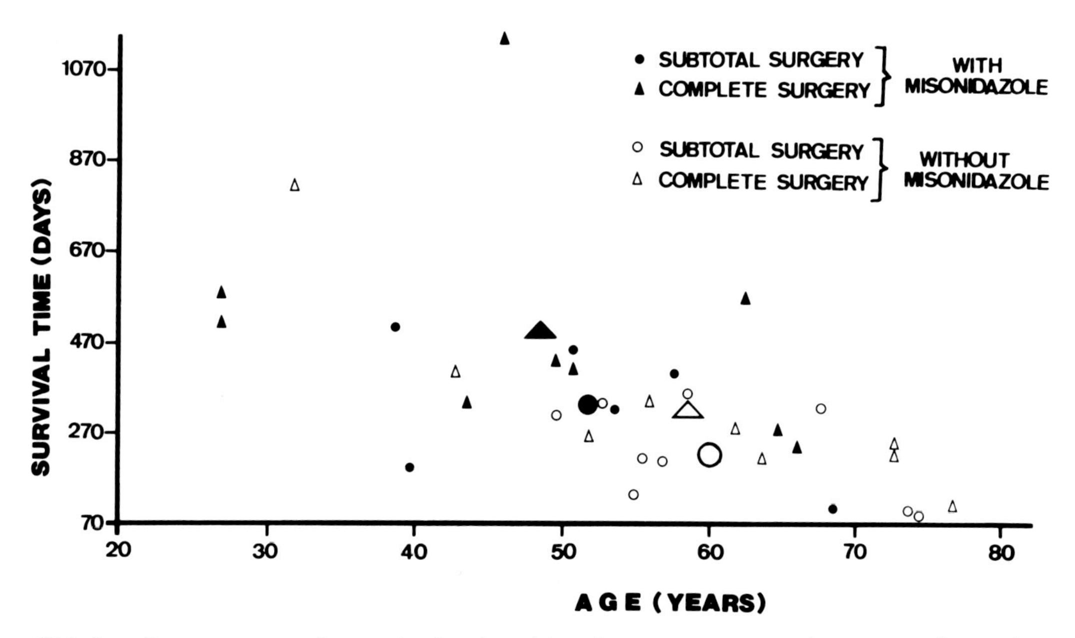

FIG.2. Scattergram of survival time (days) versus age at diagnosis (years) for all 33 patients available for analysis. The 4 larger symbols indicate the mean values for the different groups. A negative correlation between age and survival is obvious.

the Mantel-Cox test gives a significant difference in survival ($p < 0.01$) in favour of the younger group. An analysis of variance for age reveals a significant difference between patients treated with misonidazole (mean age 50.3 years) and those treated without sensitizer (mean age 59.9 years), with a mean age difference of 9.6 years ($p < 0.05$). It cannot therefore be ruled out at present that the significantly better survival for our patients treated with misonidazole is at least partly due to the differences in ages.

The rationale for our treatment schedule was to decrease the main hypoxic tumour cell mass rapidly within the first few treatment weeks, letting reoxygenation take place more effectively with the subsequent daily conventional fractionations. In the last treatment week misonidazole was given to help to destroy any remaining hypoxic tumour cells. By applying relatively high cumulative doses of misonidazole (12.6–16.4 g/m^2, mean 14.2 g/m^2) in an overall time of 7½ weeks, neurotoxic side effects were kept at a minimum and did not cause significant clinical problems [8].

The value of using hypoxic cell sensitizers in the treatment of malignant gliomas is not yet established. Favourable effects were reported by Urtasun et al. [13], who evaluated the possible enhancement of the radiation effect by metronidazole (a tumour dose of only 3000 rads was given in 9 fractions over 18 days, corresponding to 1288 rets). In the Cambridge glioma trial [14] no differences in survival could be observed with the addition of misonidazole (radiation dose 1702 rets, cumulative dose of misonidazole 12 g/m^2). More patients have to be admitted into our study before a firm conclusion about the efficacy of misonidazole can be drawn.

ACKNOWLEDGEMENTS

The authors are indebted to Hoffmann-La Roche Wien GmbH for contributing the misonidazole used in this study.

REFERENCES

[1] WALKER, M.D., ALEXANDER, E., Jr., HUNT, W.E., et al.: Evaluation of BCNU and/or radiotherapy in the treatment of anaplastic gliomas. A cooperative clinical trial, J. Neurosurg. 49 (1978) 333.

[2] JELLINGER, K., KOTHBAUER, P., VOLC, D., et al.: Combination chemotherapy (COMP Protocol) and radiotherapy of anaplastic supratentorial gliomas, Acta Neurochir. 51 (1979) 1.

[3] WALKER, M.D. GREEN, S.B., BYAR, D.P., et al.: Randomized comparisons of radiotherapy and nitrosoureas for the treatment of malignant glioma after surgery, New Engl. J. Med. 303 (1980) 1323.

[4] EORTC BRAIN TUMOR GROUP: Effect of CCNU on survival rate of objective remission and duration of free interval in patients with malignant brain glioma - Final evaluation, Eur. J. Cancer 14 (1978) 851.

[5] JENTZSCH, K., KÄRCHER, K.H., KOGELNIK, H.D., et al.: Initial clinical experience with the radiosensitizing nitroimidazole Ro 07-0582, Strahlentherapie 153 (1977) 825.

[6] KOGELNIK, H.D., MEYER, H.J., JENTZSCH, K., et al.: Hypoxic Cell Sensitizers in Radiobiology and Radiotherapy (Proc. 8th L. H. Gray Conf., Cambridge, UK, 1977), Br. J. Cancer 37 Suppl. III (1978) 281.

[7] SALAZAR, O.M., RUBIN, P., McDONALD, J.V., FELDSTEIN, M.L.: Patterns of failure in intracranial astrocytomas after irradiation: Analysis of dose and field factors, Am. J. Roentgenol. 126 (1976) 279.

[8] KOGELNIK, H.D.: Clinical experience with misonidazole - High dose fractions versus daily low doses, Cancer Clin. Trials 3 (1980) 179.

[9] ADAMS, G.E., DAWSON, K., STRATFORD, I.J.: "Electron-affinic radiation sensitizers for hypoxic cells: Prospects and limitations with present and future drugs", in Progress in Radio-Oncology (Proc. Int. Symp. Baden, Austria, 1978: KÄRCHER, K.-H., KOGELNIK, H.D., MEYER, H.-J., Eds.), Georg Thieme Verlag, Stuttgart, and Thieme-Stratton Inc., New York (1980) 84-95.

[10] SALAZAR, O.M., RUBIN, P., FELDSTEIN, M.L., PIZZUTIELLO, R.: High dose radiation therapy in the treatment of malignant gliomas: Final report, Int. J. Radiat. Oncol., Biol. Phys. 5 (1979) 1733.

[11] KRAMER, S.: Tumor extent as a determining factor in radiotherapy of glioblastomas, Acta Radiol., Ther., Phys., Biol. 8 (1969) 111.

[12] WALKER, M.D., STRIKE, T.A., SHELINE, G.E.: An analysis of dose-effect relationship in the radiotherapy of malignant gliomas, Int. J. Radiat. Oncol., Biol. Phys. 5 (1979) 1725.

[13] URTASUN, R., BAND, P., CHAPMAN, J.D., et al.: Radiation and high-dose metronidazole in supratentorial glioblastomas, New Engl. J. Med. 294 (1976) 1364.

[14] BLEEHEN, N.M.: The Cambridge glioma trial of misonidazole and radiation therapy with associated pharmacokinetic studies, Cancer Clin. Trials 3 (1980) 267.

Progress in Radio-Oncology II, edited by K. H. Kärcher et al., Raven Press, New York © 1982.

High-Dose Multiple Daily Fractionation Radiotherapy in Combination with Misonidazole as a Treatment of High-Grade Malignant Gliomas

A Pilot Study of the Radiotherapy Group of the EORTC

K.K. ANG, E. van der SCHUEREN
Academic Hospital St. Rafaël,
Leuven,
Belgium

G. NOTTER
Sahlgrenska Sjukhuset,
Göteborg,
Sweden

C. CHENAL
Hôpital Pitié la Salpétrière,
Paris,
France

J.C. HORIOT
Centre G.F. Leclerc,
Dijon,
France

J. RAPS
Middelheim Ziekenhuis,
Antwerp,
Belgium

H. van PEPERZEEL
Academic Hospital,
Utrecht,
The Netherlands

J.C. GOFFIN
Centre Hospitalier de Tivoli,
La Louvière,
Belgium

Abstract

Based on recent experimental data, a new fractionation schedule was designed for the treatment of malignant gliomas. Three irradiations, each 2 Gy at 4 h intervals, were given every day during 5 days. This scheme was repeated after a two-week rest period. A total dose of 60 Gy was thus delivered, in 30 fractions and in 10 treatment days. This fractionation schedule was combined with misonidazole in a daily dose of 1.2 g/m^2 during therapy. From March 1979 to December 1980, 97 patients were treated. The acute tolerance was very good. No major complications were observed. Only 13% of the patients experienced transient but serious discomfort during the treatment; 69% of patients had mild side effects of short duration, and 18% had no symptoms. The most frequent side effects were headache, nausea/vomiting, and somnolence. The present study has several very interesting aspects. This treatment scheme uses the same fraction size and total dose as often used in conventional treatment (60 Gy/30 fr/6 weeks) and modifies only the interval and overall time. These features make it possible to combine all irradiation sessions with high doses of anoxic cell sensitizer. It already clearly appears that the burden on the patient is reduced. Initial results are encouraging, but final assessment requires a longer follow-up. A randomized study testing this treatment modality is under way.

As the prognosis of patients with high grade malignant gliomas is very unfavourable, its treatment is an object of intensive investigation.

Conventional radiotherapy by itself or after tumor resection has been used for treating these patients for decades. The collected data from the literature (1,2,3,4,5,6,7,8,9) indicate that this treatment modality can slightly prolong the survival time of the patients, but fails to enhance the cure rate. In such a situation, the gain in the survival time has to be balanced against the burden of the treatment on the patients. One may thus conclude that the therapeutic ratio of conventional radiotherapy is rather low, for it requires 6 to 7 weeks treatment, inducing a certain discomfort to the patients and necessitating a prolonged hospitalisation, or daily visits, in order to obtain an average gain in total survival time of only about 20 weeks.

In trying to improve the therapeutic ratio, a modified time-dose-fractionation schedule was designed. Its main features are a drastic reduction in the total treatment time, without decreasing the total dose or altering the fraction size by using multiple daily fractionation (MDF), in combination with hypoxic cell sensitizer (Misonidazole).

The design of this schedule was based on the recent advances in the knowledge of the radiation tolerance of the central nervous system (CNS) and also in the cytokinetics of malignant gliomas. Experimental data suggest that the late tolerance of CNS to a fractionated radiotherapy is mainly determined by the fraction size and only minimally influenced by the overall treatment time for up to 6 weeks (10). Hence, most of the repair is carried out at the cellular level. Experimental data also suggest that this type of repair is essentially completed within a few hours. This means, using "normal" fraction size, the overall-treatment time could be reduced without decreasing the late tolerance significantly by administering several irradiations a day. However, the acute toxicity of this schedule was not yet known. If the acute side effects were reasonable, this treatment scheme clearly would reduce the burden on the patients as it only needs ten days to complete the radiotherapy.

Cytokinetic factors which have been suggested to be responsible for the resistance of malignant gliomas to ionizing radiation are the presence of a significant fraction of hypoxic cells and a very high growth rate (11,12,13). In this respect a concentrated schedule could have several advantages. It could result in an enhanced radiation effect on the tumor by reducing the repopulation of tumor cells. It is also well suited for combination with Misonidazole. Due to the neurotoxicity only 12 g/m^2 of Misonidazole can be administered. As the number of treatment days will be reduced a high dose of drug can be given each day, resulting in a high serum concentration and an optimal enhancement ratio. With the serum drug half-life of 12 h, a dose given 2 h before the first daily irradiation would also enhance the second and maybe to a lesser degree the third daily fraction which are given respectively 4 and 8 h after the first session.

The feasibility of this modified radiotherapy schedule was first evaluated at the academic hospital of Leuven. After a few months, 6 other centers of the radiotherapy group of the E.O.R.T.C. participated in this phase II trial.

MATERIAL AND METHODS

Between March 1979 and December 1980, ninety-seven patients have been accrued. Eightly percent of the patients were entered during 1980. Sixty-five patients were treated after tumor resection, 22 had an inoperable tumor and 10 had a tumor recurrence after previous surgery.

Radiation was delivered in two parallel opposing fields using photons from a Cobalt-60 source or linear accelerator. Two radiotherapy courses were given separated by a rest period, first of 4 weeks and later on down to 2 weeks. Each course consisted of 5 treatment days. The daily dose was 6 Gy delivered in three equal fractions, with 4 h interval between the irradiations. The total dose was 60 Gy of which 40 Gy was delivered to the whole brain and the remaining 20 Gy to the tumor bearing area with 2 cm margin (Fig. 1).

Misonidazole (Roche) was chosen as a hypoxic cell sensitizer. A dose of 1.2 g/m^2 was administered 2 h before the first daily irradiation during 10 treatment days resulting in a total dose of 12 g/m^2. The first twenty patients did not receive Misonidazole to test the tolerance of radiotherapy schedule.

During and after the treatment the patients were evaluated clinically to assess the acute side-effects of this therapeutic schedule. Special attention was given to signs of intracranial hypertension such as headache, nausea, vomiting, changes in the state of consciousness, deterioration of neurological status, and papilledema. The skin and mucosal (conjunctiva) reactions were also evaluated. The patients were carefully screened to detect symptoms or signs of peripheral neuropathy.

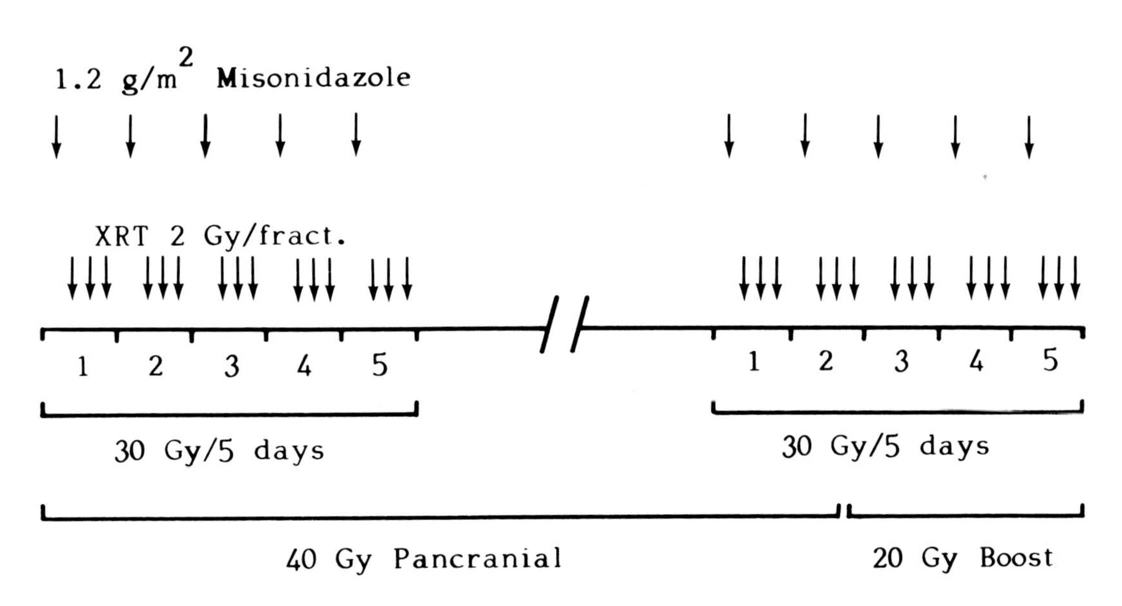

FIG.1. Time-dose-fractionation schedule.

RESULTS

No major complications were observed during or after the therapy. All patients could complete the treatment without alterations of the schedule. The type of acute side effects and the frequency at which they occurred are shown in Table I. Of the 97 patients, 67 (69 %) suffered from mild and 13 (13 %) from severe acute reactions. Seventeen patients (18 %) tolerated this therapy without any discomfort. The most frequently occurring symptoms were headache, nausea, vomiting and somnolence. Most of the patients who had mild side effects experienced 2 symptoms. The average number of symptoms per patient for the mild discomfort was 1.85. But of the 13 patients who had a severe reaction, 12 had only one symptom. In some centers steroids were given prophylactically, but in centers where the drug was only given symptomatically it has been found that less than half of the patients needed steroids. All symptoms lasted only a few days, and they regressed rapidly after treatment. The acute side effects were slightly more frequent and a little more pronounced in patients who were treated without previous tumor resection.

The skin tolerance to this therapeutic schedule was excellent. Forty-five patients (46 %) did show skin reactions ranging from mild erythema to dry desquamation. None had moist epidermolysis as sometimes occurs at the insertion of the ears at the end of conventional radiotherapy.

The initial results on disease-free survival are encouraging but longer follow-up is required for evaluation of the end results of this treatment schedule.

CONCLUSION

The feasibility of the high dose multiple daily fractionation schedule in combination with Misonidazole was tested in a phase II study

TABLE I. TYPE AND FREQUENCY OF THE SIDE EFFECTS TO MDF [a] SCHEDULE

	MDF		MDF + Miso	
	Mild	Severe	Mild	Severe
Nausea/vomiting	9	1	24	4
State of consciousness	3	0	20	3
Headache	9	2	37	1
Papilledema	4	0	10	1
Peripheral neuropathy	1	0	7	2
Total N° of patients	20		77	

[a] MDF: multiple daily fractionation.

and tolerance has been found to be excellent.

The concentrated radiotherapy already has been proven to have the advantage of a reduced burden on the patient. Whether the theoretical advantages of this schedule will be translated in better treatment results will require longer follow-up and a randomized study.

REFERENCES

[1] BOUCHARD, J., PEIRCE, C.B.: Radiation therapy in the management of neoplasms of the central nervous system, with a special note in regard to children, Am. J. Roentgenol. 84 (1960) 610-628.
[2] TAVERAS, J.M., THOMPSON, H.G., POOL, J.L.: Should we treat glioblastoma multiforme? A study of survival in 425 cases, Am. J. Roentgenol. 87 (1962) 473-479.
[3] UIHLEIN, A., COLBY, M.Y., LAYTON, D.D., PARSONS, W.R., CARTER, T.L.: Comparison of surgery and surgery plus irradiation in the treatment of supratentorial gliomas, Acta Radiol., Ther., Phys., Biol. 5 (1966) 67-78.
[4] TROUILLAS, P.: Immunologie et immunothérapie des tumeurs cérébrales, Etat Actuel Rev. Neurol. 128 (1973) 23-38.
[5] STAGE, W.S., STEIN, J.J.: Treatment of malignant astrocytomas, Am. J. Roentgenol. 120 (1974) 7-18.
[6] FIX, I.: An evaluation of the radiotherapeutic management of intracranial glial tumours without prior surgery, Am. J. Roentgenol. 96 (1966) 130-136.
[7] YUNG, W.K., STEWARD, W., MARLES, J.E., GRIEM, M.L., MULLAN, J.F.: Glioblastoma multiforme: treatment with radiation and triiodothyronine, Int. J. Radiat. Oncol., Biol. Phys. 1 (1976) 645-650.
[8] SALAZAR, O.M., RUBIN, P., McDONALD, J.V., FELDSTEIN, M.L.: High dose radiation therapy in the treatment of glioblastoma multiforme: a preliminary report, Int. J. Radiat. Oncol., Biol. Phys. 1 (1976) 717-727.
[9] WALKER, M.D., ALEXANDER, E., HUNT, W.E., MacCARTY, C.S., MAHALEY, M.S., MEALEY, J.,Jr., NORRELL, H.A., OWENS, G., RANSOHOFF, J., WILSON, C.B., GEHAN, E.A., STRIKE, T.A.: Evaluation of BCNU and/or radiotherapy in the treatment of anaplastic gliomas. A cooperative clinical trial, J. Neurosurg. 49 3 (1978) 333-343.
[10] VAN DER KOGEL, A.J.: Late effects of radiation on the spinal cord, Thesis, 1979.
[11] HOSHINO, T., BARKER, M., WILSON, C.B., BOLDREY, E.B., FEWER, D.: Cell kinetics of human gliomas, J. Neurosurg. 37 (1972) 15-26.
[12] HOSHINO, T., WILSON, C.B., ROSENBLUM, M.L., BARKER, M.: Chemotherapeutic implications of growth fractions and cell cycle time in glioblastomas, J. Neurosurg. 43 (1975) 127-135.
[13] HOSHINO, T., WILSON, C.B.: Cell kinetic analysis of human malignant brain tumours (gliomas), Cancer 44 (1979) 956-962.

Progress in Radio-Oncology II, edited by K. H. Kärcher et al., Raven Press, New York © 1982.

Radiation and Misonidazole in Children with Brain Stem Gliomas and Supratentorial Glioblastoma

A pilot study

H.J.G. BLOOM, R.D. BUGDEN
Royal Marsden Hospital and
Institute of Cancer Research,
London,
United Kingdom

Abstract

In a series of 484 children with intracranial tumours referred to the Royal Marsden Hospital for radiotherapy, there were 57 (12%) examples of inoperable pontine and medullary tumours for which the 5-year survival rate was 17%. This was in sharp contrast to the prognosis for certain other types of cerebral glioma in this series where the 5-year survival rate ranged from 30 to 70%. The limited local tumour mass in brain stem tumours, the absence of cerebro-spinal or distant metastases, and their often initial good but short-lived response to irradiation, all support the trial of a chemical radiosensitizing agent with which to try and achieve greater and more prolonged local control of the disease. Since the prognosis for cerebral hemisphere glioblastoma, which is relatively uncommon in children, is also extremely poor, such cases were included in this pilot study. The problems and possible risks associated with combined radiotherapy and a chemical radiosensitizer in children with brain tumours is discussed. Attention is given to irradiation fractionation, to factors relating to misonidazole administration, to possible sensitization of normal CNS tissue and to the potential risks of irradiation combined with a neurotoxic agent on the developing brain. So far, 8 children with brain stem tumours and 3 children with cerebral hemisphere gliomas have been treated in this study. In addition, data is also available on 3 children re-treated for recurrent medulloblastomas. Preliminary observations regarding experience with this small series will be reported including blood misonidazole levels, drug tolerance and the possible influence of anticonvulsants and steroids on toxicity.

1. INTRODUCTION

Gliomas of the brain stem (pons and medulla) are found most commonly in children. They are inoperable and are generally regarded as untreatable and inevitably fatal. Radiotherapy, in fact, produces clinical improvement in up to 75% of patients [1,2]. Unfortunately, this promising initial response is not maintained and, in the great majority of patients, there is early recurrence - with death within two years. Five-year survival rates following radical radiotherapy range between 10 and 20% [3-5]. In a UK national series of 144 children with pontine tumours, the overall 5-year survival rate was 6% [6]. Of 57 children with brain stem gliomas, representing 12% of a series of 484 children with intracranial tumours referred to the Royal Marsden Hospital for radiotherapy, 17% were still alive at 5 years, and this proportion was maintained to 20 years. These poor results are in marked contrast to our experience with radiotherapy given alone or post-operatively for certain other types of cerebral glioma in children, for which 5-year survival rates range from 30 to 70%, depending upon tumour type and site [7].

The inoperability of brain stem gliomas, their relatively small tumour burden, the absence of CNS and distant metastases, and their initial but short-lived response to irradiation represent a special challenge to the radiotherapist and suggest the use of a radio-sensitizing agent to try and achieve greater local control of the disease.

Misonidazole diffuses freely across the blood/brain barrier, with peak CSF levels of the drug in man being attained 4 to 5 hours after oral administration [8] and reaching up to 80% or more of maximum blood concentrations [8,9]. Misonidazole levels in normal brain, in glioma tissue and in tumour cyst fluid reach 50 to 70% of blood levels, and are comparable to those found in non-neurogenic tissues.

2. PROJECT AND PATIENTS

In a phase 2 study, we hope to assess the feasibility and possible value of using misonidazole in combination with radiotherapy for children aged between 3 and 15 years who have clinical and radiological evidence of inoperable brain stem tumours. We also included children with histologically verified high-grade astrocytomas (glioblastoma) of the cerebral hemispheres in whom surgical decompression or biopsy had been carried out and from which they had made a satisfactory post-operative recovery. Thus, the cases selected for this study were suffering from the most unfavourable types of cerebral glioma. We were concerned with tolerance to the drug, the possibility of early and late side-effects, and survival, compared with historical controls.

During the past 24 months, 11 consecutive cases aged between 5 and 13 have entered the study, 8 with brain stem tumours and 3 with high-grade astrocytomas of the cerebral hemispheres. Observations concerning 3 additional children with recurrent medulloblastoma during the same period were also available.

3. METHODS

3.1. Radiotherapy

We adhered to our routine radiotherapy schedule of giving a tumour dose of 5000 to 5500 rads in daily fractions, 5 days per week over 6 to 7 weeks. The volume irradiated included all known, suspected and potential tumour spread, with due consideration of possible involvement of the upper cervical cord in brain stem lesions and of the controlateral hemisphere in supratentorial glioblastoma.

3.2. Misonidazole Treatment

The prescribed total cumulative dose of misonidazole in all cases was 12 g/m^2 body surface area, given as single doses of

600 mg/m^2 daily (Monday to Friday) for 20 doses during the last 4 weeks of radiotherapy. The capsule contents were mixed with jam or honey in a teaspoon and taken after a light breakfast, 4 to 5 hours before radiotherapy.

3.3. Drug Assay

Blood samples were taken 4 hours after the administration of misonidazole (just prior to radiotherapy) on the first day of treatment with sensitizer, and on each successive Monday and Friday. In patients with drug accumulation in the blood the misonidazole dose was reduced and additional samples were analysed. Misonidazole levels were determined by polarography [10], which measures total nitroimidazoles including the parent compound and its o-demethylated metabolite (desmethylmisonidazole), itself a radiosensitizer.

3.4. Other Drugs

Of the 11 children with brain stem and cerebral hemisphere tumours, 10 received dexamethasone during treatment, generally in a step-wise reducing dosage. The total mean dose was 121.5 mg with a range of 19-319 mg. The 3 patients with cerebral hemisphere tumours also received phenytoin, 150-200 mg daily.

3.5. Clinical Assessment

General and neurological examinations were carried out weekly during treatment and thereafter, if feasible, at 2 to 4 week intervals until the patient's condition became stable, after which the assessments were made less frequently.

4. RESULTS

All patients completed the planned course of radiotherapy and all but one received the full calculated dose of misonidazole: in this patient the drug was reduced from 800→600 mg daily because the blood concentration reached 49 µg/ml. Blood levels of misonidazole in the 11 primary tumour cases are shown in Fig.1 and Table I: the range generally lies between 20 and 35 µg/ml. The mean level for all cases was 25.6 µg/ml (10.4-49.0), for 5 males 25.9 µg/ml (10.4-35.7), for 6 females 25.4 µg/ml (10.7-49.0); for 7 receiving additionally dexamethasone alone the mean level was 25.9 µg/ml (10.4-49.0), and for 3 receiving both dexamethasone and phenytoin 22.9 µg/ml (10.7-34.0). In the only patient receiving neither steroids nor anti-convulsants the mean level was 32.2 µg/ml (27-38). From data in this small series, it is apparent that sensitizer levels at 4-5 hours were not influenced by sex, dexamethasone or phenytoin.

4.1. Misonidazole Toxicity

Apart from occasional nausea and some decrease in appetite, misonidazole was remarkably well tolerated. The occasional minor

TABLE I. RADIOTHERAPY AND MISONIDAZOLE IN CHILDREN WITH BRAIN STEM TUMOURS AND CEREBRAL GLIOBLASTOMA

Patient; Sex	Age	Diagnosis[a]	Phenytoin	Total dexamethazone (mg)	Radiotherapy dose (rads)	Radiotherapy time (weeks)	Misonidazole daily (mg)	Misonidazole total (g)
PS(F)	7	BS	–	63	5000	6.5	600	12
DR(M)	11	BS	–	–	5000	6.5	700	14
CV(F)	11	BS	–	103	5000	7	800 [b]	13.4
DZ(M)	11	BS	–	178	5500	8	600	12
SS(F)	8	BS	–	28	5000	6.5	600	12
SE(M)	13	BS	–	39	5000	7	800	16
DR(M)	5	BS	–	35	5000	7	400	8
LH(F)	12	BS	–	115	5500	7	800	15.2
VL(F)	6	CG	yes	316	4850	7	400	7.6
LE(M)	6	CG	yes	19	5000	7	500	10
SP(F) [c]	12	CG	yes	319	4000	7.5	800	16

[a] BS brain stem tumour; CG cerebral hemisphere glioblastoma.

[b] Reduced to 600 mg.

[c] Previously treated for leukaemia.

gastro-intestinal disturbance was not unexpected in hospitalized children with intracranial tumours undergoing radiotherapy. Vomiting occurred in one child and was associated only with the first few doses of misonidazole. Evidence of peripheral neuropathy was rare. Only one child complained of paraesthesiae of the fingers over a 4-week period following completion of radiotherapy: this patient was the only one not receiving phenytoin or dexamethasone. In one or two patients tendon reflexes were decreased, but again this may have been due to the disease rather than to the sensitizer. There were no examples of grand mal, and only one child with a cerebral hemisphere tumour had petit mal - which was present before treatment. There were no complaints of tinnitus or of impaired hearing. In one patient with a brain stem glioma, sudden and intense irritation of the eyes together with photophobia occurred 6 hours after the first dose of 600 mg of misonidazole: it lasted a few hours, but a dislike for bright light persisted for 3 to 4 days. There were no other attacks with subsequent doses of misonidazole and no other toxic manifestations were seen: subsequent ocular sequelae were not observed during the remaining 10 months of this child's life. A maculopapular rash was observed in one patient which was almost certainly due to phenytoin and which disappeared after discontinuing the anti-convulsant.

It was appreciated from the start that misonidazole toxicity in children with intracranial tumours could prove difficult to ascertain because of changes produced by the disease itself, and because of the obvious problems in eliciting and interpreting tests

TABLE I (continued)

Toxicity gastric	Toxicity neuro.	Clinical improvement	Alive /dead; survival time (months)	Blood misonidazole levels mean (mg/ml)	Blood misonidazole levels range (mg/ml)
S1	0	+	D(11)	30.5	(26.1-38)
0	S1	+	A(18)	32.2	(27-38)
0	0	+	D(15)	33.8	(19-49)
0	0	0	?	33.0	(25.5-45.8)
0	0	+	D(8)	21.0	(17-23.9)
0	0	0	D(3)	22.2	(19.8-27.2)
0	0	+	A(7)	18.5	(10.4-23.4)
0	0	0	A(6)	22.6	(16.4-35.7)
0	0	0	D(13)	21.4	(20-24)
0	0	+	A(13)	23.4	(19.9-23.8)
0	0	+	A(3)	22.9	(10.7-34)

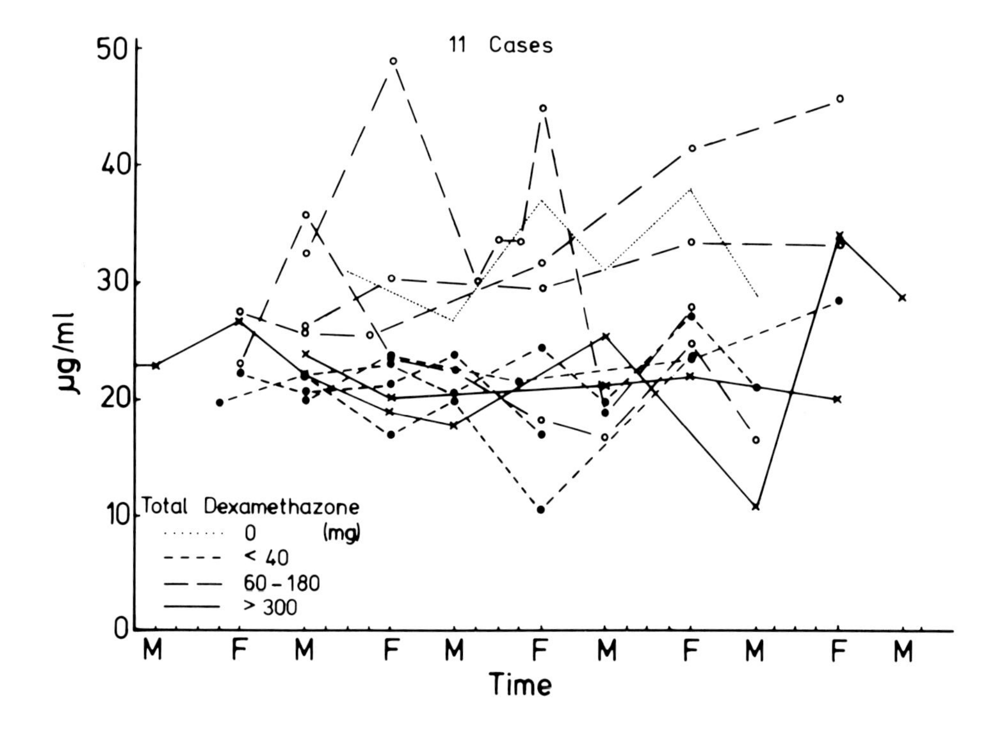

FIG.1. Blood misonidazole concentrations on Mondays and Fridays in 11 children receiving 600 mg/m^2 orally daily, 5 days/week during the last 4 weeks of radiotherapy, according to total dose of dexamethasone administered during this time.

of neurological function in young children. There were no cases, however, where misonidazole had to be discontinued on account of possible toxicity. In one child the dose of misonidazole was reduced from 800→600 mg daily because of a rather high end-of-week blood level. Misonidazole with radiotherapy was well tolerated and in several children general and neurological improvement was demonstrated during or after completion of treatment.

4.2. Survival

Of the eight children with brain stem tumours, four died at times ranging from 2.5 to 15.0 months after treatment and three are still alive at 6 to 18 months: follow-up on one is still awaited. Of the three children with cerebral glioblastoma, one is dead after 13 months and 2 are still alive at two and 13 months, respectively. The three children with recurrent medulloblastoma have only recently been treated palliatively and all had a good response.

5. DISCUSSION

In no other site of the body is it so important to consider the late effects of treatment than on the developing central nervous system of young children. Furthermore, the brain stem is believed to be one of the more vulnerable areas of the CNS to irradiation. We have, therefore, several problems to consider. First, the child's brain may be especially sensitive to large infrequent fractions of radiotherapy per se. Second, although the radio-sensitizing and cytotoxic properties of misonidazole are directed against hypoxic cells, it is possible that certain areas of the normal brain have a reduced oxygen concentration and may be rendered more susceptible to treatment. Third, a neurotoxic agent in combination with irradiation may represent an increased risk to normal nervous tissue especially in children.

Recent reports describing degenerative changes in the cerebellum and brain stem nuclei in experimental animals treated with nitroimidazole derivatives are disquieting [11-13]. The most neurotoxic nitroimidazole for the CNS seems to be Ro-07-0741, which produces haemorrhagic necrosis and demyelination in the brain stem of treated mice, changes which are associated with profound disturbance of posture, balance and locomotion.

Several observations concerning the radiosensitization of normal tissues in experimental animals have been reported in the literature (reviewed by Denekamp [14]). However, these studies involve large single doses of irradiation, and no examples of increased sensitization of normal tissue in man using fractionated radiotherapy with misonidazole have been published. Nevertheless, the CNS may be a special case and clearly one must proceed with caution, especially in young children, and be aware that any serious sequelae resulting from the use of nitroimidazoles, alone or in combination with irradiation, may not become evident for some years. However, with the present poor results for brain stem tumours following conventional treatment some risk with a new approach would seem to be justified.

As a result of these considerations we rejected hypofractionation regimens, and decided to adhere to our conventional megavoltage dose/time factors that have been used over many years in children with cerebral gliomas and found to be relatively safe. By adopting standard radiotherapy schedules and administering misonidazole with each daily fraction of irradiation, the effect of the sensitizer may be greatly reduced for the following possible reasons:

(i) the amount of drug given with each fraction will be relatively low;

(ii) the chemical radiosensitizer may be less effective when given with small rather than with large doses of irradiation;

(iii) progressive re-oxygenation of tumour tissue during irradiation leads to a decreasing need for hypoxic cell sensitizer.

There is evidence from laboratory studies that misonidazole can enhance the effect of small as well as large fractions of irradiation [15] and that 30-fraction irradiation regimens, in which a small dose of misonidazole is given with each fraction, will achieve greater cell kill than few irradiation fractions associated with larger doses of radiosensitizer [16]. In these circumstances sensitizers are most likely to be of benefit in tumours where reoxygenation is incomplete: gliomas are likely to be in this class. In the present clinical study the administration of misonidazole with daily fractions of radiotherapy was associated with blood levels of 20-35 μg/ml, concentrations which are expected to produce enhancement ratios of at least 1.3 [17,18]. Daily administration of misonidazole may also be advantageous in producing more prolonged tissue exposure for greater cytotoxicity.

The decision to use misonidazole during 4 weeks of the 6 to 7 week course of radiotherapy presented us with the option of giving the drug during the first or the last part of treatment. The degree, speed and efficiency of tumour revascularization during radiotherapy in patients with cerebral gliomas is likely to be slow and incomplete. We chose to give misonidazole during the last 4 weeks of treatment largely because of the possible onset of cerebral oedema during the initial stages of radiotherapy which, in the case of the brain stem, is particularly dangerous even without radiosensitizers.

To avoid substantial toxicity the total dose of misonidazole should not exceed 12 g/m^2 body surface area [19]. Nevertheless, even with such doses, the incidence of peripheral neuropathy may reach 20 to 30% [19,20]; the paucity of acute toxic effects in the present series is striking. Admittedly, the assessment of sensory neuropathy in young children, already with serious neurological disturbance, can be difficult, but even in older children (11-13 years) there were no significant changes. In fact, abnormal neurological signs relating to the disease often improved as radiotherapy with misonidazole progressed. An unexpectedly low incidence of misonidazole neurotoxicity has also recently been observed in adult patients treated for malignant gliomas [20-22].

There is evidence for a metabolic interaction between misonidazole and phenytoin in man [23,24] resulting in a significant reduction in the plasma half-life of the sensitizer, but no change in peak and 4-hour levels. It is believed that phenytoin, through hepatic microsomal enzyme induction, is responsible for increasing the clearance of the drug through oxidative demethylation. On the other hand, Walker et al. [22] attributed an unusually low incidence of neuropathy in adults with brain tumours treated with misonidazole during radiotherapy to the administration of corticosteroids. In collected clinical experience from two centres [20], adult patients with brain tumours receiving dexamethasone alone showed no evidence of peripheral or central neurotoxicity by misonidazole, whereas in patients receiving neither dexamethasone nor phenytoin the incidence of these two complications reached 83% and 11%, respectively.

It appears that both phenytoin and dexamethasone can each protect against misonidazole toxicity. In the present study phenytoin was a possible factor in only three patients. All but one of the 11 children with primary tumours had dexamethasone: the only example of neurotoxicity (paraesthesiae of fingers for 4 weeks) occurred in the solitary case receiving neither phenytoin nor dexamethasone.

The mechanism of possible protection from misonidazole toxicity in patients receiving corticosteroids would appear to differ from that associated with phenytoin administration. It seems that phenytoin can modify the rate of clearance of misonidazole from the blood without effecting peak blood levels [23,24]. In this way phenytoin protects against toxicity without interfering with radiosensitization. However, it reduces cytotoxicity, which is dependent more upon duration of tissue exposure than on peak drug levels. Whereas the increased rate of clearance of misonidazole by phenytoin is not thought to be a disadvantage for radiosensitization, the situation in relation to corticosteroids appears to be more complex; interference with both radiosensitization and tumour cytotoxicity leading to a reduction in therapeutic gain cannot be excluded.

So far, reported early experience with radiotherapy and misonidazole for brain gliomas in adults has not been encouraging [9,21,25], and recent studies in laboratory models have also been disappointing [26,27]. It is too early to consider treatment results in this study but, so far, the outlook for the few children with brain stem tumours treated by radiotherapy does not appear to be significantly altered by the use of misonidazole.

6. CONCLUSION

Misonidazole is well tolerated in children receiving radiotherapy for brain stem and cerebral hemisphere tumours. The interaction between nitroimidazole derivatives and other drugs used in patients with brain tumours, particularly corticosteroids, may reduce toxicity but may also interfere with radiosensitization and cytotoxicity and warrants further investigation. We hope to examine this problem further in a xenograft model consisting of human cerebral gliomas implanted subcutaneously and also intracerebrally in immune-deprived mice [28]. Although we do not have observations on possible late effects, especially on the normal developing central nervous system in long-term survivors following combined radiotherapy and radiosensitizer, the prognosis for brain stem gliomas and high-grade cerebral astrocytomas in children is so poor as to justify continued study of this treatment in such cases.

REFERENCES

[1] GREENBERGER, J.S., CASSADY, J.R., LEVENE, M.B.: Radiology 122 (1977) 463.
[2] SHELINE, G.E.: Cancer 39 (1977) 873.
[3] BLOOM, H.J.G., WALSH, L., in Cancer in Children - Clinical Management (BLOOM, H.J.G., LEMERLE, J., NEIDHARDT, M.K., VOUTE, P.A., Eds.), Springer-Verlag, Berlin (1975) 93.
[4] SCHWEISGUTH, O.: Tumeurs Solides de l'Enfant, Flammarion, Paris (1979) 191.
[5] KIM, T.H., CHIN, H.W., POLLAN, S., HAZEL, J.H., WEBSTER, J.H.: Int. J. Radiat. Oncol., Biol. Phys. 6 (1980) 51.
[6] STEWART, A.M., LENNOX, E.L., SANDERS, B.M.: Br. J. Cancer 28 (1973) 568.
[7] BLOOM, H.J.G.: Acta Neurochirurgica 50 (1979) 103.
[8] ASH, D.V., SMITH, M.R., BUGDEN, R.D.: Br. J. Cancer 39 (1979) 503.
[9] TAMULEVICIUS, P., BAMBERG, M., SCHERER, E., STREFFER, C.: Br. J. Radiol. 54 (1981) 318.
[10] BUGDEN, R.D.: Anal. Proc. 17 (1980) 283.
[11] GRIFFIN, J.W., PRICE, D.L., KNETHE, O.D., GOLDBERG, A.M.: Neurotoxicology 1 (1979) 299.
[12] ROSE, G.P., DEWAR, A.J., STRATFORD, I.J.: Br. J. Cancer 42 (1980) 890.
[13] CONROY, P.J., SHAW, A.B., McNEILL, T.H., PASSALACQUA, W., SUTHERLAND, R.M.: in Radiation Sensitizers: Their Use in the Clinical Management of Cancer (BRADY, L.W., Ed.), Masson, Paris (1980) 397.
[14] DENEKAMP, J.: Cancer Clin. Trials 3 (1980) 139.
[15] VAN PUTTEN, L.M., SMINK, T.: Br. J. Cancer 37 Suppl.III (1978) 246.
[16] DENEKAMP, J., McNALLY, N.J., FOWLER, J.F., JOINER, M.C.: Br. J. Radiol. 53 (1980) 981.
[17] KOGELNIK, H.D.: Cancer Clin. Trials 3 (1980) 179.
[18] ADAMS, G.E. (Institute of Cancer Research, London): personal communication, 1981.
[19] DISCHE, S., SAUNDERS, M.I., FLOCKHART, I.R., LEE, M.E., ANDERSON, P.: Int. J. Radiat. Oncol., Biol. Phys. 5 (1979) 851.
[20] WASSERMAN, T.H., STETZ, J., PHILLIPS, T.L.: Cancer Clin. Trials 4 (1981) 7.
[21] CARABELL, S.C., BRUNO, L.A., WEINSTEIN, A.S., RICHTER, M.P., CHANG, C.H., WEILER, C.B., GOODMAN, R.L.: Int. J. Radiat. Oncol., Biol. Phys. 7 (1981) 71.
[22] WALKER, M.D., STRIKE, T.A.: Cancer Clin. Trials 3 (1980) 105.
[23] WORKMAN, P., BLEEHAN, N.M., WILTSHIRE, C.R.: Br. J. Cancer 41 (1980) 302.
[24] GANGJI, D., SCHWADE, J.G., STRONG, J.M.: Cancer Treat. Rep. 64 (1980) 155.
[25] BLEEHAN, N.M.: Cancer Clin. Trials 3 (1980) 267.
[26] REDDY, E.K., KIMLER, B.F., HENDERSON, S.D., MORANTZ, R.A.: in Radiation Sensitizers: Their Use in the Clinical Management of Cancer, (BRADY, L.W., Ed.), Masson, Paris (1980) 457.
[27] CLENDENDON, N.R., ALLEN, N., KOBAYASHI, N., GORDON, W.A., KARTHA, M., KO, L.-W.: in Radiation Sensitizers: Their Use in the Clinical Management of Cancer (BRADY, L.W., Ed.), Masson, Paris (1980) 501.
[28] BRADLEY, N.J., BLOOM, H.J.G., DAVIES, A.J.S., SWIFT, S.M.: Br. J. Cancer 38 (1978) 263.

Progress in Radio-Oncology II, edited by K. H. Kärcher et al., Raven Press, New York © 1982.

Enhancement of the Radiation Response of Chinese Hamster Cells by Pre-Treatment with Two Platinum Complexes*

A.H.W. NIAS, Margaret LAVERICK,
Richard Dimbleby Department of Cancer Research,
St. Thomas's Hospital Medical School,
London,
United Kingdom

Irena SZUMIEL
Department of Radiobiology and Health Protection,
Institute of Nuclear Research,
Warsaw,
Poland

Abstract

Platinum co-ordination complexes are cytotoxic drugs with a mode of action analogous to alkylating agents. The original drug Cis-PDD is used in the chemotherapy of ovarian and testicular tumours, often as a single agent, but nephrotoxicity is dose limiting. A number of second generation platinum complexes have been synthesized. PAD has a very high therapeutic ratio in laboratory tumour tests but is very insoluble. CHIP has a lower therapeutic ratio but is very soluble and shows minimal nephrotoxicity in animal tests. Combinations of X-irradiation and either PAD or CHIP have been tested in vitro using CHO cell cultures. Enhancement ratios have been determined for the radiation dose/response curves of aerated cells pre-treated with different doses of drug at different time intervals. Isobolographic analysis of the data has been used to indicate the most synergistic regime. Enhancement of radiation response is also found when hypoxic cells are pre-treated with a low dose of CHIP. This produces a minimal effect on aerated cells treated with the same time/dose regime. A therapeutic strategy is therefore suggested in which low dose CHIP is used with radiotherapy to enhance the response of hypoxic cells.

INTRODUCTION

One of the many forms of combined modality treatment of malignant disease involves radiotherapy localized to the main tumour cell mass and chemotherapy to metastatic but microscopic deposits. When radiotherapy and chemotherapy are combined it is essential to know in detail the extent and manner of any interaction of ionizing radiation with the chemotherapy agent in question. This requires a fundamental study of the mode of action of each new drug and of radiation, first administered separately and then in those combinations of dose and time interval which the separate data indicate to be most likely to show interaction. The evidence for potentiation or synergism between drug and radiation requires an effect upon the biological test system which is more than

* This work has been supported in part by the International Atomic Energy Agency, the United States National Cancer Institute and the United Kingdom Cancer Research Campaign.

additive; otherwise the combination can do more harm than good in terms of general toxicity without enhancing the tumour response. This type of assay of drug / radiation interaction has been undertaken on Chinese hamster (ovary) - CHO - cells in monolayer culture *in vitro* using two different platinum co-ordination complexes.

2. PLATINUM DRUGS

The original platinum drug is Cis-diammine dichloride Platinum II - PDD. This is used as a single agent and in combination chemotherapy of ovarian, testicular and other solid tumours; but nephrotoxicity is dose limiting. A number of second generation Platinum co-ordination complexes have been synthesized.

2.1 PAD

One of these is PAD which has a therapeutic index of 235 compared with a value of 8 for PDD. (1) The structural formula of PAD is shown in Figure 1. The structure consists of cis-chlorides, planar platinum (II) and two dative co-ordinate bonds whereby the lone pairs on the primary amines are donated into the metal orbitals. Unfortunately this drug is quite insoluble and it was necessary to dissolve it in DMSO before adding to the cultured cells. Treatment with drug alone causes: growth inhibition due to an impaired G_1 - S transition and formation of a non-cycling compartment; delayed cell death (with a maximum at 72 hours after treatment); decrease of ^{14}C-TdR incorporation into DNA; non-lethal damage; no sparing effect of dose-fractionation and a minimal degree of cycle phase specificity (2).

2.2 CHIP

Because of the insolubility of PAD we turned our attention to another second generation Platinum co-ordination complex, CHIP, whose formula is shown in Figure 2. This is a Platinum IV compound which is very soluble (about eight times as soluble as PDD) and with the slightly higher therapeutic index of 12. The increased solubility of CHIP offered the hope of reduced kidney toxicity and this has been confirmed in our own test mice where gastro-intestinal damage is the dose-limiting factor. We have shown that the mode of action of CHIP is similar to that seen with bifunctional alkylating agents, as evidenced by the pattern of chromosome aberrations (3) which would place the drug into Bender's

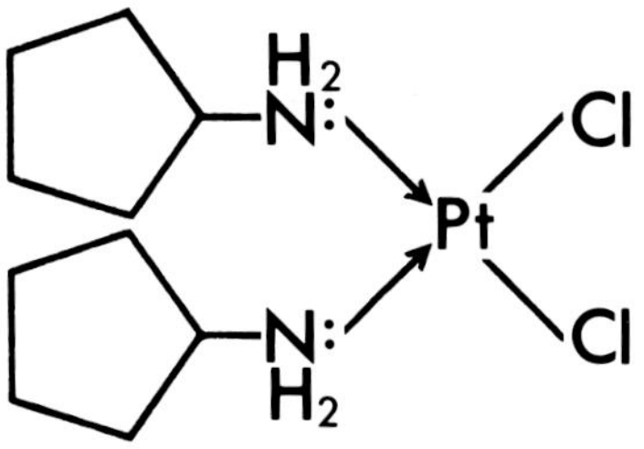

cis - dichloro-bis (cyclopentylamine) platinum (II)

FIG.1. Structure of PAD.

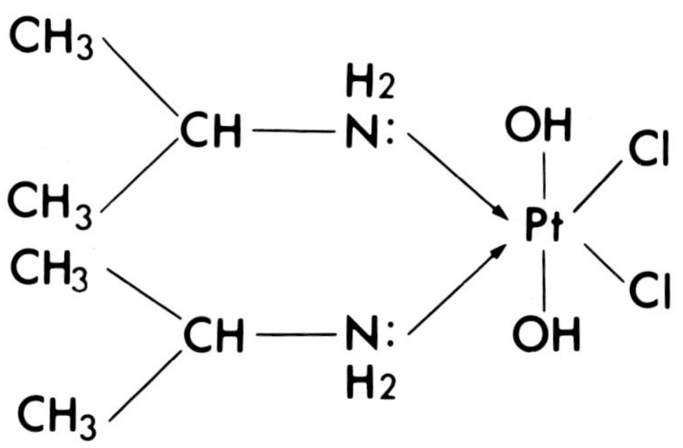

cis - dichloro-bis (isopropylamine) trans dihydroxy platinum (IV)

FIG.2. Structure of CHIP.

third class of chemicals, producing lesions which can be repaired by recombinational or post-replication repair processes (4).

2.3 CYTOTOXICITY

A comparison of the cytotoxicity of PAD and CHIP is shown in Figure 3. CHO cells were exposed to increasing concentrations of the drugs for a period of one hour at 37°C. The culture medium was then changed and the cells grown for five days before fixation, staining and colony counting (5). Roughly twice the dosage level of CHIP is required to reduce cell survival to the same level as that produced by PAD. The main difference is clearly the larger shoulder region in the curve for CHIP than that for PAD. This suggests a threshold range of drug dosage, although the subsequent cell survival curve is exponential and there is no evidence that this is a poison curve. (The parameters of these two curves, for PAD and CHIP respectively, are:- D_0 = 14 and 16 µg/ml, D_q = 28 and 88 µg/ml, n = 7 and 300).

3. COMBINATION REGIMES

3.1 PAD AND IRRADIATION IN AIR

The most effective enhancement of radiation response by PAD followed a drug dosage level high enough to reduce cell survival towards the exponential portion of the dose-response curve and given a short time interval before irradiation (6). This is shown in Figure 4 which includes a comparison of the dose-response of CHO cells treated with PAD in two different formulations of medium F12. Medium A had a lower chloride ion concentration than medium B and the half-life of the drug was 160 and 45 minutes respectively (2). A larger dosage of PAD was required in medium B than in medium A to bring about a significant dose modification of the radiation response (indicated at the top of Figure 4). In medium A a dose of 26 µg/ml of PAD reduced survival to 39% (which is on the exponential part of that curve) and it clearly enhanced the effect of radiation. However, the same dose of PAD in medium B only reduced survival to 55% (which is still on the shoulder of that curve) and there was no enhancement of the effect of radiation. Not until the much larger dose of 45µg/ml was used to reduce survival to 26% (now on the exponential part of the curve) was there any dose-modification (by a factor of 1.34).

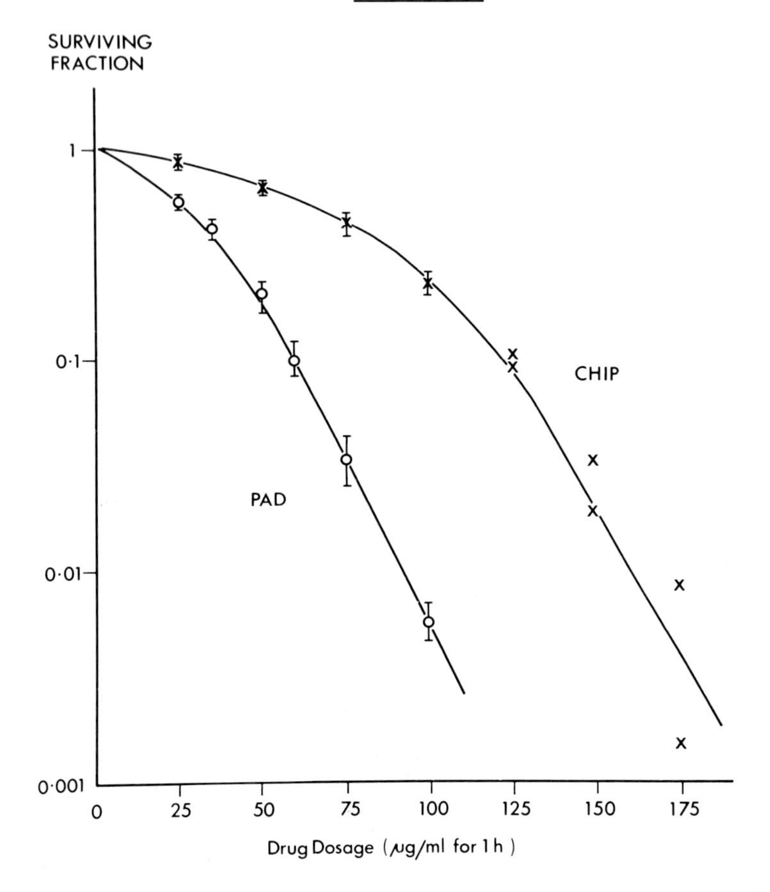

FIG.3. Dose-response curves for cells treated with PAD and CHIP (5).

3.2 CHIP AND IRRADIATION IN AIR

Since twice the dosage level of CHIP is required to reduce cell survival to the same level as that produced by PAD, a CHIP dose of 100 µg/ml was used.(5) The two survival curves in Figure 5 show the radiation dose-response of cells irradiated with or without a one-hour drug treatment given one hour beforehand. The drug pre-treatment data have been normalized. Clearly the radiation survival curve of the drug-treated cells does not fit that of the cells treated by radiation alone. After normalization the combination is not just additive, there is enhancement by a factor of 1.64. (The parameters of these two curves, for radiation alone and CHIP plus radiation respectively are:- D_0 = 230 and 140 cGy, n = 1.8). There is a simple dose-modifying effect of CHIP, just as there was for PAD, but the effect is greater after CHIP than after an equivalent pre-treatment dose of PAD.

3.3. ISOBOLOGRAPHIC ANALYSIS OF PLATINUM AND RADIATION COMBINATIONS

While difference in the slopes of dose-response curves may be taken as evidence of enhancement of radiation action by the two platinum complexes, the

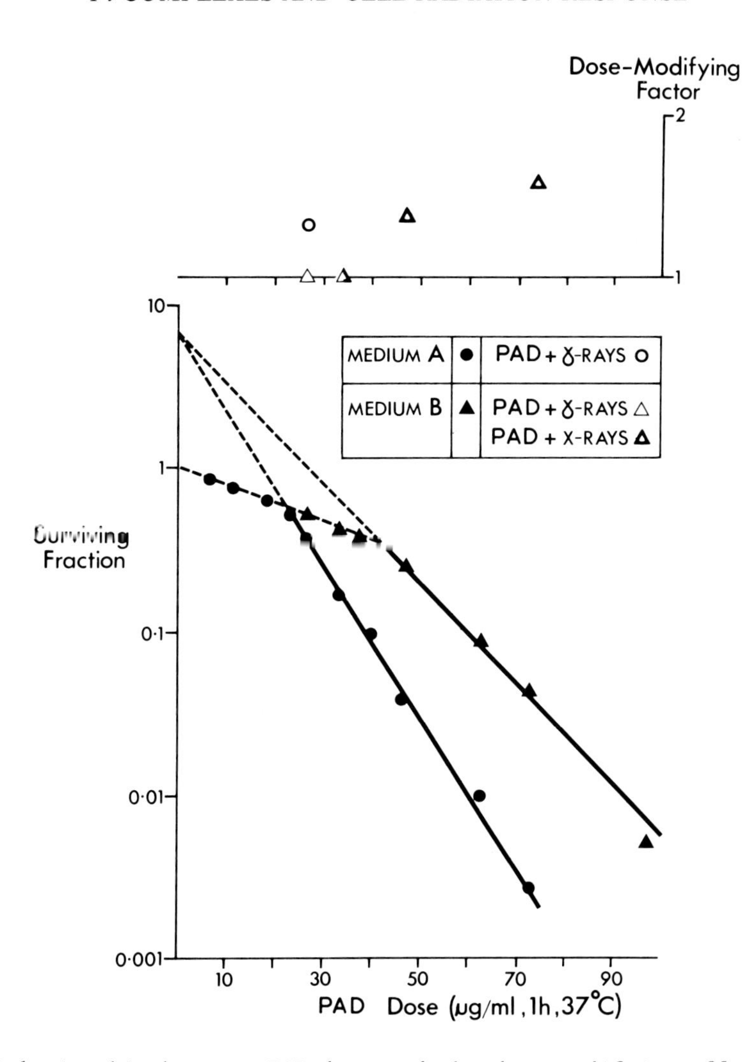

FIG.4. Relationship between PAD dose and the dose-modifying effect of combined PAD and radiation treatment in two kinds of culture medium (6).

value of enhancement ratio may be an over-simplification of the information that is available. When full dose-response curves have first been obtained for each modality of treatment and then for combined treatments the data can be analysed in more detail by the construction of isobolograms (7). "Envelopes of additivity" are derived by plotting the doses of each cytotoxic agent that give log survival values that add up to a given level of survival such as the 1 or 2 log decrements shown in Figure 6. For the upper, dotted, line (Mode I) the dose increments are plotted starting from zero and including the shoulder portions of the survival curves. The lower line (Mode II) is derived from dose increments taken from the exponential portions of the curves. Interaction between drug and radiation is assumed for Mode II but not for Mode I.

The data from combined treatments are then plotted on the isobolograms and any point lying to the right of the upper line indicates a less than additive effect while points lying to the left of the lower line indicate a

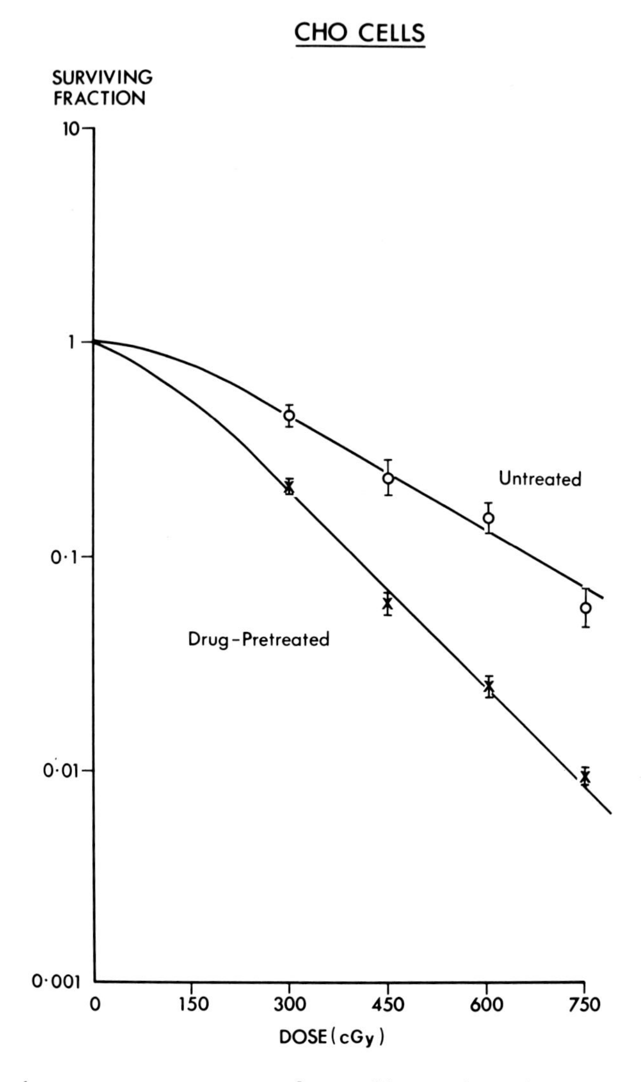

FIG.5. X-ray dose-response curves for cells with and without pretreatment with CHIP at 100 μg/ml for 1 hour at 37°C (5).

supra-additive (i.e. truly synergistic) effect. Supra-additive effects are uncommon but we have found one example with a dose-regime of 58 µg/ml CHIP on radiation (8). More commonly, the points lie within the envelope between the two lines and this merely indicates some enhancement of effect. The closer the localisation to the lower line, the more pronounced the enhancement and the more effective the regime.

The raw data used to derive the points in Figure 6 are shown in Figure 7. CHO cells were grown in medium B and the single agent survival curves are shown for (1) PAD and (2) x-rays together with the combined treatment curves for PAD at different dosages and time intervals before x-irradiation (curves 3-6) which have not been normalised in this case. The points (from curves 3, 4 and 6) for a one hour time interval between increasing dose levels of drug

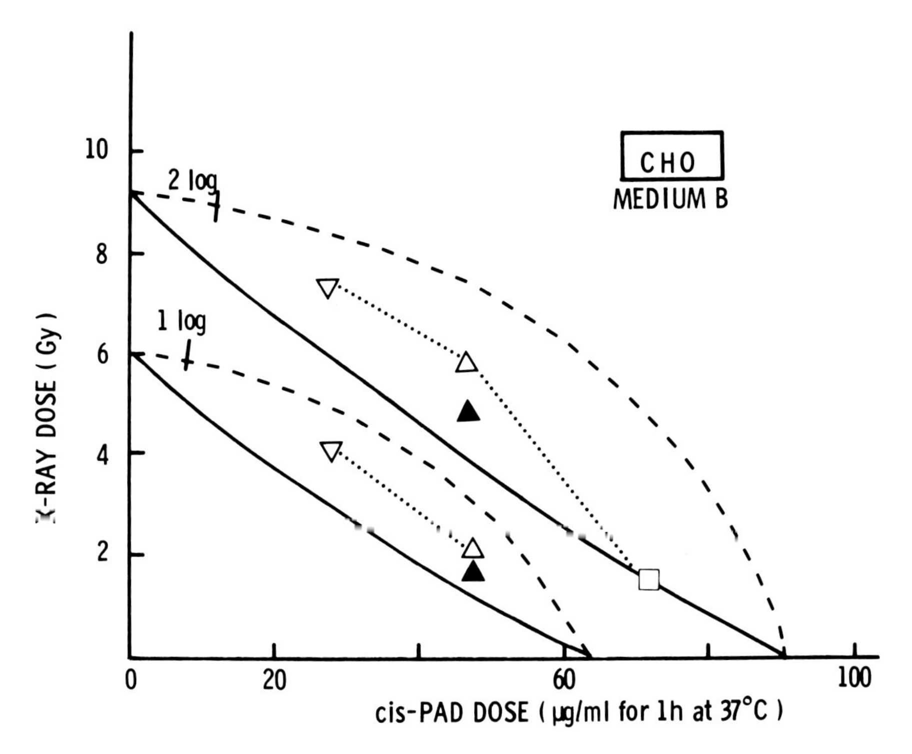

FIG.6. Isobolograms derived from curves 1 and 2 in Figure 7. Experimental points indicated with the same symbols used for curves 3-6.

treatment and then irradiation are joined in Figure 6 by a dotted line to allow for comparison. The 2 log isobologram shows that the highest PAD dosage of 72µg/ml (curve 6) gave the most effective result with the data point situated actually on the lower line of the envelope. The regime was also the most effective in terms of a simple enhancement ratio of 1.59 (as shown in Figure 4) so the two methods of analysis gave the expected result.

With the other regimes, however, there is some discrepancy between the methods. Going back along the dotted lines in Figure 6 the PAD dose levels of 46 and 26µg/ml (curves 4 and 3) gave data points in the middle of the two envelopes, indicating an average degree of enhancement by the isobolographic method. In terms of slope ratios, however, the enhancement ratios were 1.34 and only 1 respectively. On the other hand the data points for 46µg/ml given 4 hours before irradiation (curve 5) show more enhancement than the same dose given 1 hour beforehand. This is indicated by both the isobolograms and the enhancement ratio of 1.52.

There is certainly more information to be gained from the isobolographic method of analysis when full dose-response data are available. This was shown in other examples of combinations where PAD was used in medium A with time intervals ranging from 72 hours before, to 2 hours after irradiation (8). The degree of enhancement shown by the isobolograms was not always what would have been calculated from the simple enhancement ratios indicated by the survival curves. The same applied to our results with the use of CHIP and radiation. For the combination which gave the enhancement ratio of 1.64 (Figure 5)

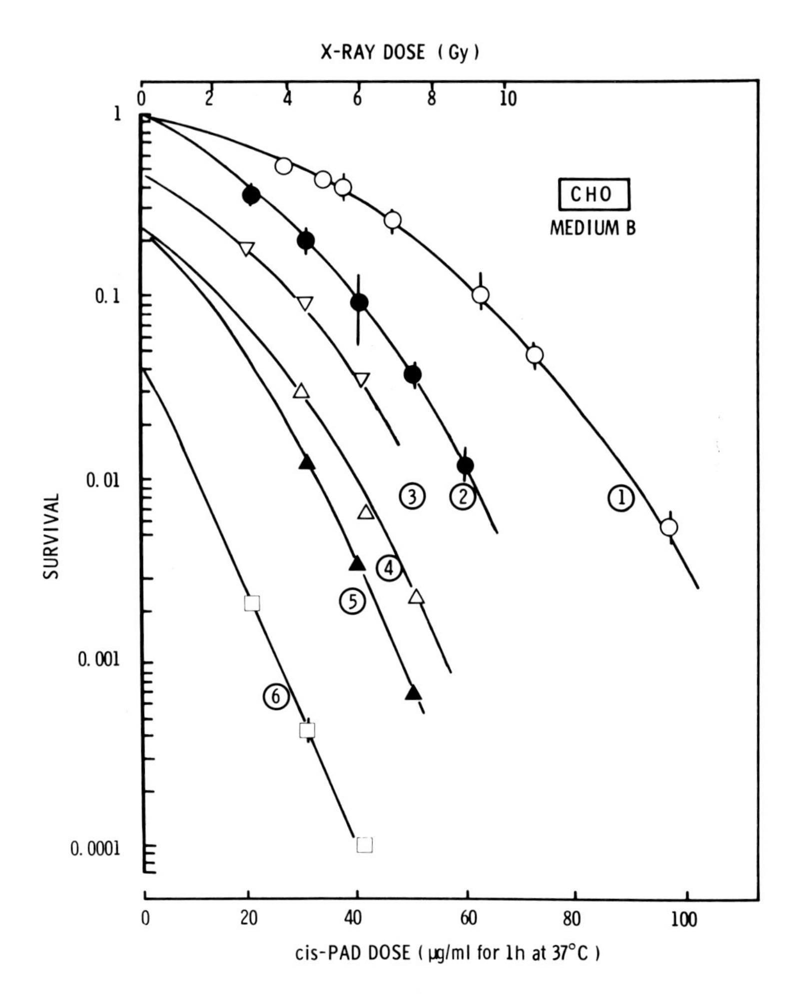

FIG.7. Dose-response curves for cells treated as follows: (1) PAD alone; (2) x-rays alone; (3) PAD at 26 µg/ml, then x-rays after 1 h; (4) PAD at 46 µg/ml, then x-rays after 1 h; (5) PAD at 46 µg/ml, then x-rays after 4 h; (6) PAD at 72 µg/ml then x-rays after 1 h.

an isobologram analysis would place the experimental point in the middle of the envelope of additivity. By contrast, the combination already referred to, which showed a supra-additive result, gave the lower enhancement ratio of 1.55. Thus one cannot predict from simple enhancement ratios whether a combination gives only enhancement or is actually supra-additive (7). This is probably because such ratios take no account of the shoulder region of survival curves.

3.4 CHIP AND IRRADIATION OF HYPOXIC CELLS

Preliminary studies of the effect of CHIP pre-treatment upon the radiation response of hypoxic cells had shown an anomalous effect in which the survival curve had a reduced extrapolation number but the same final slope after a pre-treatment with CHIP of 100 µg/ml for 1 hour and a one hour time interval before irradiation (3). Figure 8 shows the result of using half

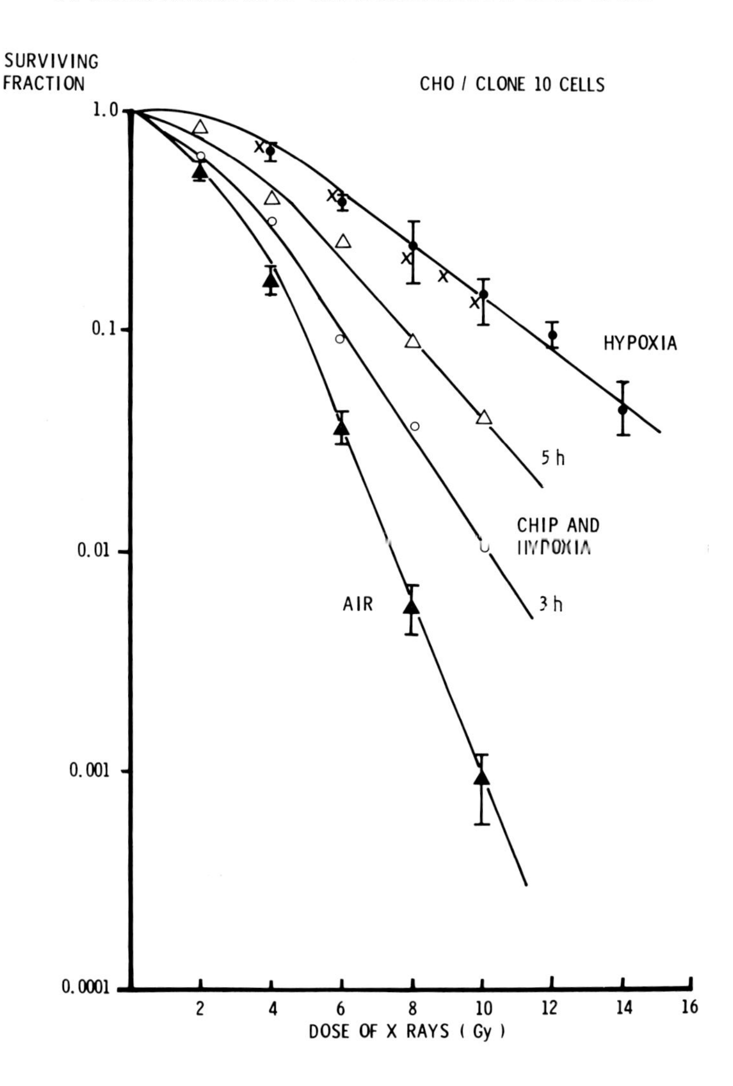

FIG.8. X-ray dose-response curves for cells treated under aerated conditions (▲), or under hypoxic conditions alone (●), or under hypoxic conditions after pretreatment with CHIP at 50 μg/ml for 1 h and then irradiated after 3 h (o), 5 h (Δ) and 24 h (×).

the drug concentration (50 μg/ml) and irradiating the hypoxic cells after time intervals of 3, 5 and 24 hours. The oxygen enhancement ratio of these cells is 2.8 so that the dose-modifying effect of CHIP pre-treatment upon hypoxic cell radiosensitivity should be compared with that figure. There was no effect after 24 hours. After 5 hours the ratio was 1.4; after 3 hours it was 1.9. Under these conditions there is minimal enhancement of the radiation response of aerated cells but clearly there is a considerable effect upon hypoxic cell radiosensitivity. Similar studies with C_3H mouse mammary tumour cells have also shown an enhancement of hypoxic cell radiosensitivity by a factor of 2.1 with no effect upon aerated cells (9).

4. CONCLUSION

These studies with combinations of platinum drugs and radiation point the way to a therapeutic strategy which may be worth clinical trial. Low dose levels of platinum which do not enhance the radiation response of aerated cells produce a very useful degree of hypoxic cell radiosensitization. By contrast, higher dose levels of platinum enhance the radiation response of aerated cells and this combination should be used with caution since normal tissues may suffer unacceptable levels of damage. By using low dose levels there may not only be useful benefit from the enhanced response of the primary tumour in the irradiated volume, but also some effect upon microscopic metastases elsewhere. It is clear that both the dose-level of platinum drug and the time interval before irradiation require to be chosen with care. These studies with CHO cells *in vitro* suggest that the optimum dose level for enhancement of hypoxic cell radiosensitivity lies on the shoulder portion of the platinum drug dose-response curve. At such a dose level, enhancement of the radiosensitivity of aerated cells is minimal although the drug is still cytotoxic. An appropriate clinical regime would involve radiotherapy to the primary tumour volume with low dosage of CHIP to provide enhancement of the radiation response of hypoxic tumour cells but no effect upon aerated normal cells other than the cytotoxic effect of the drug alone. At such low dose levels the regime will be well within the limitations of normal tissue toxicity but there should be some benefit derived from the effect of the drug upon microscopic metastases.

REFERENCES

(1) CONNORS, T.A., JONES, M., ROSS, W.C., BRADDOCK, P.D., KHOKHAR, A.R. and TOBE, M.L. New platinum complexes with anti-tumour activity. Chem-Biol. Interactions 5 (1972) 415-424.

(2) SZUMIEL, I. and NIAS, A.H.W. Action of a platinum complex (cis-dichlorobis (cyclopentylamine) platinum II) on Chinese hamster cells *in vitro* Chem-Biol. Interactions 14 (1976) 217-232.

(3) NIAS, A.H.W., BOCIAN, E. and LAVERICK, M. The mechanism of action of cis-dichlorobis (isopropylamine) trans dihydroxy platinum IV (CHIP) on Chinese hamster and C_3H mouse tumor cells and its interaction with x-irradiation. Int. J. Radiation Oncology 5 (1979) 1341-1344.

(4) BENDER, M.A., GRIPPS, H.G., BEDFORD, J.J. Mechanisms of chromosome aberration production. III Chemicals and ionizing radiation. Mutation Res. 23 (1974) 197-212.

(5) NIAS, A.H.W. and SZUMIEL, I. The effects of cis-dichlorobiscyclopentylamine Platinum II (PAD) and cis-dichlorobisisopropylamine trans-dihydroxy Platinum IV (CHIP) and radiation on CHO cells. Wadley Medical Bulletin 7 (1977) 562-567.

(6) SZUMIEL, I. and NIAS, A.H.W. The effect of combined treatment with a platinum complex and ionizing radiation on Chinese hamster ovary cells *in vitro*. Br. J. Cancer 33 (1976) 450-458.

(7) STEEL, G.G. and PECKHAM, M.J. Exploitable mechanisms in combined radiotherapy - chemotherapy: the concept of additivity. Int. J. Radiation Oncology 5 (1979) 85-91.

(8) SZUMIEL, I. and NIAS, A.H.W. Isobologram analysis of the combined effects of anti-tumour platinum complexes and ionizing radiation on mammalian cells. Brit. J. Cancer 42 (1980) 292-296.

(9) LAVERICK, M. and NIAS, A.H.W. Potentiation of the radiation response of hypoxic mammalian cells by cis-dichlorobis (isopropylamine) trans-dihydroxy Platinum IV (CHIP) Brit. J. Radiol 54 (1981) 529-530.

Progress in Radio-Oncology II, edited by K. H. Kärcher et al., Raven Press, New York © 1982.

In Vivo Analysis of the Combined Action of Ionizing Radiation and *Cis*-Diammine Dichloride Platinum (II) on a Solid Tumour

W. PORSCHEN, H. MÜHLENSIEPEN, R. PORSCHEN,
L.E. FEINENDEGEN
Institute of Medicine,
Kernforschungsanlage Jülich GmbH,
Jülich,
Federal Republic of Germany

Abstract

The present work was designed to study in vivo the mode of action of the antitumour agent cis-diammine dichloride platinum (II) (DDP) either alone or in combination with ionizing radiation on a solid experimental murine tumour. C57 BL/6J mice bearing the solid syngeneic adenocarcinoma EO 771 in one hind leg received an injection of 4 μg DDP per gram of body weight alone or 15 minutes prior to ^{60}Co gamma or cyclotron neutron irradiation (mean energy 6 MeV). The effects were assayed in vivo by volumetric measurements. In addition, 125-iodo-deoxyuridine (^{125}I-UdR) was used for tracing DNA precursor incorporation; activity loss from euoxic and average tumour cells was evaluated with a double tracer technique. Tumour growth delay was enhanced by a factor 3–5 after gamma or neutron irradiation. Dose modifying factors were identical for I-UdR incorporation 24 hours after treatment and tumour growth delay. DDP affected recovery of euoxic cells but had no lethal effect, whereas an enhanced rate of activity loss from the average tumour cell population was observed. DDP is a potent adjunct to therapy with different LET radiation causing distinct effects in euoxic and hypoxic cells.

1. INTRODUCTION

As the effect of therapy on malignant tumours is often not seen immediately, in vivo investigations on tumour kinetics are needed in order to describe the response of tumours within a reasonably short time after treatment.

It has been shown that 5-iodo-2'-deoxyuridine (I-UdR), a DNA precursor, is specifically incorporated into DNA of proliferating cells during the DNA synthesis phase [1], and investigations have shown that measurements of I-UdR uptake can be an indicator of the proliferative activity of cells [2]. The tracer is not removed from intact living cells and is only reutilized to a small extent after cell lysis because of its low initial utilization [1,3]. The use of two labelled compounds, ^{125}I-UdR and ^{131}I-UdR, allows of external determination of the incorporation and loss of radioactivity under in vivo conditions with minimum disturbance of the physiological conditions [3-10]; it measures the net flow of labelled dead or living cells as well as labelled debris from the tumour [11,12]. It can also be used for testing the relative merits of different types of therapy, because the increase of radioactivity loss is an indicator of improved therapy [4,13-16].

Recent reports have shown that platinum co-ordination complexes represent a new class of cytotoxic agents. Cis-diammine dichloride platinum (II) (DDP) has undergone extensive clinical trials. The combined action of ionizing radiation and DDP has been tested in different systems; the effect of DDP was most pronounced under hypoxic conditions. Potentiation under combined treatment was found over a wide range of drug and irradiation doses [17-21].

2. MATERIALS AND METHODS

2.1. Mice and Tumour

The adenocarcinoma EO 771 was implanted by injecting 1.7×10^5 viable cells into the right hind leg of 2 to 3-months-old male C57 BL/6J mice (30 g). The number of tumour cells which produce 50% tumour takes is about 24 cells, and there is no spontaneous tumour regression [11]. The growth fraction of a seven-day-old tumour is about 0.9, with a cell cycle time of 19 hours.

Tumour volume was calculated by measuring the largest tumour diameter. It has been demonstrated that there is a linear relationship between calculated tumour volume and weighed tumour mass in the growth range observed [11]. Groups of 10 to 20 animals were used for each experimental result.

2.2. Measurements of Radioactivity Loss

The double-tracer technique enables euoxic cells with a high proliferation rate to be labelled differentially from the 'average tumour cells', which comprise a mixture of euoxic and hypoxic cells with different proliferation activities (see Refs [14,16, 22,23].

Four days after implantation, the mice received two intraperitoneal injections of ^{125}I-UdR (Amersham-Buchler, Braunschweig) seven hours apart (2×10 μCi, specific activity 2000 Ci/mmol). Because of the high initial tumour growth rate, the labelled cells became distributed throughout the tumours during the next two days. 10 μCi of ^{131}I-UdR (specific activity 3 Ci/mmol) (Amersham-Buchler, Braunschweig) were then injected into the same animals, followed by a second injection seven hours later in order to label most of the cells that are mainly located in the perivascular euoxic area. 24 hours after the last injection the tumours were treated as described below.

It has been demonstrated autoradiographically [22] that by this labelling schedule ^{131}I-UdR is localized in the euoxic cells around blood vessels at the time of treatment, whereas ^{125}I-UdR labelled cells (termed 'average tumour cells') are distributed throughout the tumour and near necrotic regions, too (i.e. the latter contain a certain fraction of hypoxic cells).

Three days before and during the experiments in which labelled I-UdR was injected, the drinking water was substituted by a 0.1% aqueous solution of sodium iodide.

After the excretion of free radioiodine, external counting of tumour-bound radioactivity reflects the fate of tumour cell DNA [5].

The incorporated tracer is counted using a two-channel pulse-height analyser (Berthold, Wildbad) linked to a NaI(Tl) scintillation counter that is covered with a lead plate containing a hole over which the tumour is placed for counting, the rest of the body being shielded. Corrections are made for background, spill-over and physical decay.

2.3. Measurements of I-UdR Uptake

At different times after treatment, 20 μCi ^{125}I-UdR was injected intraperitoneally in order to evaluate the incorporation of the DNA precursor into tumour DNA as an indicator of DNA synthesis and/or surviving cell number, since observation is limited to proliferating cells. 48 hours after injection, the tumour-bound radioactivity was determined as described above, because at that time about 60 to 70% of the total radioactivity can be found in the DNA fraction. The results are expressed relative to tumour activity of untreated control mice.

2.4. Gamma Irradiation

During ^{60}Co gamma irradiation, the bodies of the animals were shielded by lead, permitting only exposure of the tumour-bearing leg. Doses ranged between 1 Gy and 40 Gy at a dose rate of 2.50 Gy/min. Dose measurement was by Philips ionization chamber and thermoluminescent dosimeters (TLD 100).

2.5. Neutron Irradiation

Neutrons were produced by the ^{9}Be(d,n)^{10}B reaction using 14 MeV deuterons from the CV-28 cyclotron[1] and a 'thick' water-cooled beryllium target. The neutrons had a broad energy spectrum, extending up to 16 MeV, with a mean energy of 6 MeV. Absorbed dose was determined using a tissue-equivalent ionization chamber (EG&G) inside a Lucite mouse phantom. The neutron kerma was evaluated from neutron fluence measurements with PIN-diodes and activation foils. With a neutron collimator of borated wood (field diameter 23 cm), the dose rate at a distance of 96 cm from the target was 0.4 Gy/min. The gamma component free in air, derived from measurements with film dosimeters, was about 6%. The bodies of the animals were shielded by the collimator and only the tumour-bearing legs were irradiated.

2.6. DDP

DDP was used with and without gamma or neutron irradiation in order to measure its effects on tumour volume, on I-UdR incorporation into proliferating cells, and on the activity loss of previously labelled 'average' and euoxic cells. DDP was dissolved in saline solution and 0.5-15 μg DDP per gram of body weight were injected i.p. 15 min before irradiation. In mouse

---

[1] The Cyclotron Corporation, Berkeley, California, USA.

experiments, the DDP dose required to kill 50% of the animals ranged between 13-14 μg/g body weight [19].

3. RESULTS

3.1. Growth Delay

Tumour growth was studied after giving gamma or neutron irradiation, and/or DDP. Tumour volumes were evaluated from calliper measurements.

Figure 1a gives the volume response after γ-irradiation alone and Fig.1b shows the combined effect of 0.5-15 μg DDP/g body weight and 5 Gy γ-irradiation. The treatment-induced tumour growth delay is an indicator of the efficiency of the treatment.

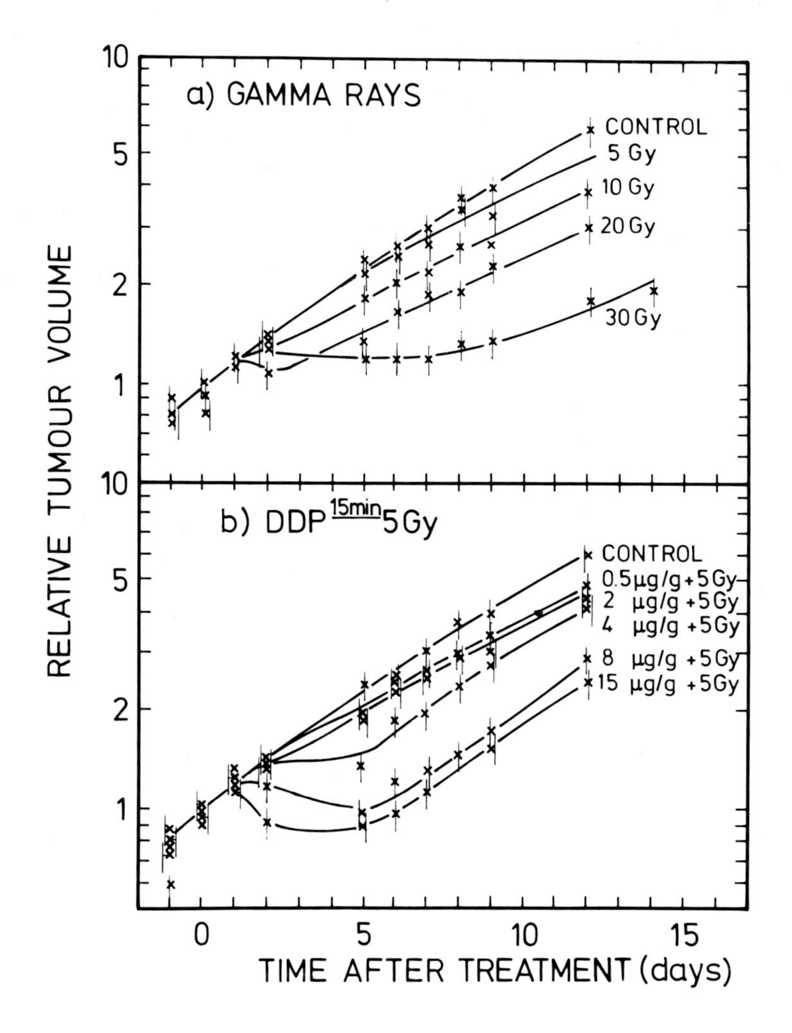

FIG.1. Relative tumour volume of adenocarcinoma EO 771 normalized to the volume at the time of irradiation (a) without or (b) with DDP.

One observes a growth delay of about 0.5 days after treatment with 5 Gy and about 8 days after irradiation with 30 Gy.

Increasing amounts of DDP enhance the growth delay induced by 5 Gy γ-rays from 0.5 to 1 day (0.5 μg DDP/g) to about 6 days (15 μg DDP/g).

Figure 2a shows the growth delay due to 0-15 μg DDP/g, 0-40 Gy γ-rays and the combined action of 0-10 Gy γ-rays 15 min after injection of 4 μg DDP/g. 4 μg DDP/g alone results in a growth delay of about 0.3 days, while 10 Gy γ-rays alone produces about 1.5-2 days delay; the combination of 4 μg DDP/g and 10 Gy γ-rays gives a delay of 4.5 days.

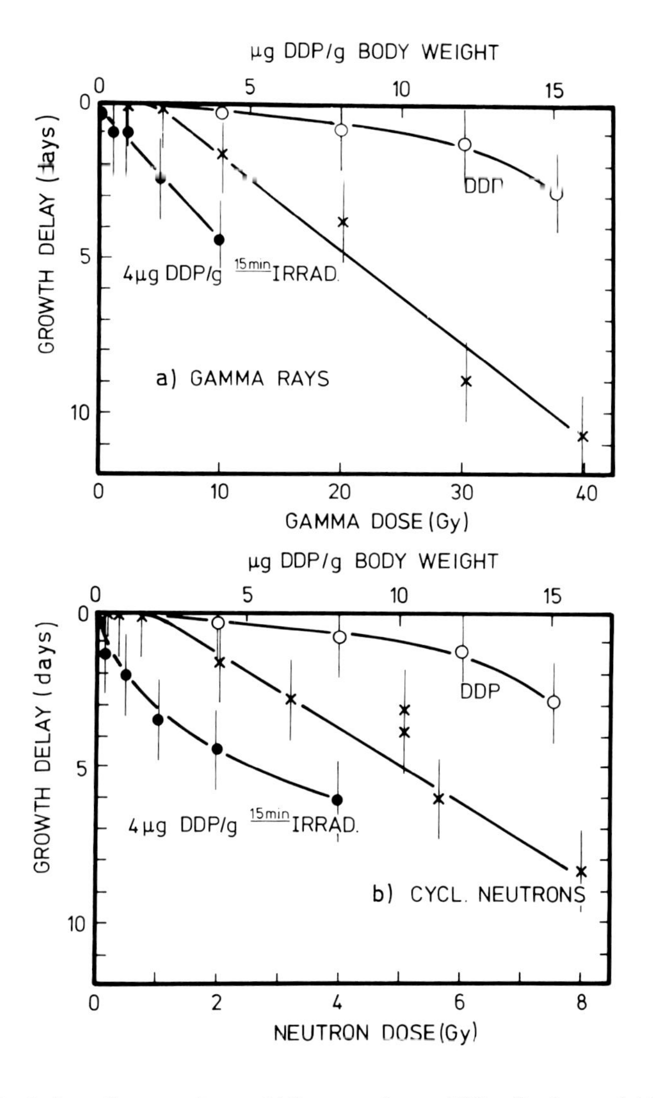

FIG.2. Growth delay from volume "1" to volume "2" of the solid experimental tumour EO 771. (a) DDP, γ-rays, and the combination of 4 μg DDP/g boy weight, 15 min before γ-irradiation; (b) DDP, cyclotron neutrons, and the combination of 4 μg DDP/g, 15 min before neutron irradiation.

Corresponding data for neutron irradiation are shown in Fig.2b. The corresponding doses of neutrons for a given level of damage are, as expected, lower; an RBE-value of about 4 is obtained.

DDP also enhances the effect of neutrons. 2 Gy neutrons gives a delay of 1.6 days, the combination of 4 μg DDP/g body weight and 2 Gy neutrons results in a delay of 4.5 days.

3.2. I-UdR Incorporation

The effect on cell production was checked by observing the uptake of ^{125}I-UdR into the tumour at various times after the various treatments. The incorporation of ^{125}I-UdR in relation to uptake in untreated tumours is an indicator of proliferation activity; it depends on the number of DNA synthesizing cells at the time of injection and on the incorporation rate per tumour cell. As seen in Fig.3a, after irradiation with 2.5 Gy of γ-rays, ^{125}I-UdR uptake becomes increasingly depressed, reaching a minimum about 4 hours after irradiation. Thereafter recovery ensues. Similar data were obtained with fast neutrons (Fig.3b).

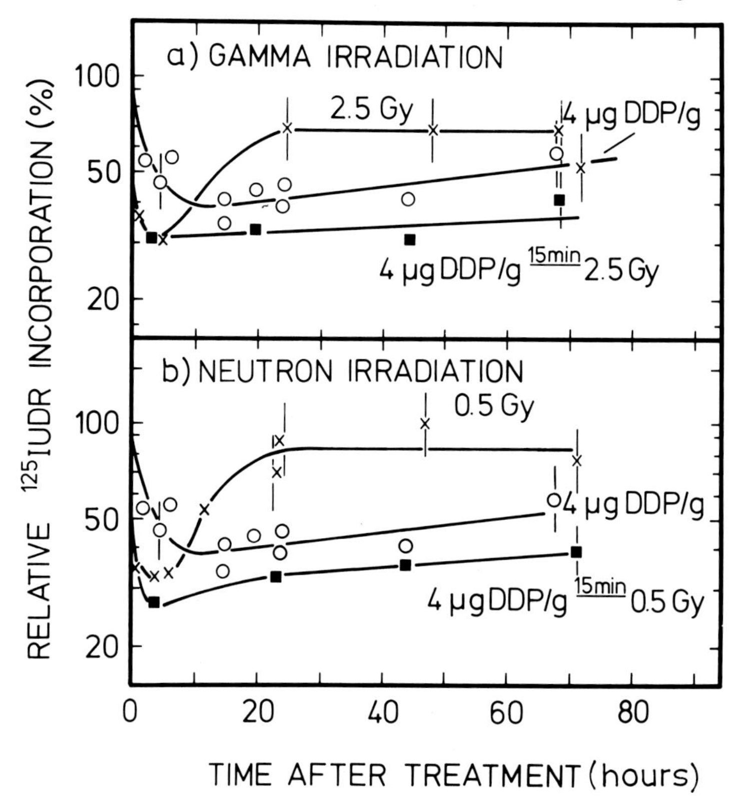

FIG.3. Incorporation of ^{125}I-UdR in solid adenocarcinoma EO 771 treated variously (irradiation alone, DDP alone, or a combined scheme with 4 μg DDP/g body weight 15 min before either (a) gamma or (b) neutron irradiation). ^{125}I-UdR was injected at the times shown; 48 h later the relative tumour-bound ^{125}I activity was evaluated.

After DDP given alone, the depression of ^{125}I-UdR uptake is similar to that after irradiation, yet recovery is delayed. After combined therapy, recovery is even more retarded.

The data indicate an interference of DDP with DNA synthesis in proliferating cells during the time of recovery.

The ^{125}I-UdR incorporation 24 h after treatment was taken as parameter for constructing dose/effect curves; Fig.4a shows the results for γ-irradiation, Fig.4b for neutrons. The effect of DDP (4 μg/g body weight) amplifies the irradiation effect on ^{125}I-UdR incorporation over a wide dose range (lower curves). 4 μg DDP/g applied in addition to irradiation is equivalent to about 7 Gy of γ-rays or to about 1.5 Gy of 6 MeV neutrons.

3.3. Treatment-Induced Activity Loss

The activity loss for euoxic and average tumour cells was evaluated after gamma, 4 μg DDP/g body weight, and combined therapy.

Figure 5a,b shows the loss of tracer from treated euoxic cells, and Fig.5c,d the loss from treated average tumour cells.

In euoxic cells, there is no effect of DDP on loss of label, irrespective of whether DDP is given alone (Fig.5a) or combined with 5 Gy γ-irradiation (Fig.5b).

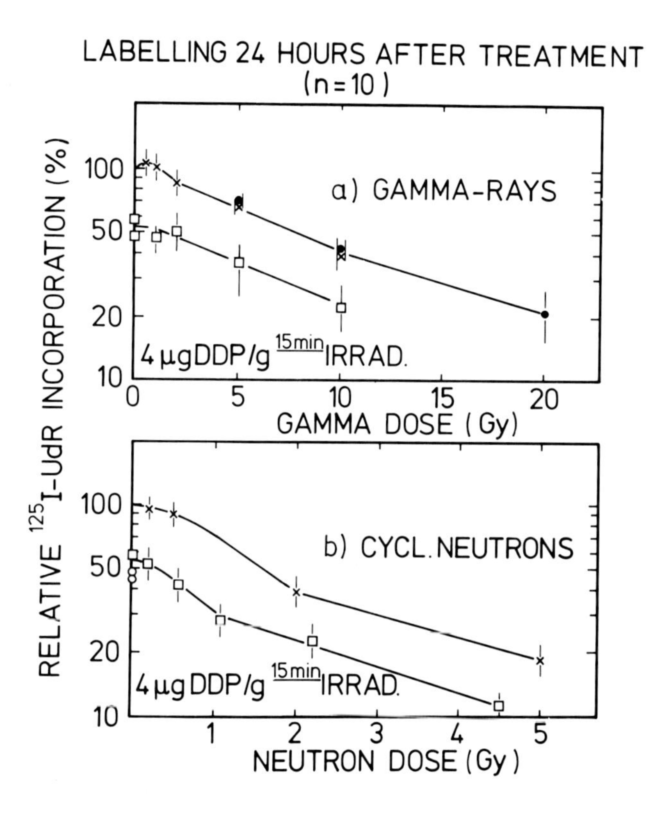

FIG.4. Relative ^{125}I-UdR incorporation after treatment for one fixed time interval of 24 h between treatment and labelling.

There is an obvious effect of DDP for labelled average tumour cells, many of which are known to be hypoxic, irrespective of whether DDP is given alone (Fig.5c) or combined with 5 Gy γ-irradiation (Fig.5d).

5 Gy of γ-rays alone enhanced the loss of tracer from average tumour cells to a lower degree than from euoxic cells. DDP alone gave a greater enhancement from average tumour cells, and this was similar to the effect when in combination with 5 Gy of γ-rays.

4. DISCUSSION

Numerous studies have been devoted to the characterization of the effect of DDP. The selective interference with DNA metabolism has been ascribed to alkylating properties, but new experimental and clinical evidence suggests that alkylation is not the sole or major mechanism of action. After injection of DDP into the peritoneal cavity of the mouse, the intact drug is rapidly excreted in the urine. The drug is passively transported across the cellular membrane and the kidney and the liver show the most significant uptake [19]. DDP is highly active against a broad spectrum of tumours, but there is no selective uptake of platinum by the tumour tissue. This suggests that tumour cells are more sensitive to the intracellular effect of DDP than are

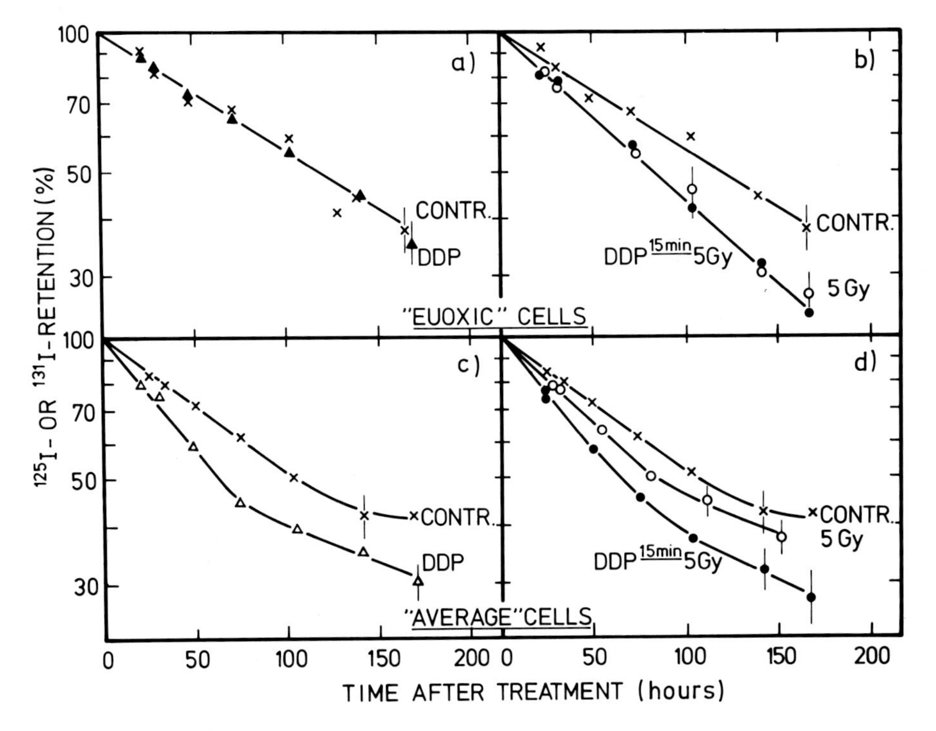

FIG.5. (a, b) Retention of ^{131}I-UdR in the solid experimental adenocarcinoma EO 771: (a) control tumours and influence of DDP alone; (b) 5 Gy γ-irradiation and 4 μg DDP/g body weight given 15 min before 5 Gy γ-irradiation.

(c, d) Retention of ^{125}I-UdR: (c) control and DDP alone; (d) 5 Gy γ-irradiation and 4 μg DDP/g given 15 min before 5 Gy γ-irradiation.

normal tissue cells. The primary reason for using DDP combined with radiation is the characterization of DDP as a radiation sensitizer [24]. The time between injection of DDP and irradiation influences the tumour response; the combination is more effective if the drug is given before rather than after irradiation [20].

This work has shown that administration of DDP 15 min before tumour irradiation results in a dose-dependent inhibition of tumour growth; increasing amounts of DDP enhance the effect of 5 Gy of γ-irradiation.

Tumour growth changes after treatment are the result of a change in the balance between cell production and cell death. Incorporation of ^{125}I-UdR may be regarded as an integral indicator of the relative rate of DNA synthesis and, therefore, as an index of cell production.

After a single injection of DDP, tumour irradiation or combined therapy, the relative incorporation of ^{125}I-UdR gradually decreased to a minimum, reached 4 hours after treatment. Recovery was delayed after injection of DDP and by combining DDP with radiation. The relative incorporation reflects the activity bound to tumour DNA. By labelling the control tumours and the treated tumours at the same time after treatment, a nearly constant ratio of uptake from 24 to 96 hours after irradiation was obtained.

After DDP treatment, the cells are 'pre-damaged' and a smaller radiation dose is sufficient to produce a given level of overall damage. Based on the uptake of ^{125}I-UdR, 4 μg DDP/g body weight may be regarded as equal to about 7 to 7.5 Gy of gamma rays or 1.5 to 2 Gy of 6 MeV neutrons.

The effect on cell loss was examined by the double-tracer technique. The in situ method measures the net radioactivity loss from the tumour. The dose of 4 μg DDP/g body weight produced no measurable effect on the treated euoxic cells and did not modify the effect of radiation on the loss of DNA-bound tracer. 5 Gy alone enhanced the rate of loss of tracer from the average tumour cells to a lower degree than observed for euoxic cells. DDP alone was more effective, but it did not significantly enhance the effect of radiation in a combined treatment (Fig.5).

The lethal action of DDP on hypoxic tumour cells and the modification of the radiation effect on recovery of euoxic cells suggests that this drug may be of value in radiation therapy.

REFERENCES

[1] HUGHES, W.L., COMMERFORD, S.L., GITLIN, D., KRÜGER, R.C., SCHULTZE, B., SHAH, V., REILLY, A.: Deoxyribonucleic acid metabolism in vivo: I. Cell proliferation and death as measured by incorporation and elimination of iododesoxyuridine, Fed. Proc. 23 (1964) 640-648.

[2] PORSCHEN, W., FEINENDEGEN, L.E.: Biologische In-vivo-Dosimetrie von 15 MeV-Neutronen bei normalen und Tumorzellen - Zellmarkierung mit Jod-125-Desoxyuridin, Strahlentherapie 145 (1973) 27-38.

[3] FEINENDEGEN, L.E., BOND, V.P., HUGHES, W.L.: Physiological thymidine reutilization in rat bone-marrow, Proc. Soc. Exp. Biol. Med. 122 (1966) 448.

[4] BEGG, A.C.: Cell loss from several types of solid murine tumour: comparison of ^{125}I-iododeoxyuridine and tritiated thymidine methods, Cell Tissue Kinet. 10 (1977) 409.

[5] DETHLEFSEN, L.A., SORENSON, J., SNIVELY, J.: Cell loss from three established lines of C3H mouse mammary tumor: A comparison of the ^{125}I-UdR and the ^{3}H-TdR autoradiographic methods, Cell Tissue Kinet. 10 (1977) 447.
[6] FEINENDEGEN, L.E., BOND, V.P., HUGHES, W.L.: 125-I-DU in autoradiographic studies of cell proliferation, Exp. Cell Res. 43 (1966) 107.
[7] FEINENDEGEN, L.E., HEINIGER, H.J., FRIEDRICH, G., CRONKITE, E.P.: Differences in reutilization of thymidine in hemopoietic and lymphopoietic tissues of the normal mouse, Cell Tissue Kinet. 6 (1973) 573.
[8] PORSCHEN, W., FEINENDEGEN, L.E.: In-vivo-Bestimmung der Zellverlustrate bei Experimentaltumoren mit markiertem Joddesoxyuridin, Strahlentherapie 137 (1969) 718.
[9] STEEL, G.G.: Cell loss from experimental tumours, Cell Tissue Kinet. 1 (1968) 193-207.
[10] FEINENDEGEN, L.E.: "Prospects and problems on the application of isotopes to pure and applied cytology", Ch. 18 in Radiotracer Techniques and Applications (EVANS, E.A., MURAMATSU, M., Eds), Marcel Dekker Inc., New York (1977) 769-821.
[11] PORSCHEN, R.: Zellkinetische Untersuchungen in vitalen und nekrotischen Tumorbereichen: Anwendbarkeit der Verlustratenmessung mit 125-Jod-desoxyuridin, Dissertation, Universität Düsseldorf, 1980.
[12] PORSCHEN, R., PORSCHEN, W., FEINENDEGEN, L.E.: "Applicability of the ^{125}IUdR in situ method for measuring cell loss in a solid adenocarcinoma", Br. J. Cancer 41 Suppl. IV (1980) 105.
[13] PORSCHEN, W., GARTZEN, J., GEWEHR, K., MÜHLENSIEPEN, H., WEBER, H.-J., FEINENDEGEN, L.E.: "In vivo assay of the radiation sensitivity of hypoxic tumour cells; Influence of gamma-rays, cyclotron neutrons, Misonidazole, hyperthermia and mixed modalities", in Hypoxic Cell Sensitizers in Radiobiology and Radiotherapy (Proc. 8th L.H. Gray Conf., Cambridge, UK, 1977), Br. J. Cancer 37 Suppl. III (1978).
[14] GEWEHR, K., PORSCHEN, W., MÜHLENSIEPEN, H., FEINENDEGEN, L.E.: Measurements of in vivo effects of an electron affinic radiosensitizer using a double tracer technique, Strahlentherapie 156 (1980) 554-560.
[15] PITTNER, W., PORSCHEN, W., FEINENDEGEN, L.E.: In-vivo-Messungen unterschiedlicher Strahlenempfindlichkeit von Tumor-Zellen während des Zellzyklus Strahlentherapie 145 (1973) 161.
[16] PORSCHEN, W., BOSILJANOFF, P., GEWEHR, K., MÜHLENSIEPEN, H., WEBER, H.-J., DIETZEL, F., FEINENDEGEN, L.E.: "In vivo assay of the radiation sensitivity of hypoxic tumour cells: influence of radiation quality and hypoxic sensitization", in Radiobiological Research and Radiotherapy (Proc. Symp. Vienna, 1976: BECK, E.R.A., Ed.) Vol.1, IAEA, Vienna (1977) 181.
[17] MURTHY, A.K., ROSSOF, A.H., ANDERSON, K.M., HENDRICKSON, F.R.: Cytotoxicity and influence on radiation dose-response curve of cis-diamminedichloroplatinum II (cis-DDP), Int. J. Radiat. Oncol., Biol. Phys. 5 (1979) 1411-1415.
[18] OSIEKA, R., SCHMIDT, G.G.: Cis-diamino-dichloro-platinum (II), Klin. Wochenschr. 57 (1979) 1249-1258.
[19] ROSENBERG, B.: Platinum coordination complexes in cancer chemotherapy, Naturwissenschaften 60 (1973) 399-406.
[20] DOUPLE, E.B., RICHMOND, R.C.: "Platinum complexes as radiosensitizers of hypoxic mammalian cells", in Hypoxic Cell Sensitizers in Radiobiology and Radiotherapy (Proc. 8th L.H. Gray Conf., Cambridge, UK, 1977), Br. J. Cancer 37 Suppl. III (1978) 98.
[21] MUGGIA, F.M., GLATSTEIN, E.: Summary of investigations on platinum compounds and radiation interactions, Int. J. Radiat. Oncol., Biol. Phys. 5 (1979) 1407-1409.

[22] BOSILJANOFF, P., PORSCHEN, W., PIEPENBRING, W., MÜHLENSIEPEN, H., FEINENDEGEN, L.E.: In-vivo-Untersuchungen über die relative Strahlenempfindlichkeit von hypoxischen Tumorzellen, Strahlentherapie 153 (1977) 178.

[23] FEINENDEGEN, L.E.: "Autoradiographische und biologische Untersuchungen der Zellproliferation in vivo", in Präoperative Tumorbestrahlung (Deutschen Röntgenkongress: HUG, O., Ed.), Urban and Schwarzenberg, Munich (1977) 12.

[24] DOUPLE, E.B., EATON, W.L., TULLOH, M.E.: Skin radiosensitization studies using combined cis-dichlorodiammine platinum (II) and radiation, Int. J. Radiat. Oncol., Biol. Phys. 5 (1979) 1383.

Progress in Radio-Oncology II, edited by K. H. Kärcher et al., Raven Press, New York © 1982.

Mechanisms of Radiosensitization and Protection Studied with Glutathione-Deficient Human Cell Lines*

L. RÉVÉSZ, Margareta EDGREN
Department of Tumor Biology II,
Karolinska Institute Medical School,
Stockholm,
Sweden

Abstract

Glutathione-deficient fibroblasts and lymphoblastoid cells, derived from patients with an inborn error of glutathione synthetase activity, and glutathione-proficient cells, derived from clinically healthy individuals, were used to investigate the importance of glutathione for radiosensitization by misonidazole. With single-strand DNA breaks as an end point, misonidazole as well as oxygen was found to lack any sensitizing effect on cells deficient in glutathione. The post-irradiation repair of single-strand breaks induced by hypoxic irradiation of misonidazole treated cells was also found to be to a great extent glutathione dependent, like the repair of breaks induced by oxic irradiation. Naturally occurring aminothiols in glutathione-deficient cells appeared to be inefficient as substitutes for glutathione. Artificial aminothiols, such as cysteamine or dithiothreitol, were found to effectively replace glutathione.

1. INTRODUCTION

Glutathione (GSH) has been shown to participate in a number of radiochemical and biochemical processes which are of a particular importance in determining the response of cells to radiation. Recently, access to GSH-deficient human cell strains opened new approaches to the study of the role GSH plays in the processes. In a preliminary communication (1), data were presented which indicate that GSH is a key substance also in the mechanism of radiosensitization by misonidazole. This paper reports further experimental observations in this respect using single-strand DNA breaks as an end-point.

2. MATERIALS AND METHODS

Cultures of EBV-transformed lymphoblastoid cells (JG) were established from an individual with an inborn error of glutathione synthetase activity. The cells had an average glutathione content of 2.3 nmol/mg protein (2). Cultures of EBV-transformed lymphoblastoid cells (CL 1) which originated from a clinically healthy individual with an average

* This work was supported by grants from the Swedish Cancer Society (project 37-B81-15XB1), the Swedish Society for Medical Sciences (projects 253 and 306), and the International Atomic Energy Agency (under Research Contract No. 1725R4/RB).

GSH content of 38.3 nmole/mg protein were used as control. Cultures of untransformed fibroblasts (VP) derived from a subcutaneous biopsy taken from another individual with a genetically defined glutathione synthetase deficiency, had an average GSH content of 0.72 nmol/mg (1,3). For these cells, untransformed fibroblasts (GP) derived from the subcutaneous biopsy of a clinically healthy brother with a GSH content of 11.55 nmol/mg protein, on the average, served as control (1,3).

Details of the tissue culturing, X-irradiation, control of hypoxic and anoxic conditions, as well as the determination of single-strand DNA breaks using the unwinding technique in weak alkali, have been described earlier (3,4).

3. RESULTS

3.1. Yield of single-strand DNA breaks

In one series of experiments we investigated the question to what extent, if any, the effect of misonidazole on the yield of radiation induced single-strand DNA breaks (ssb) is dependent upon the cellular glutathione (GSH) content.

As illustrated in FIG 1, the dose-effect relationship for the amount of radiation induced ssb in GSH-deficient JG cells, exposed to radiation under hypoxic conditions (900 ppm O_2) in the presence of 10 mM misonidazole, is identical to that seen when exposure is made in the absence of the drug. Thus, the data indicate that misonidazole has no radiosensitizing effect on hypoxic, GSH-deficient cells which, as can also be seen in FIG 1, have a sensitivity almost as high as when aerobic irradiation is made.

In contrast, the GSH-proficient control cells (CL 1) show a considerable reduction of sensitivity in regard to the yield of ssb when irradiation is made under hypoxic, instead of aerobic conditions, and a dose modifying factor (DMF) 1.93 can be calculated (FIG 2). The DMF is 2.99 when the sensitivity of oxically and anoxically irradiated CL 1 cells is compared (data not shown). In the case of these cells, treatment with 10 mM misonidazole under hypoxic conditions results in almost as much enhancement of the radiosensitivity, as the enhancement by aerobic conditions (FIG 2).

3.2. Post-irradiation repair

In a second series of experiments, the question was investigated to what extent, if any, misonidazole influences the repair of radiation induced ssb.

TABLE I shows that treatment of the fibroblasts with 10 mM misonidazole under anoxic conditions ($<$5 ppm oxygen) for 10 minutes and during a subsequent aerobic incubation period induces ssb even without irradiation. In GSH-proficient GP cells, on the average, a 40 per cent increase of ssb was calculated after a 60 minutes' period of aerobic incubation. In GSH-deficient VP cells an increase of ssb by a factor of about 2 can be noted.

FIGURES 3 and 4 illustrate the relative amount of repaired ssb in VP and GP cells at different times during an aerobic incubation following irradiation with 34 and 87 Gy, respectively. It can be noted that, in an agreement with earlier observations (3), both cell strains repair ssb to about 85 per cent within 60 minutes if irradiation was made in anoxia. A similar repair rate

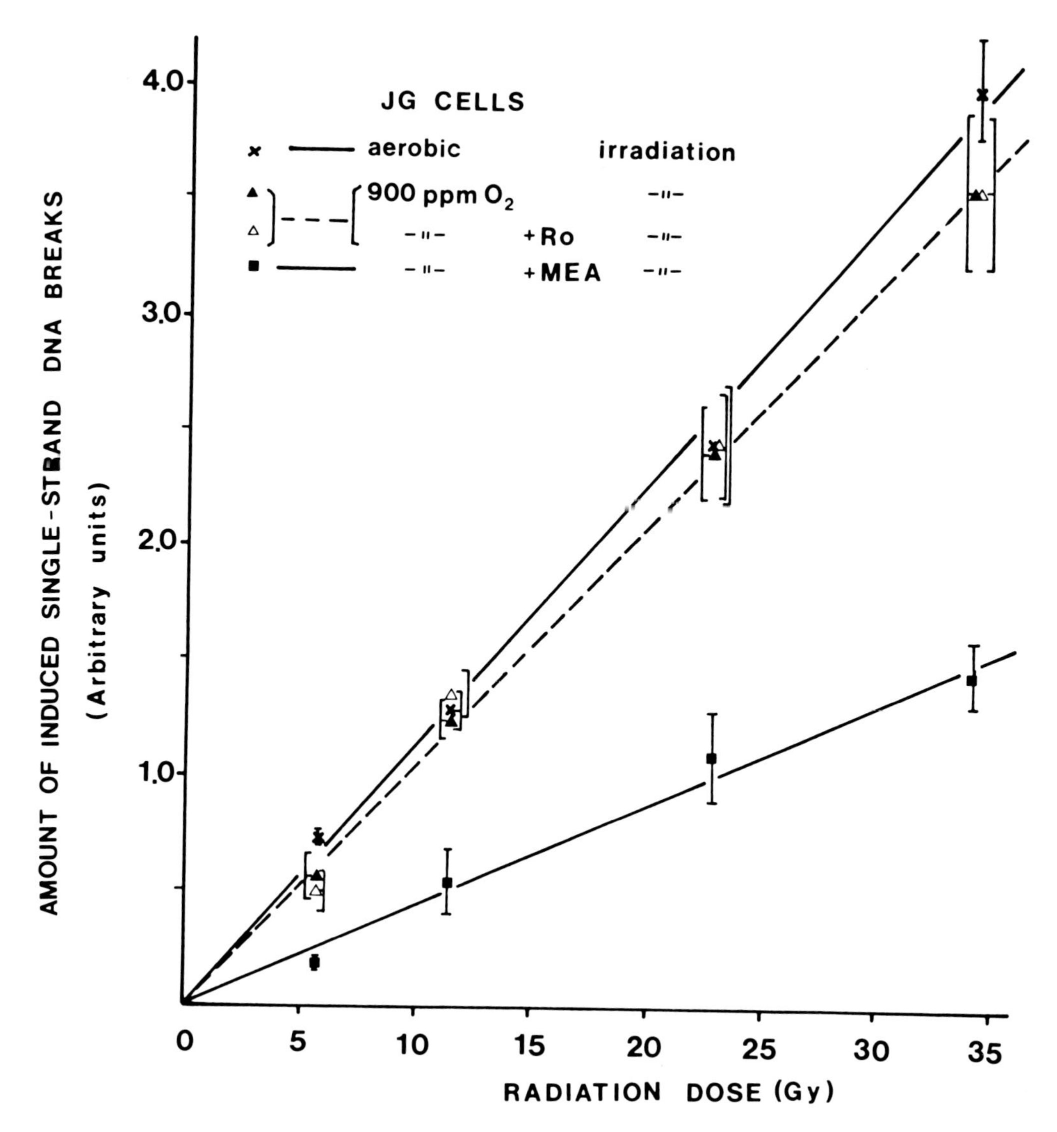

FIG.1. The yield of single-strand DNA breaks (ssb) after irradiation of glutathione-deficient JG lymphoblastoid cells, aerobically or hypoxically (900 ppm O_2) in the presence of 10 mM misonidazole (Ro), or 20 mM cysteamine (MEA) added to the medium 10 minutes before X-ray exposure. Means ± SE are indicated, as calculated from 5–6 replicate, paired experiments. Regression lines were calculated by least-squares analysis of the weighted data. The dose modifying factors (DMF), as calculated from the slope constant of the regression lines are: aerobic/hypoxic = 1.09; hypoxic/hypoxic + Ro = 1.0; hypoxic/hypoxic + MEA = 2.36.

of ssb in GP cells can be seen when misonidazole is present in 10 mM concentration during irradiation in anoxia, as well as during the aerobic post-irradiation period. In contrast, repair is strongly inhibited when VP cells were irradiated anoxically with misonidazole present. In this case, on the

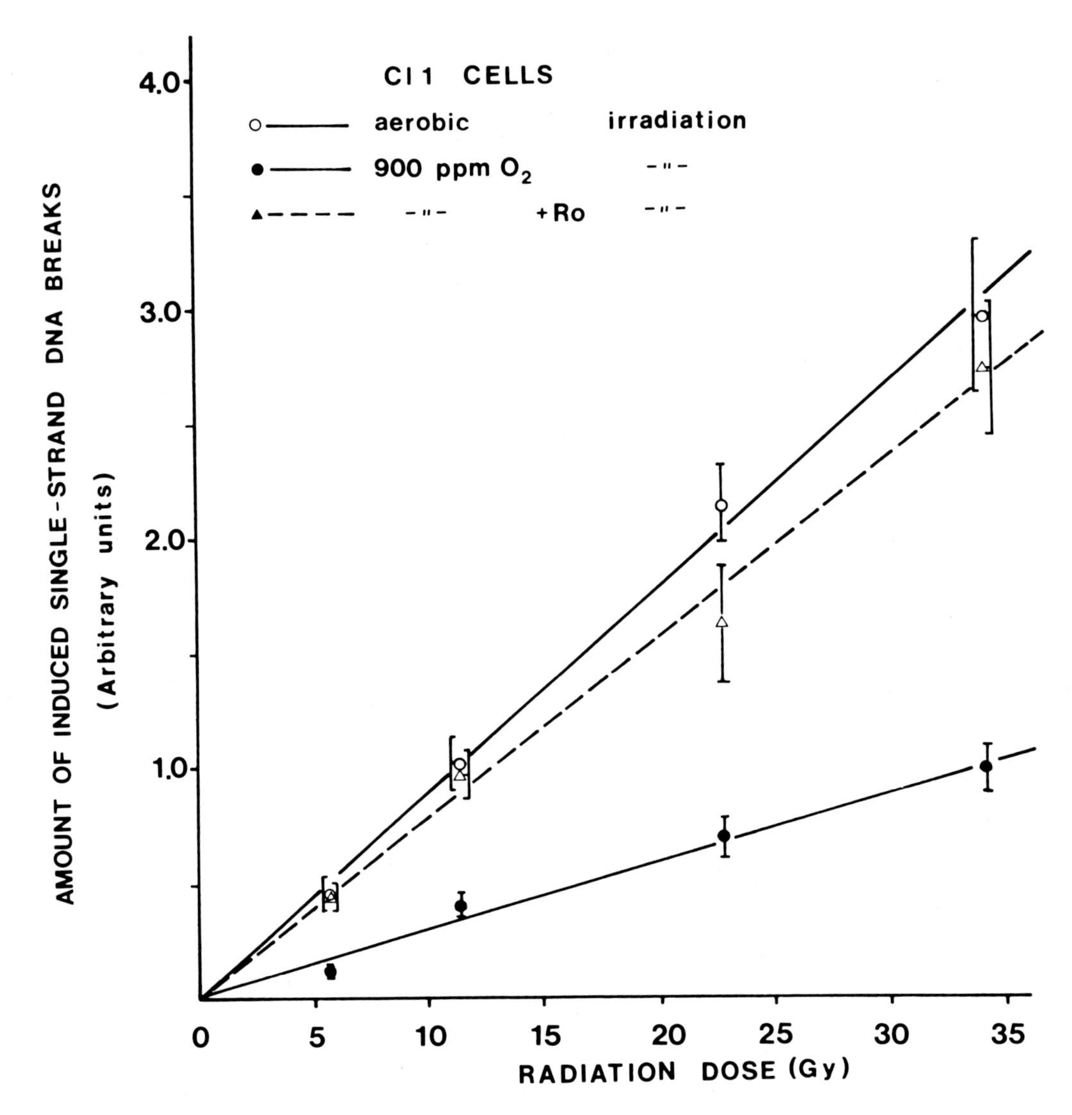

FIG.2. Yield of ssb after irradiation of GSH-proficient CL 1 lymphoblastoid cells after irradiation aerobically or hypoxically in the presence of 10 mM misonidazole (Ro) or in its absence. Means ± SE are presented from six replicate, paired experiments. The DMF calculated from the statistically defined regression lines are: aerobic/hypoxic = 1.93; hypoxic/hypoxic + Ro = 1.69.

average, not more than about 50 per cent of ssb is repaired. Thus, this repair is even lower than that seen after oxic irradiation of VP when the repair is about 75 per cent.

3.3. Substitution for GSH deficiency

In a third, complementary experimental series the GSH-deficient cells were incubated with 20 mM cysteamine (MEA) for 10 minutes before and during irradiation, or with 10 mM dithiothreitol (DTT) immediately after irradiation, in order to supplement the cellular low-molecular SH content. As illustrated

TABLE I. Relative increase of ssb in unirradiated, GSH-deficient VP cells and GSH-proficient GP cells treated with 10 mM misonidazole for 10 minutes under anoxic conditions. Determinations were made after the treatment at different times during an aerobic incubation. Means ± SE are presented as calculated from replicate paired experiments, the number of which is indicated in brackets.

Period of aerobic incubation after misonidazole treatment in anoxia	Relative increase of ssb	
	VP cells	GP cells
15 min	2.2 ± 0.8 (5)	0.9 ± 0.1 (5)
30 min	1.1 ± 0.3 (5)	1.1 ± 0.2 (5)
60 min	2.1 ± 0.7 (6)	1.4 ± 0.3 (5)

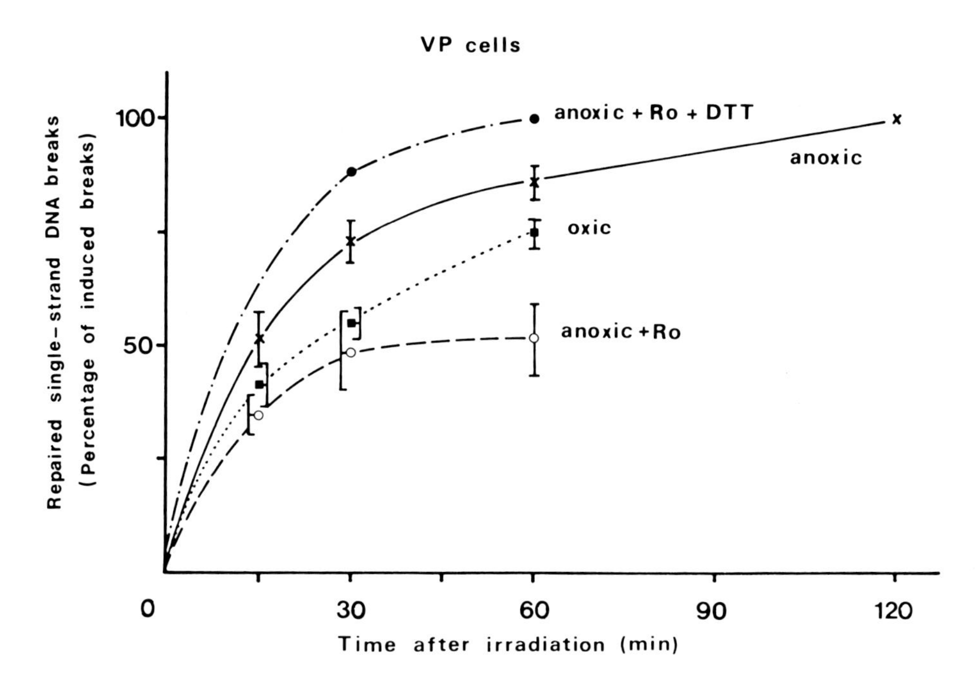

FIG.3. Relative amounts of repaired ssb at different times after irradiation of GSH-deficient fibroblasts (VP cells) under oxic (∿95% oxygen) or anoxic (< 5 ppm oxygen) conditions. When added, misonidazole (Ro) was present in 10 mM concentration during and after irradiation; dithiothreitol (DTT) was present in 10 mM concentration solely during the post-irradiation period. Means ± SE are indicated as calculated from a minimum of six replicate, paired or unpaired experiments, except the data illustrating the effect of DTT for which the points represent single determinations. Values are corrected for the induction of ssb by misonidazole toxicity.

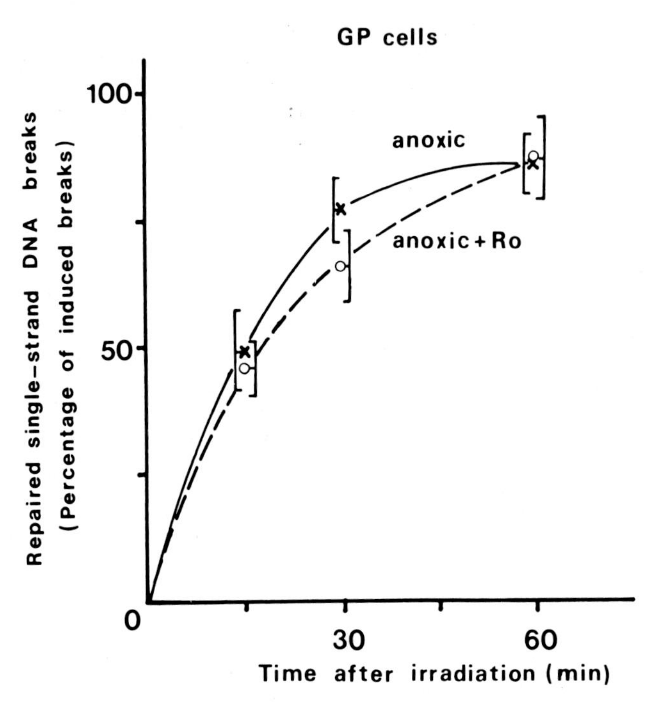

FIG.4. Relative amounts of repaired ssb at different times after irradiation of GSH-proficient fibroblasts (GP cells) under anoxic conditions with misonidazole (Ro) absent or present in 10 mM concentration during irradiation and the post-irradiation period. Results of five paired experiments are presented. The values are corrected for the induction of ssb by misonidazole toxicity.

in FIG 1, the yield of ssb after hypoxic irradiation of MEA treated JG cells decreases considerably in comparison to the untreated cells. As a result of DTT-treatment, practically all ssb induced by anoxic irradiation of misonidazole exposed VP cells is repaired within one hour (FIG 3) in contrast to the 50 per cent repair in the absence of DTT.

4. DISCUSSION

In earlier studies we have shown that misonidazole fails to sensitize to radiation hypoxic, lymphoblastoid cells (4) or fibroblasts (1) which had been derived from an individual with an inborn error of glutathione synthetase activity, and which almost totally lack glutathione. The present data obtained with lymphoblastoid cells derived from another individual with a similar genetically defined metabolic defect, confirm our earlier observations. In all these cases, the fraction of double-stranded DNA, or the number of single-strand breaks were used as an end-point. Recent studies in which clonogenic survival was chosen as the end-point are in agreement in indicating a greatly reduced, although not totally absent, enhancement of radiosensitivity by misonidazole for GSH deficient cells (5).

The dependence of the radiosensitizing effect of misonidazole on GSH is analogous to the dependence of the oxygen enhancement of radiosensitivity on this aminothiol (4). The similarity suggests that both oxygen and misonidazole may have a similar mechanism of sensitization. A competition model was suggested to explain the oxygen effect (6). According to this model oxygen reacts with radicals induced by radiation on essential target molecules, and fixes the damage there in a competition with hydrogen transfer from GSH which may repair the lesion. The observations presented here on the yield of single-strand breaks in misonidazole treated GSH-deficient and proficient cells, are consistent with a similar competition process between misonidazole and GSH.

The oxygen-mimic property of misonidazole is reflected also by the observations on the post-irradiation repair of ssb. It has been demonstrated that the repair process for a great part of the oxically induced breaks, in contrast to those induced anoxically, requires GSH probably as a co-enzyme (3). Our present data indicate that the repair of a great part of the ssb induced by radiation in the presence of misonidazole under anoxic conditions is also GSH dependent. The observation that a greater proportion of ssb remains unrepaired in GSH-deficient cells after anoxic irradiation in the presence of oxygen, may find an explanation in considering that misonidazole induces ssb in GSH-deficient cells independently of irradiation (cf. TAB. I). The repair of ssb induced by the misonidazole toxicity will also require GSH, or some other specific aminothiol. This interpretation is supported by the observation that addition of dithiothreitol (DTT) immediately after irradiation results in a nearly total repair of ssb in misonidazole treated, GSH-deficient cells.

Glutathione-deficient cells were shown to have a considerable concentration of aminothiols (3), consisting of cysteine and the dipeptides cysteinyl-glycine and glutamyl-cysteine (7). These aminothiols are, clearly ineffective as substitutes for GSH in the competition or repair processes discussed here, indicating that GSH has properties of a particular radiobiological importance. This conclusion is supported by recent studies with pulse-radiolysis (8) and cobalt-60 gamma radiolysis (9). As shown in the present and earlier reports (3, 4) some artificial aminothiols like cysteamine, dithiothreitol and mercapto-propionyl-glycine, can replace GSH effectively in processes in which other, naturally occurring aminothiols appear to be inefficient.

REFERENCES

(1) REVESZ, L., EDGREN, M., LARSSON, A., "Mechanism of radiosensitization by oxygen and misonidazole studied with glutathione deficient human fibroblasts in culture", in Radiation Research (Proc. 6th Int. Congr. Radiat. Res. Tokyo, May 1979: OKADA, S., et. al. Eds.), Japanese Assoc. for Radiation Research, Tokyo (1979) 862-866.

(2) LARSSON, A., NILSSON, K., Expression of glutathione synthetase deficiency in Ebstein-Barr virus transformed lymphoblastoid cell lines from patients with 5-oxoprolinuria, submitted to J. Inherited Metabol. Dis.

(3) EDGREN, M., LARSSON, A., REVESZ, L., Induction and repair of single-strand DNA breaks after X-irradiation of human fibroblasts deficient in glutathione, Int. J. Radiat. Biol. in press.

(4) EDGREN, M., LARSSON, A., NILSSON, K., REVESZ, L., SCOTT, O. C. A., Lack of oxygen-effect in glutathione deficient human cells in culture, Int. J. Radiat. Biol. 37 (1980) 299.

(5) DESCHAVANNE, P., Institut Gustave-Roussy, Villejuif, private communication, 1981.

(6) ALEXANDER, P., CHARLESBY, A., "Physio-chemical methods of protection against ionizing radiation", in Radiobiology Symp. Liége 1954: BACQ, Z. M., ALEXANDER, P., Eds., Butterworth, London (1955) 49-60.
(7) WELLNER, V. P., SEKURA, R., MEISTER, A., LARSSON, A., Glutathione synthetase deficiency, an inborn error of metabolism involving the gamma-glutamyl cycle in patients with 5-oxoprolinuria (pyroglutamic aciduria), Proc. Nat. Acad. Sci. USA 71 (1974) 2505.
(8) SJÖBERG, L., ERIKSEN, T. E., REVESZ, L., The reaction of the hydroxyl radical with glutathione in neutral and alkaline aqueous solution, Radiat. Res. (1981, in press).
(9) LAL, M., ^{60}Co gamma radiolysis of reduced glutathione in aerated solutions at pH values between 1-7.0, Can. J. Chem. 54 (1976) 1092.

Progress in Radio-Oncology II, edited by K. H. Kärcher et al., Raven Press, New York © 1982.

Possibilities of the Micronucleus Test as an Assay in Radiotherapy*

C. STREFFER, D. van BEUNINGEN, M. MOLLS
Institut für Medizinische Strahlenphysik und Strahlenbiologie,
Universitätsklinikum Essen,
Essen,
Federal Republic of Germany

Abstract

After treatment of cells with mutagenic agents, chromosome material appeared in the cytoplasm. It was shown that these micronuclei were formed from acentric chromosomal fragments or from whole chromosomes. After irradiation with X-rays a good correlation between the number of micronuclei and the clonogenic survival was observed in human melanoma cells. However, this did not occur after neutron irradiation. With neutrons the number of micronuclei per cell increased considerably, and therefore a simple evaluation of the test system as could be done following X-irradiation was not possible for neutrons. The mechanisms of these processes were studied with a fast proliferating cell system, the pre-implanted mouse embryo, in dependence on the radiation dose and the number of mitoses after irradiation. With moderate doses (1–5 Gy) the cells apparently had to go through one mitosis before micronuclei occurred. In contrast to chromosome aberration counts, micronuclei could be determined during interphase, the test was easy to perform and only small numbers of cells were needed. In further experiments the micronuclei were determined in human tumour biopsies before and after therapeutic irradiation and in stimulated lymphocytes from patients with non-Hodgkin lymphoma who were given whole-body irradiation later. Apparently the micronucleus test could be used as a prognostic measure for radiation response.

1. INTRODUCTION

After treatment of proliferating cells with ionizing radiation or mutagenic chemicals, chromatin particles (so-called micronuclei) can be observed in the cytoplasm [1-3]. The micronuclei are a useful indicator of cytogenetic damage. However not all chromosomal aberrations lead to the formation of micronuclei. It is generally believed that micronuclei derive from acentric chromosomal fragments or total chromosomes which are not correctly included into the daughter cell nuclei during mitosis [4].

The micronucleus test has been proposed as a simple assay to measure the cytogenetic damage of various cytotoxic agents or the combined action of such agents [1,2,5]. There has also been a description of a correlation between the number of micronuclei and the clonogenic capacity of CHO cells (for the D_0 of the survival curve) following X-irradiation [6].

In radiotherapy it is frequently found that the radiosensitivity of tumours differs between individual patients, while tumours sometimes regress rather late after irradiation. The

* The investigations were supported by the Deutsche Forschungsgemeinschaft (SFB 102) and the Bundesminister für Forschung und Technologie.

aim of the investigations described here is to see whether the micronucleus test could be of any prognostic value in this connection. Formation of micronuclei has been measured in stimulated lymphocytes of patients with non-Hodgkin lymphoma (NHL) after irradiation in vitro and in human rectum carcinoma after preoperative irradiation. In order to get a better insight into the mechanism and time course of micronucleus formation, experiments with preimplanted mouse embryos and human melanoma cells were performed after irradiation with X-rays and fast neutrons in vitro.

2. MATERIALS AND METHODS

Preimplanted mouse embryos were isolated at the two-cell stage, irradiated in the G_2-phase of this stage and cultivated in vitro as described elsewhere [7,8]. The melanoma cells were grown and irradiated as a monolayer [9]. For clonogenic survival, the cells were plated with feeder cells which were irradiated with 40 Gy in Falcon petri dishes. The colonies were counted 16 days after plating. Human lymphocytes from normal persons and patients with NHL were isolated by Ficoll-gradient centrifuging from the blood, and incubated with the mitogen phytohaemagglutinin (PHA). Biopsies from human rectum carcinoma were mechanically pressed through a nylon gauze in order to obtain single-cell preparations.

For the micronucleus test, the cells received a hypotonic treatment. After fixation with an acetic acid/ethanol mixture and air-drying, the cell nuclei and micronuclei were stained with the fluorescent dye ethidium bromide and counted under a microscope [7,8].

The preimplanted mouse embryos, the melanoma cells and the lymphocytes were irradiated either with 240 kV X-rays [8,9] or with cyclotron neutrons (D→Be, average energy: 7 MeV [10]). The radiotherapeutic treatment of the rectum carcinoma was performed in the Universitäts-Strahlenklinik Essen, while biopsies were taken in the Chirurgische Universitätsklinik Essen.

3. RESULTS AND DISCUSSION

3.1. Micronuclei in Preimplanted Mouse Embryos

The occurrence of micronuclei is dependent on cell proliferation. Preimplanted mouse embryos were chosen because only proliferating cells are present in this system and the number of cell divisions can be determined after irradiation with a rather good precision. The embryos were irradiated with X-rays at the 2-cell stage; 15 hours after irradiation the embryo cultures contain a mixture of embryos with 4-8 cells per embryo. When the number of micronuclei are measured in all embryos of these cultures, the relative increase of micronuclei per dose unit appears to be greater in the lower-dose range than in the higher-dose range (Fig.1).

This dose/effect curve is dependent on two 'overlapping' effects, which can be demonstrated by studying micronucleus

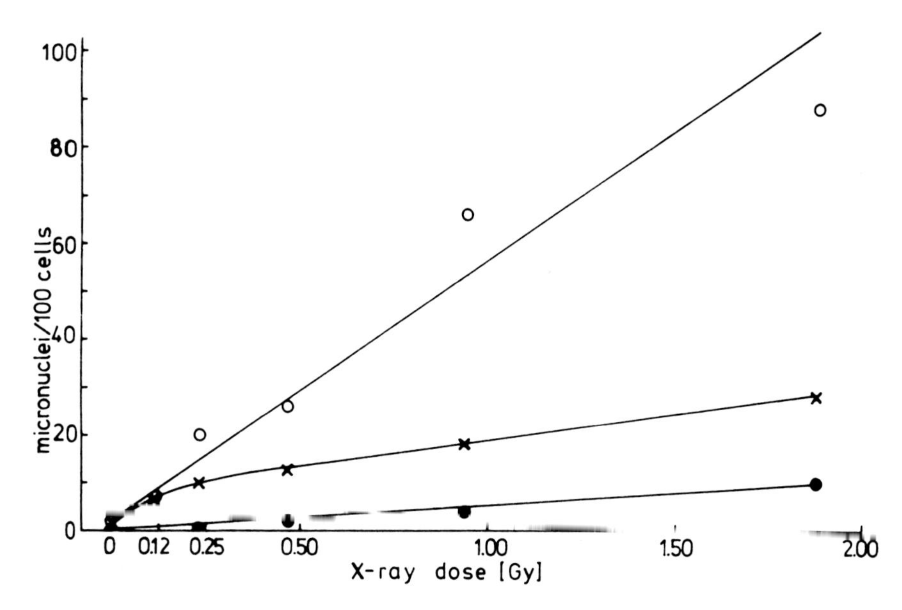

FIG.1. Micronuclei in preimplanted mouse embryos 15 hours after X-irradiation: (X) all embryos (4, 5, 6, 7, and 8-cell stage); (•) embryos in 4-cell stage; (o) embryos in 8-cell stage.

formation in embryos with 4 cells and in embryos with 8 cells separately (Fig.1). The data show that the number of micronuclei is much greater in those embryos which have already developed from the 2-cell stage to the 8-cell stage than in those which are still in the 4-cell stage. A radiation dose which induces a delay of mitosis will also postpone the formation of micronuclei. This mitotic delay increases with radiation dose and modifies the dose/effect curve when the data from all embryos (4-8 cell stage) are analysed together. At later periods (39 and 63 hours after irradiation), this effect is no longer significant and the dose/effect curve becomes linear [8].

These data clearly show that the number of mitoses and, through this parameter, the time after irradiation are important in determining the extent to which micronuclei are observed as well as the shape of the dose/effect curve.

3.2. Micronuclei in Melanoma Cells

In comparison with preimplanted mouse embryos, the situation observed with the melanoma cells is more complex, although the irradiation was performed at a time when all cells were still proliferating and an exponential growth curve was found [9]. The dose dependence of micronucleus formation is shown for different times after X-irradiation in Fig.2. It is seen that the number of micronuclei increases with time; a constant level (a plateau) is reached between 72-120 hours after irradiation. The cells have a cell cycle time of about 44 hours under these conditions. The plateau for the number of micronuclei is apparently reached after the cells have gone through somewhat more than one cell cycle, bearing in mind the mitotic delay of some hours.

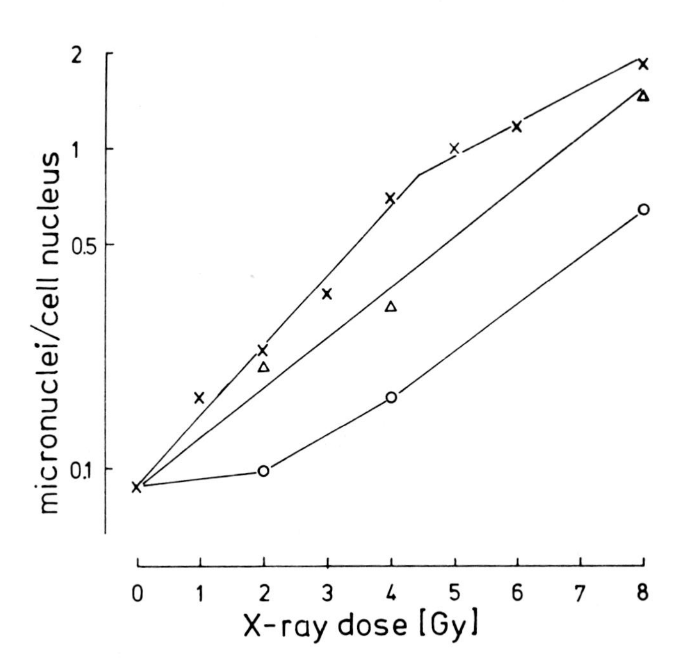

FIG.2. Number of micronuclei per cell nucleus (logarithmic scale) in human melanoma cells as a function of X-ray dose 24 hours (o), 48 hours (Δ) and 72 hours (X) after irradiation in vitro. The cells were cultivated in monolayers.

If the number of micronuclei in the plateau is compared with the clonogenic survival of the melanoma cells, a good correlation between the two dose/effect curves is observed for a certain dose range (Fig.3). Midander and Révész have reported similar data for CHO cells [6]. When a radiation dose is reached which causes on average one micronucleus per cell, the two dose/effect curves start to deviate from each other (Fig.3). Such correspondence does not exist for any dose range after irradiation with neutrons. Especially at low radiation doses, neutrons cause a much greater increase in micronuclei than do X-rays.

These data can be partly explained by experiments in which not only the total number of micronuclei is counted but also the numbers of micronuclei formed in each individual cell nucleus are determined. On the basis of such experiments, cell nuclei with 0, 1, 2, 3, 4 and ≥5 micronuclei were classified and the data plotted in Fig.4. After radiation doses which decrease the clonogenic survival to less than 1% (see Fig.3), almost 50% of the cells still have no micronucleus (Fig.4). Apparently not all 'dying' cells express their cytogenetic damage by forming micronuclei, although they have gone through a mitosis after irradiation. Comparison of the effects induced by the two radiation qualities demonstrates that neutron irradiation induces a number of micronuclei per cell greater than one in more cells than does X-irradiation, though the number of cells with more than one micronucleus per cell increases with radiation dose for both radiation qualities. In the low dose range the difference

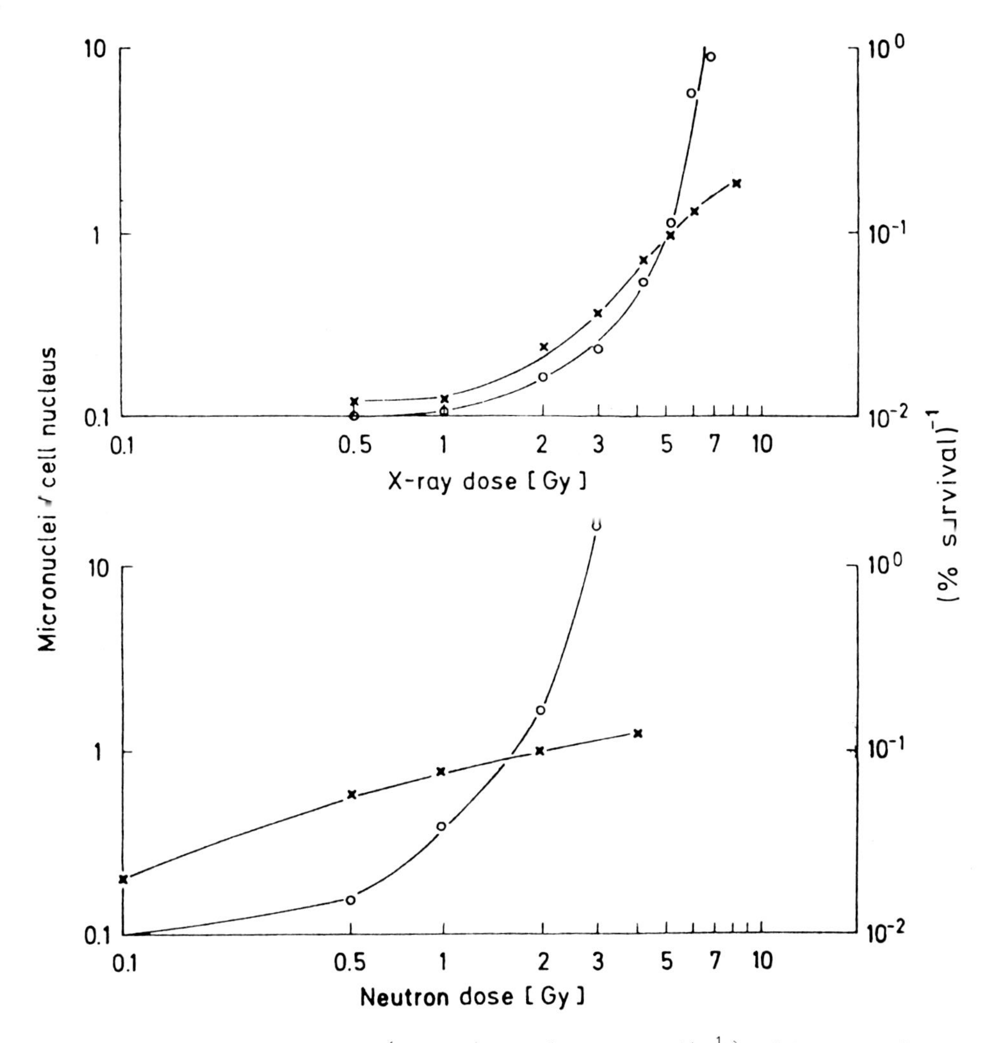

FIG.3. Clonogenic survival (plotted as (% survival)$^{-1}$) of human melanoma cells (o) and number of micronuclei per cell nucleus (X) plotted against radiation dose after irradiation with X-rays or fast neutrons. Micronuclei were determined 144 hours after irradiation.

between the qualities is very significant (Fig.4). These phenomena may be responsible for the lack of a general correlation between the number of micronuclei and the clonogenic survival obtained following neutron irradiation in general and following X-irradiation in the high-dose range.

3.3. Micronuclei in Lymphocytes

In order to obtain radiation-induced micronuclei in lymphocytes, the cells must be stimulated by a mitogen after irradiation with doses of some grays [3,11]. Patients with NHL obtained therapeutic whole-body irradiation with radiation doses of 0.15 Gy per fraction to a total of 1.5 Gy [12]. Blood samples

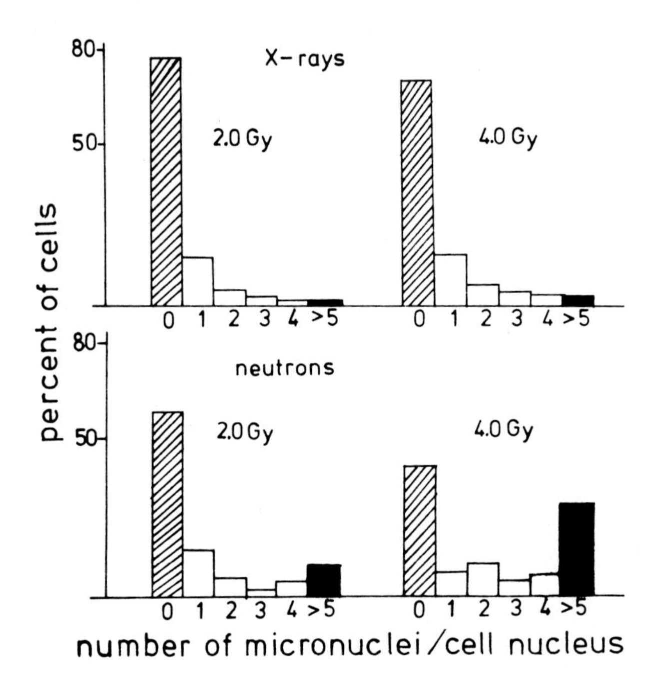

FIG.4. Human melanoma cells with 0, 1, 2, 3, 4 and ⩾ 5 micronuclei per cell (in per cent) 144 hours after X or neutron irradiation. Cells were cultivated in monolayers.

were taken before the first irradiation was performed. The isolated lymphocytes were incubated with PHA and X-irradiated directly after placing them in a culture medium.

Under these conditions micronuclei develop in the proliferating lymphocytes. The number of micronuclei in lymphocytes from patients are compared with those in lymphocytes from normal persons (Fig.5). The shaded area delimits the range of these values (mean value ± SE) for normal persons. In the lymphocytes from five out of six NHL patients, an increase in micronuclei was observed which falls within the normal range or is higher (Fig.5). These patients have been in full remission for at least one year after radiotherapeutic treatment. No micronuclei were found in the lymphocytes from a sixth patient after irradiation in vitro. This patient only had partial remission.

Investigations are being continued to see whether the micronucleus test in lymphocytes might prove to be an indicator for the degree of radiosensitivity of individual patients and whether it might be of any prognostic value for the response of certain malignancies to radiotherapy.

3.4. Micronuclei in Rectum Carcinoma

A study is being performed in our university clinics to evaluate the effect of preoperative radiotherapy on rectum carcinoma. As with other tumours, it has been observed that the tumour response varies considerably. The aim of the experiments is to investigate whether a correlation exists between the clinical observations and the formation of micronuclei. Biopsies

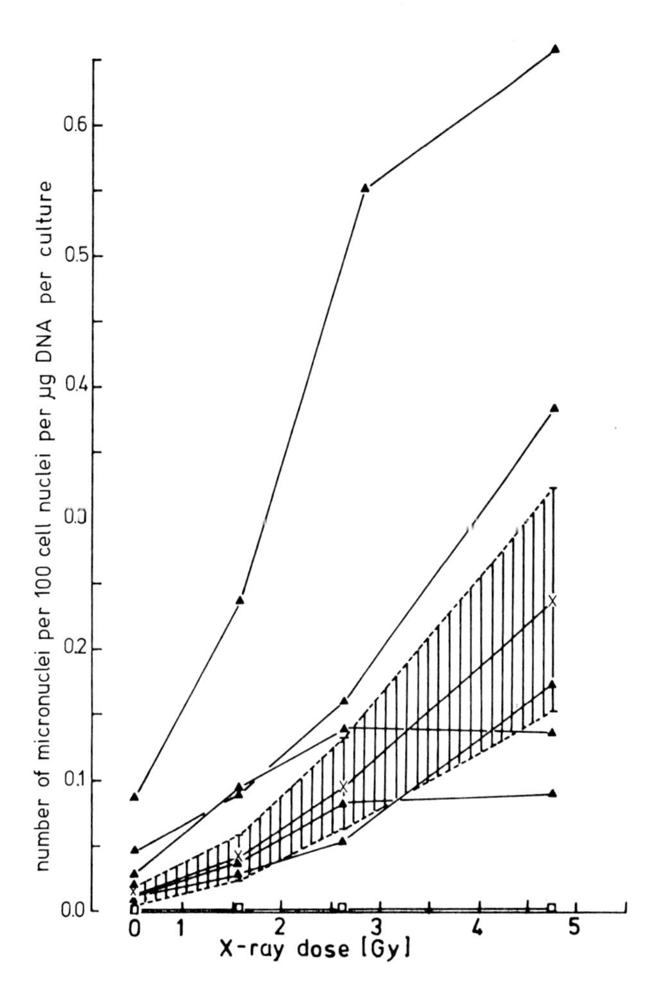

FIG.5. Micronuclei in human lymphocytes after X-irradiation in vitro and a 72-hour cultivation with PHA: normal persons ± SE (X and limits of shaded area); patients with NHL, full remission after radiotherapy (▲); patient with NHL, no remission after radiotherapy (□).

were taken before the first radiation dose, during radiotherapy and at the time of surgery.

Micronuclei were already observed in tumour cells, though to a varying degree, before irradiation. This may be a sign of dying necrotic cells. Micronuclei were found in all tumours after irradiation.

Examples for two patients are presented in Fig.6. In the tumour of patient A, the number of micronuclei increased several fold after irradiation. Tumour regression in this patient was very rapid and no tumour was found at the time of surgery - three weeks after irradiation. No radiation-induced increase of micronuclei was seen for patient B; tumour regression has been slow and tumour tissue was still observed during surgery (Fig.6). DNA measurements have shown that both tumours are diploid and

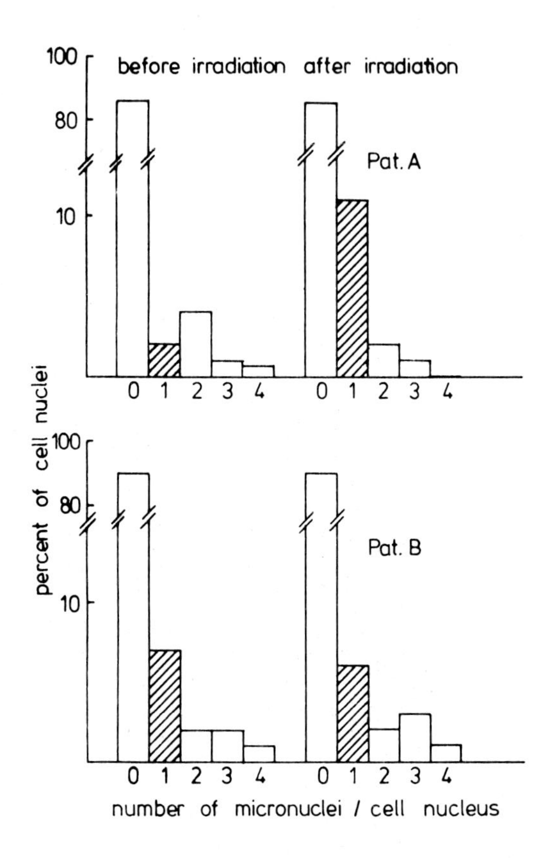

FIG.6. Cell nuclei with 0, 1, 2, 3 and 4 micronuclei per cell (per cent) in tumour biopsies from two patients (A and B) with rectum carcinoma, before and after preoperative irradiation (see text).

the degree of proliferation (number of S-phase cells) was comparable (unpublished results).

However, further data are required to indicate whether determination of micronuclei can be of any prognostic value for tumour response after radiotherapy. From the complexity of the system (its dependence on cell proliferation, etc.), it appears improbable that quantitative criteria can be developed, though estimations might be made in a qualitative way.

4. CONCLUSIONS

1. After radiation exposure, micronuclei are formed in proliferating cells. The assay is easy to perform. This cytogenetic damage can also be observed during the interphase.
2. The quantitative evaluation of dose response is dependent on the proliferation rate and time after irradiation.
3. In experimental systems a correlation exists between the number of micronuclei and clonogenic survival over a certain dose range for low-LET radiation under well-defined conditions. This is

not the case for irradiation with neutrons. It may be necessary not only to count the total number of micronuclei in a cell population but also to evaluate the number in each individual cell.

4. The preliminary data suggest that the micronucleus test might be used in a qualitative way to obtain prognostic information about the radiosensitivity of patients and tumours.

ACKNOWLEDGEMENTS

The authors gratefully acknowledge the co-operation of Professor Dr. E. Scherer, director of the Universitäts-Strahlenklinik Essen, and Professor Dr. F.W. Eigler, director of the Chirurgische Universitätsklinik Essen, and their respective staffs. The authors also express their thanks to G. Bertholdt for performing the colony-forming tests on the human melanoma cells. Finally, they would like to acknowledge the technical assistance of Mrs. Kaiser and Miss Rebmann.

REFERENCES

[1] HEDDLE, J.A.: Mutat. Res. 18 (1973) 187.
[2] SCHMID, W.: Mutat. Res. 31 (1975) 9.
[3] COUNTRYMAN, P.I., HEDDLE, J.A.: Mutat. Res. 41 (1976) 321.
[4] CARRANO, A.V., HEDDLE, J.A.: J. Theor. Biol. 38 (1973) 289.
[5] STREFFER, C., van BEUNINGEN, D., MOLLS, M., PON, A., SCHULZ, S., ZAMBOGLOU, N.: in Late Biological Effects Of Ionizing Radiation (Proc. Symp. Vienna, 1978) Vol.2, IAEA, Vienna (1978) 381.
[6] MIDANDER, J., RÉVÉSZ, L.: Br. J. Cancer 41 (1980) 204.
[7] STREFFER, C., van BEUNINGEN, D., MOLLS, M., ZAMBOGLOU, N., SCHULZ, S.: Cell Tissue Kinet. 13 (1980) 135.
[8] MOLLS, M., STREFFER, C., ZAMBOGLOU, N.: Int. J. Radiat. Biol. Relat. Stud. Phys., Chem. Med. 39 (1981) 307.
[9] van BEUNINGEN, D., STREFFER, C., ZAMBOGLOU, N., SCHUBERT, B., LINDSCHEID, K.R.: in Kombinierte Strahlen- und Chemotherapie (WANNENMACHER, M., Ed.), Urban and Schwarzenberg, Munich, Vienna, Baltimore (1979).
[10] RASSOW, J.: in Risiken und Nutzung der Strahlentherapie bösartiger Tumoren (MESSERSCHMIDT, O., SCHERER, E., Eds), Georg Thieme, Stuttgart (1978).
[11] HOLLSTEIN, M., McCANN, J.: Mutat. Res. 65 (1979) 185.
[12] BAMBERG, M., SCHULZ, U., ÖHL, S., HOLFELD, H., LENZ, W.: in Kombinierte Strahlen- und Chemotherapie (WANNENMACHER, M., Ed.), Urban and Schwarzenberg, Munich, Vienna, Baltimore (1979).

Progress in Radio-Oncology II, edited by K. H. Kärcher et al., Raven Press, New York © 1982.

Relative Toxicity of Different Nitroimidazole Sensitizers*

R.M. SUTHERLAND, P.J. CONROY, D.W. SIEMANN
University of Rochester,
School of Medicine and Dentistry,
Cancer Center,
Rochester, New York,
United States of America

Abstract

Different nitroimidazoles were compared for in vitro monolayer and multicell spheroid cytotoxicity, in vivo tumour cytotoxicity, neurotoxicity, and enhanced chemotoxicity in order to obtain information relative to mechanism(s) of these effects. The results indicate that relative effectiveness of the different nitroimidazoles is dependent in a major way on their hydrophilicity for in vitro and in vivo cytotoxicity and neurotoxicity. More hydrophilic compounds tended to be less toxic. However, significant exceptions to this relationship were found especially in experiments to evaluate enhanced chemotoxicity. The results indicate that properties other than pharmacokinetics of the nitroimidazoles must be considered in order to explain the mechanism(s) for certain nitroimidazoles and types of toxicity.

1. INTRODUCTION

Certain nitroimidazole compounds can sensitize hypoxic mammalian cells to radiation but are not toxic to aerobic cells when present for short periods of time before and during irradiation (1). However, other experiments have demonstrated cytotoxicity to both hypoxic and aerobic cells after long exposures in the absence of irradiation (2, 3, 4, 5, 6). In addition, neurotoxicity which also does not appear to be related to interaction with radiation, has occurred, both in rodents and humans, after administration of significant cumulative doses (7, 8, 9, 10, 11, 12). More recently the toxicity of certain cancer chemotherapeutic agents has been enhanced both *in vitro* and *in vivo* by addition of nitroimidazoles (13, 14, 15, 16, 17, 18, 19). The mechanism(s) of these toxicities and their possible interrelationships are unknown.

Although much is known about the relative radiation sensitizing capacities of different nitroimidazoles both *in vitro* and *in vivo*, limited data are available on the relative effectiveness of nitroimidazoles for producing cytotoxicity, neurotoxicity and enhanced chemotoxicity. Therefore, we have undertaken a comparison of different nitroimidazoles for these different toxicities using a variety of

* The research in this publication was supported by NIH Grants CA-11051, CA-20329, CA-11198, NS-16147, CM-87175, and PHS-BRSG Award, University of Rochester School of Medicine and Dentistry.

response assays for *in vitro* monolayer and multicell spheroid cultures, *in vivo* mouse tumors and nervous system functions. Specific parts of these studies and details of the methods used have been reported previously (6, 13, 18, 19, 11, 12). The nitroimidazoles evaluated were metronidazole = [1(2-hydroxyethyl)-2-methyl-5-nitroimidazole], misonidazole = [1(2-nitro-1-imidazoyl)-3-methoxy-2-propanol], and other 2-nitroimidazoles with side chains in the 1 position as follows: Ro-05-9963, R= $CH_2CHOHCH_2OH$; Ro-07-0741, R=$CH_2CHOHCH_2OF$; SR2508, R=$CH_2CONHCH_2CH_2OH$; SR2555, R=$CH_2CON(CH_2CH_2OH)_2$.

2.1 *IN VITRO* CYTOTOXICITY

Monolayer cultures of EMT6/Ro cells were exposed to 5mM concentrations of different nitroimidazoles for different times under hypoxic conditions. The cells were grown on glass dishes, placed in special aluminum chambers, sealed and evacuated to produce less than 20 ppm O_2. The glass dishes contained wells into which small volumes of nitroimidazoles at high concentration were placed prior to the evacuation procedure required to produce hypoxia. By tipping the chambers containing the dishes, the nitroimidazoles were added to the cell cultures under hypoxic conditions. At various times the dishes were removed from the chambers and the cells were harvested, counted and plated to determine clonogenic surviving fraction (CSF). Both the cells recovered per dish and the clonogenic potential of those cells were taken into account in determining the CSF. By plotting CSF versus time over a 6-hour exposure period for the different nitroimidazoles, it was possible to determine the drug exposure time required to reduce CSF to 0.5 and 0.1 (Table I). Misonidazole reduced CSF to 0.5 and 0.1 after 1.1 or 1.9 h exposure respectively, approximately equitoxic to desmethylmisonidazole (Ro-05-9963) under these conditions despite the fact that this compound is more hydrophilic (Table III). Other nitroimidazoles, SR 2508 and SR 2555, with the same electron affinity ($\sim$-390mV), but, with greater hydrophilicity were less cytotoxic than misonidazole (relative toxicity = 0.6 for SR 2508 and $\leq$ 0.3 for SR 2555).

By comparison, when EMT6/Ro cells grown as multicell spheroids of approximately 800 μm diameter and containing large necrotic centers were exposed to these compounds in spinner flasks under normal growth conditions, toxicity was markedly reduced compared with monolayer culture in the low O_2 concentration in the chamber. Misonidazole reduced CSF to 0.5 and 0.1 after 10 or 24 h respectively. These longer exposure periods required for toxicity in the spheroids are probably related to the presence of higher O_2 concentrations, perhaps 2000 to 4000 ppm in the centers of some types of spheroids (20 and Mueller-Klieser [unpublished]) despite the presence of necrosis. Other differences occur in the toxicity found in spheroids versus monolayers (Table I). Ro-05-9963 was more cytotoxic than misonidazole (relative toxicity = 1.7) and, although the toxicity of SR 2508 and SR 2555 was again reduced in correlation with increasing hydrophilicity, the reduction in toxicity was not as marked as in the monolayer experiments (relative toxicity of SR 2508 = 0.8 and SR 2555 = 0.6 to 0.7). It appears from these results that the relative toxicities of the different nitroimidazoles can be significantly influenced by the environmental and other unknown factors in the *in vitro* test systems. Clearly, spheroids differ from monolayer cultures used in these experiments, not only in the oxygen concentrations, but also due to the

TABLE I

CYTOTOXICITY OF DIFFERENT NITROIMIDAZOLE COMPOUNDS [(a)]

Nitroimidazole Compound	Exposure (h) CSF=0.5	Rel.Toxicity CSF=0.5	Exposure(h) CSF=0.1	Rel. Toxicity CSF=0.1
A. EXPONENTIAL MONOLAYER CULTURES [(b)]				
Misonidazole	1.1	1.0	1.9	1.0
Ro-05-9963	1.0	1.1	1.7	1.1
SR 2508	2.0	0.6	3.0	0.6
SR 2555	3.5	0.3	>6.0	<0.3
B. MULTICELL SPHEROID CULTURES [(b)]				
Misonidazole	10	1.0	24	1.0
Ro-05-9963	6	1.7	15	1.7
SR 2508	12	0.8	28	0.8
SR 2555	18	0.6	36	0.7

(a) Nitroimidazole concentration tested = 5mM

(b) EMT6/Ro cells

presence of necrotic products, gradients of other substances and to the presence of chronically hypoxic cells.

2.2. *IN VIVO* CYTOTOXICITY

Our previous studies were performed to determine the kinetics and magnitude of misonidazole cytotoxicity in EMT6/Ro tumors using an *in vivo - in vitro* clonogenicity assay (21). These studies showed that large single doses (>5 mmol/kg, 1 mg/g) of misonidazole in the mouse can kill up to 90% of EMT6/Ro tumor cells in 18-24 hours. No significant cytotoxicity was observed with single i.p. doses of 2.5 mmol/kg (0.5 mg/g or less). Continous venous infusion of misonidazole to mice resulted in significant cytotoxicity (30-60%) at high clinically-achievable dose levels (0.5-10 mM, 100-200 μg/ml). Because of the relatively rapid recovery in the number of clonogenic tumor cells (21, 13), these cytotoxic effects were not reflected as changes in tumor size. In general, these data suggested that with misonidazole we may expect no more than a two-fold reduction in clonogenic cells within human tumors with a hypoxic fraction of between 6 and 25%. In order to determine if the relatively large cytotoxic effects of misonidazole in EMT6/Ro tumors are observed with other sensitizers and in other tumor cell lines, we have compared the *in vivo* cytotoxicity of misonidazole with desmethylmisonidazole (Ro-05-9963), Ro-07-0741, SR 2508, SR 2555, and metronidazole in EMT6/Ro and KHT/Ro tumors using clonogenic cell survival (CSF) and tumor growth delay (TGD) assays. The results are presented in summary form in Table II. The single dose cytotoxicity of the various sensitizers compared to misonidazole is computed by the ratio of administered (mmol/kg) or tumor exposure doses (mM·h) to produce a CSF of 0.5 or 0.1 at the time of maximum response (18 to 24 hours). In EMT6/Ro tumors (350-450mg, 20-25% hypoxic fraction) misonidazole, demethylmisonidazole, and SR 2508 were equitoxic (CSF 0.5)at the relatively low dose level of 2.5 mmol/kg (1.7 mM·h); Ro-07-0741 was 4 times more effective and metronidazole 3 times less effective. With larger doses, misonidazole administration gave a CSF of 0.1 at 4.75 mmol/kg; desmethylmisonidazole and SR 2508 were approximately 2 times less effective per administered dose, but equitoxic for tumor exposure doses of ∿6.0 mM·h. Similar studies in the KHT/Ro tumor (650-850mg, 20-25% hypoxic fraction) indicate that the cytotoxic effectiveness of these sensitizers is reduced by a factor of at least 2, depending on the sensitizer studied and the route of administration (administered dose range 0.5-10.0 mmol/kg. Misonidazole reduced CSF to 0.5 at 4.75 mmol/kg (3.0 mM·h); desmethylmisonidazole (i.p.) and SR 2508 were approximately 2 times less toxic per administered and tumor exposure dose; Ro-07-0741 was 3 times more toxic, and metronidazole 2-3 times less toxic. I.v.administration of desmethylmisonidazole was more effective than i.p. injection (factor of 2), since higher peak tumor levels were achieved. Near lethal doses of misonidazole (>7.5 mmol/kg) and Ro-07-0741 (3.0 mmoles/kg) were required to reduce CSF to 0.1. None of the other sensitizers studied could be administered in doses large enough to reduce CSF to 0.1. Some of the differences in response to sensitizer treatment are related to the lower peak tumor levels (up to 50%) and hence tumor exposure dose for similar administered doses in C3H mice bearing the KHT/Ro tumor compared to Balb/c mice bearing the EMT6/Ro tumor. KHT/Ro tumors appear to be more resistant than EMT6/Ro tumors (by a factor of 2) to the cytotoxic effects of misonidazole (CSF of 0.5). At tumor exposure doses in excess of 5 mM·h, such differences are not apparent (CSF of 0.1). No growth delay for misonidazole was observed in either tumor line. In an overall sense, the results tend to indicate that the direct cytotoxic effects of these nitroimidazole compounds at clinically achievable dose levels may be relatively small or absent.

2.3 NEUROTOXICITY

Our previous studies in the mouse have shown that the quantitative functional tests (rotarod performance) when verified by clinical observation and histology are of significant value for assessing the relative neurotoxicity of nitroimidazole radiosensitizers. The rotarod performance test (the ability of a mouse to maintain locomotor coordination and balance on a rotating rod which increases linearly in speed) was used to establish dose-response curves of neurotoxicity of misonidazole, desmethylmisonidazole, Ro-07-0741, SR 2508, and metronidazole. The endpoint for comparison was the point where relative rotarod retention times declined to a value 50% of that prevailing before chronic sensitizer administration. In general, this isoeffect point preceded the appearance of severe neurological deficit or weight loss greater than 25%. The relative neurotoxicity on the basis of administered dose and clinical scoring for these compounds is summarized in Table II. SR 2508 was the least neurotoxic and Ro-07-0741 was the most neurotoxic agent studied.

TABLE II IN VIVO CYTOTOXICITY AND NEUROTOXICITY OF NITROIMIDAZOLES

Compound	Relative Toxicity Single Dose $LD_{50/2}$	Octanol:Water Partition Coefficient	Relative Cytotoxicity[b] Single Dose[c]				Relative Neurotoxicity Multiple Dose	
			EMT6		KHT		Rotarod[h]	Hearing Loss[i]
Misonidazole	1.0[a]	0.43	1.0[d]	1.0[e]	1.0[f]	1.0[g]	1.0[j]	1.0[k]
Ro-05-9963	0.5	0.13	∿0.9	1.0	0.4	0.5	0.6	0.5
Ro-05-9963[1]	-	0.13	-	-	0.7	1.0	-	-
Ro-07-0741	2.0	0.44	4.2	1.7	3.4	2.7	3.0	3.3
SR2508[1]	0.4	0.05	∿1.0	1.0	0.5	0.5	0.2	0.2
SR2555[1]	0.3	0.02	-	-	-	-	-	0.1
Metronidazole	0.4	0.96	∿0.4	∿0.2	∿0.5	0.3	0.4	0.3

(a) $LD_{50/2}$ (i.p.) for misonidazole: 8.0 mmol/kg (1.8 mg/g) in female Balb/c mice (20-25 g)

(b) Administered dose (i.p.) to produce clonogenic surviving fraction (CSF) of 0.5

(c) Dose range 0.5-12.5 mmol/kg (0.1-2.5 mg/g)

(d) Administered dose 2.5 mmol/kg (0.5 mg/g)

(e) Exposure dose 1.7 mM•h

(f) Administered dose 4.75 mmol/kg (0.96 mg/g)

(g) Exposure dose 3.0 mM•h

(h) Isoeffect is the total dose required to reduce rotarod retention in Balb/c mice to 50% of the initial value(12)

(i) Total administered dose 75 mmol/kg (15 mg/g)

(j) Isoeffect is the total dose required to produce 100% hearing loss at 32 kHz in Balb/c mice

(k) Total administered dose 60 mmol/kg (12 mg/g)

(1) I.V. injection

Ro-07-0741 produced damage to the distal areas of peripheral nerves within 1 week (total administered dose 7.5 mmol/kg, 14 mg/g) and lesions of the vestibular nuclei, cochlear nuclei, vestibulo-cerebellar and auditory tracts at 2 weeks (total dose 14 mmol/kg, 2.8 mg/g). Misonidazole administration (15-20 mmol/kg·week, 3-4 mg/g·week) produced similar peripheral lesions at 2 weeks and CNS lesions in 3-4 weeks. Desmethylmisonidazole (Ro-05-9963) treatment resulted in similar CNS lesions in approximately 5 weeks at a dose of 25 mmol/kg·week (5 mg/g·week). Further morphologic studies are in progress to compare peripheral and CNS lesion development in different mouse strains treated with misonidazole or desmethylmisonidazole for different routes of administration (oral, i.p., i.v.). The specificity of the CNS lesions in the brain stem of mice treated with nitroimidazole sensitizers is remarkable since surrounding areas appear unaffected. A study of the specific localization of ^{14}C-misonidazole in 10 areas of the CNS and peripheral nerves of mice did not indicate preferential distribution. Our initial interpretation is that lesion specificity may be better understood on the basis of differential sensitivity of nerve cells and not specific distribution of the drug.

One problem associated with the rotarod performance test is that it is laborious and an indirect measure of neurotoxicity since both sensory and motor deficit are involved in the endpoint. Current clinical trials have reported that in addition to peripheral neuropathy and convulsions, transient tinnitus and high frequency hearing loss (ototoxicity) have also occurred in patients. Therefore, we have conducted studies to assess the ototoxic potential of sensitizers in 2 strains of mice (Balb/c and C3H/heJ). The method involved modulation of the startle reflex in mice and is a direct measure of sensory deficit. High frequency hearing losses (16 or 32 kHz) can be demonstrated in both strains of mice before overt clinical symptoms, significant rotarod performance deficit, or weight loss. The time taken (weeks) to produce 100% loss of 32kHz hearing when plotted against total administered dose has been used to generate dose response curves for misonidazole, desmethylmisonidazole, Ro-07-0741, SR 2508, SR 2555, and metronidazole. These data are summarized in Table II. In broad terms, the ototoxicity data compare favorably with relative neurotoxicity as determined by the rotarod test and is considered to be a more sensitive measure. These hearing losses are reversible, as in humans, at least for the 2 sensitizers misonidazole and Ro-07-0741 studied in detail over a 2-month period after dosing stopped.

It is clear that considerably larger cumulative doses of these sensitizers are required to produce neurotoxicity in mice with chronic administration (10-40 times) compared to the doses required to demonstrate cytotoxicity following single doses. On the basis of cumulative brain exposure dose (300-400 mM·h), misonidazole, desmethylmisonidazole, and SR 2508 are equally toxic using the functional endpoints described. Ro-07-0741 has a similar electron affinity (-390mV), but is 2-3 times more toxic. Metronidazole is 2-3 times less neurotoxic per exposure dose and is also less effective as a sensitizer since its electron affinity is lower (-460mV).

2.4 <u>ENHANCED CHEMOTOXICITY</u>

Recent investigations have demonstrated that the addition of nitroimidazoles to certain cancer chemotherapeutic agents may considerably enhance the level of tumor cell cytotoxicity of these agents *in vivo* (14, 15, 16, 17, 18, 19, 22). Table III shows data from our laboratories for different anti-tumor agents combined with misonidazole. The results were obtained using either *in situ* tumor growth delay or *in vivo* to *in vitro* clonogenic cell survival assays. While misonidazole leads to little or no toxicity, the addition of this nitroimidazole to treatments with such agents as Cy, L-PAM, BCNU, and CCNU can markedly enhance the tumor response. This enhancement is time (18) and nitroimidazole dose (19) dependent. In general, administering the two agents relatively close to one another appears optimum and at least with the more effective combinations (such as CCNU plus misonidazole)enhanced tumor cytotoxicity can be observed at considerably reduced doses of the sensitizer.

The data in Table III indicate that the effectiveness of a combination is strongly dependent on the chemotherapeutic agent chosen. In particular, it should be noted that very similar anti-tumor agents, such as, for example, BCNU and CCNU, when combined with the same nitroimadazole may lead to significant differences in the level of enhancement that can be achieved. In addition, combining the same chemotherapeutic agent with closely related nitroimidazoles can also result in substantially different effects. This was initially demonstrated in the KHT sarcoma by Mulcahy et al.(18) for combinations of BCNU with either misonidazole or Ro-05-9963.

TABLE III

THE RESPONSE OF MURINE TUMOR MODELS TO COMBINATIONS OF VARIOUS CHEMOTHERAPEUTIC AGENTS AND THE RADIOSENSITIZER MISONIDAZOLE [(a)]

Tumor Model	Chemotherapeutic Agent [(b)]	DEF
EMT6/Ro Tumor	ADR	∿1.0
KHT Sarcoma	Cy	∿1.3-1.5
	L-PAM	∿1.3
	BCNU	∿1.3-1.8
	CLZ	∿1.0
	CCNU	∿1.8-2.4
RIF-1 Tumor	CCNU	∿2.0

(a) Misonidazole dose = 1.0 mg/g i.p.

(b) Chemotherapeutic agent abbreviations used are: ADR (adriamycin), Cy (cyclophosphamide), L-PAM (phenylalanine mustard), BCNU (1,3-bis(2-chloroethyl)-1-nitrosourea), CLZ (chlorozotozin), CCNU (1-(2-chloroethyl)-3-cyclohexyl-1-nitrosourea).

More recent experiments using the same tumor model have shown that, in contrast to the results observed for CCNU plus misonidazole (Table III), the addition of a range of other nitroimidazoles including metronidazole, Ro-05-9963, SR 2508, and SR 2555 to CCNU treatment is apparently fairly ineffective.

3. DISCUSSION

The data obtained in these experiments show many similarities in terms of relative toxicities of the different nitroimidazoles. Most of the nitroimidazoles studied which had similar electron affinities exhibited decreased toxicity as the hydrophilicity increased. This was true for *in vitro* monolayer and spheroid cytotoxicity, *in vivo* tumor cytotoxicity, neurotoxicity, and general toxicity ($LD_{50/2}$). These results indicate that pharmacokinetic differences caused by modifications of physical chemical properties related to chemical structure differences can play a major role in these toxic effects.

However, significant unpredictable exceptions were obtained. Ro-05-9963 was more toxic in spheroids than in monolayer cells, and SR 2508 and SR 2555 were somewhat more cytotoxic in spheroids at the concentration tested. Ro-07-0741 produced significantly greater tumor cytotoxicity and neurotoxicity when related to either administered or exposure dose of the drug. Differences in tumor cytotoxicity and neurotoxicity for different administered doses of the other nitroimidazoles were reduced when the data were related to exposure doses of the drugs in the relevant tissues, indicating again the importance of pharmacokinetics. Production of neurotoxicity required significantly greater cumulative doses than did tumor cytotoxicity.

Differences in toxicities were obtained in these studies for different tumors and for different assays. Cytotoxicity in the KHT sarcoma was, in general, less than that produced in the EMT6/Ro mammary tumor. In both types of tumor the toxicity, as measured by the *in vivo* to *in vitro* excision assay, was maximum at about 18-24 hours after administering the drugs and then rapidly (within 48 h) returned to normal. This rapid recovery is probably the reason that little or no effects on tumor growth delay were detected after administering single doses of any of the nitroimidazoles in either tumor system.

The ability of the nitroimidazoles to enhance chemotoxicity differed markedly in several respects. The ability of misonidazole to enhance chemotoxicity was very dependent on the particular chemotherapeutic agent with which it was combined. Even very similar drugs such as BCNU and CCNU differed greatly in the chemotoxicity produced when combined with misonidazole. Furthermore, for a given chemotherapeutic agent such as CCNU, the enhancement of chemotoxicity was very dependent on the particular nitroimidazole used in the combination. On the basis of these data it is unlikely that *in vitro* or *in vivo* cytotoxic properties of the nitroimidazoles alone will predict their ability to enhance chemotoxicity.

REFERENCES

(1) ADAMS, G.E., FLOCKHART, I.R., SMITHEN, C.E., STRATFORD, I.J., WARDMAN, P., and WATTS, M.E., Electron-affinic sensitization. VIII. A corre-

lation between structures, one-electron reduction potentials and efficiencies of nitroimidazoles as hypoxic cell radiosensitizers. Radiation Research 67 (1976) 9.

(2) SUTHERLAND, R.M., Selective chemotherapy of noncycling cells in an *in vitro* tumor model. Cancer Res. 34 (1974) 3501.

(3) HALL, E.J. and ROIZIN-TOWLE, L., Hypoxic sensitizers: Radiobiological studies at the cellular level. Radiology 117 (1975) 453.

(4) MOHINDRA, R.C. and RAUTH,A.M., Increased cell killing by metronidazole and nitrofurazone of hypoxic compared to aerobic mammalian cells. Cancer Res. 36 (1976) 930.

(5) MOORE, B.A., PALIC, B., and SKARSGARD, L.D., Radiosensitizing and toxic effects of the 2-nitroimidazole Ro-07-0582 in hypoxic mammalian cells. Radiat. Res. 67 (1976) 459.

(6) SRIDHAR, R., KOCH, C. and SUTHERLAND, R., Cytotoxicity of two nitroimidazole radiosensitizers in an *in vitro* tumor model. Int. J. Radiat. Oncol., Biol. Phys. 1 (1976) 1149.

(7) URTASUN, R.C., BAND, P., CHAPMAN, J.D., FELDSTEIN, J.L., MIELKE, B., and FRYER, C.G., Radiation and high dose metronidazole in supratentorial gliomas. N.Eng.J. Med. 294(1976) 1364.

(8) GRAY, A.J., DISCHE, S., ADAMS, G.E., FLOCKHART,I.R., and FOSTER, J.L., Clinical testing of the radiosensitizer Ro-07-0582. Dose tolerance, serum and tumor concentration. Clin. Radio. 27 (1976) 151.

(9) WASSERMAN, T.H., PHILLIPS, T.G., JOHNSON, R.J., GOMER, C.J., LAWRENCE, A.G., SADEE, W., MARQUES, R.A., LEVIN, T.A. and VAN RAALTIE, G., Initial clinical and pharmacological evaluation of misonidazole (Ro-07-0582), a hypoxic cell radiosensitizer. Int. J. Radiat. Oncol., Biol. Phys. 5 (1979) 775.

(10) KOGELNIK, H.D., Clinical experience with misonidazole. High dose fractions versus daily low fractions. Cancer Clin. Trials 3 (1980)179.

(11) CONROY, P.J., VON BURG, R., PASSALACQUA, W., PENNEY, D.P., and SUTHERLAND, R.M., Misonidazole neurotoxicity in the mouse. Evaluation of functional, pharmacokinetic, electrophysiologic and morphologic parameters. Int. J. Radiat. Oncol., Biol. Phys. 5 (1979) 983.

(12) CONROY, P.J., SHAW, A.B., MC NEILL, T.H., PASSALACQUA, W., and SUTHERLAND, R.M., "Radiation sensitizer neurotoxicity in the mouse" in Radiation Sensitizers, Their Use in the Clinical Management of Cancer, BRADY, L.W., Ed., Masson Publishing USA, Inc., N.Y. (1980) 397-410.

(13) SUTHERLAND, R.M., BAREHAM, B.J., and REICH, K.A., Cytotoxicity of hypoxic cell sensitizers in multicell spheroids. Cancer Clinical Trials 3 (1980) 73.

(14) CLEMENT, J.J., GORMAN, M.S., WODINSKY, I., CATANE, R., JOHNSON, R.K., Enhancement of anti-tumor activity of alkylating agents by the radiation sensitizer misonidazole. Cancer Res. 40 (1980) 4165.

(15) ROSE, C.M., MILLER, J.L., PEACOCK, J. H., PHELPS, T.A., STEPHENS, T.C., Differential enhancement of melphalan cytotoxicity in tumor and normal tissue by misonidazole. Cancer Clin. Trials, in press, (1980).

(16) TANNOCK, I., The *in vivo* interaction of anti-cancer drugs with misonidazole or metronidazole: cyclophosphamide and BCNU. Br. J. Cancer 42 (1980) 871.

(17) TANNOCK, I., The *in vivo* interaction of anti-cancer drugs with misonidazole or metronidazole: methotrexate, 5-fluorouracil and adriamycin. Br J. Cancer 42 (1980) 861.

(18) MULCAHY, R.T., SIEMANN, D.W., SUTHERLAND, R.M., *In vivo* response of KHT sarcomas to combination chemotherapy with radiosensitizers and BCNU. Br. J. Cancer 43 (1981) 93.

(19) SIEMANN, D.W., *In vivo* combination of misonidazole and the chemotherapeutic agent CCNU. Br. J. Cancer 43 (1981) 367.

(20) SUTHERLAND, R.M. and FRANKO, A.J., On the nature of radiobiologically hypoxic fraction in tumors. Int. J. Radiat. Oncol., Biol. Phys. 6 (1980) 117.

(21) CONROY, P.J., SUTHERLAND, R.M., and PASSALACQUA, W., Cytotoxicity of misonidazole *in vivo*: A comparison of large single doses with smaller doses and prolonged contact of the drug with tumor cells. Radiat. Res. 83 (1980) 169.

(22) LAW, M.P., HIRST, D.G. and BROWN,J.M., Enhancing effect of misonidazole on the response of the RIF-1 tumour to cyclophosphamide. Br. J. Cancer, in press (1981).

Progress in Radio-Oncology II, edited by K. H. Kärcher et al., Raven Press, New York © 1982.

Long-Term Results of Treatment of Patients with Metronidazole and Protracted Radiotherapy

A base for comparative randomized studies with hypoxic radiosensitizers

A.B.M.F. KARIM, K.H. NJO
Department of Radiotherapy,
Academic Hospital,
Free University,
Amsterdam,
The Netherlands

Abstract

From 1974 to 1978, a pilot study was undertaken in the Academic Hospital of the Free University of Amsterdam to evaluate the use of hypoxic radiosensitizer metronidazole (2.5 g/d in 3 divided doses; cumulative total dose 75 ($^{+19}_{-3}$)g) given with conventional protracted radiotherapy. All patients had advanced malignancies, 70 head and neck cancers being available for long-term evaluation. Only four showed evidence of (reversible) neuropathy, including one patient with two attacks of reversible psychosis. With a minimum follow-up period of 30 months, the local control rates of some of these tumours appear to be encouraging and higher (54%) than usually obtained, without evidence of any long-term enhanced late effect of radiation or carcinogesis. Although metronidazole is known to be a less-efficient radiosensitizer, the prolonged steady serum level of the drug might have played a role in accentuating the efficiency, through the known selective cytotoxic effect on the surviving fractions of hypoxic elements of the tumours. Clinical benefit has been persistently reported from metronidazole from a number of centres. Reports on other hypoxic radiosensitizers are not always clearly encouraging to date. In view of these facts, three-armed studies appear desirable and are being pursued.

1. INTRODUCTION

Interest in the clinical use of hypoxic radiosensitizers to improve the results of radiation therapy of locally advanced solid tumours has grown in the recent past. Studies with radiosensitizers under conditions of protracted radiotherapy are scarce. In such situations the residual subpopulations of hypoxic tumour cells surviving after possible re-oxygenation may be a critical factor for local failure of radiation therapy. If the radiosensitizer is effective in eliminating the residual hypoxic tumour cells, the probability of local control may be enhanced in selected clinical situations.

Such a study was conducted in the Radiotherapy Department of the Academic Hospital at the Free University of Amsterdam from 1974 through 1978. Some of the aspects of this study have already appeared in a few preliminary publications [1,2]. The purpose of this paper is to update and evaluate the long-term response of the advanced malignancies of the head and neck area and to formulate a basis for further studies with hypoxic radiosensitizers.

2. MATERIALS AND METHODS

From 1974 through 1978, a total of 70 patients with advanced malignancies of the head and neck region were treated with metronidazole and with high-dose, precision, protracted radiotherapy schemes.

2.1. Patients

All patients had tumours in late stages (stages III and IV) except for eight patients with laryngeal carcinoma, who were in stage II. However, these eight patients had a relatively large volume of tumour mass, and not more than 70% local control probability was expected in this group [3]. Almost all the other patients had large primary and/or nodal tumour masses unlikely to be easily benefitted by conventional protracted radiotherapy: that was the reason for selecting them for this study. Some had transglottic fixed laryngeal tumours, others had recurrent tumours after surgery; one had malignant multiple fistulae together with recurrence at the primary site after surgery in the oropharynx, a few had necrotic primaries or large nodal masses (> 6 cm in diameter) or bone destruction, etc. From this group of patients (stages III and IV), one could hardly expect a 30% 2-year local control probability. The tumours were all histopathologically examined: while most were of the squamous cell type, a few were adenocarcinomas, adenocystic carcinomas or acinic cell carcinomas.

2.2. Radiotherapy: Dose, Time and Technique

All patients had been treated with meticulously followed protracted radiation therapy schemes using shrinking field techniques [3-5] to deliver high doses to the tumour masses while avoiding irreversible damage to normal tissues. Masks were used in all cases. Complex multiple fields were used, in most cases with computer plans and simulator checks. All fields were treated daily to deliver 200 rads (cGy) a day, 5 fractions a week. Where the target volume was very large, 180 rads (cGy) were delivered daily, i.e. 900 rads (cGy) every week. The total dose of radiation delivered ranged between 6500 and 7800 rads (cGy) over about 7 to 8 weeks.

2.3. Metronidazole

All patients received 2.5 g of metronidazole during each treatment day. The radiosensitizer was not administered during weekends. Metronidazole was administered in 3 unequal doses: 1 g of the sensitizer was administered 2 hours before radiotherapy, and 0.75 g twice, in the morning after breakfast and in the evening before retiring. Radiation treatment was given between 11:00 and 15:00.

The cumulative dose of metronidazole varied between 72 and 94 g for almost all patients. Three patients had rather severe gastro-intestinal intolerance, and drug administration had to be stopped. They are included in this report.

3. RESULTS

Table I shows the TNM classification or the advanced stage of the cancer together with the results. The response of the tumour is indicated by local failure, local control or regrowth delay (NED[1] status for at least 15 months, with regrowth of the

TABLE I. RESPONSE OF ADVANCED HEAD AND NECK CARCINOMAS TREATED WITH RADIATION AND METRONIDAZOLE[a]

Site	Modified UICC [3] or UICC staging	Total No. patients	Local failure	Local control	Regrowth delay
OROPHARYNX	III T2N1b	4	1	2	1
	IV T2-T4N2N3 T4NO or Postop. recurr.	8	3	4	1
	Total:	12	4	6(50%)	2
LARYNX	II? IV? T2b? T4aNO	8	0	7	1
	III T2bN1b	11	4	6	1
	T3NON1	8	3	5	0
	IV T4bNO	5	2	2	1
	T2-T4N2N3	4	1	2	1
	Total:	36	10	22(61%)	4
ADVANCED:	Nasopharynx, nose-ethmoid	12	4	8(66%)	0
	Maxillary sinus, palate, oral cavity	10	8	2(20%)	0
Overall total:		70		38(54%)	

[a] Minimum follow-up period, 30 months.

[1] No evidence of disease.

tumour discovered subsequently). Patients who died from intercurrent diseases or development of distant metastases without evidence of local failure have been considered to have local control of the tumour. All patients were followed up for a minimum period of 30 months or until death: none was lost to follow-up.

An impressive spectrum of complete tumour regression has been demonstrated in a number of situations where expectations were otherwise: large tumour masses (> 6 cm) disappeared, malignant fistulae dried up, necrotic oral cancers (> 5 cm) healed, etc.

3.1. Late Radiation Reactions

No severe late radiation effects have been recorded so far. A number of arytenoid oedemas, not beyond expectations, have been observed to subside gradually. No necrosis of skin, cartilage or bone has been seen.

3.2. Toxicity of Metronidazole

Gastro-intestinal intolerance to a troublesome degree was noticed in this study. Three patients had intractable vomiting and the drug had to be stopped. In 20 patients the symptoms of nausea and vomiting could be managed reasonably well with increasing doses of changing anti-emetics. In 11 others only mild nausea occurred, manageable with the usual anti-emetics.

Sensory peripheral neuropathy of mild to moderate degree was noticed in four patients (5.7%). All but one were reversible in 6 months. The exception was one patient who had subdued symptoms of neuropathy even after 1 year; this gradually disappeared by 18 months.

Central nervous system toxicity in the form of psychosis (aphasia, loss of direction etc.) developed on two different occasions in one patient. On both occasions the psychosis was completely reversible. The patient denied the use of alcohol during treatment.

Drug rashes were seen in one patient. Fever, ototoxicity, unusual lethargy or ataxia were not noticed in this series of patients.

3.3. Second New Primary Cancer

Four new primaries have been recorded in this series between 3 and 5 years of follow-up: two patients with laryngeal carcinoma treated with metronidazole have now developed bronchogenic carcinoma, one with a similar primary has developed thyroid undifferentiated carcinoma, and one with a nasopharyngeal squamous cell carcinoma has developed a transitional cell carcinoma of the bladder. In this Institute development of lung cancer after successful treatment of laryngeal cancer is rather high, possibly above 10%, and this is being studied.

4. DISCUSSION

The results of this non-randomized study, as demonstrated by 54% local control probability with a heterogeneous group of patients with advanced head and neck carcinoma after a minimum follow-up period of 2.5 years, appear to be encouraging. It is emphasized that a conventional protracted fractionated scheme of precision high-dose radiotherapy using a shrinking field technique is meticulously followed in this Institute.

In our experience [4,5] patients treated with the above regimen without radiosensitizer have minimal severe late radiation reactions or necrosis. Fortunately, the same is true for the group treated with metronidazole. More severe degrees of acute or late skin, mucosal, cartilage or bony reactions were not observed in this series. Reactions were carefully and personally observed and recorded. Functional disability was virtually absent.

The dose of metronidazole was chosen to deliver a high cumulative dose but relatively low daily dose. The aim was to maintain relatively steady serum and tissue levels of the hypoxic radiosensitizer. The prolonged and continuous exposure of the tumour to metronidazole has been speculated [1,6] to have a double modus operandi - a selective cytotoxic effect on hypoxic cells over and above the radiosensitizing dose-modifying effect. A much lower concentration of metronidazole has been found to be effective experimentally [7]. In eliminating hypoxic cells selectively due to cytotoxicity, it has recently been emphasized by Yuhas [8] that the combined effects of radiosensitization and cytotoxicity must be considered in any data involving prolonged exposure of hypoxic tumours to the radiosensitizer. Yuhas [8] points out that the overall effectiveness of the radiation therapy might not have been increased when marginally effective fractionation schemes are used in studies with radiosensitizers. The marginally effective fractionation schemes may be changed into effective ones through the presence of the sensitizers.

The subpopulation of hypoxic tumour cells surviving the onslaught of protracted fractionation and possibly re-oxygenation may be the deciding factor for local control of a tumour. This may be the ultimate testing ground to check the efficacy of a hypoxic radiosensitizer.

4.1. The Ideal Hypoxic Radiosensitizer

The ideal radiosensitizer has not as yet been found [9-11]. However, the results of treatment of patients with metronidazole in different parts of the world appear to be persistently encouraging [1,12-14]. Apart from a study by Urtasun [15], no other study has been controlled or randomized to gain a prolonged evaluation. With the encouraging preliminary results obtained using metronidazole in advanced bladder carcinoma [13], a randomized, controlled study was started in the Radiotherapy Institute of Rotterdam in 1979. The results of these randomized studies are awaited.

Meanwhile results are being reported on randomized studies with misonidazole. Preliminary short-term results appear to be encouraging in some [16-18] while discouraging in others [17-20]. Certainly the troublesome neurotoxicity appears to be a problem when misonidazole or desmethyl-misonidazole is used [10,11,21].

The sufferings of aged patients with advanced disease should be given prior consideration in planning studies with hypoxic radiosensitizers.

4.2. A Basic Design for Controlled Studies with Sensitizers

It is to be hoped that better radiosensitizers will be available in the near future. In the meantime, it appears imperative to undertake comparative studies with existing radiosensitizers. The basic hypoxic radiosensitizer must be effective and not unacceptably toxic. On the basis of some studies, metronidazole appears to fulfil these conditions.

We have recently initiated a multi-centre protocol for a three-arm randomized controlled study on localized squamous cell carcinoma of the bronchus irradiated with administration of either metronidazole or misonidazole or without a hypoxic radiosensitizer. More studies are planned. Such studies should, it is hoped, answer some of the questions still open concerning the hypoxic cell sensitizers currently available.

REFERENCES

[1] KARIM, A.B.M.F.: Prolonged metronidazole administration with protracted radiotherapy: A pilot study on response of advanced tumours, Br. J. Cancer 37 Suppl.III (1978) 299.
[2] KARIM, A.B.M.F., FABER, D.B., HAAS, R.E., HOEKSTRA, F.H., NJO, K.H.: Metronidazole as a radiosensitizer: A preliminary report on estimation in serum and saliva, Int. J. Radiat. Oncol., Biol. Phys. 6 (1980) 1233.
[3] KARIM, A.B.M.F., SNOW, G.B., RUYS, P.N., BOSCH, H.: The heterogeneity of the T2 glottic carcinoma and its local control probability after radiation therapy, Int. J. Radiat. Oncol., Biol. Phys. 6 (1980) 1653.
[4] KARIM, A.B.M.F., SNOW, G.B., HASMAN, A., CHANG, S.C., KEILHOLZ, A., HOEKSTRA, F.H.: Dose response in radiotherapy for glottic carcinoma, Cancer 41 (1978) 1728.
[5] KARIM, A.B.M.F., RUYS, P.N., HUTCHINSON, S., BOSCH, H.: "Radiation therapy for advanced laryngeal carcinoma: The Techniques, the Dosimetry and the Results with a Clinac 4", Proc. 8th Clinac Users Meeting, Hawaii (1980) 39.
[6] FOSTER, J.L.: Differential cytotoxic effects of metronidazole and other nitro-heterocyclic drugs against hypoxic tumor cells, Int. J. Radiat. Oncol., Biol. Phys. 4 (1978) 153.
[7] SUTHERLAND, R., RICHARDSON, J., KOCH, C., SRIDHAR, R.: "Metronidazole radiation sensitization and cytotoxicity in an in vitro tumor model", in Proc. Conf. Metronidazole, Montreal, Canada (1976) 193.
[8] YUHAS, J.M.: "Chemical radiosensitization as a means of improving solid tumor radiotherapy", in Radiation-Drug Interaction in the Treatment of Cancer (SOKAL, G.H., MAICKEL, R.P., Eds.), J. Wiley and Sons, New York (1980) 137.
[9] FOSTER, J.L., WILLSON, R.L.: "Metronidazole (Flagyl) in cancer radiotherapy" in Chemotherapy (HELLMANN, K., CONNORS, T.A., Eds.), Plenum Publishing Corporation, New York (1976) 215.
[10] DISCHE, S., FOWLER, J.F., SAUNDERS, M.I., STRATFORD, M.R.L., ANDERSON, P., MINCHINTON, A.I., LEE, M.E.: A drug for improved radiosensitization in radiotherapy, Br. J. Cancer 42 (1980) 153.
[11] DISCHE, S., SAUNDERS, M.I., STRATFORD, M.R.L.: Neurotoxicity with desmethylmisonidazole, Br. J. Radiol. 54 (1981) 156.
[12] URTASUN, R., BAND, P., CHAPMAN, J.D., FELDSTEIN, M.L., MIEKE, B., FRYER, C.: Radiation and high dose metronidazole (Flagyl) in supratentorial glioblastomas, New Engl. J. Med. 294 (1976) 1364.

[13] SOUREK, Z.V.: personal communications, 1978, 1979.
[14] LEE, D.J., WHARAM, M.D., KASHIMA, H., ORDER, S.: Short course high fractional dose irradiation in advanced and recurrent head and neck cancer, Int. J. Radiat. Oncol., Biol. Phys. 5 (1979) 1829.
[15] URTASUN, R.: personal communication, 1978.
[16] PHILLIPS, T.L.: personal communication, 1980.
[17] DISCHE, S.: personal communication, 1981.
[18] BATAINI, J.P.: personal communication, 1980.
[19] SAUNDERS, M.I.: personal communication, 1981.
[20] EORTC controlled trial on 'glioblastoma and misonidazole': preliminary results, April 1981, Rotterdam.
[21] WASSERMAN, T.H., PHILLIPS, T.L., JOHNSON, R.J., GOMER,C.J., LAWRENCE, G.A., SADEE, W., MARQUES, R.A., LEVIN, V.A., VANRAALTE, G.: Initial United States clinical and pharmacologic evaluation of misonidazole (Ro-07-0582), a hypoxic radiosensitizer, Int. J. Radiat. Oncol., Biol. Phys. 5 (1979) 775.

Progress in Radio-Oncology II, edited by K. H. Kärcher et al., Raven Press, New York © 1982.

Clinical Experience with the Dihydroquinoline Type of Hypoxic Sensitizer MTDQ

I.L. RODÉ, Z. POLLÁK, V. BÄR
Department of Oncoradiology,
Postgraduate Medical School,
Budapest
and
National Oncological Institute,
Budapest,
Hungary

Abstract

MTDQ is a peroxide-decomposing and radical-scavenger anti-oxidant. According to in vitro and in vivo examinations, MTDQ increases the radiation sensitivity of hypoxic cells without affecting the radiosensitivity of oxygenized cells. Administration of 2.3 g/m^{2}MTDQ for 100 days is void of any toxic effect, as has been confirmed by clinical findings of a combined treatment of MTDQ with radiation therapy in more than 200 patients. MTDQ accumulates in the tumour tissue. Autoradiographic examinations revealed that concentrations of ^{14}C-labelled MTDQ in tumour tissue is three times that found in blood plasma 26 hours after a single administration. MTDQ meets the requirements for continuous and repeated administration. Metabolism is delayed, while its biological half-life in human plasma varies between 48 and 57 hours. MTDQ permitted the successful radiotherapy of several tumours resistant to radiation, i.e. synoviomas and soft-tissue sarcomas, fibrosarcomas, neurofibrosarcomas etc. Evidence of the effect has been obtained in patients with liver metastases. Objective and subjective symptoms (enlargement of the liver, vomitus, meteorism, loss of weight) confirmed the presence of metastasis at an advanced stage of a verified primary tumour. MTDQ was administered to the patients before, during, and for 6 weeks after radiotherapy – if considered necessary for a longer time – at daily doses of 2 to 6 tablets. The combined MTDQ and radiation therapy yielded satisfactory clinical regression of the tumour. Stagnation was confirmed by scintigraphy.

The chemical name of the active substance known as MTDQ is:

6,6'-methylene-bis(2,2,4-trimethyl-1,2-dihydroquinoline)$_n$

where n = 1, 2 or 3. The structural formula of the radiosensitizer is shown in Fig.1. Its structure and the in vitro studies performed hitherto render it probable that the main factors in the mechanism of action are peroxide decomposition, hydrol formation and polymerization, and the release of oxygen 'in situ' in the irradiated tissue.

Biopharmacological, pharmacological, and toxicological investigations of MTDQ have been carried out in Hungarian scientific institutes. The hypoxic sensitizing action has been verified in in-vitro studies on V79 hamster cell culture at the Columbia University by Astor and Hall[1].

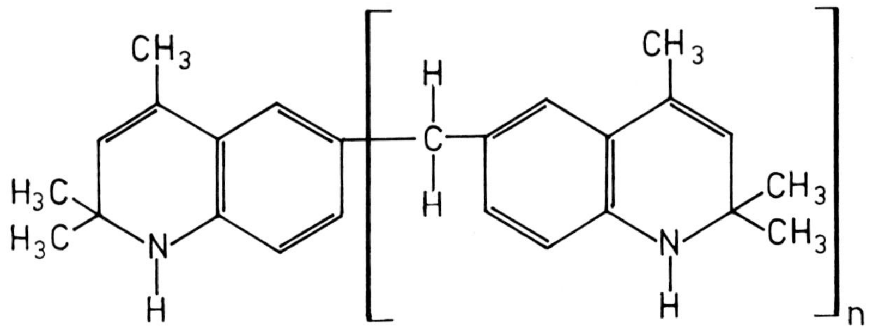

FIG.1. Chemical structure of MTDQ.

The present paper reports on the results of pharmacokinetic and clinical examinations performed at the National Institute of Oncology, Budapest.

The clinicopharmacological investigation of the compound revealed that:

A. A dose of 2.3 g/m^2 body surface area per day of MTDQ administered perorally for 100 days to patients with stage 3 and 4 tumours of various localizations did not cause undesired effects, except for a mild initial nausea observed in less than 10% of the patients treated. Nor did it affect the function of the vital organs. MTDQ is void of teratogenic action [2].

B. The enrichment and distribution of ^{14}C-labelled MTDQ in basospinaloma of the neck and lips was recorded 26 h after oral administration. It was found that the concentration of the labelled substance in tumour tissue was three times that observed in plasma. More than 90% of the orally administered MTDQ dose is eliminated with the faeces.

V79 Chinese hamster cells were exposed to graded doses of ^{60}Co gamma-rays under both aerated and hypoxic conditions in the presence of various concentrations of misonidazole, of MTDQ, or of a combination of them. It was found that MTDQ has a radiosensitizing action equal to or slightly superior to misonidazole for the same concentration of each. Its sensitizer enhancement ratio (SER) amounted to 1.2 in this experiment. The sensitizer enhancement ratios of these two drugs are additive.

The pharmacokinetic parameters and the results of the preclinical studies point to the importance of a premedication lasting for 8 to 14 days before radiotherapy, in order to obtain an effective level of MTDQ in tissue.

A therapy comprising MTDQ given in combination with radiation has been carried out in more than 200 patients. With simultaneous administration of MTDQ and radiation therapy, and with continuing administration of MTDQ during the post-irradiation period for at least 6 weeks, tumour regression has been achieved in stage 3 and 4 neoplasms of relatively low radiation sensitivity, such as malignant synovioma, fibrosarcoma, neurofibrosarcoma, epithelial malignant tumours of the head and neck, malignant melanoma, and carcinoma of the breast. The compound has proved suitable for repeated and prolonged administration [3,4]. In addition, in patients with squamous cell carcinoma of the tongue or with Enzinger's sarcoma, combined treatment has resulted in incomplete, though clinically appreciable and lasting regression in 40 to 60% of the cases examined. The sensitizer enhancement ratio was found to be 1.3 in a clinical trial.

Results obtained with MTDQ combined with ^{60}Co irradiation in two patients with planocellular cancer of the maxillary sinus that had been found to be inoperable and radiation resistant are shown in Figs 2 and 3. After delivering 30 Gy combined with MTDQ administration, both patients became free of clinical symptoms and the size of tumours decreased. In the patient shown in Fig.3, the dislocated right eye reopened.

It should be emphasized that a combined treatment comprising MTDQ administration and low-dose grid X-ray or ^{60}Co irradiation of metastatic liver tumours deriving from breast cancer or melanoblastoma has also been performed. The primary tumour was cancer of the breast in eight patients and melanoblastoma in seven. In 14 patients perihepatic pain, vomiting, nausea and disturbances of the digestion were eliminated or substantially decreased; hepatic enlargement became reduced by more than 3 cm. In 11 patients regression lasted for 3 months. The treatment remained unsuccessful in one patient. No radiogenic impairment of the parenchymatous organs and/or the intact irradiated region of the liver could be observed after the combined treatment. The treatment was well tolerated, and in no case had it to be interrupted in consequence of undesired side-effects.

The addition of MTDQ inhibits the cytotoxicity which misonidazole shows towards hypoxic cells, even when the concentration of MTDQ is one-fiftieth (1/50) of the nitroimidazole. If, indeed

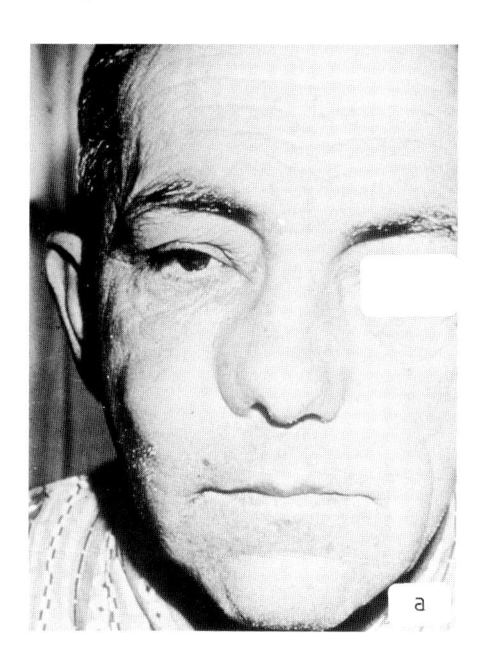

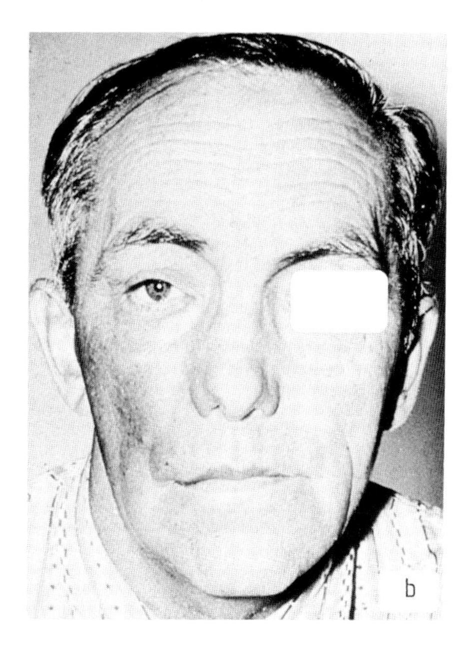

FIG.2. 54-year-old patient. Diagnosis - planocellular cancer of the maxillary sinus. (a) Before treatment; (b) after treatment.

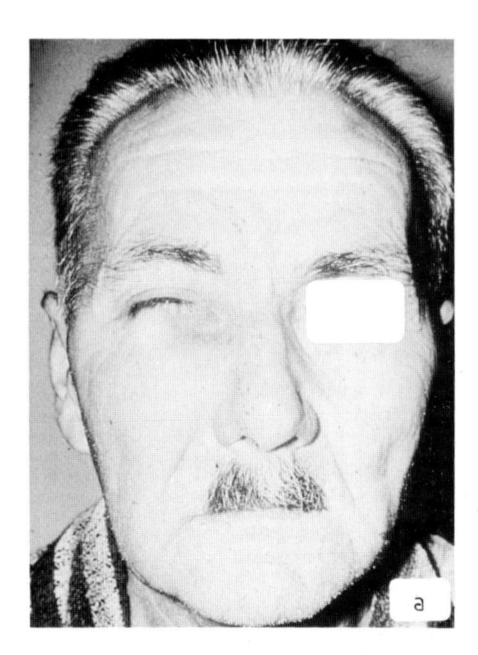

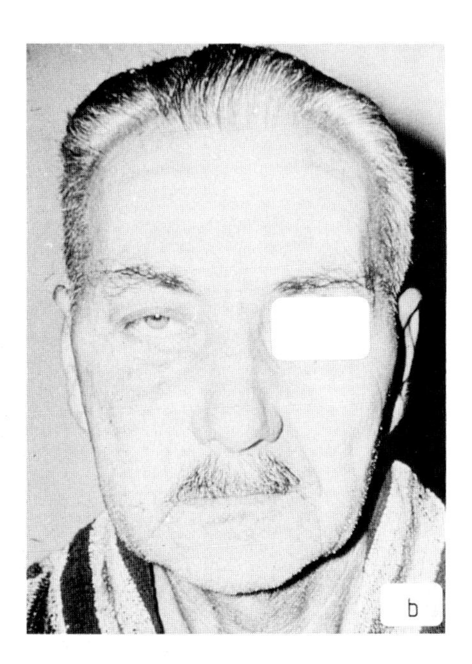

FIG.3. 57-year-old patient. Diagnosis - planocellular cancer of the maxillary sinus. (a) Before treatment; (b) after treatment.

the cytotoxicity of misonidazole towards hypoxic cells is related to the neurological toxicity of large doses of this drug in man, the exciting possibility might be considered that MTDQ may be able to reduce or eliminate this troublesome side-effect.

It is also worthwhile mentioning that the water soluble disodium sulphonic acid salt of MTDQ (carrying the code name MDS) has recently been developed. It has been found to exert a radiation-protective action on well oxygenated cells without affecting the radiosensitivity of hypoxic tumour cells. Trials under way that involve the combined use of MTDQ and MDS are encouraging and make use of an additional treatment modality, i.e. the radiation protection of well oxygenated cells by means of a protective drug applied together with the hypoxic cell sensitizer, according to the principle suggested by Chapman in 1979 as a possible means of improving results in the radiotherapy of cancer [5].

REFERENCES

[1] ASTOR, M., HALL, E.J.: Br. J. Cancer 39 (1979) 510-512.

[2] ERDÉLYI, V., et al.: Strahlentherapie 156 (1980) 198-200.

[3] RODÉ, I., et al.: in Radiobiological Research and Radiotherapy (Proc. Symp. Vienna, 1976: BECK, E.R.A., Ed.) Vol.1, IAEA, Vienna (1977) 265-271.

[4] RODÉ, I., et al.: in Proc. 14th Int. Congr. Radiology, Rio de Janeiro, 1977, Book of Abstracts, National Organizing Committee of the Congress, p.272.

[5] CHAPMAN, J.D.: New Engl. J. Med. 26 (1979) 1429-1431.

Progress in Radio-Oncology II, edited by K. H. Kärcher et al., Raven Press, New York © 1982.

How Do We Find Better Radiosensitizers?

G.E. ADAMS, P.W. SHELDON, I.J. STRATFORD
Radiobiology Unit,
Physics Department,
Institute of Cancer Research,
Sutton, Surrey,
United Kingdom

Abstract

The hypoxic cell radiosensitizer, misonidazole, currently in trial, is neurotoxic, and this limits its clinical applicability. The development of new sensitizers with improved therapeutic ratio depends on identifying the chemical and structural properties associated with neurotoxicity or increasing the sensitization efficiency without increasing this neurotoxicity. The paper discusses firstly some current approaches aimed at reducing neurotoxicity, including the study of the possible relevance of lipophilic properties and, secondly, methods for increasing drug efficiency. The latter includes exploitation of the electron-affinity relationship, the synthesis of compounds with both sensitizing and alkylating activity, the study of an unusual series of 'super-active' sensitizers and exploration of non-nitro-containing electron-affinic compounds.

1. INTRODUCTION

There is little doubt that the largest single factor influencing the response of experimental tumours to radiation is the presence of radiation-resistant hypoxic cells. Such cells must also occur in many human tumours. However, definitive evidence that such cells are a <u>major</u> reason for local failure in the treatment of human cancer by fractionated radiotherapy is still lacking. Nevertheless, the present state of knowledge of the radiobiology of experimental tumours must reinforce the view that, notwithstanding the beneficial effects of re-oxygenation, elimination of hypoxic cell radiation resistance would result in substantial benefit in the clinical situation. Of the various approaches to the hypoxia problem, the use of chemical sensitizers which increase tumour response without increasing radiation response in normal tissues is the most attractive on several grounds.

Many chemical compounds, particularly the nitroimidazoles, are known to act as differential hypoxic cell sensitizers in experimental tumour systems. Of these, one compound, misonidazole, is undergoing extensive clinical trial in various centres. Although there is evidence already available that benefit can be obtained in some clinical situations, it is now clear that the dose-limitation imposed by the neurotoxic properties of this drug will seriously limit its value and that sensitizers with higher therapeutic ratios are required. This paper describes some of the steps that are being taken to develop such improved drugs.

2. IMPROVEMENT OF THERAPEUTIC RATIO

Therapeutic ratio can be increased either by reduction of the neurotoxic properties of sensitizers or by increasing their sensitizing efficiencies. Advances have been made in both directions.

2.1. Reduced toxicity

The proposal that the lipophilic properties of sensitizers may influence their potential neurotoxicological properties was supported by experimental evidence on the uptake of various drugs in brain and tumour tissue in experimental mice *[1]*. Over a wide range of octanol-water partition coefficient, P, drug-uptake in brain decreased with decreasing values of P under conditions where uptake in tumour was unaffected.

Investigations of the neurotoxic properties of sensitizers administered to experimental mice utilise various methods involving morphological, behavioural and biochemical end-points. Some of these studies in which neurotoxicity of sensitizers have been associated with the increase of lysozomal enzyme levels in peripheral nerve preparations provide support for the lipophilicity hypothesis *[2]*.

2.1.1. Desmethylmisonidazole

Desmethylmisonidazole, the O-demethylated product of misonidazole, is the major metabolite of misonidazole in man and other species.

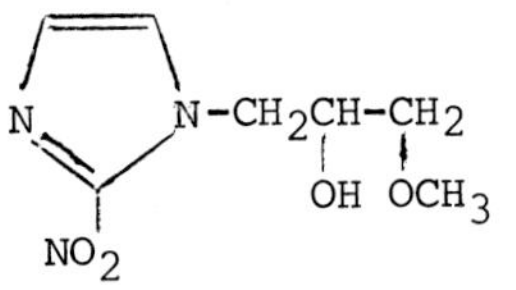

N N-CH_2CH-CH_2 (OH, OCH_3); NO_2

misonidazole

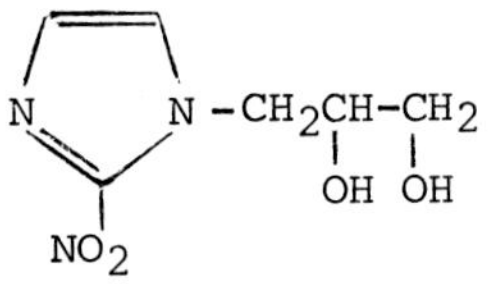

N N-CH_2CH-CH_2 (OH, OH); NO_2

desmethylmisonidazole

Its sensitizing efficiency *in vivo* is approximately the same as that of misonidazole and this, together with the fact that it is less lipophilic (the octanol-water partition coefficient, P, is 0.11 compared with 0.43 for misonidazole) prompted its clinical investigation. Dische *[3]* has reported preliminary data showing good tumour penetration following oral administration of the drug. The proportion of the patients developing peripheral neuropathy suggested, though, that this drug is not substantially less neurotoxic than misonidazole. However, more data are required from these and other studies with desmethylmisonidazole which are currently in progress.

2.1.2. The drug SR 2508

The possibility that the neurotoxicity of sensitizers may be related to their lipophilic properties has led to the synthesis and testing of several compounds of low lipophilicity. One of these is the compound SR 2508 *[4]*.

N N $-CH_2CONHCH_2CH_2OH$

NO_2

SR 2508

Sensitizing efficiency *in vivo* is slightly less than that of misonidazole but the drug is less toxic. Its low lipophilicity (P = 0.046) explains the relatively low uptake in mouse brain. Although at an advanced stage of pre-clinical development, this drug has not yet been administered to patients.

2.1.3. *Dexamethasone protection*

Clinical studies with misonidazole have shown that the neurotoxicity of this drug is reduced in patients receiving dexamethasone during radiotherapy treatment. The mechanism of this effect is unknown but recent laboratory studies have shown that this steroid protects against the cytotoxicity of misonidazole *in vitro [5]*. It is disturbing, however, that these studies showed that dexamethasone reduced the radiosensitivity of the particular cell line used, although the sensitizing efficiency of misonidazole was unaffected. If this radioprotective effect proves to be a general phenomenon, it has obvious implications in regard to the administration of steroids during radiotherapy treatment.

Other membrane-active agents, e.g. flurbiprofen, also protect against the *in vitro* toxicity of misonidazole without affecting its radiosensitizing efficiency *[6]*. Significantly, however, flurbiprofen had no effect on cellular radiosensitivity in this system. Investigations such as this hold promise for the eventual improvement of the therapeutic ratios of sensitizers by reduction of toxicity.

2.2. Increased sensitization efficiency

It is now well-established that electron affinity is one of the major factors influencing sensitization efficiency, although other properties are certainly involved. This, together with the possibility that the lipophilicity hypothesis may not apply in man *[3]* has widened the search for sensitizers more effective than misonidazole. Application of the electron affinity relationship has helped identify many compounds that are more efficient than misonidazole *in vitro*. We have examined the effectiveness of some of these compounds *in vivo* using experimental murine tumour systems. Some examples of the results are given in Table I.

The sensitizing action of the misonidazole analogue benznidazole (Ro 07-1051) has been known for many years *[7]*. However, its low solubility and relatively high lipophilicity has tended to reduce interest in studying its effectiveness *in vivo*. It is, however, a potent sensitizer, at least as effective as misonidazole, and we believe it warrants further investigation.

TABLE I. SENSITIZATION IN VIVO BY SOME NITROIMIDAZOLES

Compound	Tumour	Assay	Route	$LD_{50}/2D$ mg/g	t^{a} min	SER^{b} ¼ LD_{50}	SER^{b} max
Misonidazole	MT (in WHT ♀ mice)	agar	i.p.	1.8	60	++++	+++++
Desmethylmisonidazole	"	"	"	2.9	45	++++	+++++
Benznidazole	"	"	p.o.	-	180	-	+++++[c]
RSU 1047	"	"	i.p.	2.5	45	+++	++++
RSU 3048 (NSC 38087)	"	"	i.v.	-	0	-	+[d]
"	"	regrowth delay	p.o.	> 0.4	30-90	-	o[e]
H007	"	agar	i.p.	1.8	30	++++	+++++
Misonidazole	Lewis lung (in C57 ♀ mice)	lung colony	i.p.	1.2	60	++++	+++++
RSU 1047	"	"	"	1.7	45	++++	+++++
Ro 03 8800	"	"	"	2.6	75	+++	-
Ro 03 8799	"	"	"	1.2	60	+++	-
Misonidazole	B16 melanoma (in C57 ♀ mice)	lung colony	"	1.2	60	++++	+++++
RSU 1047	"	"	"	1.7	45	++++	+++++

Notes a) t = optimum irradiation time after drug administration.

b) SER = sensitizer enhancement ratio
1.0 = o
1.1-1.25 = +
1.26-1.50 = ++
1.51-1.75 = +++
1.76-2.0 = ++++
> 2.0 = +++++

c) at 1.0 mg/g.

d) at 0.05 mg/g.

e) at 0.4 mg/g.

The compound RSU 1047 is a member of a homologous series of 2-nitroimidazoles substituted at N1 by β-hydroxyalkylamines and of general formula

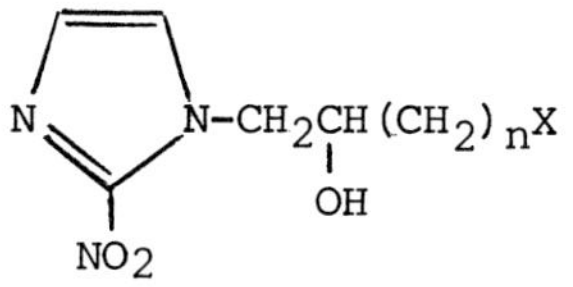

N, $N-CH_2CH(CH_2)_nX$, OH, NO_2 where X = various bases.

In 1047, n=2 and X=morpholine, and the compound appears to be optimal in this particular homologous series. It is more efficient than misonidazole *in vitro*, less toxic *in vivo* and, in the mouse system, less neurotoxic. However, in the tumour systems given in Table I, its sensitizing efficiency *in vivo* does not exceed that of misonidazole. Nevertheless, overall, its properties are such as to warrant further development.

The Roche compounds 8789 and 8800 *[8]* are closely related to the above structures. In particular, 8799 is the β-hydroxy-propanolamine (piperidine) derivative. While its effectiveness is slightly lower than that of misonidazole in the tumour systems investigated here, similar effectiveness has been observed with some other tumours (Williams and Denekamp, private communication). A significant feature of this drug is that, in experimental primates, 8799 is less neurotoxic than misonidazole (Smithen, private communication). Clinical studies with this drug will be starting shortly.

The compound RSU 3048 (NSC 38087) is a member of an unusual and highly active series of 5-substituted 4-nitroimidazoles. These are discussed later.

An example of a compound where sensitizing effectiveness *in vivo* is much greater than would be predicted on electron affinity grounds is the Hoechst compound H.007. This is a metronidazole analogue with the nitro group in the 5-position and accordingly its electron affinity is substantially lower than that of misonidazole. Nevertheless, the data in Table I show that in the MT tumour it is at least as effective as misonidazole. If high electron affinity of sensitizers is a contributory factor to their neurotoxic properties, then the Hoechst compound may be substantially less neurotoxic. Appropriate laboratory investigations are in progress in our laboratory.

2.2.1. Sensitizers with alkylating properties

In 1979, Chapman and colleagues reported that the alkylating agent CB 1954 also possessed strong radiation sensitizing ability *in vitro [9]*.

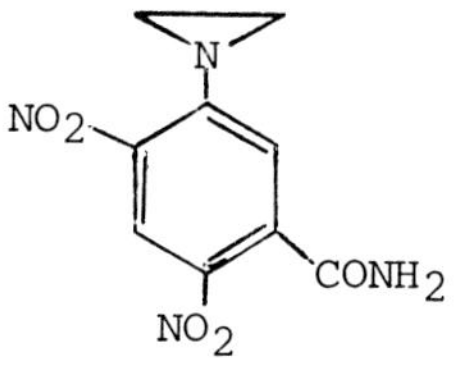

N, NO_2, $CONH_2$, NO_2

Recently *[10]* we have confirmed this and shown also that CB 1954 is considerably more efficient than misonidazole *in vitro*, despite their similarity in one-electron reduction potential.

The compound phenyl AIC, 2-phenyl-4(5)-amino-5(4)-carboxamide is known to protect against the cytotoxic (presumably alkylating) action of CB 1954 [11,10]. When cells were irradiated in the presence of both 1954 and phenyl AIC, the sensitization efficiency of the former was reduced to that of misonidazole, indicating that phenyl AIC can protect against part of the sensitization seen with 1954 alone. Since phenyl AIC has no effect on the sensitization efficiency of misonidazole, it was concluded that the higher sensitization efficiency of 1954 may be associated with its alkylating activity.

Figure 1 shows the results of some *in vitro* experiments with hypoxic Chinese hamster cells. The enhancement ratios are plotted as a function of drug concentration in the medium. CB 1954 is more efficient than misonidazole. Further, a derivative of this compound with a similar electron affinity shows an even greater increase in sensitizing efficiency.

Some of our current synthetic work involves inserting various alkylating functions into the 2-nitroimidazole nucleus and other non-nitro containing sensitizers of high electron affinity, e.g. quinones. Since CB 1954 is known to sensitize *in vivo* [9] this approach should hold some promise for the development of better radiosensitizers.

2.2.2. An unusual series of super-active nitroimidazoles

Of the various types of nitroimidazoles whose sensitizing properties show departures from those expected on electron affinity grounds, a series of 4-nitroimidazoles, substituted in the 5-position with sulphone or sulphonate groups, shows the greatest deviation [12,13]. Some of these compounds are two orders of magnitude more efficienct than misonidazole *in vitro*. Their

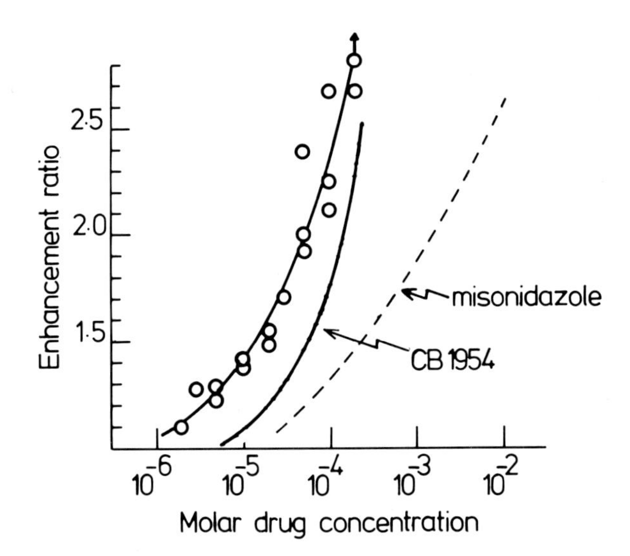

FIG.1. Radiosensitization of hypoxic Chinese hamster cells by a derivative of CB 1954.

general structure is as follows:

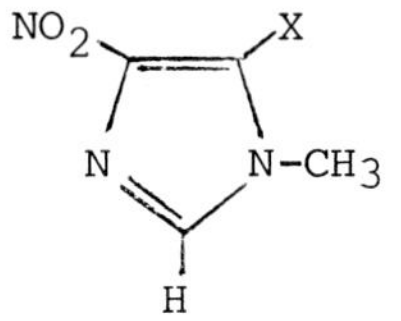

NO_2 X

N $N\text{-}CH_3$ where X = SO_2R

or $SO_2.OR$

H

Table II lists some *in vitro* sensitization efficiencies, defined as the concentration required to give an enhancement ratio of 1.6 ($C_{1.6}$). (The lower the value of $C_{1.6}$, the higher the sensitization efficiency).

Values of $C_{1.6}$ range from 3 x 10^{-4}, the value for misonidazole in this system, to $\sim 10^{-6}$, i.e. a 300-fold improvement of efficiency over misonidazole for only a modest change in electron-affinity ($\sim$50 mV). The reasons for the abnormally high efficiencies are not known, although it is interesting that reversal of the position of the 4-nitro and the 5-substituent reduces sensitizing efficiency while actually tending to increase the electron affinity.

So far, only one compound in this series (NSC 38087) has been evaluated *in vivo* (side chain substituent, SO_2.O.phenyl, see Table II). It is fairly toxic *in vivo* and at the low drug doses used for the sensitizer studies (Table I) little radiosensitization was observed. However, appreciable chemosensitization has been noted in mice bearing the MT tumour and treated with both 38087 and melphalan (Sheldon, unpublished).

The compounds in Table II vary widely in sensitization efficiency, partition coefficients and other properties. Investigations are in progress on both the radio- and chemosensitization properties *in vivo* of some of the compounds in this unusual series.

2.3. Non-nitro compounds

The possibility that the neurotoxic properties of some of the current radiosensitizers are linked with the presence of the nitro group has renewed interest in other types of compounds that can function as hypoxic cell sensitizers. Some compounds, e.g. the nitroxyl stable free-radical structures, function as radiosensitizers *in vitro* by mechanisms other than those of the purely "electron-affinic" type. Although most of these are inactive *in vivo* due to their high chemical reactivities, stereo-chemical protection of the reactive nitroxyl group may render some compounds more metabolically stable. This rationale forms the basis of some current synthetic programmes.

In the purely "electron-affinic" group of sensitizers, it has been known for many years that a wide range of chemical compounds containing conjugate electrophors in the structure show sensitizing properties. Much of the early data on such compounds referred only to bacterial systems [e.g. ref. 14], although there are no grounds for believing that some of these agents or their derivatives would not show sensitizing properties in mammalian systems - *in vitro* or *in vivo*. Structures include simple unsaturated molecules containing halogens, conjugate keto groups

TABLE II. SOME UNUSUALLY EFFICIENT 4-NITROIMIDAZOLES[a]

Substituent	E^1_7 (mV)	P	$C_{1.6}$
SO_2.O.naphthyl	-340	-	1.0×10^{-6} [b]
SO_2.O.4-nitrophenyl	-302	8	1.2 "
SO_2.O.4-methoxyphenyl	-335	10	1.5 "
SO_2.O.2-methoxyphenyl	-365	20	1.5 "
SO_2.morpholine	-406	0.69	1.5 "
SO_2.O.4-chlorophenyl	-345	-	2.6 "
SO_2.O.phenyl	-342	12.3	2.6 "
SO_2.phenyl	-376	10.8	3.6 "
SO_2CH_3	-355	0.33	4 "
$SO_2N(CH_3)_2$	-409	0.49	15 "
SO_2NH phenyl	-398	5.2	18 "
SO_2NH.2-methoxyphenyl	-408	6.5	30 "
SO_2NH_2	-395	0.23	35 "
SO_2NHCH_2morpholine	-402	0.27	35 "
$SO_2NHCH_2N(CH_3)_2$	-402	0.66	60 "
SO_2NH.2-toluyl	-426	7.0	100 "
SO_2naphthyl	-453	14	100 "
SO_2NH.2-chlorophenyl	-444	2.2	300 "
[Misonidazole]	-389	0.43	300 "

[a] E^1_7 is the one-electron reduction potential;
P is the octanol–water partition coefficient at pH 7.4;
$C_{1.6}$ is the concentration required to give an enhancement ratio of 1.6.

[b] For an enhancement ratio of 1.55 (i.e. not 1.6) in this case.

or keto-esters, aromatic or heterocyclic compounds substituted with electrophilic groups and the quinones. Compounds containing the quinonoid nucleus are particularly relevant because of the usually high electron affinities associated with compounds of this type.

REFERENCES

[1] BROWN, J.M., WORKMAN,.P., Partition coefficients as a guide to the development of radiosensitizers which are less toxic than misonidazole, Rad. Res. 82 (1980) 171.

[2] ADAMS, G.E., CLARKE, C., DAWSON, K.B., SHELDON, P.W., STRATFORD, I.J., "Nitroimidazoles as hypoxic cell sensitizers and cytotoxic agents", in Biological Bases and Clinical Implications of Tumour Resistance (Proc. 2nd Int. Symp., Rome, Sep. 1980) (1981) in press.

[3] DISCHE, S., SAUNDERS, M.I., STRATFORD, M.R.L., Neurotoxicity with desmethyl misonidazole, Br. J. Radiol. 54 (1981) 156

[4] BROWN, J.M., LEE, W.W., "Pharmacokinetic considerations in radiosensitizer development", in Radiation Sensitizers (BRADY, L.W., Ed.), Masson Publishing USA Inc. (1980) 2-13.

[5] MILLAR, B.C., JINKS, S., The effect of dexamethasone on the radiation survival response and misonidazole-induced hypoxic cell cytotoxicity in Chinese hamster cells, V79-753B, *in vitro*, Br. J. Radiol. (1981) in press.

[6] MILLAR, B.C., JINKS, S., POWLES, T.J., Flurbiprofen, a non-steroidal anti-inflammatory agent, protects cells against the cytotoxic effects of hypoxic cell radiosensitizers *in vitro*, Br. J. Cancer (1981) submitted for publication.

[7] ASQUITH, J.C., WATTS, M.E., PATEL, K., SMITHEN, C.E., ADAMS, G.E., Electron-affinic sensitization. V. Radiosensitization of hypoxic bacteria and mammalian cells *in vitro* by some nitro-imidazoles and nitropyrazoles, Rad. Res. 60 (1974) 108.

[8] SMITHEN, C.E., CLARKE, E.D., DALE, J.A., JACOBS, R.S., WARDMAN, P., WATTS, M.E., WOODCOCK, M., "Novel (nitro-1-imidazolyl)-alkanolamines as potential radiosensitizers with improved therapeutic properties", in Radiation Sensitizers (BRADY, L.W., Ed.), Masson Publishing USA Inc. (1980) 22-32.

[9] CHAPMAN, J.D., RALEIGH, J.A., PEDERSON, J.E., NGAN, J., SHUM, F.Y., MEEKER, B.E., URTASUN, R.C., "Potentially three distinct roles for hypoxic cell sensitizers in the clinic", in Proc. 6th Int. Congress Radiat. Res., Tokyo, Japan (1979) (OKADA, S., IMAMURA, M., TERASIMA, T., YAMAGUCHI, H., Eds.) JARR, 885-892.

[10] STRATFORD, I.J., WILLIAMSON, C., HOE, S., ADAMS, G.E., Radiosensitizing and cytotoxicity studies with CB 1954 (2,4-dinitro-5-aziridinyl benzamide), Radiat. Res. (1981) in press.

[11] HICKMAN, J.A., MELZACK, D.H., Protection against the effects of the antitumour agent CB 1954 by certain imidazoles and related compounds, Biochem. Pharmacol. 24 (1975) 1947.

[12] ADAMS, G.E., AHMED, I., FIELDEN, E.M., O'NEILL, P., STRATFORD, I.J.,"The development of some nitroimidazoles as hypoxic cell radiosensitizers", in Radiation Sensitizers (BRADY, L.W., Ed.), Masson Publishing USA Inc. (1980) 33-38.

[13] ADAMS, G.E., FIELDEN, E.M. HARDY, C.R., MILLAR, B.C., STRATFORD, I.J., WILLIAMSON, C., Radiosensitization of hypoxic mammalian cells *in vitro* by some 5-substituted 4-nitroimidazoles, Int. J. Radiat. Biol. (1981) in press.
[14] ADAMS, G.E., COOKE, M.S., Electron-affinic sensitization. I. A structural basis for chemical radiosensitizers in bacteria, Int. J. Radiat. Biol. 15 (1969) 457.

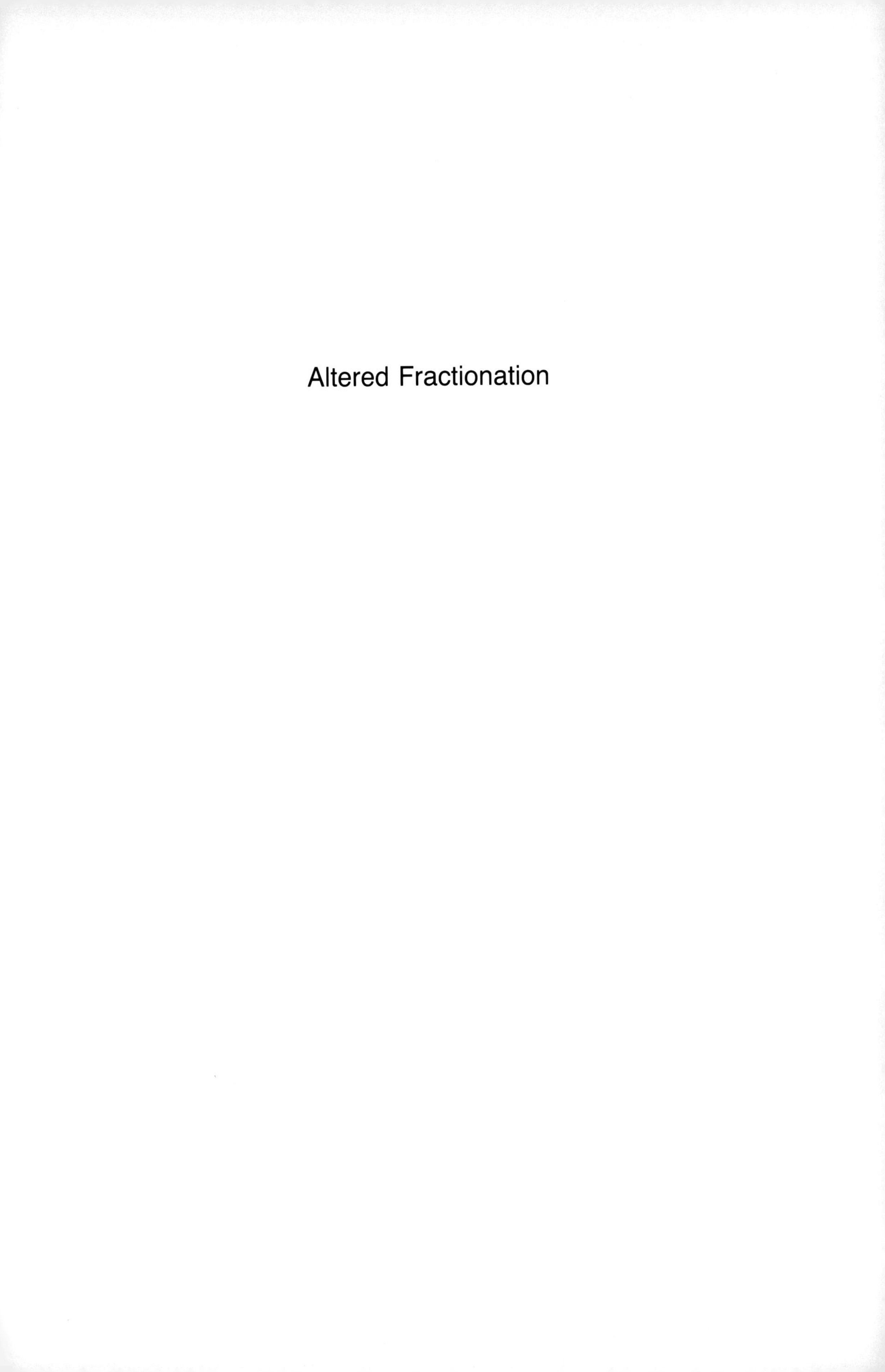

Altered Fractionation

Progress in Radio-Oncology II, edited by K. H. Kärcher et al., Raven Press, New York © 1982.

Differences in the Fractionation Response of Acutely and Late-Responding Tissues*

H. Rodney WITHERS
Department of Radiation Oncology,
UCLA Center for Health Sciences,
Los Angeles, California,
United States of America

Howard D. THAMES Jr.
Department of Biomathematics,
University of Texas M.D. Anderson Hospital
and Tumor Institute,
Houston, Texas,
United States of America

Lester J. PETERS
Institute of Oncology and Radiotherapy,
Prince of Wales Hospital,
University of New South Wales,
Sydney,
Australia

Abstract

There is a spectrum of rates of development of radiation injury in various normal tissues. These differences reflect differences in the proliferation kinetics of the target cells: basal cells of skin, or crypt cells of intestinal mucosa turn over rapidly and express acute injury, oligodendrocytes and renal tubule cells turn over slowly and manifest injury late. In modern radiotherapy, late effects are usually dose-limiting but this results, in part, from adopting dose-fractionation patterns aimed at by-passing serious acute effects. From clinical experience and recent radiobiological experiments, it has become abundantly clear that the form of the dose/survival relationship is different for the target cells of acute and late normal tissue effects. The difference is that late effects change in their severity more rapidly with change in size of dose: changes in acute reactions from a modified dose-fractionation regimen do not predict for the changes in late effects: the changes in late effects are always greater. With increase in dose per fraction, the change in late effect is disproportionately greater than the change in acute effect and vice versa. If the clonogens of rapidly responding tumours have the same relative dose/survival characteristics as the target cells in acutely responding normal tissues, it is clear that a potential exists for worsening or improving the therapeutic ratio by merely manipulating the size of dose per fraction: the therapeutic ratio should increase continuously as the size of dose per fraction decreases. Thus, the questions become, what is the ***practical,*** lower limit for the size of dose per fraction in the treatment of tumours showing early responses to radiation, and what are the total doses required for various late effects as a function of dose per fraction.

* This work was supported in part by United States Public Health Service grant CA-29644 and CA-29026 awarded by the National Cancer Institute.

There are many clinical studies which have demonstrated that when conventional, approximately 200 cGy fractions are replaced by fewer, larger dose fractions and the total dose adjusted to give equivalent or even less severe acute effects, the late effects become more severe (1-13). A second observation on the relationship between acute and late effects comes from neutron radiotherapy where it has been noted that, for a comparable level of acute injury from neutrons and x-rays, the late injuries are more severe in patients exposed to neutrons (14, 15). Both these clinical observations have counterparts in experimental animal studies (16-21) and may be explained by differences in the x-ray dose/survival relationships of the relevant target cells for acute and late injury (19, 22).

Before considering the quantitative aspects of acute and late responses, it is helpful to document our concept of these two responses (19, 23). There is clearly a spectrum of rates of development of radiation-induced normal tissue injury - some appearing within hours, others taking years to become fully manifest. It is now appreciated that the rate of development of a radiation effect in a tissue is not primarily a function of the radiosensitivity of the target cells, but rather of their proliferation kinetics. Rapidly turning-over cell populations show an early response to, and, if doses are not too high, recovery from, radiation insult. Very slowly proliferative tissues show a slow response to radiation and it is uncertain at what rate, and to what extent, the surviving cells regenerate. In general, however, it would be expected that, during a five to eight week course of radiotherapy, significant regeneration of surviving cells would occur in acutely responding (early effects) tissues, but not in slowly responding, late effects tissues.

The target cells for acute injury have been identified in most tissues (e.g. basal cells in skin, crypt cells in small bowel), but the clear acknowledgment of the target cells for late effects has been clouded by the concept that vascular damage was the basis for such effects. It simplifies an understanding of late effects to realize that vascular damage is a late effect in its own right, but that it is not the primary cause of other late effects. The target cells for late effects need clearer identification, but, for example, they are oligodendrocytes and renal tubule cells in the central nervous system and kidney, respectively. It is the slow death of these slowly turning-over cell populations that leads to late effects, just as it is the rapid death of rapidly turning-over cells that causes acute reactions, e.g. in intestinal mucosa or skin (19, 23).

The relative shapes of dose/survival curves for the cells of acutely- and late-responding tissues will determine their relative responses to dose fractionation. (For the present discussion we will not consider, in a quantitative way, the role of regeneration of surviving target cells during a course of fractionated irradiation.) There are several methods by which responses in the low dose (shoulder) region of the survival curve may be determined.

The shape of the dose/survival curve for cells _in vivo_ can sometimes be measured directly at low doses by a transplantation assay (24-26) but for similar direct measurements of cell survival _in situ_, high doses are necessary to reduce cellular survival to levels low enough to permit counting of discrete colonies regenerating from about 1 surviving cell (27-30). The shape of the survival curve at the low doses of clinical interest (e.g. between 0-500 cGy) can be deduced from multifraction experiments if it is known that each equal-sized dose fraction kills the same proportion of target cells. In two such assay systems (jejunal mucosa and spermatogenic epithelium) it has been shown that there is a high probability that this is indeed the case (29, 30). Based on the assumption of equal effect per fraction, dose/survival curves for several tissues have been reconstructed from multifraction experiments (29-32). Even if a survival curve cannot be reconstructed, inferences about its shape can be drawn from the doses needed for an isoeffect when the radiation is given in different numbers of various-sized fractional doses, and when it is assumed once again that an equal effect is obtained per fraction (33).

In Figure 1 are shown hypothetical dose-survival curves for the target cells for acute and late injury, and for each curve the repeated initial shoulders that represent the responses to repeated doses of sizes A and B when repair of sublethal injury is complete in each interfraction interval (for simplicity the effects of repopulation are ignored). The curves differ in their "curviness" in a way consistent with the differences in the acute and late dose-fractionation responses with changes in size of dose per fraction, as described above, and with the high RBE for late effects with low doses of neutron irradiation. Thus, when the total dose C is delivered with the larger dose per fraction B, the increase in late effects (resulting from increased killing of target cells) greatly exceeds the increase in acute effects. If the total dose given in the large fractions of size B were decreased so that the same level of acute response occurred as after multiple doses of size A, it is clear that late effects would exceed those resulting from the smaller dose per fraction A. In this model then we find a possible explanation for the many clinical studies that have shown that when conventional 200 cGy fractions are replaced by fewer, larger dose fractions and the total dose adjusted to give equivalent or less severe acute effects, the late effects become more severe. Put otherwise, the slope of the response to fractionated doses (repeated initial shoulders in Figure 1) increases more rapidly with increasing dose per fraction for late effects: the "effective D_o" for a regimen of repeated doses changes faster with dose per fraction for late responses.

The sterilization of mammalian cells can be fairly well described mathematically by combining a linear (α) and a dose-squared (β) coefficient of dose. Thus, the proportion of cells surviving (surviving fraction, S.F.) may be described mathematically as:

$$S.F. = e^{-(\alpha D + \beta D^2)}$$

The relatively greater susceptibility of late effects tissues to killing from accumulated sublethal injury, and, conversely, the relatively greater susceptibility of acutely responding tissues to single hit injury, as illustrated in Figure 1, represents a higher β/α ratio in the late effects cell survival equation (19, 22).

A characteristic of the response of mammalian cells to neutrons is that single-hit killing predominates at all doses, overshadowing the lethality from accumulation of sublethal injury (Figure 1). In other words, the β/α ratio characteristic of neutron dose/survival curves is small and the survival curve is essentially exponential to fairly low levels of cell survival. Thus, the $RBE_{n/x}$ for late effects may be higher than that for acute effects because target cells for the former are relatively insensitive at low x-ray doses (Figure 1), and are characterized by a larger $_1D_o$. The higher $RBE_{n/x}$ for late effects thus signifies no special sensitivity of these tissues to neutrons but rather their relative insensitivity to low doses of x-radiation. Perhaps of equal clinical relevance is the fact that the $RBE_{n/x}$ for late effects may be expected to increase more rapidly with decreasing dose per fraction than that for acute effect, by virtue of the larger β/α ratio.

What are the data that support the notion that the β/α ratio is higher for the target cells for late injury than for the target cells for acute injury and, therefore, that the survival curves shown in Figure 1 are appro-

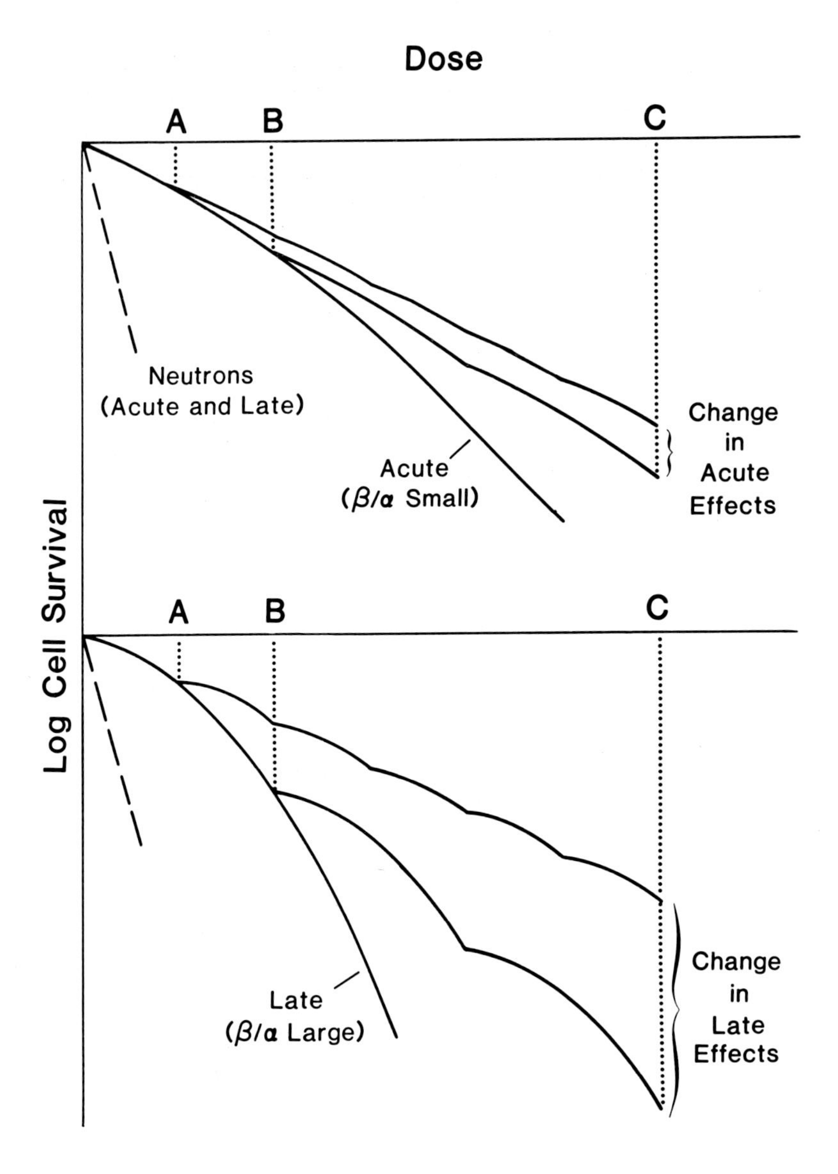

FIG.1. Hypothetical survival curves for the target cells for acute and late effects in normal tissues exposed to x-rays or neutrons. The β/α ratio in the equation for surviving fractions, $S.F. = \exp(-\alpha D-\beta D^2)$ is higher for late effects than for acute effects in x-irradiated tissues resulting in a greater rate of change in effect in late-responding tissues with change in dose: at dose A, survival of target cells is higher in late-effects than acute-effects tissues, whereas at dose B, the reverse is true. Therefore, increasing the dose per fraction from A to B will result in a relatively greater increase in late than acute injury. In the case of neutrons, the β/α ratio is low with no detectable influence of the quadratic function βD^2 over the first two decades of reduction in cell survival - implying that accumulation of sublethal injury plays a negligible role in cell killing by doses of neutrons of clinical interest: at these doses, the $RBE_{n/x}$ is higher for late than for acute effects.

priate? If the rate of sterilization of the target cells for late effects increases more rapidly with increase in dose than does the rate of sterilization of target cells for acute effects, then reducing the size of dose fractions should preferentially "spare" late effects tissues. In general terms, if the rate of change of effect with change in size of dose is greater for late than for acutely-responding tissue cells, then the slope of an isoeffect curve, in which the total dose for a certain effect is plotted as a function of size of dose per fraction, should be steeper for late than for acute endpoints. Such isoeffect curves for a variety of acutely and late responding tissues in experimental animals (18, 29, 30, 31, 34-43) are shown in Figure 2.

It is clearly apparent that the slopes of the isoeffect curves for four late effects (in spinal cord, kidney, lung, skin) are consistently steeper than similar curves for acute effects in a variety of tissues (skin epithelium, jejunal and colonic mucosa, spermatogenic epithelium, bone marrow stem cells). These isoeffect curves support the concept of a higher β/α ratio in the survival curve equation for the target cells for late effects and are consistent with the x-ray survival curves shown in Figure 1.

The curves in Figure 2 resemble traditional isoeffect curves in which the total doses for an isoeffect are plotted as a function of the number of dose fractions, N. For any given isoeffect (e.g. skin desquamation), N and dose per fraction are linked inversely. But when a large number of different isoeffects are plotted (e.g. for effects in skin, bone marrow, gut, testis, etc.), and the total doses for isoeffects vary greatly, a certain value of N on the abscissa does not represent the same size of dose per fraction for the various isoeffects. For example, if an isoeffect in skin (e.g. moist desquamation) required 4000 cGy in 10 fractions and a bone marrow isoeffect (e.g. $LD_{50/30}$) required 1000 cGy in 10 fractions, then at a value of 10 for N on the abscissa, the slope of the isoeffect curve for skin would be that appropriate for 400 cGy dose fractions while that for bone marrow would be appropriate for 100 cGy dose fractions. It is obvious from Figure 1 that the important variable in determining the total dose for a certain effect is the size of dose per fraction and not the number of fractions per se. Accordingly, when isoeffect curves for many different endpoints are plotted, as they are in Figure 2, the use of N as a variable becomes meaningless and comparisons can only be drawn on the basis of size of dose per fraction. (The same criticism applies to the use of N in isoeffect formulae purported to apply to all normal tissues.) Thus, the isoeffect curves plotted in Figure 2 are different from the usual Strandqvist-type isoeffect curves and are more "correct."

The similarity of the slopes of the acute isoeffect curves in Figure 2 suggests that the β/α ratio is the same for all acutely-responding tissues. While the survival curves for the stem cells of the jejunal crypt and spermatogenic epithelium are not detectably different from one another (19), it should be emphasized that the isoeffect curves in Figure 2 only reflect the β/α ratio and do not permit estimates to be made of the absolute values of β and α.

If the isoeffect curves for acute and late responding experimental animal tissues (Figure 2) are appropriate for man, implying that the relative shapes of the survival curves for acute and late effects are as shown in Figure 1, and if most tumors respond as do acutely responding normal tissues, then it is possible that hyperfractionation may improve the therapeutic ratio in clinical radiotherapy. For example, if a large number of doses of size A (Figure 1) were to replace a smaller number of the larger dose fraction B, the ultimate survival of the late effects target cells would change to being higher rather than lower than the acutely-responding tumor clonogens. Thus, the difference in the responses of tumors and late effects

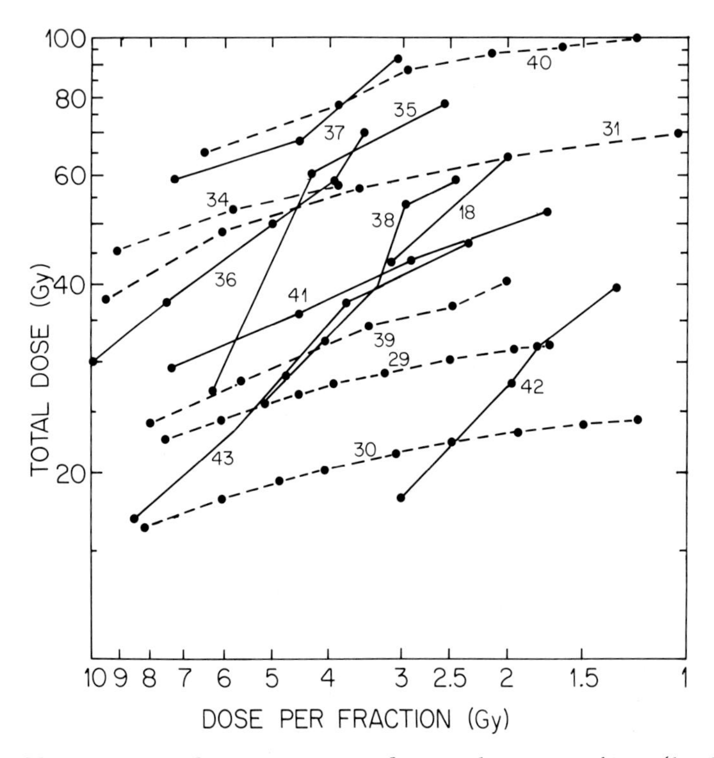

FIG.2. Isoeffect curves for a variety of acutely responding (broken lines) and late responding (full lines) tissues in experimental animals. The numbers identify, in the bibliography, the source of the data for each curve. Note that the slopes of the curves for the late-responding tissues are consistently greater than those for acute effects. The curves indicate that when the size of dose per fraction is increased, the total dose for a late isoeffect must be reduced more than would be the case if the aim was to achieve equal acute isoeffects. Note that the abscissa plots size of dose per fraction rather than number of fractions. Curve number 40 is displaced to 10-fold higher total doses for convenience of presentation. (Modified from Refs [19] and [22].)

tissues would be exaggerated (to the benefit of the normal tissue), other things being equal. (In order for one other effect, regeneration of tumor clonogens during therapy, to be equal in conventional and hyperfractionated regimens, the frequency of exposure must be increased in the hyperfractionated schemes (to multiple daily fractions) to ensure that the total treatment duration is not extended significantly.)

The next question is, what dose schedules should be used in hyperfractionated regimens. The shapes of several of the curves for acute isoeffects are fairly well defined to doses of 200 cGy or less (Figure 2), but there is a need for more precise data between doses of 50-250 cGy to help in the design of clinical hyperfractionation schedules using doses of approximately 100 cGy. While the isoeffect curves for acute effects are well defined at high doses per fraction and fairly well defined at lower doses, the late isoeffect curves are imprecise at whatever dose one chooses. Essentially no useful data exist for comparison of dose/response relationships for late effects at doses ranging between 50 and 250 cGy. Thus, at present, it is not possible to estimate precisely what doses should be used to replace conventional 200 cGy dose fractions in order to achieve the same late effects with hyperfractionated regimens. Estimates from the meager data available, using elegant analytical methods (19, 22, 31, 33, 44, 45), suggest that tolerable late effects doses for some tissues may be up to 20-30% higher when conventional 200 cGy schemes are further fractionated (44, 45) into doses of about 100 cGy. If these estimates for tolerance of late responding tissues are correct, the total doses used in previous trials of hyperfractionation may have been too low.

In practice there are many possible hyperfractionation schemes (46), but one example would be 100 cGy given 12 or 13 times per week for the same overall time as in a conventional, 200 cGy, regimen. Such a regimen may give equal late effects but would be expected to yield a more severe acute response. In many instances, the more severe acute responses may be masked by regeneration of surviving target cells during the course of therapy: if not, sufficient amelioration of acute normal tissue responses could be achieved by slight protraction of the overall treatment duration. It should be mentioned in passing, that, in hyperfractionated regimens, small differences in size of dose per fraction (e.g. 5 cGy) amplify into large differences in total dose.

In conclusion, it is clear that there are two major deficits in our knowledge of multifraction dose responses: the quantitation of the response of various dose-limiting late effects target cells especially at low fractional doses (ranging, say, from 0-300 cGy) and the variation in response to doses between 0-300 cGy of various human tumor clonogens as a function of their histology and proliferation kinetics. Since it is unlikely that late effects have their origin in vascular damage (19, 23), it will be necessary to quantitate the fractionation responses of the target cells in many tissues such as spinal cord, kidney, liver, skin, lung, etc. in order that optimum doses per fraction and total doses may be chosen for hyperfractionation trials. Because of the diversity of tumor types and proliferation kinetics, the hyperfractionation responses of human tumors may have to be evaluated as an "incidental" result of the application of hyperfractionation regimens designed to deliver doses as high as tolerable by slowly responding tissues.

REFERENCES

[1] ANDREWS, J.R.: Dose time relationships in cancer radiotherapy. A clinical radiobiology study of extremes of dose and time, Am. J. Roentgenol. 93 (1965) 56.

[2] ARCANGELI, G., FRIEDMAN, M., PAOLUZI, R.: A quantitative study of late radiation effect on normal skin and subcutaneous tissues in human beings, Br. J. Radiol. 47 (1974) 44.

[3] ATKINS, H.L.: Massive dose technique in radiation therapy of inoperable carcinoma of the breast, Am. J. Roentgenol. 91 (1964) 80.

[4] BATES, T.D., PETERS, L.J.: Dangers of the clinical use of the NSD formula for small fraction numbers, Br. J. Radiol. 48 (1975) 773.

[5] BENNETT, M.R.: The treatment of stage III squamous carcinoma of the cervix in air and hyperbaric oxygen, Br. J. Radiol. 51 (1978) 68.

[6] CHU, F.C.H., GLICKSMAN, A.S., NICKSON, J.J.: Late consequences of early skin reactions, Radiology 94 (1970) 669.

[7] GAUWERKY, F., LANGHEIM, F.: Der Zeitfaktor bei der strahleninduzierten subkutanen Fibrose, Strahlentherapie 154 (1978) 608.

[8] KIM, J.H., CHU, F.C.H., HILARIS, B.: The influence of dose fractionation on acute and late reactions in patients with postoperative radiotherapy for carcinoma of the breast, Cancer 35 (1975) 1583.
[9] MONTAGUE, E.D.: Experience with altered fractionation in radiation therapy of breast cancer, Radiology 90 (1968) 962.
[10] SAUSE, W.T., STEWART, J.R., PLENK, H.P., LEAVITT, D.D.: Late skin changes following twice-weekly electron beam radiation to post-mastectomy chest walls, Int. J. Radiat. Oncol., Biol. Phys. (1981, in press).
[11] SINGH, K.: Two regimes with the same TDF but differing morbidity used in the treatment of stage III carcinoma of the cervix, Br. J. Radiol. 51 (1978) 357.
[12] STELL, P.M., MORRISON, M.D.: Radiation necrosis of the larynx, Arch. Otolaryngol. 98 (1973) 111.
[13] DISCHE, S., MARTIN, W.M.C., ANDERSON, P.: Radiation myelopathy in patients treated for carcinoma of bronchus using a six fraction regime of radiotherapy, Br. J. Radiol. 54 (1981) 29.
[14] STONE, R.S.: Neutron therapy and specific ionization, Am. J. Roentgenol. 59 (1948) 771.
[15] HUSSEY, D.H., FLETCHER, G.H., CADERAO, J.B.: "A preliminary report of the MDAH-TAMVEC neutron therapy pilot study, in Proc. 5th Int. Congr. Radiation Research, Academic Press, New York (1975) 1106-1117.
[16] BERRY, R.J., WIERNIK, G., PATTERSON, T.J.S., HOPEWELL, J.W.: Excess late subcutaneous fibrosis after irradiation of pigskin consequent upon the application of the NSD formula, Br. J. Radiol. 47 (1974) 277.
[17] TURESSON, I., NOTTER, G.: "Late effects of various dose-fractionation regimens", in Biological Bases and Clinical Implications of Tumor Radioresistance (Proc. 2nd Int. Symp. Rome, 1980: FLETCHER, G.H., NERVI, C., WITHERS, H.R., Eds), Masson Publishing Co. (1981, in press).
[18] WITHERS, H.R., THAMES, H.D., FLOW, B.L., MASON, K.A., HUSSEY, D.H.: The relationship of acute to late skin injury in 2 and 5 fraction/week γ-ray therapy, Int. J. Radiat. Oncol., Biol. Phys. 4 (1978) 595.
[19] WITHERS, H.R., THAMES, H.D., PETERS, L.J., FLETCHER, G.H.: "Normal tissue radioresistance in clinical radiotherapy", in Biological Bases and Clinical Implications of Tumor Radioresistance (Proc. 2nd Int. Symp. Rome, 1980: FLETCHER, G.H., NERVI, C., WITHERS, H.R., Eds), Masson Publishing Co. (1981, in press).
[20] HUSSEY, D.H., GLEISER, C.A., JARDINE, J.H., RAULSTON, G.L., WITHERS, H.R.: "Acute and late normal tissue effects of 50 $MeV_{d/Be}$ neutrons", in Radiation Biology in Cancer Research (MEYN, R.E., WITHERS, H.R., Eds), Raven Press, New York (1980) 471-488.
[21] WITHERS, H.R., THAMES, H.D., HUSSEY, D.H., FLOW, B.L., MASON, K.A.: RBE of 50 MV (Be) neutrons for acute and late skin injury, Int. J. Radiat. Oncol., Biol. Phys. 4 (1978) 603.
[22] THAMES, H.D., WITHERS, H.R.: Changes in early and late radiation responses with altered dose fractionation: Implications for dose survival relationships, Int. J. Radiat. Oncol., Biol. Phys. (1981, in press).
[23] WITHERS, H.R., PETERS, L.J., KOGELNIK, H.D.: "The pathobiology of late effects of irradiation", in Radiation Biology in Cancer Research (MEYN, R.E., WITHERS, H.R., Eds), Raven Press, New York (1980) 439-448.
[24] TILL, J.E., McCULLOCH, E.A.: Early repair processes in marrow cells irradiated and proliferating in vivo, Radiat. Res. 41 (1970) 450.
[25] GOULD, M.N., CLIFTON, K.H.: The survival of rat mammary cells following irradiation in vivo, Radiat. Res. 72 (1977) 343.
[26] DE MOTT, R.K., MULCAHY, R.T., CLIFTON, K.H.: The survival of thyroid cells following irradiation, Radiat. Res. 77 (1979) 395.

[27] WITHERS, H.R.: The effect of oxygen and anaesthesia on the radio-sensitivity in vivo of epithelial cells of mouse skin, Br. J. Radiol. 40 (1967) 335.
[28] WITHERS, H.R., ELKIND, M.M.: Microcolony survival assay for cells of mouse intestinal mucosa exposed to radiation, Int. J. Radiat. Biol. 17 (1970) 261.
[29] THAMES, H.D., WITHERS, H.R.: Test of equal effect per fraction and estimation of initial clonogen number in microcolony assays of survival after fractionated irradiation, Br. J. Radiol. 53 (1980) 1071.
[30] THAMES, H.D., WITHERS, H.R., MASON, K.A., REID, B.O.: Dose survival characteristics of mouse jejunal crypt cells, Int. J. Radiat. Oncol., Biol. Phys. (1981, in press).
[31] DOUGLAS, B.G., FOWLER, J.F.: The effect of multiple small doses of x-rays on skin reactions in the mouse and a basic interpretation, Radiat. Res. 66 (1976) 401.
[32] MASUDA, K., WITHERS, H.R.: Method of constructing a dose-response curve for normal cells in situ, Br. J. Radiol. 49 (1976) 351.
[33] DUTREIX, J., WAMBERSIE, A.: "Cell survival curves deduced from non-quantitative reactions of skin, intestinal mucosa and lung", in Cell Survival After Low Doses of Radiation: Theoretical and Clinical Applications (ALPER, T., Ed.), John Wiley and Sons, Bristol (1975) 335-341.
[34] FOWLER, J.F., DENEKAMP, J., DELAPEYER, C., HARRIS, S.R., SHELDON, P.W.: Skin reactions in mice after multifraction X-irradiation, Int. J. Radiat. Oncol., Biol. Phys. 25 (1974) 213.
[35] HOPEWELL, J.W., FORTE, J.L., YOUNG, C.M.A., WIERNIK, G.: Late radiation damage to pig skin, Radiology 130 (1979) 783.
[36] VAN DER KOGEL, A.J.: Radiation tolerance of the rat spinal cord: Time dose relationships, Radiology 122 (1977) 505.
[37] WHITE, A., HORNSEY, S.: Radiation damage to the rat spinal cord: The effect of single and fractionated dose of x-rays, Br. J. Radiol. 51 (1978) 515.
[38] CALDWELL, W.L.: "Time-dose factors in fatal post-irradiation nephritis", in Cell Survival After Low Doses of Radiation: Theoretical and Clinical Applications (ALPER, T., Ed.), John Wiley and Sons, Bristol (1975) 328-332.
[39] WITHERS, H.R., MASON, K.A.: The kinetics of recovery in irradiated colonic mucosa of the mouse, Cancer 34 (1974) 896.
[40] WITHERS, H.R.: "Isoeffect curves for various proliferative tissues in experimental animals", in Proc. Conf. Time-Dose Relationships in Clinical Radiotherapy (Time-Dose Conference, Madison, 1974: CALDWELL, W.L., TOLBERT, D.D., Eds), Madison Printing and Publishing Co., Madison (1975) 30-38.
[41] FIELD, S.B., HORNSEY, S., KUTSUTANI, Y.: Effects of fractionated irradiation on mouse lung and a phenomenon of slow repair, Br. J. Radiol. 49 (1976) 700.
[42] HOPEWELL, J.W., WIERNIK, G.: "Tolerance of the pig kidney to fractionated X-irradiation", in Radiobiological Research and Radiotherapy (Proc. Symp. Vienna, 1976: BECK, E.R.A., Ed.) Vol.1, IAEA, Vienna (1977) 65-73.
[43] WARA, M.W., PHILLIPS, T.L., MARGOLIS, L.W., SMITH, V.: Radiation pneumonitis: A new approach to the derivation of time-dose factors, Cancer 32 (1973) 547.
[44] THAMES, H.D., WITHERS, H.R., PETERS, L.J., FLETCHER, G.H.: Hyper-fractionation strategies based on late normal tissue responses, Int. J. Radiat. Oncol., Biol. Phys. (1981, in press).
[45] THAMES, H.D., WITHERS, H.R., DISCHE, S.: Effect of dose per fraction on the incidence of late radiation myelopathy: an estimate of β/α,

Br. J. Radiol. 54 (1981, in press).
[46] PETERS, L.J., WITHERS, H.R., THAMES, H.D. Jr.: "Radiobiological bases for multiple daily fractionation", these Proceedings, p.317.

Progress in Radio-Oncology II, edited by K. H. Kärcher et al., Raven Press, New York © 1982.

Repair of Late and Early Radiation Injury in Lungs of Experimental Mice

J.F. FOWLER
Gray Laboratory of the Cancer Research Campaign,
Mount Vernon Hospital,
Northwood, Middlesex,
United Kingdom

E.L. TRAVIS
Radiation Oncology Branch of the
National Cancer Institute,
Bethesda, Maryland,
United States of America

Abstract

The measurement of breathing rate in mice after irradiation of the thorax enables both an ***acute phase*** of pneumonitis (2–5 months) and a ***late phase*** of fibrosis (after 8 months) to be distinguished. These types of damage can be 'spared' by different procedures. LD_{50} experiments only investigate the acute pneumonitis, not the late fibrosis. Breathing rate measurements can be made on the same animals for more than a year after irradiation, so that late and early injury in lung can be directly compared.

The increase in breathing rate after thoracic irradiation of mice has been correlated with histological changes [1-3]. Two distinct waves of injury can be identified. The 'early' phase consists of pneumonitis at 2-5 months and the 'late' phase of progressive fibrosis later than 8 months [2]. LD_{50} experiments measure the 'early' phase only, of course, and are no guide to late fibrosis.

Repair between daily fractions has been measured as similar to that in skin or slightly less, i.e. F_r about 65% instead of 75% at doses per fraction of 400-500 rads of X-rays. This rapid 'Elkind'-type repair is similar for early and late injury [4].

Slow repair in our mice is present but smaller than that reported originally by Field, Hornsey and Kutsutani [5], and occurs only for the 'early' phase of injury, not beyond 30 weeks. We have also observed slow repair after neutron irradiation - in contrast to the lack of slow repair reported by Field and Hornsey [6] - although the magnitude is small, about 1 gray of neutrons. This widens the possibilities for explaining the mechanism of slow repair.

The radioprotective drug WR 2721 protects against the early phase of lung injury after X-irradiation (DMF $\sim$1.2), but protects to a larger extent against the late injury (DMF $\sim$1.5).

The question of whether repair capability decreases after multiple small fractions has been investigated by "partial treatment and top-up" experiments: 4, 7 or 10 equal daily fractions of 0.25 or 0.3 Gy of X-rays were followed by a series of neutron top-up doses, given as single doses on the day after the last X-ray fraction. The difference between the 10 and 7 fraction top-up doses ($D_{10} - D_7$) was the same as that between the 7 and 4 fraction top-up doses ($D_7 - D_4$), thus demonstrating no evidence of less repair after 7 fractions of this size than before 7 fractions were reached.

REFERENCES

[1] TRAVIS, E.L., VOJNOVIC, B., DAVIES, E.E., HIRST, D.G.: A plethysmographic method to test lung function in locally irradiated mice, Br. J. Radiol. 52 (1979) 67-74.
[2] TRAVIS, E.L.: The sequence of histological changes in mouse lungs after single doses of X-rays, Int. J. Radiat. Oncol., Biol. Phys. 6 (1980) 345-347.
[3] TRAVIS, E.L., DOWN, J.D., HOLMES, S.J.: Breathing frequency as a measure of acute and late radiation damage in mouse lungs, Int. J. Radiat. Oncol., Biol. Phys. 5 Suppl.2 (1979) 85.
[4] TRAVIS, E.L., DOWN, J.D., HOLMES, S.J., HOBSON, B.: Radiation pneumonitis and fibrosis in mouse lung assayed by respiratory frequency and histology, Radiat. Res. 84 (1980) 133-143.
[5] FIELD, S.B., HORNSEY, S., KUTSUTANI, Y.: Effects of fractionated irradiation on mouse lung and a phenomenon of slow repair, Br. J. Radiol. 49 (1976) 700-707.
[6] FIELD, S.B., HORNSEY, S.: Slow repair after X-rays and fast neutrons, Br. J. Radiol. 50 (1977) 600-601.

Progress in Radio-Oncology II, edited by K. H. Kärcher et al., Raven Press, New York © 1982.

Rate of Repopulation in a Slow and a Fast Growing Mouse Tumour

J. KUMMERMEHR
Abteilung Strahlenbiologie,
Gesellschaft für Strahlen- und Umweltforschung mbH,
Neuherberg

K.-R. TROTT
Strahlenbiologisches Institut,
Universität München,
Munich

Federal Republic of Germany

Abstract

A slow growing adenocarcinoma (doubling time 9 d) and a fast growing fibrosarcoma (doubling time 1.5 d) were compared for their capacity to repopulate, using both regrowth delay and local tumour control as assays. Experiments included split-dose irradiation with priming doses of 10 and 20 Gy followed by hypoxic test doses within 7 d. In the carcinoma further protocols of 5 fr./8 d versus 5 fr./26 d and 10 fr./9 d versus 10 fr./27 d were tested. Repopulation in the carcinoma was remarkably high, corresponding to a 'repopulation dose' of 2 Gy/d after 10 Gy. In the 5 and 10 fraction schedules, 'repopulation doses' during the extra 18 d of treatment time were 1.5 Gy/d and 1.4 Gy/d if assayed by regrowth delay, but 2.6 Gy/d and 3.9 Gy/d if assayed by tumour control. Derived doubling times during repopulation were between 1.8 and 0.65 d. In the fibrosarcoma, repopulation within 7 d after 10 or 20 Gy was equivalent to 1.3 Gy/d when measured by regrowth delay, while it was nil in tumour cure experiments. It is concluded that the capacity for repopulation is predicted better by the potential doubling time (i.e. at theoretical maximum rate of repopulation) than by the volume doubling time at the time of treatment.

1. INTRODUCTION

The response of tumours to fractionated radiotherapy depends on various processes taking place during the intervals between fractions; the processes were summarized by Withers [1] as the 4 "R"s. Of these, recovery from sublethal damage, reoxygenation and redistribution have received much attention in experimental studies. Repopulation has received much less attention: probably based on the observation of a low clinical growth rate in most human tumours (a doubling time in the range of the usual overall treatment time), this time-dependent factor has usually been given low priority [2].

The aim of the present study is to compare the rate of repopulation in two murine tumours with widely differing growth rates, using both regrowth delay and local tumour cure as assays.

2. MATERIALS AND METHODS

2.1. Tumours

Experiments were carried out on two systems that are isogenic to our C3H mouse colony. A limited number of passages was used, retransplanted from deep-frozen material.

Mammary carcinoma AT 17 originally arose within the irradiated field in an animal permanently cured of a sarcoma. Over 13 animal passages, the tumour has maintained its well differentiated morphology and its initial slow growth rate. At the treatment size chosen (100 mg) the volume doubling time (T_d) is 9-11 d, the potential doubling time (T_{pot}) is 1.8 d and the median cell cycle time (T_c) is 16 h.

Fibrosarcoma SSK 2 has been derived from a methyl-cholanthrene-induced sarcoma (SSK 31) by in-vitro cloning. For this solid tumour, T_d is 1.5 d and T_{pot} is 1.1 d.

Transplantation of tumours was done by inserting a 1 mm^3 tumour fragment subcutaneously into the right flank of a mouse.

2.2. Growth Measurements and Quantization of Regrowth Delay

Prior and subsequent to treatment, tumour size was regularly recorded using a calibrated Perspex stencil. From growth curves plotted for individual tumours, the median growth delay was calculated as the geometric mean ± its standard error (SEM). End points used were 2 times treatment size for AT 17 and 3 times treatment size for SSK 2.

The rationale of growth-delay experiments designed to measure repopulation is to compare tumours that during 'inter-treatment time' have attained different numbers of clonogenic cells. This difference can be quantified if tumours are submitted to a range of test doses given under local hypoxia and if the dose/response curves arising are related to the day of the test dose. From isoeffect doses, a dose increment is then derived that is a measure of the repopulation afforded.

Local tumour control was ascertained by the absence of local regrowth over an appropriate observation period; this was 6 months for SSK 2 or 12 months for AT 17. Actuarial data were corrected following Kaplan and Meier, and doses giving a 50% cure probability (TCD 50) were calculated by a program proposed by Porter [3].

2.3. Irradiation

Animals were locally irradiated with 300 kV X-rays (collimating tube 25 mm dia.; filtering 0.5 mm Cu and 1.0 mm Al; dose rate 4 Gy/min) under hexobarbitone anaesthesia. For irradiation under the condition of local hypoxia, tumours were clamped off prior to and during irradiation by a plastic clamp.

Tumours were submitted to a split-dose regime consisting of a priming ambient dose of 10 or 20 Gy followed by test doses under local hypoxia after 1, 3 or 4, or 7 d. Carcinoma AT 17 was further tested using fractionated schedules, comparing 5 fractions given over 8 or 26 d with 10 fractions given over 9 or 27 d, all doses being delivered under local hypoxia.

3. RESULTS

Data on local tumour control are listed in Table I. 'Repopulated doses', calculated from differences between TCD 50 values or derived from regrowth-delay curves, are given in Table II for carcinoma AT 17 and in Table III for sarcoma SSK 2.

3.1. Split-Dose Experiments with AT 17

Regrowth-delay curves for AT 17 arising from split-dose experiments are shown in Fig. 1. Following a priming dose of 10 Gy, delay curves to subsequent test doses are significantly different for the 1, 3 and 7 d intervals. Incremental doses, evaluated at a delay level of 65 d, are given in Table II.

Following 20 Gy, a dose virtually only survived by hypoxic cells, the response curves are steeper, reflecting the greater damage imparted by the conditioning dose. The dose difference between the 3 and 7 d intervals, measured at a delay level of 100 d, is 5 Gy.

Data on local tumour control derived over an observation period of 12 months are listed in Table I. Disregarding the overlap in the 95% confidence intervals, the differences between the TCD 50 values are notably consistent with the delay data (Table II).

TABLE I. LOCAL TUMOUR-CONTROL DATA FOR AT 17 AND SSK 2

TCD 50 values are total doses, including priming dose

Tumour system	Regime tested			TCD 50 value (Gy)	95% Confidence limits (Gy)
	Priming dose (Gy)	Time to test dose (d)	Fraction-ation scheme		
AT 17	10	1		67 [a]	
		3		70.6	68.1–74.7
		7		78.5	74.4–91.6
	20	1		60.6	54.2–78.9
		3		65.5	62.2–70.8
		7		70.5	66.3–78.0
AT 17			5 fr./8 d	119	110–133
			5 fr./26 d	166	150–198
			10 fr./9 d	147	140–159
			10 fr./27 d	218	202–253
SSK 2	10	1		61.6	57.1–67.2
		4		63.8	61.4–69.4
	20	1		65.3	61.0–71.0
		4		61.1	56.2–67.6
		7		65.0	58.5–71.2

[a] Evaluable only at the 46% cure level.

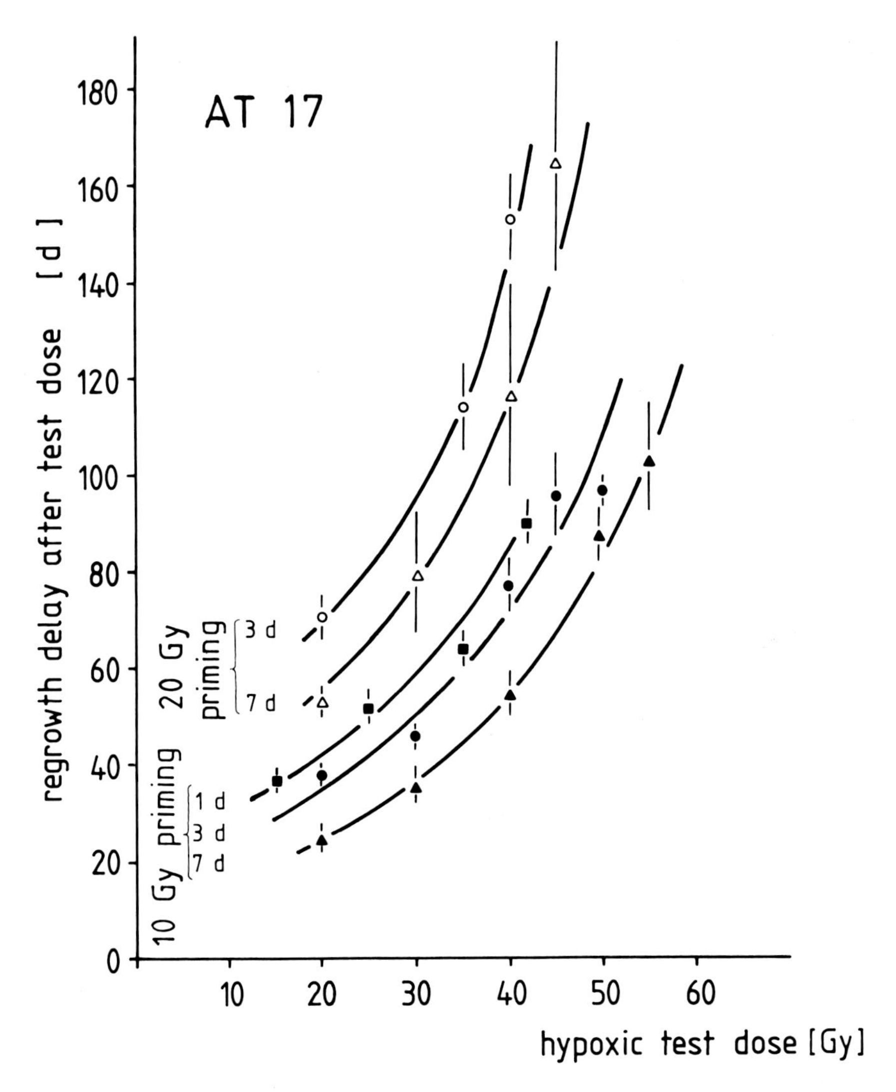

FIG.1. Regrowth delay curves for carcinoma AT 17 treated with priming doses of 10 Gy (full symbols) or 20 Gy (open symbols) under ambient conditions, followed by hypoxic test doses (see Table I). Intervals between treatments are 1 day (squares), 3 days (circles), and 7 days (triangles).

3.2. Multifractionation Experiments with AT 17

Regrowth delay curves for the 5 and 10 fraction regimes with AT 17 are given in Fig. 2. As a continuation of the argument on evaluating split-dose experiments, curves have been normalized to the last treatment day, thus taking the last dose fraction as a test dose. Dose differences thus measured at the 120 d delay level are 28 Gy (5 fr.) and 25 Gy (10 fr.).

In contrast to growth-delay data, the TCD 50 values from cure experiments indicate much higher dose differences for the corresponding schedules, viz. 47 Gy for the 5 fraction regime and 71 Gy for the 10 fraction regime.

TABLE II. DATA ON REPOPULATION IN CARCINOMA AT 17

Regime and assay	Interval (d)	Repopulated dose (Gy)	Repopulated dose per day (Gy/d)	Doubling time[a] (d)
10 Gy + test dose				
- Regrowth delay (assessed at 65 d)	1 to 3	4	2.0	1.3
	3 to 7	8	2.0	1.3
	1 to 7	12	2.0	1.3
- Tumour cure (assessed at 12 months)	1 to 3	3.6 [b]	1.8	1.4
	3 to 7	7.9	1.95	1.3
	1 to 7	11.5 [b]	1.92	1.3
20 Gy + test dose				
- Regrowth delay (assessed at 100 d)	3 to 7	5	1.25	2.0
- Tumour cure (assessed at 12 months)	1 to 3	4.9	2.45	1.0
	3 to 7	5.0	1.25	2.0
	1 to 7	9.9	1.65	1.5
5 fr./8 d versus 5 fr./26 d				
- Regrowth delay (assessed at 120 d)		28	1.55	1.6
- Tumour cure (assessed at 12 months)		47	2.6	0.96
10 fr./9 d versus 10 fr./27 d				
- Regrowth delay (assessed at 120 d)		25	1.39	1.8
- Tumour cure (assessed at 12 months)		71	3.95	0.63

[a] Doubling time during repopulation has been calculated assuming a D_0 of 3.6 Gy.
[b] Evaluable only at the 46% cure level.

3.3. Split-Dose Experiments with SSK 2

Regrowth-delay curves for SSK 2 obtained in the split-dose experiments are shown in Fig. 3. In spite of some scatter of the data (10 Gy series) and large error bars (20 Gy series), the curves are significantly different, at least for the 1 d and 7 d intervals. Dose differences evaluated at delay levels of 20 d or 35 d, respectively, are given in Table III. In contrast to regrowth-delay data, local tumour control experiments gave no consistent difference of curability between the 1, 4 and 7 d intervals. The TCD 50 figures are given in Table I.

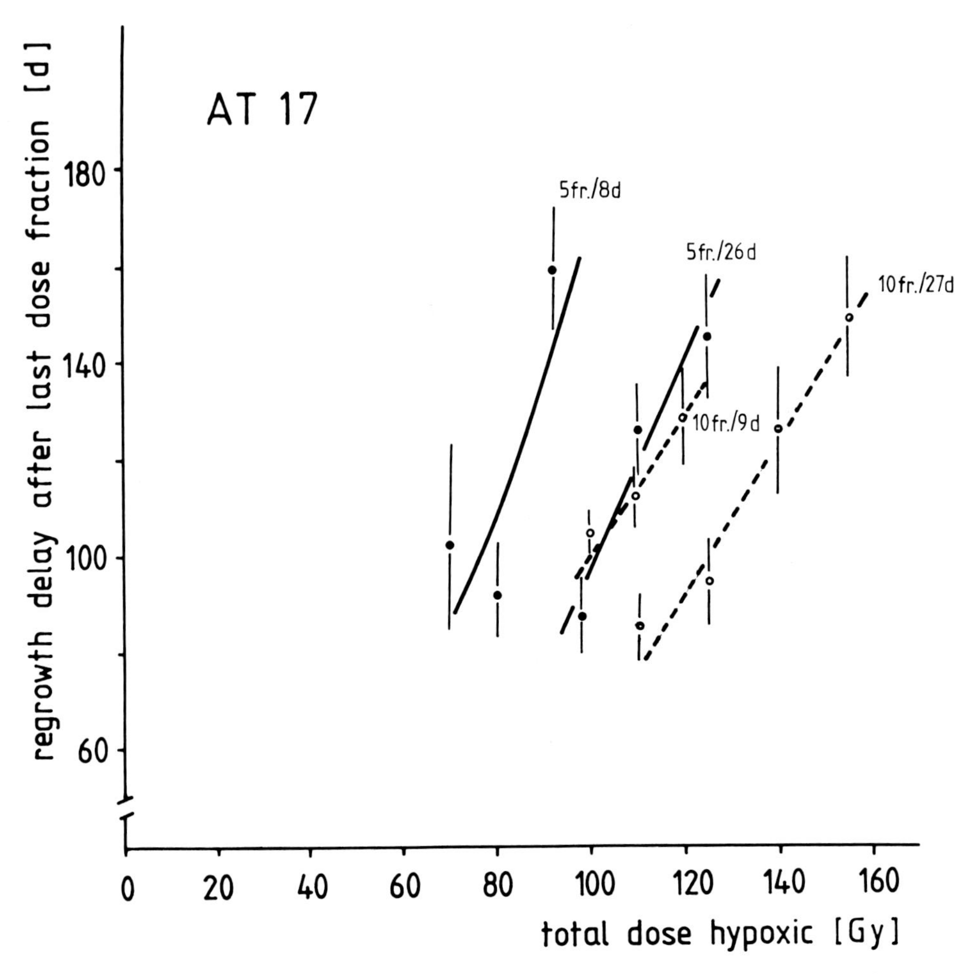

FIG.2. Regrowth delay curves for carcinoma AT 17 treated by the fractionation protocols indicated on the curves (● 5 fractions; o 10 fractions). Delay has been normalized to the day of the last dose fraction.

4. DISCUSSION

The most remarkable finding in the present experiments is an unexpectedly high potential of repopulation in the slow growing adenocarcinoma.

A compilation of the repopulation data of this tumour is listed in Table II, which gives a survey of doses necessary to compensate for repopulation in the various experimental sets and assays. It also gives estimates of the average doubling times during repopulation, assuming a D_0 for acutely hypoxic cells of 3.6 Gy.

For the split-dose experiments, the agreement between regrowth-delay and tumour-control data is exceptionally good. Repopulation between 1 and 7 d after a dose of 10 Gy is compensated for by an average daily dose of 2.0 Gy/d, and the onset of repopulation as determined from the 3 d interval is rapid. The calculated

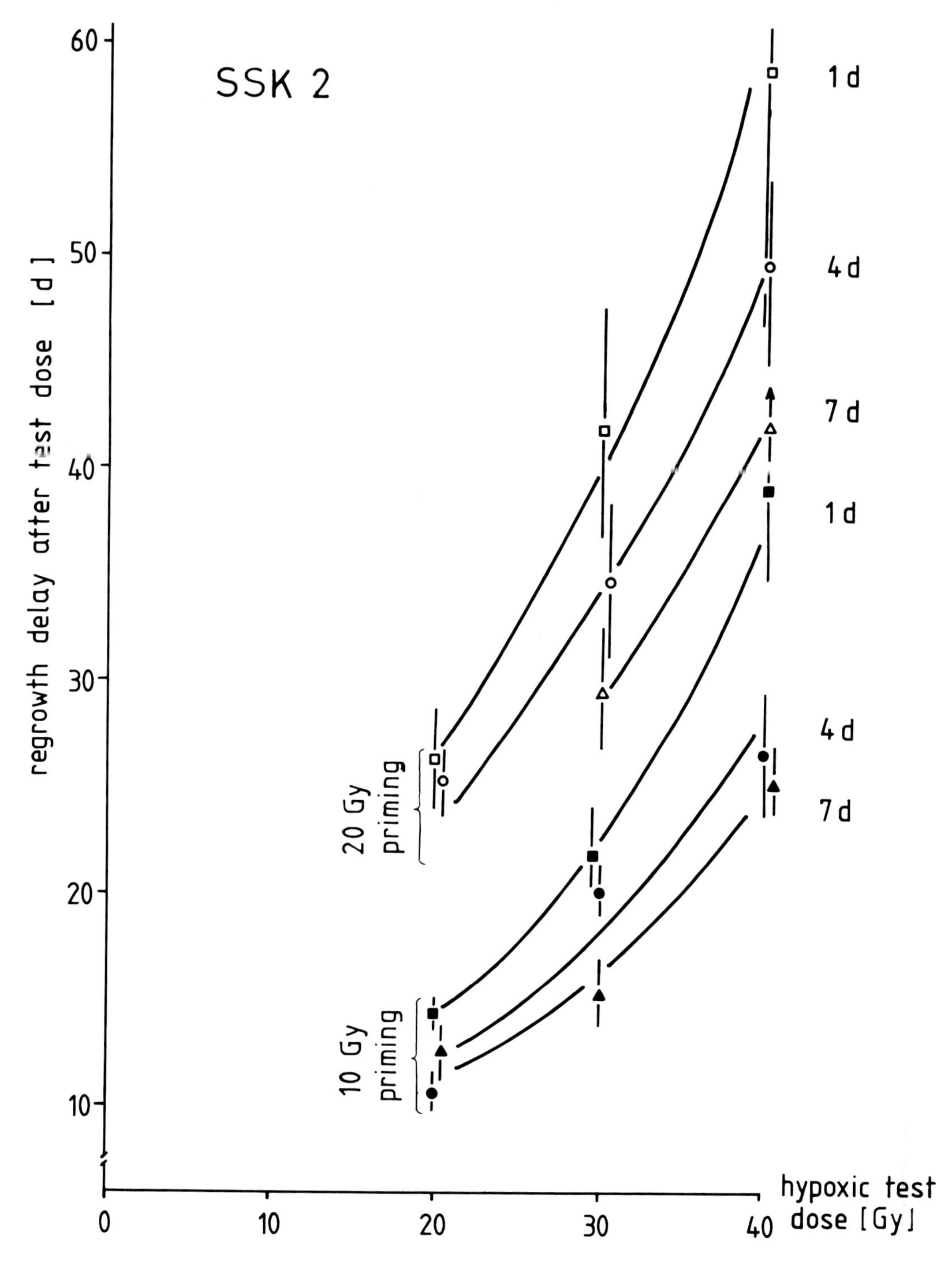

FIG.3. Regrowth delay curves for sarcoma SSK 2 treated with priming doses of 10 Gy (full symbols) or 20 Gy (open symbols) under ambient conditions, followed by hypoxic test doses (see Table I). Intervals between treatments are 1 day (squares), 4 days (circles), and 7 days (triangles).

TABLE III. DATA ON REPOPULATION IN SARCOMA SSK 2

Regime and assay	Interval (d)	Repopulated dose (Gy)	Repopulated dose per day (Gy/d)	Doubling time[a] (d)
10 Gy + test dose				
- Regrowth delay (assessed at 20 d)	1 to 4	4.5	1.5	1.4
	4 to 7	2.8	0.93	2.3
	1 to 7	7.3	1.22	1.7
- Tumour cure (assessed at 6 months)	1 to 4	2.2	0.7	2.9
20 Gy + test dose				
- Regrowth delay (assessed at 35 d)	1 to 4	3.9	1.3	1.6
	4 to 7	4.1	1.36	1.5
	1 to 7	8.0	1.32	1.6
- Tumour cure (assessed at 6 months)	1 to 7	-0.3	-	-

[a] Effective doubling time during repopulation based on a D_0 of 3.0 Gy, as measured by in-vitro plating.

doubling time is, therefore, as short as 1.6 d. This is far less than the volume doubling time of the tumour at the time of treatment (9 d), and comes close to the potential doubling time (1.8 d). Any prediction of the rate of repopulation based on the clinical doubling time would, therefore, have led to a disastrous underestimation. A slightly smaller rate of repopulation was found after 20 Gy, pointing to a rapid reoxygenation of the hypoxic cells surviving this treatment.

Measurements of repopulation in the 5 and 10 fraction schedules gave greatly differing results, depending on the assay. Using regrowth delay, the mean repopulated dose per day during the extra 18 d of the extended treatments was 1.56 Gy/d (5 fr.) and 1.4 Gy/d (10 fr.), comparable to the figures obtained in the split-dose experiments. In contrast, cure data gave much higher figures, viz. 2.6 Gy/d (5 fr.) and 3.95 Gy/d (10 fr.), doses that call for effective doubling times of 0.96 d and 0.63 d, the latter being identical to the median cell cycle time of the tumour cells.

The discrepancy observed between the assays does not necessarily imply a difference deriving from the methodological approach. Rather, it is the fact that the range of doses applied for the two assays is widely different that must be taken into account. As pointed out by Suit et al. [4], a critical interdependence may exist between the size of dose per fraction and the fractionation interval for maximum repopulation. His data on a fast-growing mouse carcinoma (T_d = 2.5 d) tested using the same 10 fr./9 d and 10 fr./27 d regime as ours yielded a repopulated dose of 3.37 Gy/d, and for a 5 fr./4 d and 5 fr./20 d regime a dose of 1.84 Gy/d, in accord with our figures.

For sarcoma SSK 2 a compilation of the data measured in the split-dose experiments is given in Table III; it is clear that the results from the two assays are entirely contradictory. No repopulation was detected in the tumour-cure experiments, with the reservation that a 7 d interval for a 10 Gy dose was not tested. Regrowth-delay data, however, indicate considerable repopulation, both after 10 and 20 Gy, amounting to an average repopulated dose of 1.27 Gy/d. At present a decision on which assay gives the more relevant results cannot be made. Preliminary data on in-vitro cloning experiments with the SSK 2 tumour indicate that, after 20 Gy, the number of clonogenic cells stays constant until day 4, and then increases by a factor of 2.5 at day 7. After a hypoxic dose of 20 Gy, the number of clonogenic cells again remains constant until day 4, but then increases by a factor of 5 at day 7. A comparable delayed onset of proliferation was found by Hermens and Barendsen [5] in the R 1 rhabdomyosarcoma, and a reduced rate of proliferation until day 7 was recently reported by Rowley et al. [6] for a rat hepatoma.

The reasons for this relatively slow repopulation in the fast sarcoma, particularly in relation to the carcinoma, are not clear. However, there may be a connection between the histological changes seen in the two tumours and the rate of repopulation. Carcinoma AT 17 reacts to radiation with rapid cell lysis and depletion, which possibly serves as a stimulus to repopulation; sarcoma SSK 2 maintains both its structure and cellularity during the first week after the doses tested.

REFERENCES

[1] WITHERS, H.R.: "Capacity for repair in cells of normal and malignant tissues", in Time and Dose Relationships in Radiation Biology as Applied to Radiotherapy (Carmel Conference Report), Brookhaven National Laboratory Rep. BNL-50203 (C-57) (1970) 54-69.

[2] ELLIS, F.: "The relationship of biological effect to dose-time-fractionation factors in radiation therapy", in Current Topics in Radiation Research (EBERT, M., HOWARD, A., Eds) Vol. 4, North Holland Publishing Co., Amsterdam (1968) 359-397.

[3] PORTER, E.H.: The statistics of dose/cure relationships for irradiated tumours, Part I, Br. J. Radiol. 53 (1980) 210-227.

[4] SUIT, H.D., HOWES, A.E., HUNTER, N.: Dependence of response of a C3H mammary carcinoma to fractionated irradiation on fractionation number and intertreatment interval, Radiat. Res. 72 (1977) 440-454.

[5] HERMENS, A.F., BARENDSEN, G.W.: Changes of cell proliferation in a rat rhabdomyosarcoma before and after X-irradiation, Eur. J. Cancer 5 (1969) 173-189.

[6] ROWLEY, R., HOPKINS, H.A., BETSILL, W.L., RITENOUR, E.R., LOONEY, W.B.: Response and recovery kinetics of a solid tumour after irradiation, Br. J. Cancer 42 (1980) 586-595.

Progress in Radio-Oncology II, edited by K. H. Kärcher et al., Raven Press, New York © 1982.

Recovery from X-Ray Damage in Plateau-Phase Human Tumour Cells

R.R. WEICHSELBAUM, A. MALCOLM, J.B. LITTLE
Joint Center for Radiation Therapy,
Department of Radiation Therapy,
Harvard Medical School,
Boston, Massachusetts
and
Laboratory of Radiobiology,
Department of Physiology,
Harvard School of Public Health,
Boston, Massachusetts
United States of America

Abstract

Cells derived from radiocurable tumours demonstrate less potentially lethal damage repair in vitro than cells derived from tumours considered not radiocurable. Investigation of a human melanoma line shows that although large fractions (1200 rads) induce large amounts of initial damage, much of this is 'potentially lethal' and is repaired. Detailed investigations of early phases of potentially lethal damage repair in this human melanoma line show a very rapid phase and slower phases of repair, with marked recovery after relatively short times (hours).

1. INTRODUCTION

In some human tumours the delivery of multiple daily fractions of X-rays has been shown to enhance local control when standard fractionation has not proven effective [1-3]. Theoretical radiobiological advantages of hyperfractionation include enhanced 'single-hit killing', redistribution of tumour cells to a more oxygenated and/or radiosensitive phase of the cell cycle, and minimization of the effect of radioresistant or hypoxic tumour clones, since differences in surviving fraction between resistant and normal cells are less magnified in the low-dose region of the survival curve [1-4].

Density-inhibited plateau-phase cultures of human tumour cells have several of the characteristics of human tumour cell populations in vivo, including slowly proliferating and non-proliferating clonogenic cells and a cell loss factor [5,6]. The enhancement in survival after delay in subculture in plateau-phase cells after X-ray treatment is referred to as potentially lethal damage (PLD) repair. PLD repair has been demonstrated in animal and human solid and ascites tumours [5-7].

We have previously examined the repair of potentially lethal X-ray damage in plateau-phase cultures of human tumour cells and found that some cell lines derived from non-radiocurables showed more PLD repair than either normal cells or cells derived from tumours considered radiocurable [8]. We have recently examined the early time course of PLD repair in human melanoma line C143 as well as PLD repair after small and large fraction sizes. We present herein the results of these experiments and update our data on PLD repair in human tumours. We discuss these results in terms of their possible applications to fractionated radiotherapy.

2. MATERIALS AND METHODS

The human tumour lines and methods for their establishment and maintenance have been described in a previous publication [9]. Cells are grown in Eagle's Minimal Essential Medium supplemented with 10% foetal calf serum, 90 mg/l of glucose, 0.6 mg/l of sodium pyruvate and 15 μg/l of gentamycin in an atmosphere of 95% air - 5% CO_2. Radiations are carried out with a 220 kVp GE Maximar unit operating at 15 mA and yielding a dose rate of 50 rad/min. The protocol for the measurement of PLD repair is as follows. Cells are plated into 60 mm dishes and grown to confluency. The culture medium is renewed for three days and the experiment is performed on the fourth. Cells are irradiated at room temperature and returned to the incubator. Single plates are removed and the cells trypsinized and replated at low density at regular intervals. The medium is changed at 5 to 7 days after irradiation and 12 to 18 days later the dishes are rinsed, fixed and stained. Only colonies observed under a dissecting microscope to be composed of 50 cells or more are scored as survivors. Results are plotted as recovery ratio versus time cf explant. The enhancement in survival after explant is interpreted as being due to the repair of potentially lethal X-ray damage. The recovery ratio is determined by dividing the surviving fraction at each time by the zero-hour survival. The recovery ratio is employed in order to normalize the data from all experiments because of some variability in the initial zero-hour survival fractions.

Experiments on rapid PLD recovery were carried out using a cobalt-60 source at 37°C at a dose rate of approximately 44 rad/s. Recovery was measured at 2, 6 and 10 minutes as well as at 2, 4, 6, 8, 10 and 24 hours. No differences have been noted in rapid PLD recovery at room temperature or 37°C.

3. RESULTS

Potentially lethal damage recovery was studied by measuring survival as a function of the time between X-irradiation of density inhibited plateau-phase cultures and their subcultures at low density to measure colony-forming ability.

Results for cell lines C32 and C143 (melanomas) and TX4 (osteosarcoma) irradiated with 700 rads are shown in Fig.1. The recovery ratio is shown on the ordinate, and the interval between irradiation and subculture on the abscissa. The enhancement seen in survival reflects the recovery from potentially lethal X-ray damage (an 8 to 2.1-fold recovery). The X-ray PLD recovery seen with lines GBM (glioblastoma), SAOS (osteosarcoma) and PAS (hyper-

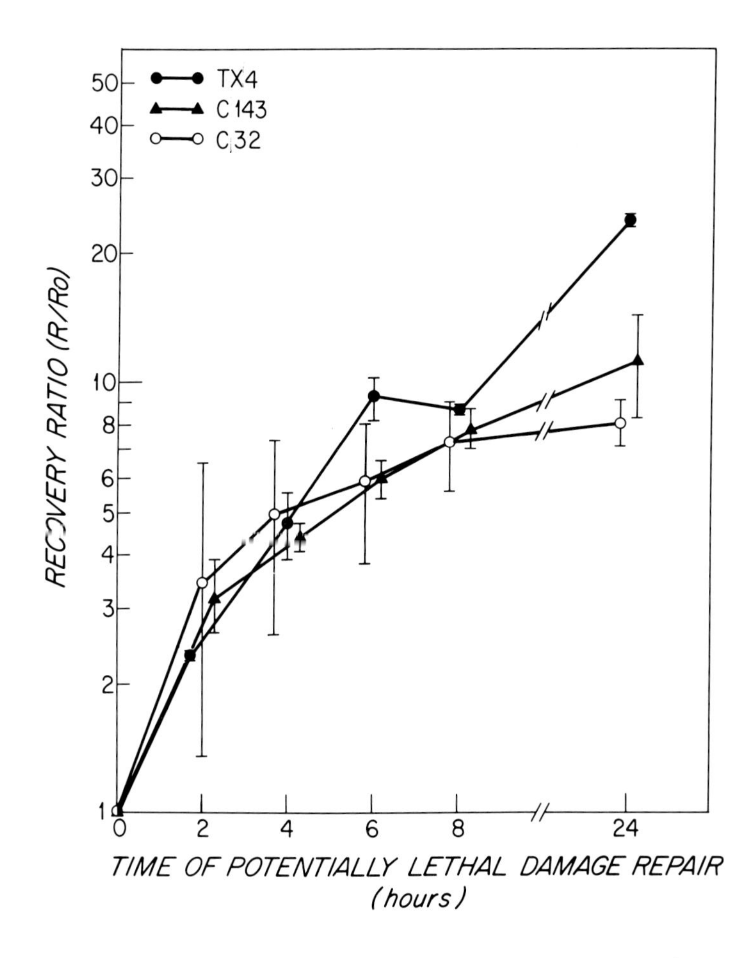

FIG.1. Potentially lethal damage repair of cell lines TX4 (osteosarcoma), and C143 and C31 (melanomas).

nephroma) is shown in Fig.2. The recovery ratios varied from 2.5 to 4.4. Figure 3 shows PLD recovery in lines MCF7 and MDA (breast cancers) and LAN-1 (neuroblastoma). These tumours are considered radiocurable; as can be seen in Fig.3, they exhibit significantly less PLD repair than the non-radiocurable tumour cell lines shown in Figs 1 and 2 (1.5 to 2-fold recovery), although there is a heterogeneity among and between tumour cell types.

Figure 4 shows PLD recovery in human melanoma cell line C143 after irradiation with 350 or 1200 rads. The zero-time survival was 0.37 after 350 rads and 0.001 after 1200 rads. However, over a 50-fold recovery takes place after 6 hours at the larger doses, whereas only a 2-fold recovery is seen at 350 rads.

The occurrence of a rapid phase of PLD recovery between 2 and 10 min, as well as a slower phase of recovery between 30 min and 24 hours is shown in Fig.5. A significant enhancement in survival (recovery ratio of 2.5) was seen after only a 6 min recovery interval.

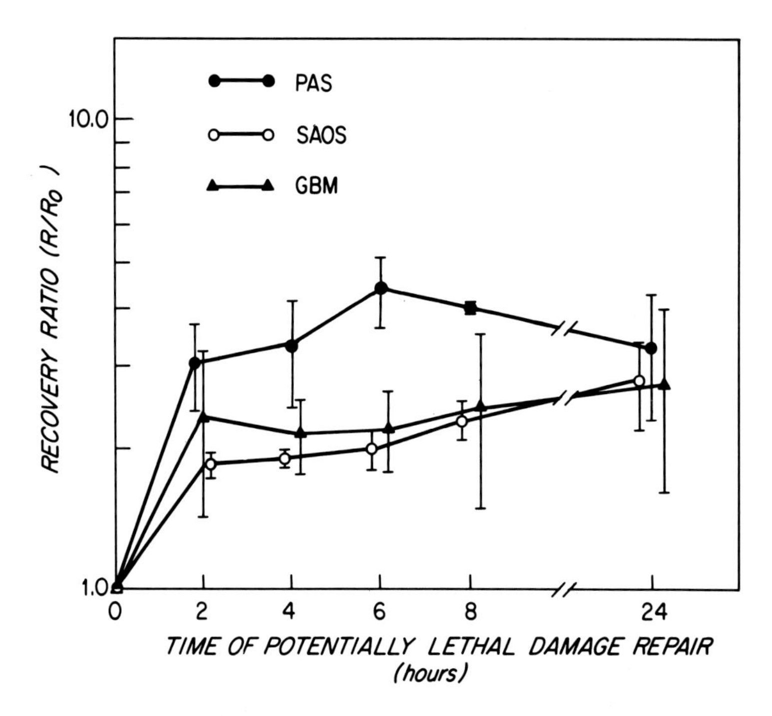

FIG.2. Potentially lethal damage repair of cell lines PAS (hypernephroma), SAOS (osteosarcoma) and GBM (glioblastoma).

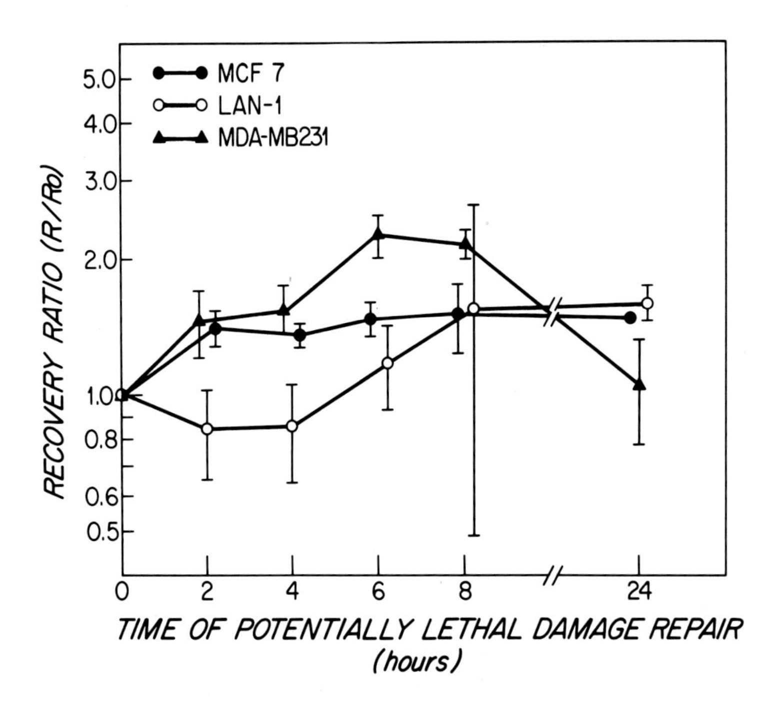

FIG.3. Potentially lethal damage repair of cell lines MCF7, MDA (breast cancers) and LAN-1 (neuroblastoma).

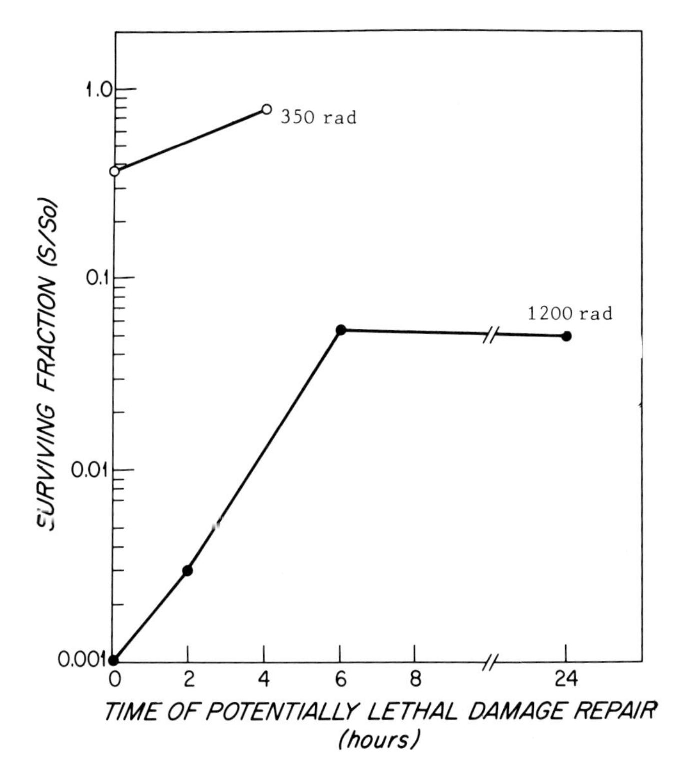

FIG.4. Potentially lethal damage repair in human melanoma line C143 following irradiation with 350 or 1200 rads.

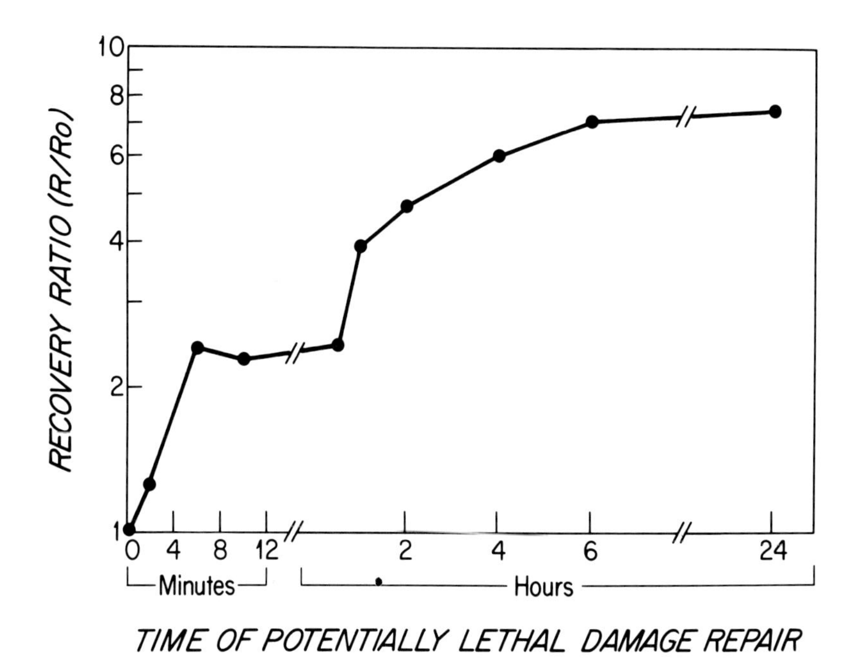

FIG.5. Early potentially lethal damage repair time examined in human melanoma line C143.

4. DISCUSSION

Potentially lethal damage repair has a well-defined molecular basis in bacteria and yeast [10]. More recently, Weichselbaum et al. [11] suggested a molecular basis for PLD repair in human cells. Thus, we considered this an important repair process for investigation in human tumour cells. Cells derived from tumours considered not clinically radiocurable (melanoma, osteosarcoma) demonstrated significantly more PLD recovery than all human tumour and normal cell lines examined in our laboratory. Three cell lines derived from tumours considered radiocurable exhibited less PLD recovery than other cell lines examined under our conditions (with the exception of fibroblast strains from ataxia telangiectasia patients). Other tumour cell lines derived from tumours generally considered non-radiocurable were intermediate in their capacity for PLD repair. Even a small recovery from a 30-fraction scheme might cause enough enhancement in the ultimate surviving fraction to decrease the local radiocurability of a tumour. If some constituent tumour cells have PLD repair characteristics of tumour cell lines TX4, C132 or C143, this might render them quite 'radio-incurable'. Thus, factors that affect PLD repair might be important in clinical radiotherapy.

Large fraction radiotherapy has been proposed as a treatment for malignant melanoma based on the fact that the survival curve for melanoma cells grown and irradiated in vitro has been reported to have a large shoulder [12]. We have not confirmed this finding. Furthermore, the results in Fig. 4 indicate that large fractions may not be the optimal method to eradicate melanoma, since any non-cycling cells in a condition analogous to the stationary-phase tumour cells in the present experiment may be extremely efficient in the repair of potentially lethal damage. Large fractions may induce much 'initial' damage (analogous to the zero-hour time), much of which may be repaired by 6 or 24 hours. The extrapolation from data based on exponentially growing cells may thus be misleading when determining the ultimate surviving fraction in cells proficient in PLD repair. Recently, several clinical trials have shown conventional fractionation to be equal to large fraction radiotherapy in the treatment of human melanoma [13,14].

Our data may have implications for other tumours besides melanoma. Recently, for example, patients with bronchial carcinoma were treated with low daily doses (200-250 rads), compared with higher daily fractions (440-1100 rads) at less frequent time intervals. The smaller daily doses proved more effective in local control [15]. Large fractions (above 500 rads) may cause serious long-term normal tissue sequelae, especially if a substantial volume of normal tissue is irradiated, and may actually lessen the therapeutic ratio. This observation suggests that long-term normal tissue damage may not be based on PLD repair. The ultimate tumour radiocurability may depend upon the number of constituent tumour cells analogous to plateau-phase cells; the amount of PLD repair performed by these cells may determine the ultimate surviving fraction. Large fractions might be more effective in tumours not proficient in PLD repair.

Investigation of rapid PLD recovery in human melanoma line C143 indicates that PLD recovery occurs as early as 2 minutes after irradiation, and suggests that it may be divided into a very rapid phase and a slower second phase, which occurs between approximately 1 and 24 hours and in a more gradual fashion. This PLD recovery may be analogous to the PLD recovery seen by Little et al. [7] in NCTC tumour cells, although in their study very early PLD repair times were not investigated. Thus an attractive hypothesis is that twice-a-day radiotherapy may deliver a second fraction before PLD repair is complete; it is tempting to speculate that a dose delivered almost immediately after the initial dose might inhibit the very rapid PLD recovery. Furthermore, relatively less PLD repair may occur after multiple doses, and we are currently investigating this in vitro. Zinninger and Little [16] demonstrated that small fraction, protracted irradiation preferentially kills cells in stationary cultures (compared to exponentially growing cells).

In summary, potentially lethal damage repair may be an important cellular process in human tumour radiocurability. Cells derived from tumours that appear radiocurable demonstrate less PLD repair than cells derived from tumours that are not radiocurable. Investigations of a C143 human melanoma line show that, although large fractions induce large amounts of initial damage, much of the damage is only 'potentially lethal' and is repaired. Detailed investigation of early phases of PLD repair in line C143 showed some repair at very early times after irradiation.

REFERENCES

[1] LITTBRAND, B., EDSMYR, F.: Preliminary results of bladder carcinoma irradiated with low individual doses and a high total dose, Int. J. Radiat. Oncol., Biol. Phys. 1 (1976) 1059-1062.

[2] NORIN, T., ONYANGO, J.: Radiotherapy in Burkitt's lymphoma: conventional versus superfractionated regime - early results, Int. J. Radiat. Oncol., Biol. Phys. 2 (1977) 399-406.

[3] CHOI, C.H., SUIT, H.D.: Evaluation of rapid radiation treatment schedules utilizing two treatment sessions per day, Radiology 116 (1975) 703-707.

[4] SUIT, H.D.: Superfractionation, Int. J. Radiat. Oncol., Biol. Phys. 2 (1977) 591-592.

[5] LITTLE, J.B.: Repair of sublethal and potentially lethal radiation damage in plateau phase cultures of human cells, Nature (London) 224 (1969) 804-806.

[6] HAHN, G.M., LITTLE, J.B.: Plateau phase cultures of mammalian cells: An in vitro model for human cancer, Curr. Top. Radiat. Res. 8 (1972) 39-83.

[7] LITTLE, J.B., HAHN, G.M., FRINDEL, E., TUBIANA, M.: Repair of potentially lethal radiation damage in vitro and in vivo, Radiology 106 (1973) 689-694.

[8] WEICHSELBAUM, R.R., SCHMIT, A., LITTLE, J.B.: Cellular repair factors influencing radiocurability of human malignant tumors, submitted for publication.

[9] WEICHSELBAUM, R.R., NOVE, J., LITTLE, J.B.: "Radiation response of human tumor cells in vitro," in Radiation Biology in Cancer Research (MEYN, R.E., WITHERS, H.R., Eds), Raven Press, New York (1980) 345-351.

[10] PATRICK, M.H., HAYNES, R.H., URETZ, R.B.: Dark recovery phenomena in yeasts: I, Radiat. Res. 21 (1964) 144-161; II, Radiat. Res. 23 (1964) 564-579.

[11] WEICHSELBAUM, R.R., NOVE, J., LITTLE, J.B.: Deficient repair of potentially lethal damage in ataxia telangiectasia and xeroderma pigmentosum fibroblasts, Nature (London) 291 (1978) 261-262.
[12] BARRANCO, S.C., ROMSDAHL, M.M., HUMPHREY, R.M.: Irradiation response of malignant melanoma cells grown in vitro, Cancer Res. 31 (1971) 830-833.
[13] TROTT, K.R., von LIEVEN, H.R., KUMMERMEHR, J., et al.: Radiosensitivity of malignant melanoma, Part 2. Clinical Study, Int. J. Radiat. Oncol., Biol. Phys. 7 (1981) 15-20.
[14] LOBO, P.A., LIEBNER, E.J., CHAO, J.J., KANJI, A.M.: Radiotherapy in the management of malignant melanoma, Int. J. Radiat. Oncol., Biol. Phys. 7 (1981) 21-26.
[15] EICHHORN, H.J.: Different fractionation schemes tested by histological examination of autopsy specimens from lung cancer patients, Br. J. Radiol. 54 (1981) 132-135.
[16] ZINNINGER, G.F., LITTLE, J.B.: Fractionated radiation response of human cells in stationary and exponential phases of growth, Radiology 108 (1973) 423-428.

Progress in Radio-Oncology II, edited by K. H. Kärcher et al., Raven Press, New York © 1982.

Radiobiological Bases for Multiple Daily Fractionation*

Lester J. PETERS
Institute of Oncology and Radiotherapy,
Prince of Wales Hospital,
University of New South Wales,
Sydney,
Australia

H. Rodney WITHERS
Department of Radiation Oncology,
UCLA Center for Health Sciences,
Los Angeles, California,
United States of America

Howard D. THAMES Jr.
Department of Biomathematics,
University of Texas M.D. Anderson Hospital
and Tumor Institute,
Houston, Texas,
United States of America

Abstract

The use of multiple daily dose fractions (MDF) may be advantageous for one or more of the following reasons: increase in number and reduction in size of dose fractions, reduction in overall time of treatment, and facilitation of other aspects of treatment which indirectly relate to dose fractionation. The rationales underlying each of these changes from conventional daily radiotherapy are reviewed, and a strategy is developed to increase the probability of achieving a favourable therapeutic ratio between tumour control and late normal tissue injury. For a given level of late injury, severe acute reactions limit MDF if large volumes are treated too rapidly. However, the use of split course MDF schedules to overcome this problem is logically unsound if the total duration of treatment then exceeds that which could be achieved with a less intensive continuous MDF course. A theoretically optimal regimen of MDF involves: (i) the use of a fractional dose below which little further sparing from fractionation can be expected with regard to late effects in the dose-limiting normal tissue; (ii) a total dose determined by the tolerance of this tissue at the specified dose per fraction; and (iii) an overall time of treatment which is the minimum consistent with acceptable acute reactions, provided that sufficient time is allowed between dose fractions for Elkind repair in the normal tissues determining the late effects.

---

* This work was supported in part by United States Public Health Service grant CA-29644 and CA-29026 awarded by the National Cancer Institute.

TERMINOLOGY

Multiple daily fractionation (MDF) implies simply that more than one radiotherapy treatment is given per day. Since there are different radiobiological rationales for different versions of MDF we suggest the following definitions to avoid ambiguity:

1. Hyperfractionation (Ultrafractionation): A regimen of MDF in which the major change from conventional therapy is an increase in number (and reduction in size) of dose fractions rather than a shortening of overall treatment time.

2.(a) Accelerated fractionation: A regimen of MDF in which the major change from conventional therapy is a shortening of overall treatment time rather than an increase in the number of dose fractions.

(b) Concomitant boost: A special case of accelerated fractionation in which a reduced field boost is delivered as a second daily dose fraction in addition to the scheduled conventional treatment.

3. Adjunctive MDF: A regimen in which more than one dose fraction is given on certain treatment days to facilitate other aspects of treatment rather than primarily to modify either total fraction number or duration of treatment.

Some regimens of MDF may combine elements of more than one of the above strategies but for the purpose of discussion of rationale we will first consider them separately.

HYPERFRACTIONATION

There are sound radiobiological reasons why increasing the number of dose fractions and reducing the incremental fraction size might be expected to improve the therapeutic ratio in radiotherapy. These are not arguments for hyperfractionation *per se*. However, to increase the number of dose fractions without MDF would require protraction of treatment, with the undesirable consequence of increasing the time available for tumour cells to regenerate during the course of irradiation. The following considerations are therefore based on the premise that overall treatment time is the same as that for conventional radiotherapy, so that effects of regeneration can for practical purposes be factored out.

Radiosensitisation by redistribution: Since cellular radiosensitivity varies as a function of position in the division cycle, the cells surviving after each dose of irradiation will be grouped predominantly in the more resistant phases. Allowing these cells to redistribute themselves in cycle will improve the efficiency of killing by the next dose, and the more fractions used, the greater the opportunity for redistribution to take place[1]. Redistribution occurs in all cycling cells - normal and neoplastic - so that no therapeutic gain may occur with respect to normal epithelial cells which determine acute radiation reactions. However, the cells generally considered responsible for late radiation sequelae are very slowly cycling. They will therefore be much less sensitised by redistribution, resulting in less severe late reactions for a given level of tumour control, or conversely, better tumour control for a given level of late reactions.

Possible reduction of OER: As the size of dose per fraction is reduced, relatively more cell killing results from "single hit" lethal events than from accumulated sublethal injury[2], or in terms of the linear-quadratic model of cell survival the α component of cell killing predominates. Biophysically, single hit killing can be explained by the presence of foci of high density ionisation in beams of low average LET, in which case it might be predicted that the OER would be lower in these regions. Experimentally, there is no unanimity of results, partly because of the technical problems associated with measurement of OER at low doses [3]. However, in several mammalian cell systems, significantly reduced values of OER have been found with low doses or low dose rates, and any influence of hyperfractionation on OER could only be beneficial.

Considerations of survival curve shape: Analysis of the slope of isoeffect curves that have been determined experimentally or deduced clinically for a variety of acute and late normal tissue reactions has resulted in the constant finding of a higher exponent of fraction number for late effects[4]. This suggests a basic difference in the survival characteristics of cell populations determining acute and late responses, wherein the latter are more critically influenced by dose fractionation. This difference may be related to the age distribution of cells in actively cycling tissues compared with more quiescent ones, and to the relative lack of redistribution in the latter as discussed above. Regardless of the explanation however, the general statement can be made that in terms of the linear-quadratic model of cell survival, the nett β/α ratio is greater for late effects.The higher the β/α ratio, the more the survival curve bends in the dose range of clinical interest and the greater the relative sparing achieved by reduction in size of dose per fraction down to the limit at which essentially all cell killing results from single hit (α type) lethal events. Inasmuch as the response of tumours to irradiation is more akin to acute than late normal tissue reactions (since both are determined by depopulation of actively proliferating cells) it can be predicted that the therapeutic ratio with respect to late normal tissue injüry would be improved by hyperfractionation (Fig.1). This does not of course guarantee a favourable therapeutic ratio for all categories of tumour, but does increase the probability of its occurrence provided the β/α ratio for tumour cell survival is less than that for cells of the dose limiting normal tissue. Insufficient data are presently available for human tumours to know how commonly this condition exists.

ACCELERATED FRACTIONATION

In contrast to hyperfractionation the principal aim of accelerated fractionation is to reduce the overall time of treatment without increasing the incremental dose per fraction. Such a strategy is indicated when there is reason to believe that the tumour being treated is capable of a vigorous regenerative response or when the clinical situation demands as rapid a rate of tumour regression as possible.[1]

---

[1] Tumour regression rates depend on many factors other than cell kill, such as proliferation kinetics, tumour architecture, and rate of removal of dead cells and debris. However the only factor over which the therapist has influence is the rate of cell killing.

The rate of regeneration of clonogenic cells in human tumours can only be inferred from clinical experience using different overall times of treatment without a change in fraction number or size. Some of the estimates obtained in this way indicate a significant regenerative potential[5]. On first principles, the doubling time of clonogenic cells in a tumour must be shorter than that for the tumour as a whole. Thus, even in tumours having an average volume doubling time of about 2 months [6], regeneration of clonogenic cells could amount to several doublings over the period of a conventional course of treatment, while in tumours which are clinically rapidly growing, the extent of regeneration would probably be even greater. Since as a first approximation each doubling of clonogenic cell number would require an extra dose of 200-300 rads to offset, it is easily seen that the success of a course of treatment in some circumstances could be decided by the parameter of overall time.

The limitation to accelerated fractionation is the severity of acute reactions in proliferating tissues which normally themselves regenerate during treatment. Late effects, on the other hand, are relatively insensitive to overall treatment time provided the size of dose per fraction is unchanged. This is reflected in isoeffect plots by a higher exponent of time for acute reactions than for late effects. Gauging the shortest overall time of treatment consistent with acceptable acute reactions in different clinical situations calls for careful judgement. It is important to stress that the major rationale for accelerated fractionation is lost when a break in treatment is introduced to allow for acute reactions to subside — if this results in the total duration of treatment being as long as for a conventional continuous course.

CONCOMITANT BOOST

This is a special case of accelerated fractionation in which the boost to gross disease (which would normally be given through reduced fields after completion of a basic course of treatment) is given <u>during</u> the basic course as a second dose on certain treatment days. The rationale for this technique is again the reduction in overall time required to deliver the total dose and the usual indication is when a tumour mass appears to be growing under conventional treatment.

FIG.1. Effect of reduction in size of dose per fraction (d) in terms of the ratio β/α using the survival curve equation $SF_D = [\exp(-\alpha d-\beta d^2)^N]$, where $D = N \cdot d$, and N is the number of fractions. Higher values of β/α apply to late injury rather than to acute reactions in normal tissues. Kinetically, tumour response is analogous to an acute normal tissue reaction since both are determined by depopulation of actively proliferating cells. Reduction in size of dose per fraction from d_1 to d_2 allows the total dose for a given level of late injury to beincreased from D_1 to D_2 (Fig.1a). The same changes in dose per fraction and total dose would, however, result in a greater degree of depopulation in tumour cells having a lower a/a ratio (Fig.1b) since the sparing effect of dose fractionation is less marked. The insets show the shoulder region of the single-dose survival curves in more detail.

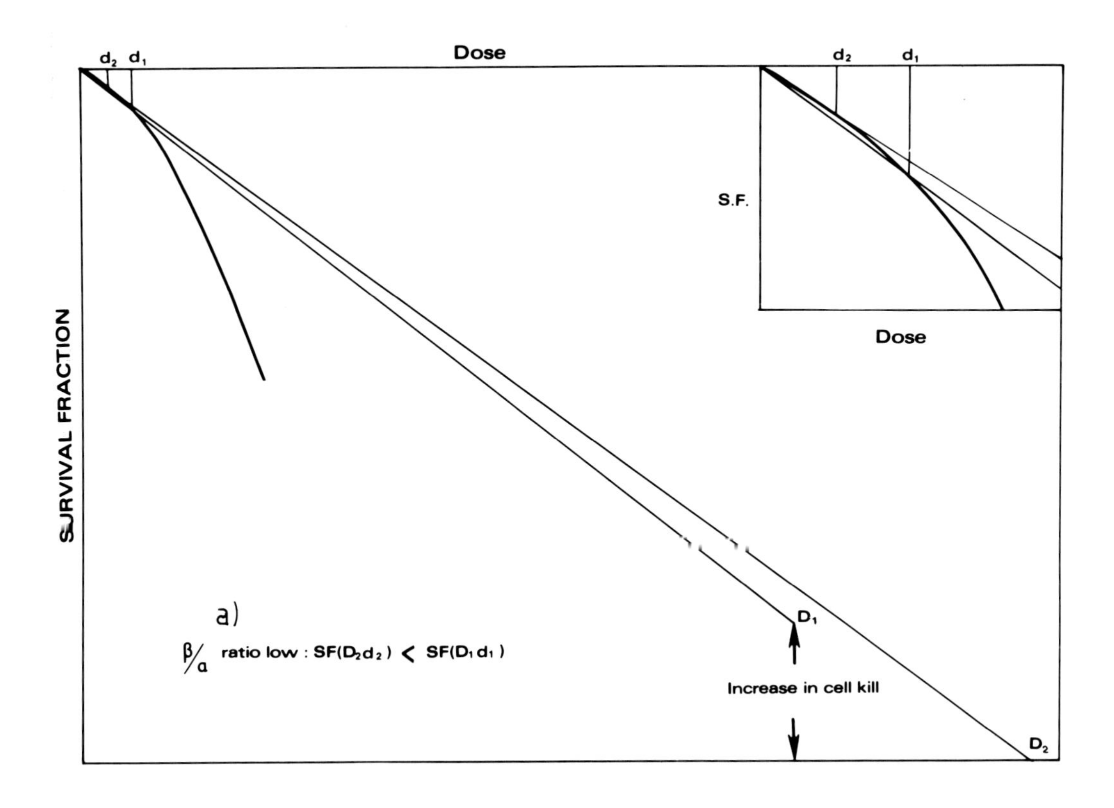
Dose
d_2 d_1
SURVIVAL FRACTION
S.F.
Dose
a)
β/α ratio low : SF(D_2d_2) < SF(D_1d_1)
D_1
Increase in cell kill
D_2

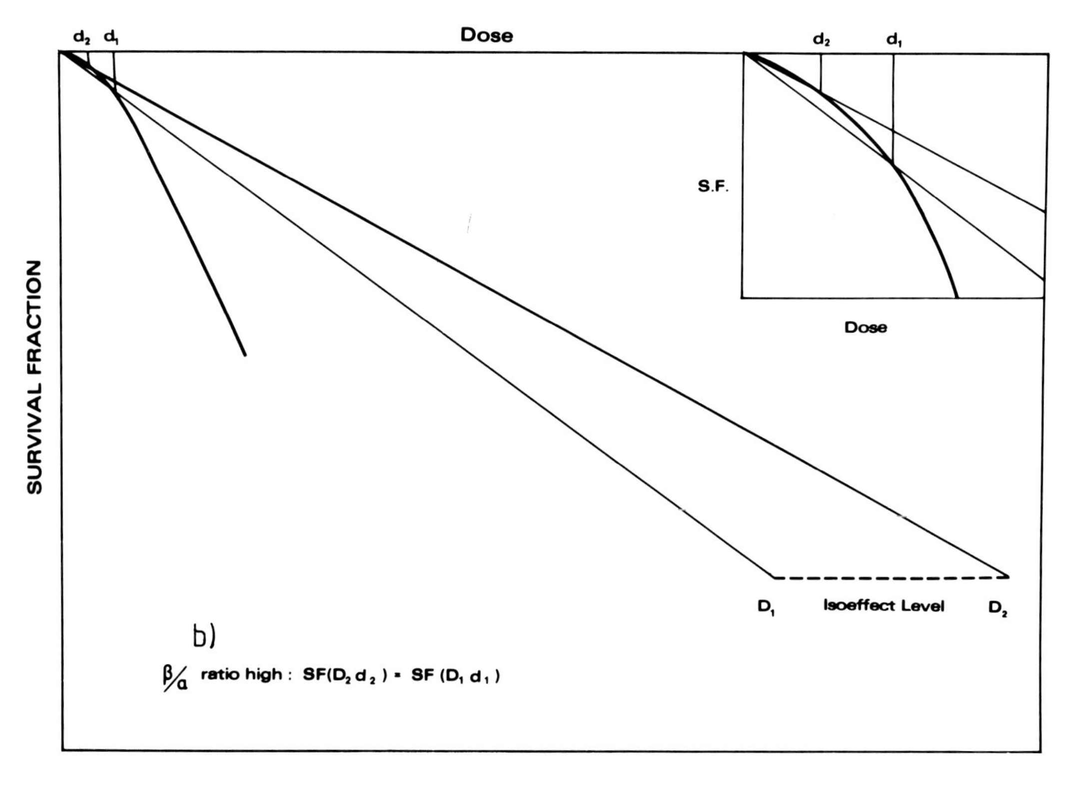
Dose
d_2 d_1
SURVIVAL FRACTION
S.F.
Dose
D_1
Isoeffect Level
D_2
b)
β/α ratio high : SF(D_2d_2) = SF (D_1d_1)

ADJUNCTIVE MDF

In this mode of treatment, the use of MDF is secondary to other considerations. For example, when using the hypoxic cell radiosensitiser Misonidazole (whose toxicity is limited by its cumulative total dose) it is rational to give more than one radiation dose on each day of drug administration to exploit its long plasma half life. An inverse of this strategy is to use MDF in regimens combining cytotoxic chemotherapy or hyperthermia and radiotherapy when one wishes to minimise the risk of interaction by giving the two modalities on different days, while still avoiding the use of large radiation dose fractions. Adjunctive MDF may also be used in situations where access to a particular therapy machine is limited to certain days of the week, or purely for reasons of convenience.

From the viewpoint of time/dose relationships, the use of MDF in this way is of little consequence provided there is no major change in the number and size of dose fractions or in overall time, and provided the dose fractions are adequately spaced (see below). If significant changes in either fraction number or time are made, the considerations discussed under Hyperfractionation and Accelerated Fractionation, respectively, would apply.

SPACING OF DOSES WITH MDF

It is implicit in all the foregoing discussion that sufficient time is allowed between doses of MDF to permit effective repair of sublethal injury - at least in normal late effect tissues. Some advantage may accrue by not greatly exceeding the minimum time required (which has yet to be accurately established for human late effects) since repair may be slowed, reduced, or even absent in hypoxic tumour cells [7]. To the extent that accumulated sublethal injury contributes to tumour cell kill, this would be expected to improve the therapeutic ratio. However, as the size of dose per fraction is decreased, relatively less cell killing of this type occurs and the potential advantage of accurately timed fractionation would decrease as the size of dose per fraction is reduced.

STRATEGY FOR OPTIMAL "COMBINED RATIONALE" MDF

It is possible to exploit some of the potential advantages of both hyperfractionation and accelerated fractionation together by using multiple small daily dose fractions over a shortened total duration of treatment. Since from the logistic point of view, it is rarely possible to give more than 3 fractions per day, it is clear that mutual limitations exist between fraction number (and thus fraction size) and time.

Under these circumstances, one has to balance the potential benefit of increased dose fractionation against the potential detriment of more prolonged treatment. In general terms, the point of diminishing return with respect to dose fractionation is reached when killing of normal late effect tissue cells is essentially all by single hit events. This probably occurs with doses of the order of 100 rads, and further reduction in size of dose per fraction would not appear to be justified. Having chosen a fractional dose, the number of fractions needed to reach a prescribed level of late tissue injury must be decided, and then the treatment should be given in the shortest overall time consistent with acceptable acute reactions.

CONCLUSION

Multiple daily fractionation offers a new range of possibilities for optimisation of radiotherapy using readily available low LET beams. There are good radiobiological reasons for predicting that a reduction in the size of dose per fraction and shortening of overall time of treatment might improve the therapeutic ratio between tumour eradication and late normal tissue injury. In addition, MDF may allow more efficient use of chemical radiosensitisers and combined modality treatments. The limitation to MDF is the severity of acute reactions produced in normal proliferating tissues if treatment is given too rapidly. However, split course MDF regimens should logically be avoided.

REFERENCES

[1] WITHERS, H.R.: Cell cycle redistribution as a factor in multifraction irradiation, Radiology 114 (1975) 199.
[2] ELKIND, M.M.: Fractionated dose radiotherapy and its relationship to survival curve shape, Cancer Treat. Rev. 3 (1976) 1.
[3] ALPER, T., Ed.: Cell Survival after Low Doses of Radiation: Theoretical and Clinical Implications, John Wiley & Sons, London (1975) 107-175.
[4] THAMES, H.D., WITHERS, H.R.: Changes in early and late responses with altered dose fractionation: Implications for dose-survival relationship, Int. J. Radiat. Oncol., Biol. Phys. (1981, in press).
[5] FLETCHER, G.H.: "Basic clinical parameters", in Textbook of Radiotherapy, 3rd ed., Lea and Febiger, Philadelphia (1980) 180-218.
[6] CHARBIT, A., MALAISE, P., TUBIANA, M.: Relation between pathological nature and the growth rate of human tumours, Eur. J. Cancer 7 (1971) 307.
[7] ALPER, T.: "Dose fractionation and dose rate", in Cellular Radiobiology, Cambridge Univ. Press, London (1979) 164-187.

Progress in Radio-Oncology II, edited by K. H. Kärcher et al., Raven Press, New York © 1982.

Unconventional Fractionation in the Treatment of Primary and Secondary Lung Tumours*

L.R. HOLSTI, M. SALMO
Department of Radiotherapy and Oncology,
University Central Hospital,
Helsinki,
Finland

M.M. ELKIND
Division of Biological and Medical Research,
Argonne National Laboratory,
Argonne, Illinois
and
Department of Radiology,
University of Chicago,
Chicago, Illinois,
United States of America

Abstract

When daily fractions with equal doses and equal intervals are used in radiotherapy, a sufficient time is obviously not allowed for reoxygenation to exert a significant effect. Reasonably large initial doses may be the method of choice. Once tumour shrinkage and reoxygenation have started, optimal treatment might imply smaller dose fractions and shorter intervals, the latter to ensure that contributions from repopulations remain ineffective. Based on these ideas of Elkind, three unconventional fractionation schemes with decreasing individual tumour doses were tested in a clinical series consisting of 17 primary lung carcinomas and 38 lung metastases from different primaries. The results suggested the existence of a rapid and a slow regression pattern of irradiated tumours. In the rapidly shrinking, sensitive group, tumours disappear totally in most instances. The half-regression time correlated with the disappearance rate. Rapidly shrinking radiosensitive and slowly shrinking radioresistant tumours can be identified within a week following an initial large dose fraction. Lung fibrosis correlated with the half-regression time. The more rapidly the tumour disappeared, the higher was the degree of fibrosis.

1. INTRODUCTION

When daily fractions with equal doses and equal intervals are used in radiotherapy, sufficient time is obviously not allowed for reoxygenation to exert a significant effect. According to Elkind et al. [1] and Elkind [2,3], a reasonably large degree of cell killing may be required before cell lysis and attendant

* The Helsinki group were supported by the Sigrid Jusélius Foundation, Helsinki, Finland.

reoxygenation can set in to a significant extent. Accordingly, it was suggested that reasonably large initial doses separated by intervals of three to four, or possibly more, cell cycle times could be more effective in allowing reoxygenation to play a part. Once tumour shrinkage and reoxygenation have started, optimal treatment might comprise smaller dose fractions and shorter intervals, the latter to ensure that contributions from repopulations remain ineffective.

2. FRACTIONATION SCHEMES AND EVALUATION OF RESULTS

Based upon the above considerations, three unconventional fractionation schemes (UFS) with decreasing individual doses were tested in a clinical series consisting of 17 measurable primary lung carcinomas and 38 lung metastases from various primaries. The control group (CFS) was given conventional fractionation. The various fractionation schemes are specified in Table I. The UFS/2 and UFS/3 regimens were designed to reduce fibrosis [4].

TABLE I. UNCONVENTIONAL (UFS) AND CONVENTIONAL (CFS) FRACTIONATION SCHEMES USED

Week	Fraction dose (rad)				
	Monday	Tuesday	Wednesday	Thursday	Friday
SCHEME 1 (UFS/1): Total dose 5000 rad in 14 fractions					
1	1000				
2	700				500
3	300		300		300
4	300		300		300
5	200	200	200	200	200
SCHEME 2 (UFS/2): Total dose 4400 rad in 14 fractions					
1	800				
2	600				400
3	300		300		250
4	250		250		250
5	200	200	200	200	200
SCHEME 3 (UFS/3): Total dose 5000 rad in 13 fractions					
1	800				
2			700		
3			600		
4	500			500	
5	300		300		300
6	200	200	200	200	200
CONVENTIONAL SCHEME (CFS): Total dose 5000 rad in 25 fractions					
1 – 5	200	200	200	200	200

The shrinkage rate of the tumours, based upon the reduction in area, was calculated by the method of least squares, fitted to an exponential regression.

The calculation of the half-regression time was based on measurements of tumour size from weekly chest X-rays. As a criterion of tumour regression, we used a 5 mm decrease in diameter within 7 days, irrespective of tumour size. Using this criterion, it is possible to calculate the maximum half-regression time for tumours of any size.

Lung fibrosis was evaluated roentgenologically.

3. TUMOUR REGRESSION AND LOCAL CONTROL

Our observations indicate that the proportion of tumours disappearing is high in all UFS groups, especially when the half-regression time is less than 40 days (Table II). In the UFS/1 and UFS/3 groups, disappearance of tumours was also noted in cases where the half regression time was longer, in four cases even more than 80 days. In the CFS group, none of the eight tumours disappeared, and the half-regression time was more than 40 days in all cases.

It was only possible to estimate the growth rate, in terms of the doubling time, in some of the tumours. Our observations indicated that the half-regression time did not correlate with the doubling time of those tumours (Table II). This observation differs from those reported by Breur for fractionated radiotherapy of lung metastases with total doses of 2000-5000 rad [5] and by van Peperzeel for single-dose treatment with 1000 rad [6]. Both of these authors stressed that growth rate is one of the most important parameters in determining radiosensitivity with regard to volume reduction.

TABLE II. TUMOUR REGRESSION AND DISAPPEARANCE FOR THE THREE UNCONVENTIONAL FRACTIONATION SCHEMES (see Table I)

Half-regression time (days)	Disappearance	Doubling time (days)				
		≤40	41–80	81–160	161–240	>240
SCHEME 1 (UFS/1)						
≤40	10/12			1		
41–80	3/9	1	1			
>80	2/4		1			
SCHEME 2 (UFS/2)						
≤40	5/6	2		1	1	
>40	0/5		1		2	
SCHEME 3 (UFS/3)						
≤40	12/13		1	2		6
41–80	2/6			1	2	
>80	2/7		1	3		

According to Malaise et al. [7], the half-regression time for different types of lung metastases after a single dose of 1000 rad is of the order of 14 days. The interval after irradiation before the minimum volume was reached was longer when the doubling time before irradiation was larger.

Dutreix et al. [8] have used a concentrated split-course modification consisting of 2 fractions of 850 rad 48 hours apart, a rest interval of 3 weeks and a second series of 3000 rad in 3 weeks. Tumour regression was faster with concentrated irradiation than with fractionated irradiation.

The mean survival time of our lung cancer patients is shown in Table III. The longest survival was noted in the UFS/3 group. However, the stage and performance distribution was not uniform between the groups. Nevertheless, the impression is that the survival of these lung cancer patients does not differ significantly from that achieved with other forms of radiotherapy. Some of the patients with primary lung cancer, as well as some of those with pulmonary metastases are alive at more than 800 days after treatment.

4. NORMAL TISSUE REACTION

Observation of lung fibrosis in the various treatment groups showed that fibrosis developed mainly in those cases where the tumour disappeared, and that there was, in those cases, no great difference between the different schemes with regard to the frequency of fibrosis. The UFS/2 scheme was somewhat better than the other two in the latter respect, but the probability of tumour disappearance with this regimen is lower than with UFS/1 and UFS/3. It seems as if the smaller initial dose and the longer time elapsing between the first and second fractions reduced the percentage of fibrosis.

When the tumour shrinks very slowly, no heavy fibrosis occurs, i.e. fibrosis is only of slight or medium degree. If the half-regression time of the tumour is short, fibrosis will develop early and is severe [4]. This supports our earlier view concerning the importance of reducing the field size for rapidly shrinking tumours during treatment. As shown by us in another study [9], roentgenological evaluation of lung fibrosis seems insufficient. We therefore now use different pulmonary function tests and CT scans to assess the early and late injury in the lung tissue.

TABLE III. SURVIVAL OF LUNG CANCER PATIENTS

Treatment scheme	Patients alive	Mean survival (days)
UFS/1	0/5	226
UFS/2	0/2	129
UFS/3	2/7	474
CFS [a]	1/3	459

[a] Including one on chemotherapy.

5. IDENTIFICATION OF RADIORESISTANT TUMOURS

From Table II it can be seen that the shrinkage rates and disappearance rates differ within a treatment group. Broadly speaking, response to the UFS schemes falls into two main classes [4,10]. Tumours that do not start to shrink during the first week are "radioresistant", while those that do are "radiosensitive". If the half-regression time for different types of lung metastases after a single dose of 1000 rad is of the order of 14 days [7], it might be reasonable to prolong the interval between the first and second dose fractions to 10 or 14 days in order to produce greater shrinkage of the tumour [3].

Arcangeli et al. [11] used the same approach and gave an initial dose of 800-1000 rad to patients with advanced or recurrent head and neck tumours. After 10 days of rest, tumour shrinkage was estimated and the tumour was classified as a responder or a non-responder. The remaining part of the treatment was given as conventional fractionation. 2/3 or 3/4 of the responders exhibited complete tumour shrinkage, while none of the non responders showed a complete response.

This approach might serve as a method of empirical identification of radioresistant tumours, but experience is still limited. The slowly responding tumours might respond to radiosensitizers. We have, therefore, recently started to include misonidazole on a randomized basis in our study.

6. CONCLUSIONS

6.1. Fractionation with unequal (i.e. decreasing) fraction sizes and intervals between fractions appears to have some advantages.

6.2. Preliminary experience has suggested the existence of two main types of regression pattern in irradiated tumours, the rapidly shrinking (radiosensitive) and the slowly shrinking (radioresistant).

6.3. In the rapidly shrinking, sensitive group, tumours disappear totally in most instances. A short half-regression time correlated with a high disappearance percentage.

6.4. The half-regression time did not correlate with the doubling time of the tumour.

6.5. Rapidly shrinking, radiosensitive and slowly shrinking radioresistant tumours can be identified within a week after an initial large-dose fraction.

6.6. Lung fibrosis following unconventional fractionation correlated with the half-regression time. The more rapidly the tumour disappears, the higher is the degree of fibrosis.

REFERENCES

[1] ELKIND, M.M., WITHERS, H.R., BELLI, J.A.: Intracellular repair and the oxygen effect in radiobiology and radiotherapy, Front. Radiat. Ther. Oncol. 3 (1968) 55-87.

[2] ELKIND, M.M.: "Reoxygenation and its potential role in radiotherapy", in Time and Dose Relationships in Radiation Biology as Applied to Radiotherapy (Carmel Conference Report), Brookhaven National Laboratory Rep. BNL-50203 (1970) 318-333.

[3] ELKIND, M.M.: "Recovery, reoxygenation and a strategy to improve radiotherapy", in Biological and Clinical Basis of Radiosensitivity (FRIEDMAN, M., Ed.), Charles C. Thomas, Springfield, Illinois (1974) 343-372.

[4] HOLSTI, L.R., SALMO, M., ELKIND, M.M.: "Non-standard fractionation: Clinical observations", in Biological Bases and Implications of Tumor Radioresistance (Proc. 2nd Int. Symp., Rome, 1980: FLETCHER, G.H., NERVI, C., WITHERS, H.R., ARCANGELI, G., MAURO, F., TAPLEY, N. duV., Eds), Masson Publishing USA, Inc., New York (1981, in press).

[5] BREUR, K.: "Effectiveness of radiation in relation to growth characteristics of tumors and normal tissues", in Biological and Clinical Basis of Radiosensitivity (FRIEDMAN, M., Ed.), Charles C. Thomas, Springfield, Illinois (1974) 502-527.

[6] van PEPERZEEL, H.A.: Effects of single doses of radiation on lung metastases in man and experimental animals, Eur. J. Cancer 8 (1972) 665-675.

[7] MALAISE, E.P., CHARBIT, A., AHAVAUDRA, N., COMBES, P.F., DOUCHEZ, J., TUBIANA, M.: Change in volume of irradiated human metastases, Investigation of repair of sublethal damage and tumour repopulation, Br. J. Cancer 26 (1972) 43-52.

[8] DUTREIX, J., SCHLIENGER, M., CHAUVEL, C., DAGUIN, R.: Concentrated irradiation. Concentrated palliative radiotherapy for tumours affecting the oesophagus, brain, bones and mediastinum, Ann. Clin. Res. 3 (1971) 9-15.

[9] MATTSON, K., HOLSTI, L.R., KORHOLA, O., SALMO, M.: "Radiation injury to the lung following intensive split-course therapy with three vs five weeks rest intervals; a radiological and physiological study" (Proc. 2nd Int. Symp., Rome, 1980: FLETCHER, G.H., NERVI, C., WITHERS, H.R., ARCANGELI, G., MAURO, F., TAPLEY, N. duV., Eds), Masson Publishing USA, Inc., New York (1981, in press) *Poster abstract.*

[10] HOLSTI, L.R., SALMO, M., ELKIND, M.M.: "Unconventional fractionation in clinical radiotherapy", in Hypoxic Cell Sensitizers in Radiobiology and Radiotherapy (Proc. 8th L.H. Gray Conf., Cambridge, UK, 1977), Br. J. Cancer 37, Suppl. III (1978) 307-310.

[11] ARCANGELI, G., MAURO, F., NERVI, C., STARACE, G.: A critical appraisal of the usefulness of some biological parameters in predicting tumor radiation response of human head and neck (H&N) cancer, Br. J. Cancer (1980, in press).

Progress in Radio-Oncology II, edited by K. H. Kärcher et al., Raven Press, New York © 1982.

Size/Dose Relationships and Local Control of Oro- and Hypopharyngeal Cancer Treated by Radiotherapy

J.P. BATAÏNI, J. BERNIER, C. JAULERRY, F. BRUNIN
Institut Curie,
Paris,
France

Abstract

A retrospective study of a large series of megavoltage treated head and neck cancers is under way at the Institut Curie. Control of subclinical disease in the neck with elective irradiation is confirmed (1.9% failure). Clinical nodes of up to 3 cm are controlled in 90% of the cases with doses higher than 65 Gy, whereas 3 out of 4 nodes of larger volume are controlled with doses higher than 70 Gy. No dose/response curve is elicited for control of the primary except for T1 and T2 cancer of pyriform sinus. T1 oropharyngeal cancer is controlled in 90% of the cases at all dose levels and T2 cases in about 75–80%. For the large majority of cases of oropharyngeal carcinoma (T3 and T4), there was no evident correlation between dose and control. Approaches other than the use of conventional radiotherapy should be looked for if better local control is aimed at.

1. INTRODUCTION

Radiotherapy plays an important role in the management of head and neck carcinoma and especially so at the Institut Curie (I.C.), where all referred cases are primarily treated by radical radiotherapy. Salvage surgery, either on the primary or for node failure, is performed whenever possible in certain persistant or recurrent tumours.

Over the past 10 years, many reports tended to demonstrate close correlations between the size of the tumour mass and the dose necessary to produce permanent control by radiation [1,2]. Total doses were to be increased to obtain the same percentage of control in the more advanced stages [3-6].

This paper presents an analysis of size/dose relationships and local control - which is the principal cause of failure in head and neck cancer - based on a retrospective study of a large series of patients treated by megavoltage radiotherapy at I.C. since 1958.

2. CASES TREATED AND METHODS

In this study, the UICC T and N staging is utilized. The series analysed concern mainly 715 cases of carcinoma of the tonsillar and soft palate regions (661 males and 54 females; the site of origin was the tonsillar region including the glosso-palatine sulcus in 80% of the cases, the soft palate in 15%, and it was undetermined in the remaining 5% of the cases). They also include 434 cases of carcinoma of the pyriform sinus occurring in male patients, the female incidence at this site accounting only for 2 to 3%.

Radical treatment was attempted for all patients. Tumour doses averaged 65 Gy, given over 40 days for cancer of the oropharynx and over 45 days for cancer of the hypopharynx, delivered in 5, or rarely 6, treatments per week. However, an appreciable number of patients (about 10%) received less than 55 Gy because of their poor general condition.

The weekly tumour dose was 10 to 11 Gy, but was raised in oropharyngeal carcinoma to 12 or even 13 Gy when the target volume was reduced for the last part of the treatment. No predetermined total tumour doses were applied, the final dose being adjusted according to the response of the tumour in every individual case. Doses higher than 75 Gy were rarely used for cancer of the pyriform sinus, but were applied in 87 cases of carcinoma of the tonsil and soft palate.

Analysis of control of neck disease was extended to other sites in the head and neck region, especially in 359 cases of glosso-epiglottic carcinoma. Clinically negative necks received tissue doses of 50 to 55 Gy over 5 weeks, whereas involved nodes received booster doses to reduced volumes, up to 75 or even 90 Gy.

3. CONTROL OF SUBCLINICAL DISEASE IN THE NECK

For many decades, elective irradiation of the neck has been systematically performed at I.C. With supervoltage irradiation, tissue doses of 50 to 55 Gy given over about 5 weeks were employed.

In 698 patients out of a series of 1829 cancers of the supraglottis, and oro- and hypopharyngeal cancers, isolated failure in the neck occurred in 13 patients only, i.e. in 1.9% of cases; moreover, most of these failures occurred outside or were marginal to the irradiated volumes.

4. CONTROL OF NECK NODES

In the same series, isolated neck failure occurred in 35 patients (6.3%) out of 556 patients staged N1 and N2, and in 62 patients (10.8%) out of 575 N3 patients.

Nodal control according to size and radiation dose is analysed (Tables I and II) for a series of 950 cancers of the head and neck (434 pyriform sinus, 157 epilarynx and 359 glosso-epiglottic cancers). In the determinate group of patients, 90% control was obtained for nodes up to 3 cm for dose levels above 65 Gy. For larger nodes, control was, for doses above 70 Gy, 84% for nodes of 3 to 5 cm and 73% for nodes larger than 5 cm.

5. CONTROL OF OROPHARYNGEAL CARCINOMA

The 3-year actuarial local control rate in a consecutive series of 661 male patients irradiated for carcinoma of the soft palate and tonsil, including the glosso-palatine sulcus, was 91% in 56 T1, 74% in 140 T2, 52% in 233 T3, and 31% in 232 T4 cases.

TABLE I

CANCER OF PYRIFORM SINUS, EPILARYNX AND GLOSSO-EPIGLOTTIC REGION

NODAL CONTROL ACCORDING TO SIZE AND RADIATION DOSE *
(primary controlled)

Dose	Node single ≤3cm			Node >3 cm		
Gy	N°	failure	control in det. group **	N°	failure	control in det. group **
-55	9	3	1 / 4	5	4	0/ 4
60	23	6	8/ 14	11	4	1/ 5
65	15	2	6/ 8	7	3	1/ 4
70	32	3	11/ 14	6	3	2/ 5
75	52	2	22/ 24	12	2	6/ 8
80	31	1	16/ 17	9	2	6/ 8
85	23	1	12/ 13	9	1	2/ 3
90	28	1	14/ 15	16	1	6/ 7
90+	18	1	11/ 12	7	0	2/ 2
Total	231	20	101/121 (83.5%)	82	20	26/46 (56.5%)

<······ $p < 0.001$ ·······>

* at 2 yr for pyriform sinus, at 3 yr for other sites.

** patients lost to follow-up or dead from dist. metast., intercurrent disease, and 2nd primary are excluded

TABLE II

NODAL CONTROL ACCORDING TO SIZE AND RADIATION DOSE

Primary controlled
(determinate group)

Dose Gy	3 cm	Dose Gy	3 - 5cm	5 cm
≤65	15/26 57%	≤70	3/9 33%	1/9 11%
	$p < 0.001$		$p < 0.05$	$p < 0.02$
>65	86/95 90%	>70	11/13 84%	11/15 73%
Total	101/121 83.5%		14/22 63.6%	12/24 50%

Local control as well as absolute and determinate survival were higher in the much smaller female population.

The 3-year actuarial control rate in relation to radiation dose is shown in Table III. For a closer analysis, patients receiving less than 55 Gy and patients having an associated failure in the neck were not included in the analysis (evidently improving the control rate data).

TABLE III

SQUAMOUS CANCER OF TONSIL AND SOFT PALATE

Actuarial 3 yr control rate of Primary according to stage and radiation dose

565 M and F patients **

Dose Gy	T1 Nº	T1 Control	T2 Nº	T2 Control	T3 Nº	T3 Control	T4 Nº	T4 Control
-60	10	100%	24	81%	32	70%	32	49%
65	15	92%	37	72%	54	62%	37	50%
70	14	91%	41	85%	63	59%	53	35%
75	4	75%	19	73%	32	50%	29	47%
75+	7	100%	6	80%	25	44%	31	24%

** Patients receiving doses < 55 Gy and having associated failure in the neck are not included.

For 50 patients with T1 disease, control was excellent at all dose levels used in all but 3 patients.

In the 127 patients with T2 disease, local control was obtained in about 80% of the cases, again at all dose levels. For the 388 patients staged T3 and T4, the control rate did not increase with increase of dose from 60 to over 75 Gy. As no predetermined dose was delivered, and because complete response at the end of radiotherapy is a good predictor of the definitive outcome [6], (76% with 3-year control versus 33% in immediate and in complete regression, respectively, in our series) [7], analysis was performed separately for those who had complete regression and those who still had residual disease at the completion of radiotherapy.

Again no correlation was found between control and dose in the two subpopulations (Tables IV and V). The same conclusions were obtained when data were considered in terms of tumour NSD.

Analysis of local control in the series of 359 glosso-epiglottic cancers yielded similar results but at a lower level - except for T1 lesions.

6. CONTROL OF CANCER OF THE PYRIFORM SINUS

Correlations between dose and control according to stage of the disease have been analysed (Table VI). For T1 and T2 tumours, local control of 60 to 65% is obtained at 2 years with doses higher than 65 Gy, whereas no dose/response curve could be derived for the more advanced tumours.

Comparative control of small primary T1 and T2 tumours with nodes up to 3 cm was attempted, assuming similarities in the number of tumour cells and their status. With doses up to 65 Gy, control of the primary was obtained in 36% of the cases (10/28) and node control in 40% (6/15); with higher doses, control of the primary was obtained in 65% of the cases (22/34) and node control in 93% (50/54), since higher doses could be delivered with impunity.

TABLE IV

SQUAMOUS CANCER OF TONSIL AND SOFT PALATE

3 yr actuarial control rate of Primary according to dose and status at end of RT

565 M and F patients

DOSE Gy	T 1 Complete regression N°	Control	T 2 Complete regression N°	Control
-60	9	100%	21	84%
65	15	92%	33	73%
70	12	100%	38	87%
75	4	75%	17	76%
75+	7	100%	6	80%
	Incomplete regression N° 3 1 failure > 65 Gy		Incomplete regression N° 11 4 failures : 1< 60 Gy 1< 65 Gy 1< 70 Gy 1< 75 Gy	

TABLE V

SQUAMOUS CANCER OF TONSIL AND SOFT PALATE
3 yr control rate of Primary according to dose and status at end of RT

565 M and F patients **

DOSE Gy	T 3 Complete regression N°	Control	T 3 incomplete regression N°	Control	T 4 Complete regression N°	Control	T 4 incomplete regression N°	Control
-60	23	73%	9	59%	9	62%	23	45%
65	37	73%	17	37%	22	67%	15	24%
70	47	67%	16	33%	24	51%	29	24%
75	18	67%	14	26%	8	100%	21	30%
75+	I0	85%	15	18%	6	25%	25	24%

** Patients receiving doses < 55 Gy and having associated failure in the neck are not included.

A dose/response curve is still observed for large nodes, but none could be elicited for T3 and T4 tumours [8].

TABLE VI

CANCER OF THE PYRIFORM SINUS

2 yr control of Primary according to stage and radiation dose (determinate group)

DOSE Gy	T1 +T2 Control		T3 + T4 Control	
-50	1/ 5		1/ 9	
55	0/ 2		6/20	30%
60	3/ 9		10/22	45%
65	6/12	50%	14/35	40%
70	14/21	66%	30/75	40%
75	6/10	60%	19/49	39%
75 +	2/3		7/18	39%
TOTAL	32/62	52%	87/228	38%

$p < 0.06$

7. DISCUSSION AND CONCLUSIONS

Control of subclinical disease in the neck with doses of 50 to 55 Gy is well confirmed by this study and will not be discussed further.

Control of clinical nodes is dose dependent; doses as high as 80 to 90 Gy have often been delivered to the node volume.

As regards control of the primary, raising doses from 60 to 75 Gy, which is, in most situations, the maximum permissible dose, did not seem to increase the control rate for T3 and T4 tumours of the oropharynx, either when tumour had completely disappeared or, more particularly so, when there was residual disease at completion of the course of radiotherapy.

Tumour control for T1 and T2 cases was excellent or 'almost excellent' at all dose levels.

As regards cancer of the pyriform sinus, a dose/response curve was elicited for T1 and T2 cases, though it is less significant than that obtained in another study on supraglottic cancer [9].

Further analysis will be made using 6-week dose equivalents [10], also taking account of the fact that clinical and microscopical exophytic tumours are more readily controlled than infiltrating or deeply ulcerated tumours.

A number of factors governing tumour response and control - other than the O_2 factor, the importance of which has been overemphasized [11] - as, for example, the presence in large primary tumours of heterogeneities in the tumour cell populations, seem to be of importance.

In advanced tumours approaches other than radical radiotherapy used alone must be looked for if better tumour control is to be obtained.

ACKNOWLEDGEMENTS

The authors are grateful to Rodney Withers M.D. and Lester Peters M.D. for reading the manuscript and for their positive criticism, and to B. Asselain M.D. and Mrs M. Lavée for their help with the statistics.

REFERENCES

[1] FAYOS, J.J., LAMP, I.: Radiation therapy of carcinoma of the tonsillar region, Am. J. Roentgenol., Radium Ther. Nucl. Med. 111 (1971) 85-94.

[2] PEREZ, C.A., LEE, F.A., ACKERMAN, L.V., KORBA, Purdy J., POWERS, W.E.: Carcinoma of the tonsillar fossa: significance of dose of irradiation and volume treated and the control of the primary tumor and metastatic neck nodes, Int. J. Radiat. Oncol., Biol. Phys. 1 (1976) 817-827.

[3] SHUKOVSKY, L.J., FLETCHER, G.H.: Time, dose and tumor volume relationships in the irradiation of squamous cell carcinoma of the tonsillar fossa, Radiology 107 (1973) 621-626.

[4] SPANOS, W.J., SHUKOVSKY, L.J., FLETCHER, G.H.: Time, dose and tumor volume relationships in irradiation of squamous cell carcinomas of the base of tongue, Cancer 37 (1976) 2591-2599.

[5] SHUKOVSKY, L.J., BAEZA, M.R., FLETCHER, G.H.: Results of irradiation in squamous cell carcinomas of the glossopalatine sulcus, Radiology 120 (1976) 405-408.

[6] FLETCHER, G.H.: Squamous cell carcinomas of the oropharynx, Int. J. Radiat. Oncol., Biol. Phys. 5 (1979) 2073-2090.

[7] BATAÏNI, J.P., et al.: Radiotherapy of oropharyngeal carcinoma. Experience at the Institut Curie (1981, in preparation).

[8] BATAÏNI, J.P., et al.: Radical radiotherapy of cancer of the pyriform sinus (1981, in preparation).

[9] GHOSSEIN, N., BATAÏNI, J.P., ENNUYER, A., STACES, P., KRISHASWAMY, V.: Local control and site of failure in radically irradiated supraglottic laryngeal cancer, Radiology 112 (1974) 187-192.

[10] FLETCHER, G.H., SHUKOVSKY, L.J.: Isoeffect exponents for the production of dose-response curve in squamous cell carcinomas treated between 4 and 8 weeks, J. Radiol., Electrol., Med. Nucl. 57 (1976) 825-827.

[11] KAPLAN, H.S.: On the relative importance of hypoxic cells for the radiotherapy of human tumours, Eur. J. Cancer 10 (1974) 275-280.

Progress in Radio-Oncology II, edited by K. H. Kärcher et al., Raven Press, New York © 1982.

Management of Advanced Head and Neck Squamous Carcinoma by Multiple Daily Sessions of Radiotherapy and Surgery

A. RESOULY, V.H.J. SVOBODA
St. Mary's General Hospital,
Portsmouth,
United Kingdom

Abstract

Fifty nine patients, mostly with advanced squamous carcinoma of the head and neck, were treated in Portsmouth between 1973 and 1980. A ^{60}Co machine was used, and a tumour dose between 50 and 55 Gy was delivered in 24 to 34 sessions, so that three fractions were given every day, with a minimum interval of three hours and an overall time of 10 to 14 days. When a large volume was irradiated prophylactically, the dose was 48 Gy in 30 sessions over 12 days, and the primary volume was boosted three weeks later by another 12 to 18 Gy in 9 to 12 sessions over 3 to 4 days. Full response of the primary tumour was achieved in 51 out of the 59 patients. Lymphatic masses responded similarly to the primary tumour, but the dose of 50 Gy in 24 to 30 fractions was too low. The authors recommend multiple-daily-session radiotherapy as a primary treatment of choice for most advanced squamous cell carcinomas of the head and neck, with elective surgery at 3 to 4 months after completion of radiotherapy.

1. INTRODUCTION

Radical radiotherapy is the treatment of choice for early carcinomas of the oral cavity and larynx. As surgery for advanced tumours carries high morbidity, some patients are referred for palliative radiotherapy. During the last generation, considerable progress was made by the introduction of standard block dissection and of commando operations. Megavoltage radiotherapy reduced the need for interstitial treatment and at least one author, Henk [1], claimed success with the hyperbaric oxygen tank. Pierquin [2] tried to imitate the low dose rate of curietherapy by external beam therapy. Backstrom [3] explained his success with fast, small fractions by the reduced protection which hypoxia offers under such circumstances.

In the years 1973–1980 we had encouraging experience with fast fractionation radiotherapy [4], i.e. multiple-daily-session radiotherapy (MDS RT), and this led to substantial changes in our policy.

2. MATERIAL

This paper discusses all patients with histologically confirmed squamous cell carcinoma of the head and neck, treated by MDS RT in Portsmouth. In the years 1973–1977 we used this method to treat 18 patients because they were too frail to travel daily or because of advanced tumours. When the late damage was found acceptable, we irradiated by MDS RT most advanced patients with poor prognosis so that, before 1 January 1981, altogether 59 patients had been treated. All ages and general conditions were included, as well as six patients with second primary carcinoma which led in all these six cases to death. Table I lists all patients by UICC TNM stages.

3. METHODS

Until 1977 we treated limited primary volume with only the nearest ipsilateral draining area (6-9 cm^3) by a pair of wedged cobalt-60 fields. The tumour dose of 50-55 Gy was delivered in 24 to 34 sessions over 10 to 16 days. Three fractions, each

TABLE I. HEAD AND NECK SQUAMOUS CARCINOMA

Multiple-daily-session radiotherapy, 1973-1980

Site	TNM classification	Number
Maxillary antrum	T3,4 NO MO	3
Tonsil	T2 NO MO	1
	T2 N2,3 MO	3
Nasopharynx	T2 N3 MO	1
Tongue	T3 NO MO	2
Submand. metast.	-	1
Hypopharynx	T3 NO MO	1
	T2,3 N3 MO,1	3
Valeculla	T3 NO,1 MO	2
Aryepiglottis	T1,2,3,4 N1 MO,1	7
Pyriform fossa	T2 NO MO	2
	T3 NO MO	2
	T2,3 N1 MO	3
	T2,3 N2,3 MO	3
Larynx	T2,3 NO MO	6
	T2,3 N1,2 MO	5
	T4 NO MO	1
	Neck rec., 8 cm dia.	1
Vocal cord	T1,2 NO MO	9
	T3 NO MO	1
Postcricoid	T1,3 NO MO	2
Total		59

between 175 and 230 cGy, were given every treatment day, with a minimum interval of three hours. Later the treatment always included a larger prophylactic volume. Since 1979, in all patients who were in the EORTC pilot study, we delivered 48 Gy in 30 fractions given over 12 days to the large prophylactic volume, and gave a boost of 12-20 Gy in 3 to 4 days to the residual and primary site. The fractions were smaller than 200 cGy. Nine patients were also given 1 g/m^2 Misonidazole during each of the ten treatment days. There was no modification of the acute response and there were no complications.

4. RESULTS

4.1. Normal Tissue Response

4.1.1. EARLY REACTIONS

The first side-effects appeared at the end of the second week of treatment. The reaction culminated a week later by stomatitis or dysphagia and cough. This required readmission of five frail or old patients. The skin reaction was negligible, but there was patchy or confluent fibrin mucositis. The discomfort in all early lesions started to improve during the second week after completed radiotherapy and the patient was symptom-free four weeks later.

4.1.2. LATE DAMAGE

Six patients with early tumours of the larynx had no signs of late damage three years after radiotherapy.

4.1.3. POST-RADIATION CHANGES IN ADVANCED TUMOURS

The large primary tumour impedes normal tissue healing of the acute changes and increases the risk of late damage and complications.

Overlap of anterior neck fields often produces oedema of the anterior neck skin, and mucosal oedema is common in the larynx. Both these changes are semi-permanent, but the symptoms often improve or disappear during 2 to 12 months.

Some advanced tumours heal without late damage, namely T3 tumours of the larynx and tongue. Moderate damage in the form of neck or laryngeal fibrosis or oedema were observed in four of our patients. In contrast to them, severe late damage was usually disabling.

One form was laryngeal fibrosis, which we observed in three patients. One was treated for post cricoid carcinoma and is still dysphagic 84 months after the treatment and has permanent tracheostomy. Both the anterior neck, larynx and upper oesophagus are fibrosed. Another patient, whose advanced tumour of the ventricular band required tracheostomy before radiotherapy, developed laryngeal fibrosis and the tracheostomy had to be reopened. He is now without recurrence at 14 months. The third

TABLE II. PRIMARY TUMOUR FAILURE AND LOCAL RECURRENCES

Neck, squamous cell carcinoma, multiple-daily-session radiotherapy, 59 patients, closing date 1981-01-01

Site	Dose (rads)	No. of fractions	Early response	Comments
Larynx, 8 cm neck metast.	5250	30	partial	extent underestimated
Tonsil, T2 N3, 81 years	4250	24	partial	refused to continue; neck masses - full resp.
Vocal cord, T4 NO	5000	26	partial	laryngectomy at 6 months; survived 45 months; no sign of recurrence
Postcricoid, 82 years	5250	30	full	local rec. at 5 months
Vocal cord T3, stoma, 81 years	5250	30	full	local rec. at 5 months
Subglottis, T2, 77 years	5000	24	full	local rec. at 8 months
Subglottis, stoma, 77 years	5500	30	full	local rec. at 12 months
Subglottis, T2	5000	34	full	local rec. at 10 months
Epiglottis	5000	27	full	local rec. at 13 months

patient also had tracheostomy before radiotherapy and his primary tumour was cleared but the larynx was functionally inadequate, which led to aspiration. At 11 months the patient is fed by nasogastric tube.

Another form was cartilage necrosis, which we observed in two patients. One, with T3 N1 tumour of the vocal cord and tracheostomy before radiotherapy, had laryngectomy and neck dissection before we suspected residual tumour. Instead there was only cartilage necrosis in the larynx, although metastasis was present in the lymph nodes. The patient is now without recurrence 69 months after treatment. Finally, a patient with T4 tumour of the aryepiglottis died nine months after radiotherapy of aspiration pneumonia. The post-mortem revealed no residual tumour, but only cartilage necrosis leading to perforation of the epiglottis.

4.2. Tumour Response

Full response of the primary tumour within three months after radiotherapy was achieved in 51 out of 59 patients. The response was not assessed in five patients, and in three cases only partial response was observed. Table II lists these three and also six further patients where the primary tumour recurred.

Table III describes all the courses where the total tumour dose was only 50 Gy or less. The tumour control is inadequate. Large primary tumours not only tend to recur, but the risk of metastasis is also greater (Table IV). Neck lymphadenopathy responded equally as well as the primary tumour of a comparable site, or better (Table V). With advanced primary tumours the risk of lymphatic metastasis outside the irradiated area is considerable (Table VI). Some of these patients were salvaged by block dissection, some had further radiotherapy. In some patients haematogenous metastases developed affecting lung, bone or liver. Table VII summarizes the early results of the EORTC Pilot Study. The closing date is 1 January 1981.

TABLE III. LOW DOSE RESULTS

Head and neck, squamous cell carcinoma, advanced disease, multiple-daily-session radiotherapy

Dose (rads)	No. of fractions	(days)	Result
2000	10	14	prophyl.; lymph node metast. in treated area
4250	24	10	residual tumour
5000	30	16	large volume; no sign of recurrence at 24 months
5000	30	12	residual tumour
5000	27	11	local recurrence at 13 months
5000	26	14	residual tumour
5000	24	10	local recurrence at 8 months

TABLE IV. RESPONSE OF LARGE TUMOURS OF PRIMARY DIAMETER 5 cm OR MORE

Head and neck, squamous cell carcinoma, multiple-daily-session radiotherapy, 8 patients

Site	No. of patients	Lymph node recurrence at: (months)	No sign of recurrence at: (months)
Larynx	3	5	5; 24
Max. antrum	2	5	19
Tongue	2	12	11
Tonsil	1	-	5

TABLE V. RESPONSE OF NECK LYMPHADENOPATHY

Squamous cell carcinoma, multiple-daily-session radiotherapy, 25 patients

Stage	Number	No local recurrence at: (months)	Residual or recurrent (months)
N1	12	5; 5; 6; 6; 9; 10; 12; 14; 24; 30	7; 5 (block dissection, now 69 months with no sign of recurrence)
N2	5	9; 12; 17; 19	residual (mass 8 cm dia.)
N3	8	5; 5(6 cm dia.); 5(6 cm dia.); 7; 11; 11	residual (4 cm dia.); residual (6 cm dia.)

5. DISCUSSION

We were primarily interested in the response of squamous carcinoma to MDS RT and in the tolerance of upper respiratory and oropharyngeal mucosa. We did not try to show improved survival. We have therefore excluded no patients from this study.

Frequent fractionation shortens the hospitalization and, since the acute reaction is moderate and short, brings faster relief of symptoms than conventional regimes. The late damage in early tumours is nil and it is not inevitable even in advanced tumours. The response, both of the primary tumour and the lymphatic metastases, was very good providing that the dose given by a single course was not less than 52 Gy in 30 sessions. Even large and fixed lymphatic masses can probably be controlled by such treatment. We are convinced that MDS RT offers better therapeutic ratio than when less frequent sessions are used. The risk of lymphatic neck metastases in large tumours demands the inclusion of substantial prophylactic volume in the first part of the treatment, and the addition of a boost to the primary and residual site after three weeks. During the pause the mucositis clears and the second course does not cause any discomfort. In the EORTC fractionation and Misonidazole trial

TABLE VI. LYMPH NODE RECURRENCE OUTSIDE TREATED AREA

Head and neck, squamous carcinoma, multiple-daily-session radiotherapy

Primary site	Site of recurrence and time of recurrence (months)		Surgery[a]
Antrum	submandibular	2	-
Tonsil	left low cervical	7	-
	right mid-cervical	13	-
Tongue	mid-cervical	10	block dissection; after 12 months NSR
Pyriform fossa	upper cervical	2	-
Aryepiglottis	upper cervical	3	-
Aryepiglottis	upper cervical	7	-
Epiglottis	upper cervical	8	-
Larynx	upper cervical	5	block dissection; after 69 months NSR
Vocal cord	upper cervical	27	block dissection; after 56 months NSR

[a] NSR: no sign of recurrence.

48 Gy tumour dose (TD) is given in 30 sessions over 12 days, followed three weeks later by 19-22 Gy in 4 to 5 days, 160 cGy per fraction.

Our eight years of experience encouraged us to change our management. MDS RT will now represent our first attempt to control the advanced or prognostically unfavourable tumours of head and neck. Laryngectomy would be considered imperative only in advanced tumours of the larynx that lead to airway obstruction. In any other case we recommend mandatory examination under general anaesthetic three months after radiotherapy with a biopsy from the primary area or residual tumour. Post-radiation fibrosis starts to develop after four months and radical surgery should be performed before this time. However, we had minimal difficulties even much later. Close co-operation between the radiotherapist and surgeon is essential, both before any decision is taken about the primary treatment and then during radiotherapy and the follow-up.

The MDS RT plan should consist of two parts. In the first course the primary tumour, with a safe 'healthy' margin, should be treated together with at least whole ipsilateral neck regardless of clinical involvement. Three weeks are then allowed to clear the mucositis. The second course treats only the residual tumour and the primary site. Alternatively, after 44.25 Gy in 25 sessions, a further 5 sessions are given to a limited volume within the 12 day course (dose per fraction 175 cGy, tumour dose 53 Gy).

Stridor or airway obstruction after radiotherapy for T3 or T4 tumour of the larynx is an indication for laryngectomy, even if

TABLE VII. EORTC PILOT STUDY 1979: CLOSING DATE 1981-01-01

Head and neck, squamous cell carcinoma, advanced disease, multiple-daily-session radiotherapy (MDS RT)

Diagnosis	Surgery before radiotherapy	TNM	Type of radiotherapy [a]	Response	Surgery	Follow up and comment[b] (months)
Tongue, 8 cm dia.	Nasogastric tube	T3 NO	MDS RT + boost	full	–	>12 NSR, alcoholism
Hypopharynx	Tracheotomy	T3 NO	MDS RT + Miso	full	–	> 3 NSR, ? cause
Neck met., 8 cm dia.	Laryngectomy	rec.	MDS RT + Miso	residual	–	> 3 residual tumour
Vocal cord	–	T3 NO	MDS RT + boost	full	–	11 NSR
Pyriform fossa, 4 cm	–	T2 NO	MDS RT + Miso	full	–	14 NSR
False cord	Tracheotomy	T2 NO	MDS RT + Miso	full	–	14 NSR, larynx fibrosed
Epigl., valeculla	–	T3 N1	MDS RT + Miso	full	–	14 NSR
False cord	–	T3 N1	MDS RT + boost + Miso	full	block dissection neg.	12 NSR
Epiglottis, diab.	–	T2 N1	MDS RT + boost	full	block dissection for node, recurrence positive	14 nodal recurrence
Pyriform fossa	–	T2 NO	MDS RT + Miso	full	–	10 NSR
Tongue	–	T3 NO	MDS RT + Miso	full	block dissection for node, recurrence positive	12 NSR
Hypopharynx	–	T3 N2	MDS RT + boost	full	–	9 NSR
Tonsil	Local exc. neck mass	T2 N3	MDS RT + boost	full	–	7 NSR
Aryepiglottis	–	T3 N1	MDS RT + boost + Miso	full	–	6 NSR
Hypopharynx, neck	–	T2 N3 M1	MDS RT + boost	full	–	5 with bone metast.
Valec., epigl. tongue	Tracheotomy	T3 NO	MDS RT + boost	full	–	5 NSR, oedema
Pyriform fossa	–	T3 N3	MDS RT + boost	{tumour: full node(s): residual}	–	5 nodal residuum
Pyriform fossa	–	T2 N3	MDS RT + boost	tumour: full node(s): residual	block dissection, necrotic tumour	5 NSR

[a] MDS RT indicates single course; MDS RT + boost indicates additional boost; Miso indicates misonidazole therapy also.

[b] NSR: no sign of recurrence.

the tumour has completely regressed. Any lymph node mass that persists 6 to 8 weeks after radiotherapy should be removed by block dissection if operable, provided the primary tumour is controlled. A similar policy should be used in other lesions of the head and neck.

6. CONCLUSIONS

MDS RT is well tolerated and offers better therapeutic ratio for advanced squamous carcinomas of the head and neck than does conventional fractionation. The authors recommend it as the primary treatment of choice in the majority of patients. A large prophylactic volume must be treated to 70% to 90% of the total tumour dose. Mandatory examination under general anaesthetic and biopsy should be performed three months later, followed by surgery within the next month if residual tumour is present and operable. Such a policy offers good tumour control with little morbidity and mutilation. Good results need close co-operation between the surgeon and radiotherapist, who must together take any decision in each individual case.

REFERENCES

[1] HENK, J.M., KUNKLER, P.B., SMITH, C.W.: Radiotherapy and hyperbaric oxygen in head and neck cancers, Lancet ii (16 July 1977) 101-105.

[2] PIERQUIN, B., BAILLET, F., BROWN, O.: L'effet différentiel de l'irradiation continue ou semi-continue à faible débit des carcinomes épidermoides, J. Radiol. d'Electrol., Med. Nucl. 51 (1974) 533-536.

[3] BACKSTROM, A., JAKOBSSON, P.A., LITTBRAND, B., WERSALL, J.: Fractionation scheme with low individual doses in irradiation of carcinoma of the mouth, Acta Radiol., Ther., Phys., Biol. 12 (1973) 401-406.

[4] SVOBODA, V.H.J.: Radiotherapy by several sessions a day, Br. J. Radiol. 48 (1975) 131-133.

Progress in Radio-Oncology II, edited by K. H. Kärcher et al., Raven Press, New York © 1982.

Superfractionated Radiotherapy of Carcinoma in the Mouth*

B. LITTBRAND
Department of Oncology,
University of Umeå,
Umeå

P. JAKOBSSON
Department of Oncology,
University of Uppsala,
Uppsala

Sweden

Abstract

There are several theoretical reasons for using a fractionation scheme comprising multiple small doses in a range close to 1 gray. The main reason for this fractionation scheme was the result of the radiobiological experiment in vitro which showed an oxygen enhancement ratio which was dose dependent and low at low doses. This trial was started at Radiumhemmet in Stockholm in 1970 according to these assumptions. A series of operable epidermoid cancer in the mouth was randomized to one of the following two treatment schemes: 3 X 1.0 Gy each day to a total dose of 42 Gy; or 1 X 2.0 Gy each day to a total dose of 40 Gy. The treatments were given using ^{60}Co γ-radiation, and the field included the tumour region and the lymph nodes in the upper neck. Three weeks after the termination of radiotherapy, the patients were operated upon. All patients have been followed up for at least 8 years. The survival rate shows a significant difference in favour of the superfractionated group. The post-operative complications were low and the same in both groups. In no case was there any late injury. Thus the superfractionated treatment gave a better therapeutic ratio than the conventional one.

1. INTRODUCTION

The oxygenation of the cells in neoplastic tissues varies greatly as a result of the deficient vascular system. The lack of homogeneity of the tumour cell population as regards oxygenation, and hence radiation sensitivity, constitutes a major problem in radiation therapy. Observation in experiments with a variety of biological materials suggests that the cellular repair of sublethal damage is oxygen dependent [1]. Furthermore, the difference in repair indicates a difference in extrapolation numbers of the survival curves of cells irradiated in the absence or in the presence of oxygen and a dose-dependent oxygen enhancement ratio (OER).

In the 1974 L.H. Gray Conference [2], various authors presented results indicating a dose-dependent OER [3-6]. Most found an OER close to 1 in the region of 1 Gy, but one result in

* Supported in part by the Swedish Cancer Society.

disagreement with the others was also presented [7]. A fractionation scheme with low doses could, therefore, possibly circumvent the problem during radiotherapy of the particular radioresistance of poorly oxygenated tumour cells. Other theoretical considerations also favour a fractionation scheme of small multiple doses (as indicated elsewhere in these Proceedings [8]). In the light of this experience, it was deemed of interest to ascertain whether a fractionation schedule of 1 Gy at each treatment would be applicable in the treatment of carcinoma of the mouth [9].

2. CASES AND TREATMENT

The case series comprised patients from Radiumhemmet in Stockholm with carcinoma of the mouth. They were treated with preoperative radiotherapy and surgery. All the tumours were classified according to the TNM classification (UICC, Geneva 1968). The distribution of tumour site and tumour stage are presented in Tables I and II respectively. A diagnosis of epidermoid carcinoma was always confirmed histologically before the treatment began.

TABLE I. THE DISTRIBUTION OF PATIENTS IN THE TREATMENT GROUPS ACCORDING TO SITE OF TUMOUR.

FRACTIONATION SCHEME PER DAY (TOTAL DOSE)	GINGIVA	TONGUE	FLOOR OF THE MOUTH
3 x 1 Gy (42 Gy)	1	7	4
1 x 2 Gy (40 Gy)	8	7	3

TABLE II. THE DISTRIBUTION OF PATIENTS IN TREATMENT GROUPS ACCORDING TO TNM CLASSIFICATION (UICC, GENEVA 1968). (ALL THE TUMOURS WERE M_0).

FRACTIONATION SCHEME PER DAY (TOTAL DOSE)	T_1N_0	T_1N_1	T_2N_0	T_2N_1	T_3N_0	T_3N_1
3 x 1 Gy (42 Gy)	3	0	4	1	2	2
1 x 2 Gy (40 Gy)	4	2	9	2	1	0

The patients were randomized to be treated according to either one of two fractionation schemes. Scheme I comprised 3 X 1 Gy per day (at 4-hourly intervals), 5 days a week, to a total dose of 42 Gy given over 3 weeks. Scheme II comprised 1 X 2 Gy per day, 5 days a week, to a total dose of 40 Gy given over 4 weeks. External irradiation was the rule, with individual planning of the dose. All fields were irradiated at each session. The radiation source was a Siemens gammatron, a ^{60}Co unit, giving a dose rate of 0.4-0.6 Gy/min at a distance of 60 cm. Three weeks after the termination of radiotherapy, the patients were operated upon (tumour resection and cervical dissection on the same side). The mean age, 60 years (range 28-80), was the same in the two treatment groups.

3. RESULTS AND DISCUSSION

The patients have been followed up for at least eight years and the survival in the two treatment groups is illustrated in Fig.1. After treatment with fractionation scheme I, eight patients out of 12 (67%) were still living after eight years, compared with four patients out of 18 (22%) after scheme II. According to a χ^2-analysis, the difference is significant ($p<0.05$). Furthermore the differences are supported by an analysis of the causes of death. One patient died of cancer and three patients of intercurrent diseases in group I; in group II, eight patients died of cancer and six of intercurrent diseases. The microscopy of the surgical specimens disclosed no sign of residual malignancy in 10 out of 12 patients treated with scheme II despite a lower cumulative radiation effect (CRE) value in scheme I (12.30 and 13.60 CRE units in schemes I and II, respectively[1] [cf. Table III]). This could be explained by the shorter treatment time in scheme I - three weeks, compared with four weeks in scheme II. There was no problem with the healing of the mucosa or with the increased time interval before surgery in any of the patients. After surgery, one patient in group I had an osteitis, and one patient in group II had an osteitis and fistula. No late injury was recorded in any patient in either treatment group.

A comparison between the two treatment groups is complicated by a difference in the tumour site distribution because of the randomization (Table I). However, neither does this difference nor do the differences in the distribution of tumour stage (Table II) favour the superfractionated group (scheme I).

As a result of the shorter treatment time in scheme I (3 weeks) as compared with scheme II (4 weeks), it is natural that the maximum early reaction in the mucosa is more pronounced in group I (Table III). However, more importantly, the postoperative complications and the late injuries do not differ between the two groups. One could argue that radiotherapy was not the only treatment and that the patients were also operated upon but, bearing in mind that the fractionation schemes were the only variable in this trial and that the postoperative and late complications were

---

[1] CRE unit, sometimes termed reu. For information about the CRE concept, see, for example, Ref. [10].

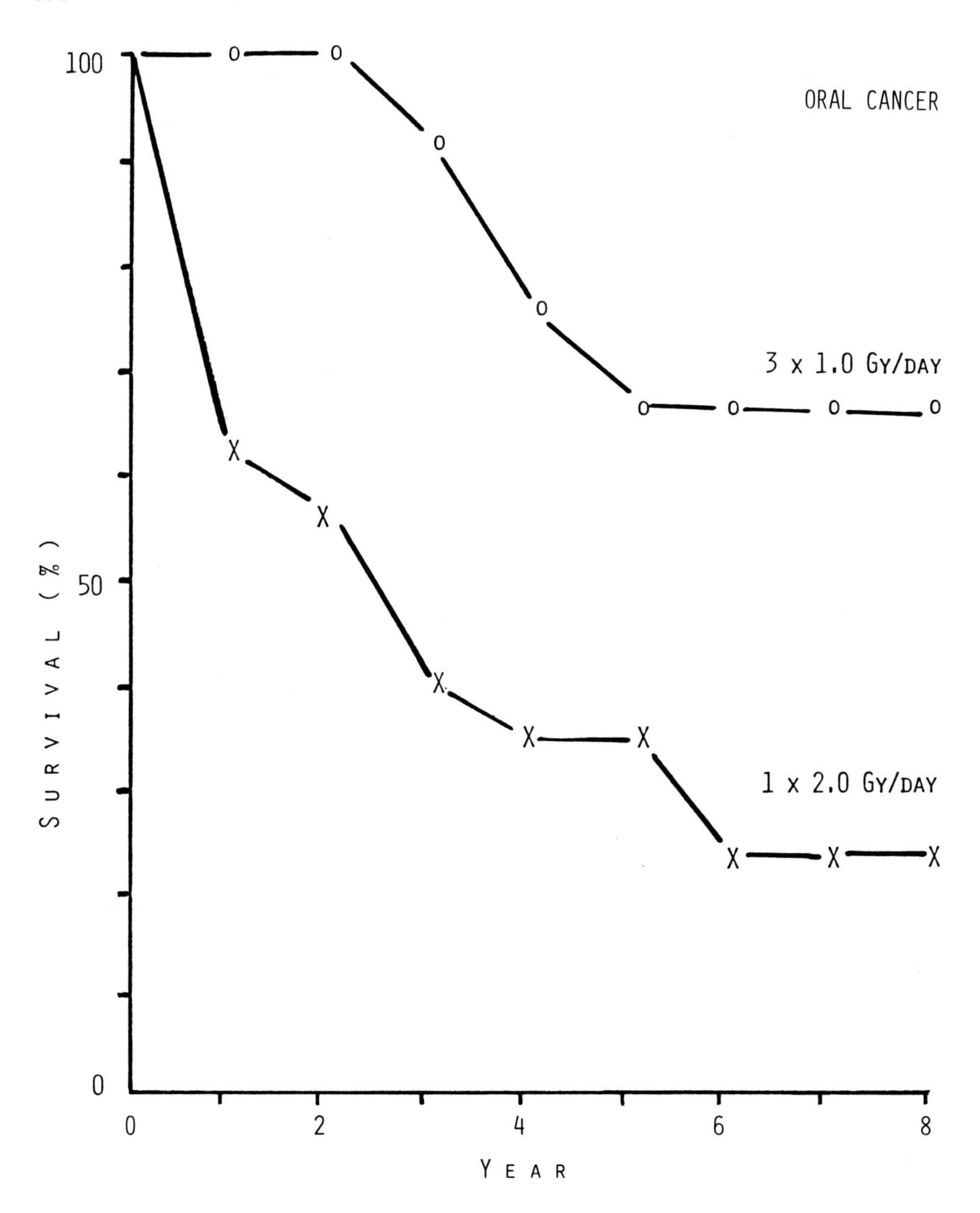

FIG.1. The survival of patients treated with preoperative radiotherapy and surgery. The patients were randomized to one of two fractionation schemes. Scheme I: 3 × 1 Gy each day to a total dose of 42 Gy; Scheme II: 1 × 2 Gy each day to a total dose of 40 Gy.

the same in the two groups, the results indicate an increased therapeutic ratio after superfractionated radiotherapy as compared with conventional radiotherapy.

TABLE III. THE DISTRIBUTION OF MAXIMUM EARLY EFFECT IN THE MUCOSA AFTER RADIOTHERAPY OF THE PATIENTS IN THE TREATMENT GROUPS.

FRACTIONATION SCHEME PER DAY (TOTAL DOSE)	REACTION IN THE MUCOSA		
	ERYTHEMA	SPOTTED MUCOSITIS	CONFLUENT MUCOSITIS
3 x 1 Gy (42 Gy)	2	4	6
1 x 2 Gy (40 Gy)	4	11	3

REFERENCES

[1] HALL, E.J.: The effect of hypoxia on the repair of sublethal radiation damage in cultured mammalian cells, Radiat. Res. 49 (1972) 405.

[2] Cell Survival After Low Doses of Radiation: Theoretical and Clinical Implications (Proc. 6th L.H. Gray Conf., London, 1974: ALPER, T., Ed.), The Institute of Physics, and John Wiley and Sons Ltd, London, New York (1975).

[3] NIAS, A.H.W., GILBERT, C.W.: "Response of HeLa and Chinese hamster cells to low doses of photons and neutrons", ibid, pp. 93-99.

[4] McNALLY, N.J.: "The effect of repeated small doses of radiation on recovery from sublethal damage by Chinese hamster cells irradiated in oxic or hypoxic conditions in the plateau phase of growth", ibid, pp. 119-125.

[5] CHAPMAN, J.D., GILLESPIE, C.J., REUVERS, A.P., DUGLE, D.L.: "Radioprotectors, radiosensitizers, and the shape of the mammalian cell survival curve", ibid, pp. 135-140.

[6] RÉVÉSZ, L., LITTBRAND, B., MIDANDER, J., SCOTT, O.C.A.: "Oxygen effects in the shoulder region of cell survival curves", ibid, pp. 141-149.

[7] KOCH, C.J.: "Measurement of very low oxygen tensions in liquids: does the extrapolation number for mammalian survival curves decrease after X-irradiation under anoxic conditions?", ibid, pp. 167-173.

[8] PETERS, L.J., WITHERS, H.R., THAMES, H.D. Jr.: "Radiobiological bases for multiple daily fractionation", these Proceedings, p.317.

[9] BÄCKSTRÖM, A., JAKOBSSON, P.A., LITTBRAND, B., WERSÄLL, J.: Fractionation scheme with low individual doses in irradiation of carcinoma of the mouth, Acta Radiol., Ther., Phys., Biol. 12 (1973) 401.

[10] DUNCAN, W., NIAS, A.H.W.: Clinical Radiobiology, Churchill Livingstone, Edinburgh, London, New York (1977) 180-181.

Progress in Radio-Oncology II, edited by K. H. Kärcher et al., Raven Press, New York © 1982.

Biological and Therapeutic Studies of Multifractionation

J. DUTREIX, J.M. COSSET, F. ESCHWEGE
Institut Gustave—Roussy,
Villejuif,
France

A. WAMBERSIE
UCL — Clinique Universitaire St. Luc,
Brussels,
Belgium

Abstract

After short comments on the radiobiological rationale of multifractionation, the authors report three preliminary clinical studies: 16 fractions per day for 3 days (11 patients); 8 fractions per day for 5 days (56 patients); 5 fractions per day for 10 days (49 patients). No dramatic improvement was observed for advanced tumours with a poor prognosis with conventional schedules. However, early tolerance is improved, particularly when very extended volumes (in 23 patients) are treated. Conclusions are restricted to the particular schedules used.

1. INTRODUCTION

Multifractionated irradiation with two or more fractions per day allows a reduction in the number of treatment days. This offers a practical advantage which is commonly used when an adjuvant therapy with hypoxic cell sensitizers or hyperthermia [1] is used in association with the irradiation. It also allows a reduction of the overall treatment time, which seems advisable for fast-growing tumours, without using large fractions - considered by most radiotherapists as detrimental for late reactions. However, radiobiology does not provide clear evidence that additional therapeutic benefit could arise from the unusual time/dose factors of multifractionated irradiation [2,3].

The fraction size can be smaller than the standard 2 Gy fraction. This would increase the differential effect between tumour and normal tissues if the curvatures of the cell survival curves were different in this range of small doses [4].

The spacing between the fractions is smaller than the standard 24 h, so that repair of sublethal injuries may be incomplete between fractions. This could result in a therapeutic benefit if the repair in tumour cells were slower than in normal cells; however, there is no biological evidence for such a difference.

Reoxygenation is a fast mechanism that can occur within a short temporal spacing of the fractions. It is, however, unlikely to be more active than for standard fractionation.

Finally, the repetition of the fractions at short intervals may interact with the cell kinetics, mitotic delay and progression in the cycle. The therapeutic significance of this effect is unknown.

The same biological mechanisms are involved in low-dose-rate irradiation. The therapeutic efficiency of interstitial and endocavitary irradiation is considered by some authors to be mainly related to the time/dose distribution. Based on this, Pierquin [5] has stimulated a clinical study of low-dose-rate external radiotherapy. Multifractionation should be equivalent to a low dose rate if the spacing of the fractions is small enough with respect to the time constants of the biological mechanisms involved.

2. PRELIMINARY STUDY WITH 16 FRACTIONS PER DAY

A preliminary biological study was carried out to compare different irradiation sequences with early skin reactions (dry desquamation) used as the biological test on a normal tissue. The technique, presented in another paper [6], consists in matching the doses necessary for causing a given skin reaction on two symmetrical fields treated on the same patient with the two different schedules that are being compared (250 kV X-rays). Equivalence was found for:

Conventional fractionation: 3 × 3 Gy per week for 3 weeks = 27 Gy

Short-course irradiations given over 3 days:
- Concentrated irradiation, 2 × 8.5 Gy = 17 Gy
- Multifractionation, 16 × 0.45 Gy at 30 min intervals given each day, repeated for 3 consecutive days = 22 Gy
- Low dose rate (0.92 Gy/h), 2 × 4 h each day, repeated for 3 consecutive days = 22 Gy

The multifractionated protocol has been applied as a palliative treatment for advanced head and neck tumours and lymph nodes (11 patients). A single 250 kV X-ray beam was used with a total skin dose of 26 to 30 Gy.

The skin reaction (moderate to brisk dry desquamation) peaks at 25 days and is repaired at 40 days. Mucositis peaks at 10 days and is repaired at 20 days. Tumour regression was appreciable at 10-20 days. The effects on skin, mucosa and tumours were similar to those observed for a concentrated irradiation of 2 × 9 Gy in 3 days.

A second treatment was given to 5 patients 3-6 weeks later; two of them showed no evidence of local disease at the time of death (20 months).

3. THERAPEUTIC STUDY WITH 8 FRACTIONS PER DAY

Treatments with 16 daily fractions could not be undertaken with high-energy beams on account of the machine time required.

A study was carried out by Castera et al. [7] with ^{60}Co radiation using 8 daily fractions (0.9 Gy) at 2 h intervals in 2 series of 5 consecutive treatment days with a 2-week rest period (7.20 Gy per treatment day, 36.0 Gy per series, 72.0 Gy for full treatment).

This was used for 56 patients with head and neck tumours ($9\ T_2$; $43\ T_3$; $4\ T_4$: of these 34 were N_0, N_1, 22 were N_2, N_3).

All primary tumours remain clinically controlled during follow up except for 4 cases with clinically suspected recurrence.

In 9 cases lymph-node remnants necessitated a surgical excision, which healed without any complications or later recurrence.

Brisk mucositis is repaired at 2 months in 75% of the cases; it has lasted for over 4 months in 3 patients. Larynx oedema was observed in 4 patients; one required tracheotomy.

This series of patients was compared with a series of 54 patients previously treated with conventional fractionation (70 Gy given over 7 weeks) and with the same field size and arrangement.

For multifractionated irradiation (MFI), the early mucosal reaction is greater but its duration is shorter, mouth dryness is also worse, while the late cervical fibrosis is more frequent but asymptomatic. The fatal complications are similar (haemorrhagia: 4 MFI versus 5 conventional; extended necrosis: 3 MFI versus 5 conventional). Local control of the primary tumour is better with MFI (4 failures versus 14) but the control of the lymph nodes is worse (9 failures versus 3). The survival rate is higher for MFI (Table I). This comparison suggests some benefit with the multifractionated irradiation.

4. THERAPEUTIC STUDY WITH 5 FRACTIONS PER DAY

Two studies are currently under way at our Institute to check whether a smaller number of daily fractions, compatible with machine availability, could still bring a therapeutic benefit.

For technical reasons the daily fraction number has been reduced to 5 given at 2 h intervals. Irradiations are carried out with the same beams as are used for conventional treatment, ^{60}Co γ-rays, Linac high-energy photons and electrons, and with the same geometry.

4.1. First Study

Patients with advanced head and neck epitheliomas and with a poor prognosis have been entered in the first study.

The multifractionated protocol consists of 5 fractions of 0.75 Gy per fraction (3.75 Gy per treatment day) given alternately

TABLE I. COMPARISON OF SURVIVAL RATES FOR MULTIFRACTIONATION AND CONVENTIONAL FRACTIONATION

Months	Multifractionated irradiation		Conventional fractionation	
6	47/56	84%	44/54	81%
12	34/51	67%	28/52	54%
18	12/19	63%	16/49	33%
23	5/8	63%	14/47	30%

on two opposed lateral fields, repeated on 5 days per week (18.75 Gy weekly dose). A first series is being given for 10 days in 2 weeks (37.5 Gy). A second series is being given after a 3-week rest interval.

The total dose to the posterior part of the cervical lymphatic chains (irradiated by 8-10 MeV electrons when the dose to the spinal cord reaches 30 Gy) is limited to 50 Gy (no palpable nodes) or 65 Gy (palpable nodes).

The multifractionated protocol is being applied only to the upper half of the target volume. The lower neck is treated by anterior and posterior ^{60}Co ports in the conventional way.

Twenty-six patients have been included (Table II). Follow up has been for more than one year for 23 patients.

Skin reactions are minimal. Moderate subcutaneous fibrosis develops during the following months. Mucous membrane reaction, according to the score system in use at our Institute, is considered above acceptable tolerance for 7 (out of 26) patients. This proportion is similar to that observed with the conventional treatment. The reaction is not increased by previous chemotherapy.

Tumour regression (Table III) is greater for tumours of the oropharynx than for those of the oral cavity; most of the latter were very extensive. At 6 months, 2 patients were dead (with metastases), the tumour had recurred in 11 cases, and a necrosis was observed in 4 patients. Eight patients have remained free of local disease without complications.

There may be some moderate benefit in the treatment by multifractionation of these advanced tumours. However, the bad local prognosis is not dramatically improved.

TABLE II. MULTIFRACTIONATED IRRADIATION: HEAD AND NECK PROTOCOL

A first series comprising 5 fractions of 0.75 Gy per fraction given each day on 5 days each week for 2 weeks (37.5 Gy). After a 3-week rest, a second identical series is given, to result in an overall dose of 75 Gy. 26 patients were entered (oropharynx 9, oral cavity 17). Unsuccessful chemotherapy before these schedules on 11 of the patients.

T \ N	N_0	N_1	N_2	N_3	
T_3	5	4	4	2	15
T_4	3	2	1	5	11
Totals	8	6	5	7	26

TABLE III. MULTIFRACTIONATION: TUMOUR REGRESSION

Minimum tumour volume (%)	Head and neck (see Table II)	Extended volumes (see Table IV)
0	6	5
20	11	2
40	5	2
60	0	4
80	0	4
Unchanged	0	2
Not analysable	4	4
	26	23

TABLE IV. MULTIFRACTIONATION: EXTENDED VOLUME PROTOCOL

5 fractions of 0.7 Gy per fraction given each day on 4 days each week for 2 weeks (28 Gy). After a 2-week rest, a second series is considered.
23 patients were entered.

Tumour type		Irradiated volume	
Embryon. sarcomas	6	Abdomen	14
Soft-tissue sarcomas	5		
Malignant schwannomas	3	Thorax	7
Melanomas	2		
Miscellaneous	7	Head–neck	2

4.2. Second Study

The second study is aimed at investigating the clinical feasibility of multifractionation for very large tumours, the irradiation of which is usually not considered.

The protocol consists of 5 fractions of 0.70 Gy per fraction (3.5 Gy/d) given at 2 h intervals, repeated on 4 days per week (14 Gy weekly dose). A first series is being given for 8 days in 2 weeks (28 Gy). After a 2-week rest period, a second series is considered. The treatment is being carried out with 5.5 MV X-rays.

Twenty-three patients with various types of tumour were entered in this protocol (Table IV).

The region treated was the abdomen in 14 cases (total abdomen in 6 cases), and the thorax (mediastinum or hemi-thorax)in 7 cases.

No skin reaction was observed except in patients treated by chemotherapy (in particular Actinomycin D) prior to irradiation.

The digestive tolerance of the 14 patients treated on the abdomen was satisfactory in 5 cases, acceptable in 6 cases and poor in 3 cases. Nausea and vomitting occurred during the first treatment days and usually disappeared afterwards. They were not observed when the coeliac area was spared. Diarrhoea was observed only in 2 cases at the end of the treatment.

The tolerance of the 7 patients treated on the thorax was excellent or acceptable: of 4 patients who were irradiated over the whole length of the oesophagus, 2 complained of moderate dysphagia at the end of the treatment.

On account of the good tolerance of the treatment, the dose was slightly increased to 0.80 Gy per fraction, with a total of 32 Gy for the 2-week treatment, for the last 6 patients of this series.

Tumour regression has been analysed for 19 patients (Table III). It was variable, even within each pathological group. In one third of the cases, the regression has been large enough, with a definite symptomatic improvement.

However, the survival rate is low. Three patients (out of 23) have survived for one year; only one (at 19 months) is without evidence of recurrence (entered protocol with extensive recurrence in the abdomen of a nephroblastoma previously locally treated with radiation; was given total abdominal multifractionated irradiation).

5. CONCLUSIONS

From the preliminary biological studies and this limited clinical experience, we can conclude:

(a) There is no radiobiological evidence for a definite therapeutic advantage of multifractionation;

(b) Our limited clinical experience does not offer evidence of a dramatic advantage for advanced cases with poor prognosis for conventional schedules;

(c) The early tolerance seems satisfactory and probably better than with conventional treatment for extended volumes;

(d) Longer follow up is needed to assess the late tolerance;

(e) Our conclusions apply to the particular schedules we have used; multifractionation offers a large diversity of schedules which are worth investigating.

REFERENCES

[1] ARCANGELI, G., BAROCAS, A., MAURO, F., NERVI, C., SPANO, M., TABOCCINI, A.: Multiple daily fractionation (MDF) radiotherapy in association with hyperthermia and/or Misonidazole. Experimental and clinical results, Cancer 45 (1980) 2702-2711.

[2] DUTREIX, J., LELLOUCH, J., WAMBERSIE, A.: Mécanismes radiobiologiques différentiels entre l'irradiation continue de la curiethérapie et l'irradiation fractionnée de la radiothérapie externe, J. Radiol., Electrol. Med. Nucl. 53 (1972) 221-226.

[3] DUTREIX, J., WAMBERSIE, A.: "Multifractionated irradiation - biological bases," in Biological Bases and Clinical Implications of Tumor Radioresistance (Proc. 2nd Int. Symp. Rome, 1980: FLETCHER, G.H., NERVI, C., WITHERS, H.R. et al., Eds), Masson Publishing USA, Inc., New York (1981, in press).
[4] ALPER, T., Ed.: "Cell survival after low doses of radiation", Proc. 6th L.H. Gray Conference (London, 1974), The Institute of Physics and J. Wiley and Sons, London (1975) 16-21.
[5] PIERQUIN, B., BAILLET, F.: The differential effect of continuous irradiation at low dose rate, Br. J. Radiol. 44 (1971) 236.
[6] DUTREIX, J., WAMBERSIE, A., BOUHNIK, C.: Cellular recovery in human skin reactions: Application to dose fraction number overall time in radiotherapy, Eur. J. Cancer 9 (1973) 158-167.
[7] CASTERA, D., LEGROS, M., MOUILLET, J.: Etude de la radiothérapie hyperfractionnée chez 56 patients atteints de tumeurs de la tête et du cou, J. Radiol., Electrol., Med. Nucl. 59 (1978) 611-614.

Progress in Radio-Oncology II, edited by K. H. Kärcher et al., Raven Press, New York © 1982.

Local Treatment of Malignant Brain Tumours by Removable Stereotactically Implanted Radioactive Sources

P.H. GUTIN, T.L. PHILLIPS, Y. HOSOBUCHI,
W.M. WARA, A.R. MACKAY, K.A. WEAVER,
Sharon LAMB, S. HURST
University of California San Francisco School of Medicine,
San Francisco, California,
United States of America

Abstract

Thirty-one patients harbouring malignant primary or metastatic brain tumours were treated with 34 implantations of ^{198}Au or ^{125}I sources using stereotactic neurosurgical techniques. Most tumours had recurred after surgery, whole-brain irradiation, and treatment with all feasible chemotherapeutic agents. All of the sources were mounted in catheters for removal after the desired dose had been delivered. One or more sources were placed in each tumour to deliver 3.5–7.2 krad for ^{198}Au or 3.0–12.5 krad for ^{125}I. Seven patients received ^{198}Au sources; four are evaluable. One responded for 5 months and one deteriorating patient with a recurrent tumour stabilized for 6 months. One patient with an unbiopsied tumour that was radiographically consistent with a diagnosis of anaplastic astrocytoma received a boost dose with ^{198}Au after whole-brain irradiation and is stable more than 16 months after implantation. One patient with a glioblastoma deteriorated despite treatment. Twenty-four patients underwent 27 implantations with removable high-activity ^{125}I sources; 15 patients (18 implantations) are currently evaluable. There were 13 responses for 2+ to 6 months, two occurring in patients who received interstitial boosts in combination with external irradiation. One patient with an anaplastic astrocytoma has stabilized for more than 10 months, while one patient with a metastatic melanoma failed to respond to two successive ^{125}I implants. Two patients with glioblastoma multiforme deteriorated within 2 months after implantation and were found to have metastases from these tumours in distant regions of their brain despite apparent local control of the primary mass.

1. INTRODUCTION

More than 90% of anaplastic astrocytomas and glioblastomas are localized [1]. Metastases from these tumors within the central nervous system are uncommon [2], and systemic metastases are rare [3,4]. Most solid tumors that have not metastasized by the time they are detected can be cured by surgery and radiation therapy [5]. Even though most malignant brain tumors are localized, the principal experimental treatment has been systemic chemotherapy, for which the traditional target has been metastatic disease. The use of systemic chemotherapy for brain tumors seems misguided when the obstacles to its efficacy are considered. Hyperthermia, intratumoral and intra-arterial chemotherapy, and interstitial brachytherapy have been suggested as local therapies for brain tumors. Because radiation therapy has been the most effective modality against malignant brain tumors [6,7] and because radiation-induced necrosis has prevented delivery of more than 6000-7000 rad at conventional dose-rates [8], brachytherapy is a logical local treatment for brain tumors.

Over the past 4 years we have been acquiring experience with the interstitial irradiation of brain tumors and have used stereotactic neurosurgical techniques to place permanent or removable radiation sources into tumors [9,10]. We prefer the removable sources for malignant or metastatic tumors.

2. MATERIALS AND METHODS

2.1 Patients

From April, 1979, to April, 1981, 31 patients ranging in age from 14 to 69 years were implanted 34 times with either Au-198 or I-125 sources that were later removed. Sixteen patients harbored primary anaplastic astrocytomas or glioblastomas that recurred after surgery, whole brain irradiation, and treatment with all feasible chemotherapeutic agents. Because of compromised bone marrow reserves, some patients could not tolerate continued chemotherapy. Four patients, 1 with malignant melanoma and 3 with breast carcinoma, had brain metastases that recurred after surgery and whole brain irradiation.

Six patients had implants for boosts just before or just after external brain irradiation. Two harbored glioblastomas and three harbored anaplastic astrocytomas; one harbored an unbiopsied thalamic tumor that was radiographically consistent with anaplastic astrocytoma. One patient with a solitary brain metastasis from melanoma had an implant boost.

2.2 Implantation Technique

Sources were implanted under local anesthesia through a burr hole using the Leksell stereotactic system modified for use with the GE 8800 CT scanner [11]. A base plate is fixed at four points to the skull's outer table and a CT scan is performed with a plastic replica of the conventional metal stereotactic frame attached to the base plate. With the plastic frame in position, scan artifacts are avoided, and the tumor target(s) can be visualized and related precisely to the frame's center by the computer program intrinsic to the GE 8800 scanner (FIG. 1). The coordinates for the target sites are calculated, the patient is taken to the operating room, and, with the metal frame in position, the radioactive sources are implanted.

Implants were performed with an afterloaded silastic catheter developed in cooperation with the American Heyer-Schulte Corporation (Goleta, California). An outer cannula is passed to the target using a stylet to push it through the guides of the stereotactic frame and through the interposed brain; the position of the catheter is fixed at the burr hole with an adjustable plastic collar. The stylet is removed and the cannula is afterloaded with a coaxial inner cannula containing the radioactive source(s). The wound is then reapproximated such that the catheter(s) are completely covered. Patients are isolated, with proper precautions observed, for the interval necessary to deliver the desired dose, after which catheters and sources are removed under local anesthesia.

2.3 Sources and Dosimetry

Au-198 was supplied by Nuclear Sources and Services, Inc., Houston, Texas, and I-125 by 3M Company, St. Paul, Minnesota. A well ionization chamber with a sensitivity of 2.3×10^{11} A/mg Ra equivalent for the higher photon energies was used for calibration of the sources, with a precision of $\pm$ 1% for Au-198. For I-125, an absolute calibration factor is not yet available; the chamber reading for each source is related to the activity stated by the manufacturer.

Doses are calculated by assuming that each implant source is a point source of isotropic radiation. The Au-198 sources were assumed to produce the same dose

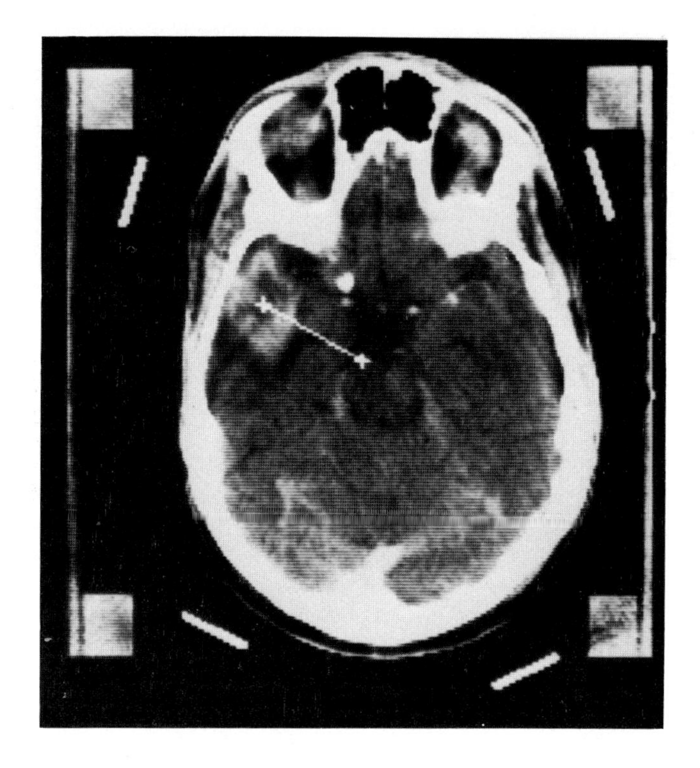

FIG.1. CT scan taken with the plastic stereotactic frame in position. The 4 vertical (square) posts of the frame are visible. The geometric center of the frame (small cross) is related by the scanner's computer to a target for ^{125}I placement in the tumor. The coordinates of the frame's center and the target are normally also displayed.

distributions as Ra-226 of the same mg equivalency. For I-125, a dose rate constant of 1.2 rad $cm^2/h \cdot mCi$ at 0.5 cm is assumed [12]. Anisotropies in the radiation fields produced by the I-125 sources are currently not accounted for in the computer's dose calculation.

A CT scan is used to plan treatment. For roughly spherical tumors, all sources are positioned at the tumor's center, and sources are positioned along the axis of elongated (prolately ellipsoidal) tumors to make the dose more uniform (FIG. 2). After stereotactic placement of sources, orthogonal radiographs are taken to determine source relationships, and a computer program converts position data, source strengths, and implantation duration into total-dose contours in any plane. The resulting dose plot is scaled to match the magnification of the radiographs to allow superimposition.

2.4 Evaluation of patients

Corticosteroid doses were adjusted as needed to improve neurologic function and reduce the symptoms of increased intracranial pressure. Because improvement caused by steroids can mimic response to interstitial radiation, doses were increased only when required to treat clearcut clinical deterioration. In addition, attempts were made to reduce the steroid dose every 6 to 8 weeks if the patient was clinically stable or improving. Anticonvulsants were used when medically indicated. Patients were evaluated by neurological examinations and CT scans at intervals of 8 weeks, when

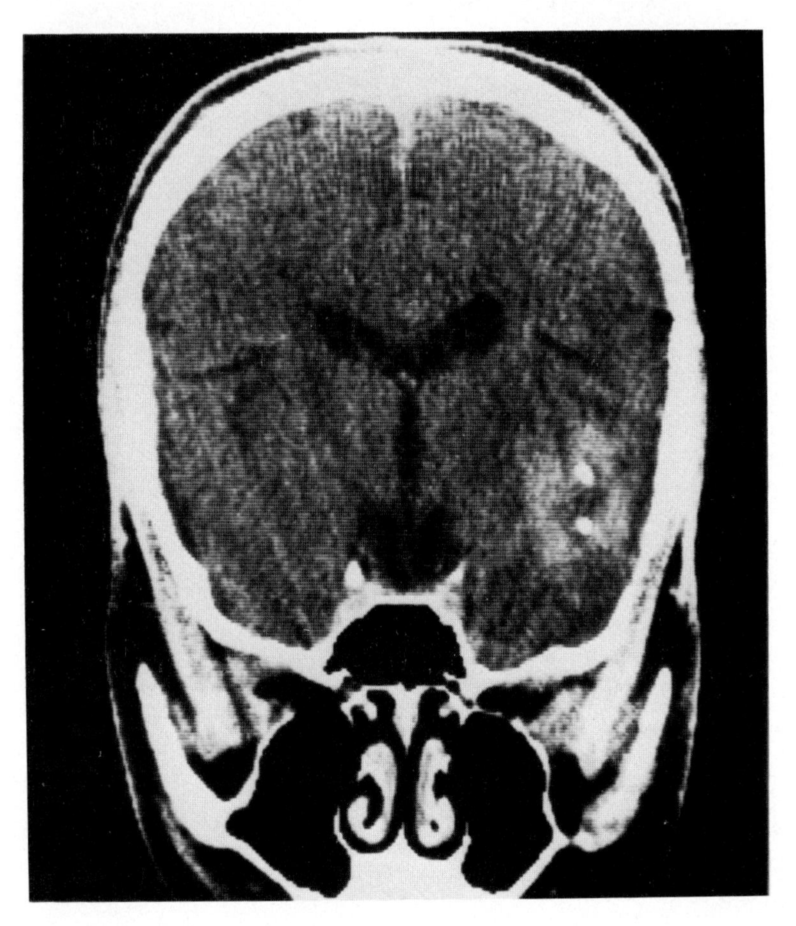

FIG.2. CT scan localizing the position of ^{125}I sources in two catheters in a recurrent right temporal lobe glioblastoma.

possible, and graded using the criteria developed by Levin et al. [13]. Patients were considered evaluable if they were alive and available for their first evaluation 8 weeks after implantation.

3. RESULTS

3.1 Gold-198

Seven patients were treated with removable Au-198 implants to minimum tumor doses of 3500-7200 rad at minimum dose rates of approximately 40-80 rad/h. Three were not evaluable; 1 died from meningeal carcinomatosis and another from a systemic fungal infection before the first evaluation. The third patient refused followup.

One patient treated with a removable Au-198 implant (5500 rad) for a recurrent glioblastoma responded dramatically for 5 mo,[1] and survived 11 mo before dying. A

[1] mo is used as symbol for month(s).

patient with melanoma deteriorating with a recurrent, previously-irradiated brain metastasis stabilized for 6 mo after Au-198 implantation (6000 rad). Another patient with an unbiopsied thalamic tumor (most likely an anaplastic astrocytoma) is stable 16+ mo after 3500 rad interstitial and 5000 rad whole brain irradiation. One patient with a recurrent glioblastoma continued to deteriorate and died 2 mo after implantation.

3.2 Iodine-125 (high activity)

Twenty-four patients underwent 27 implantations of removable high-activity I-125 sources. Dose rates of 25 to 100 rad/h were achieved at the periphery of sizable tumors when single or multiple sources in single or multiple well-positioned catheters were used. Minimum tumor doses were 3000-12 500 rad.

Nine patients are not evaluable. One patient developed a brain abscess at the implant site that required craniotomy (with an incidental subtotal resection). One died of a pulmonary embolism, 1 with breast carcinoma died of liver metastases, and 1 with anaplastic astrocytoma died from massive hepatic necrosis of unknown cause. Three patients have not returned for their first evaluation. As patients receive progressively higher interstitial radiation doses, we are noting clinical and radiographic signs of deterioration at the early evaluations. We believe this is caused by reactive cerebral edema from the intense local irradiation, but the syndrome is indistinguishable from tumor regrowth. Two patients with recurrent malignant primary brain tumors are currently in this situation and are therefore not evaluable.

Fifteen patients treated with 18 high-activity I-125 sources are evaluable. Nine patients with recurrent tumors responded for 2+ to 6 mo. Two of these, whose tumors recurred after a first implant, responded to a second treatment. Recurrent tumors that responded included anaplastic astrocytomas, glioblastomas, and metastases from breast carcinoma. One patient with an anaplastic astrocytoma and another with a brain metastasis from malignant melanoma received interstitial radiation boosts before external brain irradiation and have responded for 4+ and 5 mo (FIG. 3).

One patient with a recurrent anaplastic astrocytoma has stabilized for 10+ mo, while 1 patient with a recurrent melanoma metastasis failed to respond to 2 successive I-125 implants. Two patients with glioblastomas deteriorated within 2 mo after implantation and were found to have metastases from these tumors in distant regions of their brains, despite apparent control of the primary mass.

4. DISCUSSION

Removable sources are superior for the brachytherapy of brain tumors because they afford greater control over the dose delivered, prevent migration of the sources from necrotic tumor regions, and make possible the removal of sources if emergency decompressive surgery is required. In addition, a removable implant system allows the use of long-lived, high-activity isotopes (I-125) without risking brain toxicity or compromising the safety of those close to the patient after discharge from the hospital.

Our experience suggests that low-activity I-125 sources implanted in reasonable numbers are sufficient only for treating low grade astrocytomas [10]. The relatively recent availability of high-activity I-125 sources, which give higher dose rates, has made it possible to treat brain tumors with malignant growth characteristics. Dose rates in excess of 30 rad/h, commonly used in the brachytherapy of cancer at other sites, are necessary to treat primary and metastatic malignant brain tumors [10,14].

A comparison of the effectiveness of Au-198 and I-125 was not possible in this study; the I-125 implants were done late in the study when our increased aggressiveness and our greater technical facility guaranteed that higher radiation

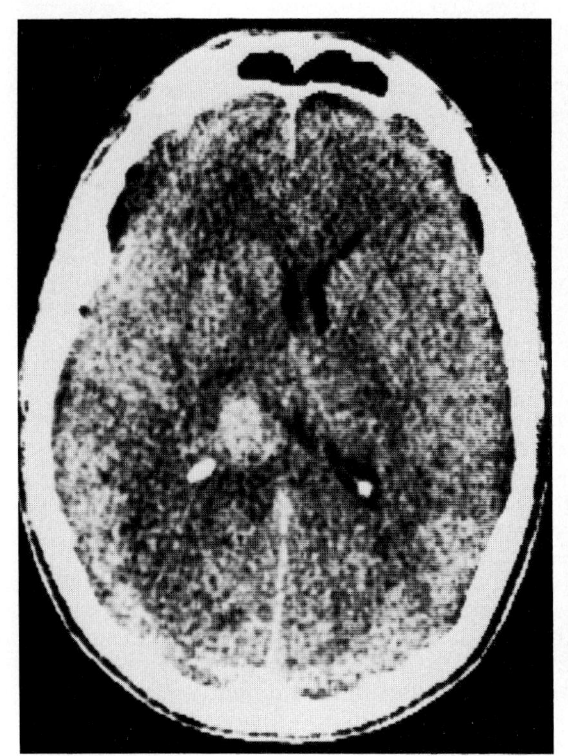
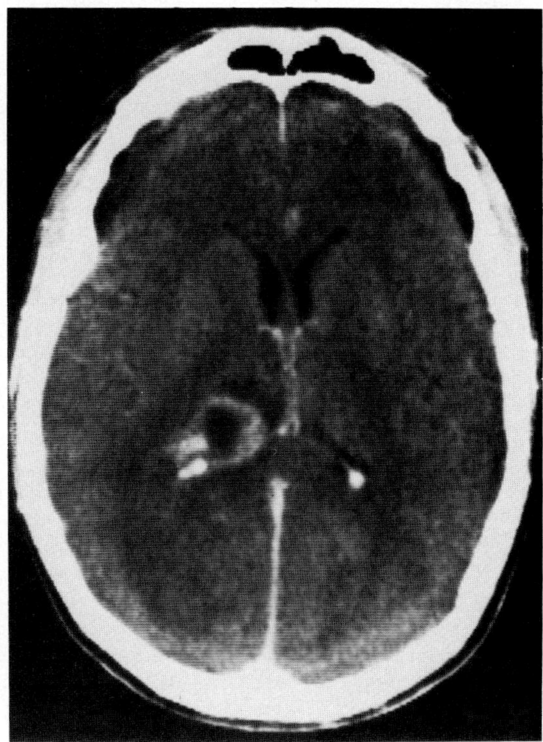

FIG.3. CT scans showing an anaplastic astrocytoma of the posterior thalamus before (left) and 2 months after (right) treatment with removable high-activity ^{125}I sources (3900 rad) and limited field external irradiation (4000 rad). Tumor volume and midline shift are reduced.

doses could be delivered more accurately. However, I-125 seems to be the superior isotope for the interstitial irradiation of brain tumors. It emits relatively low energy (27 to 35 keV) characteristic X-rays, which limits its tissue half-value layer to 2 cm. The amount of radiation penetrating to surrounding brain is far less than that from Ir-192 or Au-198, and "leakage" through the skull is reduced, which affords better protection for families and medical personnel.

It is clear that interstitial brachytherapy has the potential to afford significant palliation in patients with recurrent malignant brain tumors. Because survival in patients with malignant gliomas has been shown to increase stepwise with radiation dose (up to 6000 rad) [7], it may be that interstitial radiation boosts along with conventional external radiation will produce long-term remissions or even cures while limiting the exposure of normal brain to radiation.

5. ACKNOWLEDGMENTS

We thank the Medical Product Division of the 3M company for their cooperation in providing high-activity I-125 sources, the Special Products Division of the American Heyer-Schulte Corporation for their help in developing removable implant catheters, and Neil Buckley for editing and preparing the manuscript.

REFERENCES

[1] HOCHBERG, F.H., PRUITT, A.: Assumptions in the radiotherapy of glioblastoma, Neurology (Minneapolis) 30 (1980) 907.

[2] ERLICH, S.S., DAVIS, R.L.: Spinal subarachnoid metastasis from primary intracranial glioblastoma multiforme, Cancer 42 (1978) 2854.

[3] ALVORD, E.C.: Why do gliomas not metastasize? Arch. Neurol. 33 (1976) 7.

[4] SMITH, D.R., HARDMAN, K.M., EARLE, K.M.: Metastasizing neuroectodermal tumors of the central nervous system, J. Neurosurg. 31 (1969) 50.

[5] ZUBROD, C.G.: Chemical control of cancer, Proc. Natl. Acad. Sci. U.S.A. 69 (1972) 1042.

[6] WALKER, M.D.: Chemotherapy: Adjuvant to surgery and radiation therapy, Semin. Oncol. 2 (1975) 69.

[7] WALKER, M.D., STRIKE, T.A., SHELINE, G.E.: An analysis of dose-effect relationship in the radiotherapy of malignant gliomas, Int. J. Radiat. Oncol., Biol. Phys. 5 (1979) 1733.

[8] SHELINE, G.E., WARA, W.M., SMITH, V.: Therapeutic irradiation and brain injury, Int. J. Radiat. Oncol., Biol. Phys. 6 (1980) 1215.

[9] HOSOBUCHI, Y., PHILLIPS, T.L., STUPAR, T.A., GUTIN, P.H.: Interstitial brachytherapy of primary brain tumors: Preliminary report, J. Neurosurg. 53 (1980) 617.

[10] GUTIN, P.H., PHILLIPS, T.L., HOSOBUCHI, Y., WARA, W.M., MACKAY, A.R., WEAVER, K.A., LAMB, S., HURST, S.: Permanent and removable implants for the brachytherapy of brain tumors, Int. J. Radiat. Oncol., Biol. Phys. submitted for publication.

[11] MACKAY, A., GUTIN, P., HOSOBUCHI, Y., NORMAN, D.: "CT stereotaxis and interstitial irradiation for brain tumor", in Interventional Radiologic Techniques: Computerized Tomography and Ultrasonography (MOSS, A.A., GOLDBERG, H.I., Eds), University of California Printing Department, Berkeley (1981) 93-99.

[12] KRISHNASWAMY, V.: Dose distribution around an I-125 seed source in tissue, Radiology 126 (1978) 489.

[13] LEVIN, V.A., CRAFTS, D.C., NORMAN, D.M., HOFER, P.B., SPIRE, J.-P., WILSON, C.B.: Criteria for evaluating patients undergoing chemotherapy for malignant brain tumors, J. Neurosurg. 47 (1977) 329.

[14] PIERQUIN, B.: The destiny of brachytherapy in oncology, Am. J. Roentgenol. 127 (1976) 495.

Progress in Radio-Oncology II, edited by K.
Kärcher et al., Raven Press, New York © 1

Stereotactic Interstitial Therapy of Non-Resectable Intracranial Tumours with Iridium-192 and Iodine-125

F. MUNDINGER
Abteilung Stereotaxie und Neuronuklearmedizin,
Neurochirurgische Universitätsklinik,
Freiburg im Breisgau,
Federal Republic of Germany

Abstract

Interstitial or intracavitary Curie therapy of intracranial tumours and tumour cysts has been carried out using the stereotactic computer technique including computed tomography (CT stereotaxy). All tumours are bioptically classified intraoperatively in a smear preparation. Today non-resectable tumours of grade I and II have iridium-192 and iodine-125 sources implanted permanently. The dose distribution (tumour surface dose of 120 Gy), the optimum implantation sites, and the implantation activities are calculated using the programs developed together with W. Birg. Anaplastic gliomas are treated with the brachy-Curie-therapy (GammaMed® iridium-192 contact radiation device, or temporary implantation of iodine-125) usually in combination with percutaneous irradiation. In the case of germinomas, malignant ependymomas, and medulloblastomas, the brachy-Curie-therapy is combined with percutaneous irradiation of the neuroaxis. Technique and dosimetry are reported. From a series of more than 1200 patients (by March 1981), the results of 739 cases that were locally treated with iridium-192 and 146 cases treated with iodine-125 are given.

1. INTRODUCTION

The stereotactic local-irradiation treatment of intracranial tumours is carried out with the computer-aided version of the stereotactic device developed by Riechert and Mundinger [1,2] (Fig. 1) and, more recently, by computed-tomography stereotaxy (CT stereotaxy) [3,4]. The radiopharmaceutica that have been used at Freiburg for implantation since 1952 are listed in Table I, and the clinical indications in Table II.

2. INTERSTITIAL THERAPIES

2.1. Brachy-Curie Therapy

In brachy-Curie therapy, the dose of the interstitially or intracavitarily implanted emitter is delivered within minutes. Such therapy is carried out with the aid of a GammaMed® iridium-192 contact-irradiation device that I constructed together with Sauerwein [5] (Fig.2). Alternatively, catheter systems loaded with iodine-125 seeds may be implanted temporarily for a period of several days [6,7].

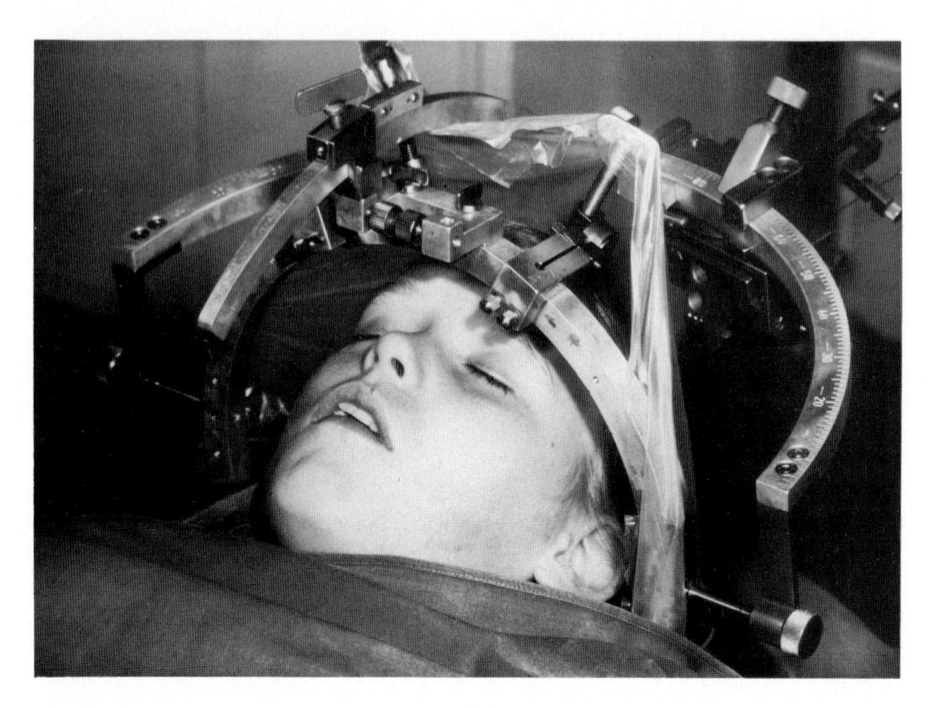

FIG.1. Stereotactic apparatus developed by Riechert and Mundinger in the modification by Mundinger and Birg for computed stereotaxy. The base ring is fixed with fixers on the head. The half-ring with probe holder in the previously calculated position is attached to the base ring. The implantation cannula is inserted through a small drilled hole (6 mm dia.) into the tumour.

TABLE I

Radionuclides and number of stereotactic interstitial irradiation procedures (1952 – March 1981)

P-32	6
Co-60	179
Y-90	44
I-125	146
Ta-182	21
Ir-192 GammaMed®	247
Ir-192	492
Au-198	128
Total	1263

TABLE II

Interstitial Curie therapy procedures (1952 - March 1981)

Glioblastoma	170
Astrocytoma	370
Oligodendroglioma	93
Ependymoma	28
Pinealoma	14
Meningioma	15
Metastasis	18
Sarcoma	11
Pituitary adenoma	246
Craniopharyngioma	21
Other lesions	41
Hypophysectomy	55
Pallidotomy	21
	1103

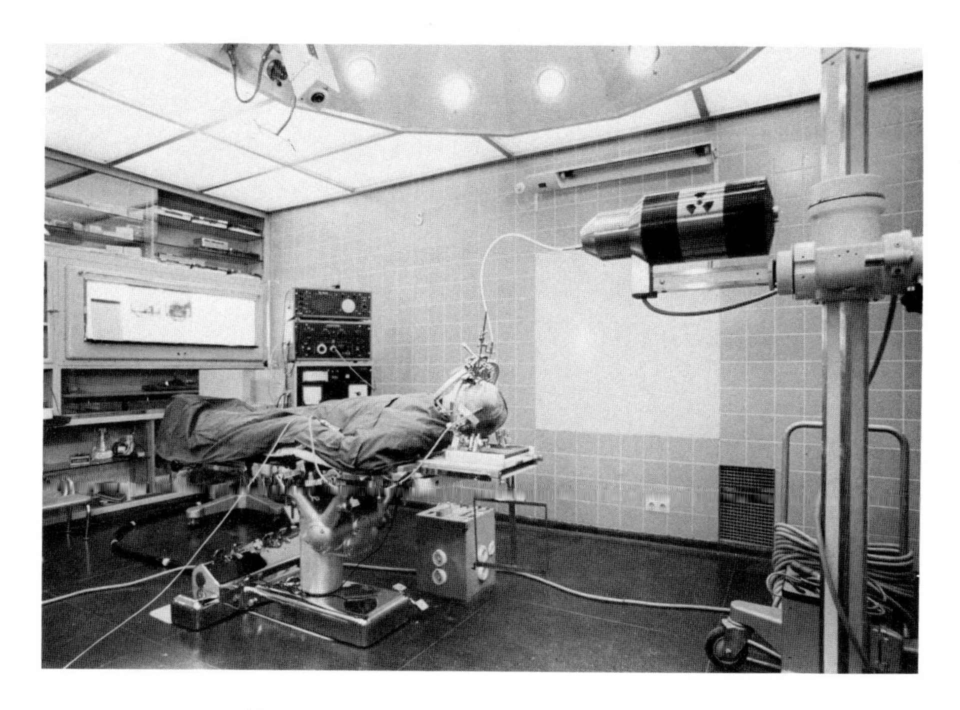

FIG.2. The GammaMed® iridium-192 contact radiation device. The connecting tube is attached to the stereotactically introduced application cannula at one end and to the exit channel of the shielding container at the other. With the aid of automatic remote control, the emitter is introduced into the intertumourally inserted application cannula and is automatically withdrawn into the shielding container after the calculated exposure has been completed.

2.2. Curie Therapy

Curie therapy is applied in the form of a permanent implantation, according to a concept that I developed in 1957. This method involves long-term irradiation, i.e. for periods of several months, using ^{192}Ir [8-11] and, since February of 1979, also ^{125}I-filled titanium seeds [6,7] absorbed on Dowex [12].

All tumours are initially biopsied intra-operatively and confirmed with the aid of smear preparations and, additionally, in sections fixed in paraffin wax [10] (Fig.3). Biopsies can even be performed in the diencephalon, mesencephalon and brain stem–pons areas with relatively very little risk (∿1%) using computer-aided stereotaxy. Nuclide implantation is undertaken immediately after biopsy.

2.3. Technique Chosen

Which form of irradiation is to be used depends on the grade of the tumour. Brachy-Curie therapy must be given preference with any anaplastic or malignant tumour; such therapy can, if necessary, be given in combination with local long-term irradiation or percutaneous irradiation.

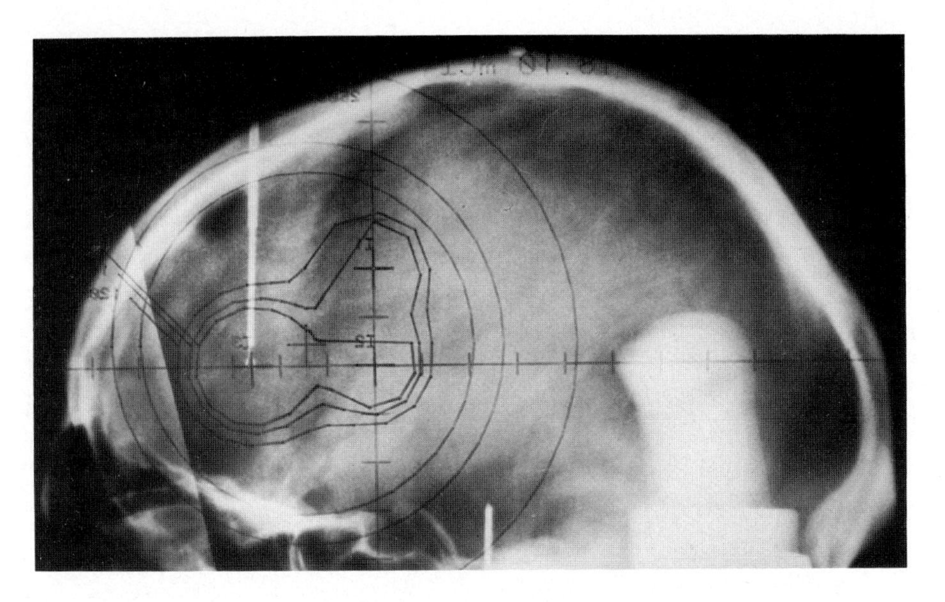

FIG.3. Astrocytoma in the fronto-precentral area. Corresponding to the CT scan, the localization and activity of the ^{125}I seeds are calculated (3 targets in this case). The biopsy probe (1 mm dia.) lies in the frontal implantation target.

Curie therapy is indicated for low-grade tumours. Our experience has shown that, even when percutaneous irradiation techniques are carried out perfectly, no local regression can be effected on these tumours because of their great radioresistance, even to doses as high as 70 Gy.

Smaller, peripherally situated tumours (metastases) are resected as completely as possible, using the approach through the bone, with a cortical opening of only 5-6 mm. Subsequently intracavitary tumour wall irradiation is given using ^{125}I implants.

Cysts or cystic tumours are first drained and subsequently implanted.

Our CT-stereotaxy technique is a combination of computer-aided stereotaxy and X-ray transmission computed tomography [7,13]. Using this technique, the location, configuration and volume of the tumour, and thus the dosimetric data (activity, number of implants, sites of implants and their relation to one another, isodose distribution) can immediately be calculated on the computer of the X-ray tomograph (Fig.4). The target points for multiple implants and the angular parameters of the stereotactic device can also be computed. Hence optimum placement and distribution of the radioisotopic implants can be achieved (Figs 5,6). Biopsy and implantation are both carried out in the operating theatre.

3. RESULTS

The data in Tables I and II include 247 ^{192}Ir GammaMed® irradiation treatments and 492 permanent ^{192}Ir implants. Also

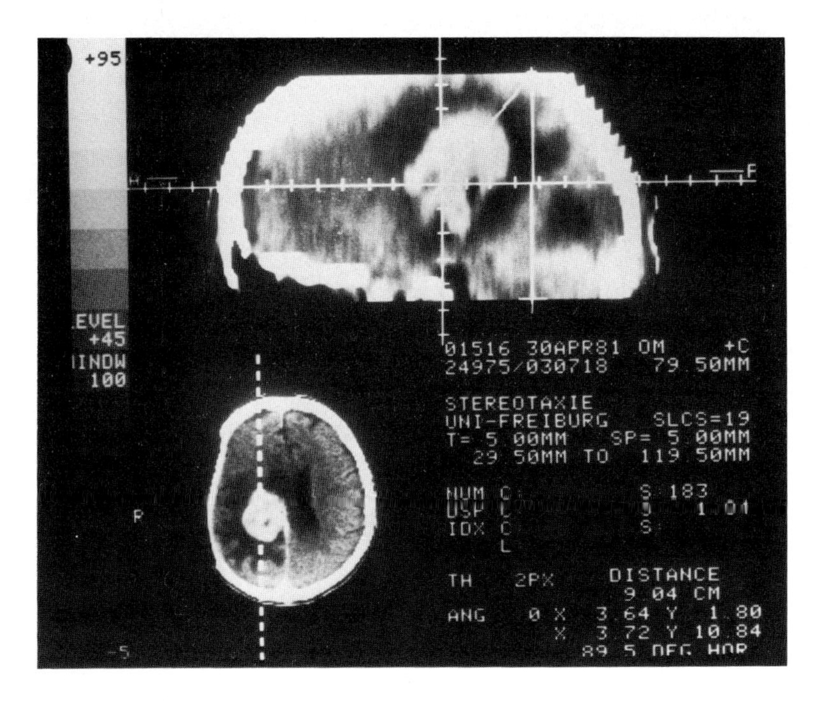

FIG.4. Malignant melanoma. The bony trepanation point can be determined in the sagittal and coronal (vertical) reconstructions so that, for example, a pin-shaped tumour can be axially implanted.

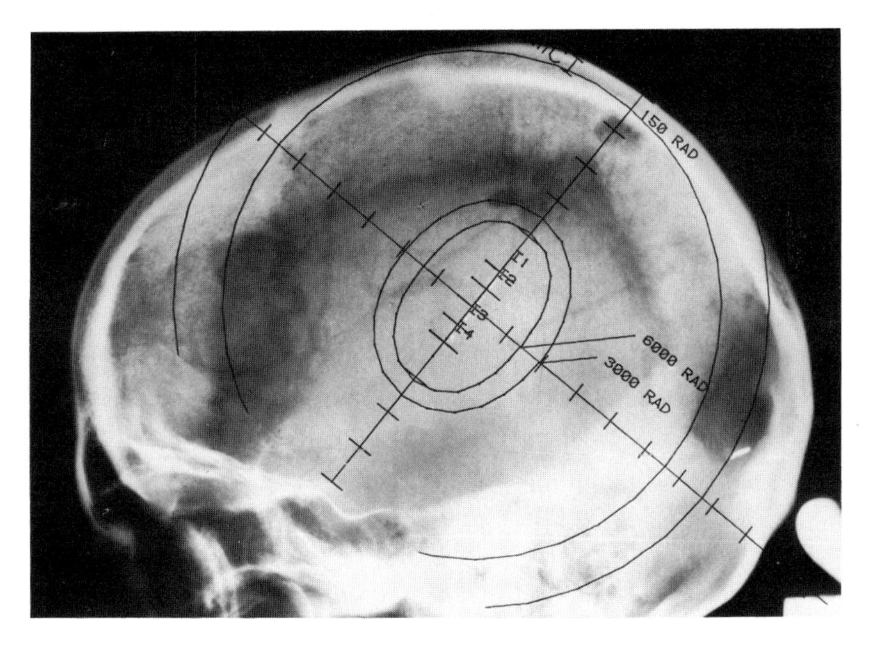

FIG.5. Malignant melanoma of Fig 4. X-ray lateral view for control of catheter position for brachy-Curie therapy and tracing of the previously calculated tumour surface dose (40 Gy).

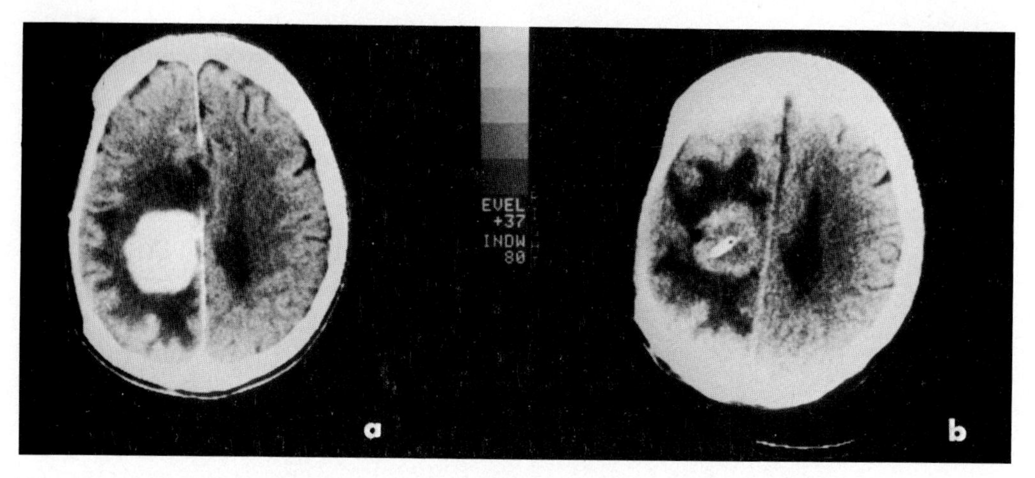

FIG.6. Malignant melanoma of Figs 4 and 5. (a) CT-scan before the brachy-Curie therapy. (b) The control CT-scan with the implanted catheter loaded with ^{125}I seeds. The previously calculated target point in the centre of the tumour has been 'hit' exactly.

included are the 146 permanent ^{125}I-seed implantations that were performed between 2 February 1979 and 31 March 1981. The long-term results of the variably long series of ^{192}Ir brachy-Curie therapy and permanent implants are reported elsewhere [8,9,11,14,15].

The results of the GammaMed® ^{192}Ir brachy-Curie therapy given with or without a preliminary 6-day-infusion treatment with radiosensitizers (bromodeoxyuridine, 5-fluorouracil, methotrexate) have been evaluated; Table III shows the average survival times. The 12-year long-term control achieved with the permanent ^{192}Ir implants have also been analysed, and Table IV shows the 3 and 5-year survivals for gliomas of grades I and II. The results of ^{125}I Curie therapy that have already been evaluated are listed in Tables V and VI.

4. DISCUSSION

The 3 and 5-year survivals of patients receiving ^{192}Ir Curie therapy indicate the good palliative as well as curative effect of this type of irradiation therapy. The 26-month follow-up period for ^{125}I cases is too short to warrant a similar assessment. Up to now, our follow-up controls (using regular CT scans - usually every 3 months) have, particularly with tumours of grades III and IV, shown a rather limited radiobiological reaction at the 120 Gy isodose (tumour surface). The generally better tolerance is probably due to the only slight perifocal oedema. Advantages in such treatment also lie in the fact that radiation protection is easier to manage during the operation, and that practically no protection measures have to be taken for the surroundings. The last-mentioned points, in particular, pose a great advantage over ^{192}Ir Curie therapy. Our experience to date leads me to believe that, especially for the larger tumours, ^{125}I can replace ^{192}Ir in Curie therapy because of the improved positioning, dosimetric and control techniques of CT stereotaxy, as well

TABLE III

Survival after interstitial brachy-Curie therapy with/without radio-sensitization (179 patients)

	n+	n-	t_s +	t_s -	$t_{max.}$ +	$t_{max.}$ -
Glioblastoma	62	9	9.6± 6.9	8.3± 6.8	35.3*	23.6
Astrocytoma	43	4	22.1±16.9	24.1±16.7	65.3*	47.5
Oligo-dendroglioma	25	7	37.2±33.4	30.1±18.5	111.5*	63.3
Metastasis	6	2	24.4±35.2	9.1± 0.8	95.5*	9.6
Ependymoma	6	2	24.1±22.1	49.3±59.9	55.8*	93.1
Undiff. meningioma	2	1	16.7±16.5	28.8	28.4	28.8
Other tumors	7	3	10.1±10.9	20.3±19.2	32.5	16.1
	151	28				

* still living;
n number of cases;
t_s survival (months);
$t_{max.}$ maximum survival (months);
+ / - with / without sensitization.

TABLE IV

3- and 5-year survival time of 108 patients with intracerebral gliomas after permanent Iridium-192 implantation. Evaluation period 12 years (1965-1977)

	3-year survival (%)	5-year survival (%)
Hemispheric gliomas (43 patients)	69.6	38.7
pilocytic astrocytoma (I)	50.0	50.0
fibrillary astrocytoma (II)	53.6	32.1
oligodendroglioma (II)	27.3	9.0
Midline gliomas (65 patients)	63.1	45.8
pilocytic astrocytoma (I)	74.0	63.4
fibrillary astrocytoma (II)	70.8	36.8
oligodendroglioma (II)	-	-

as the possibility of early detection of tumour recurrence or of under-dosed parts of tumours, thus allowing for an early re-implantation.

5. INDICATIONS

Stereotactic interstitial irradiation today, commonly using ^{192}Ir and/or ^{125}I, either for brachy-Curie therapy or for permanent

TABLE V

Cerebral gliomas treated with interstitial ^{125}I Curie therapy (Febr. 1979 - March 1981)

	number of cases	
	treated	still living
Pilocytic Astrocytoma I	33	30
Fibrillary Astrocytoma II	49	46
Anaplastic Astrocytoma III	17	9
Oligodendroglioma II	9	8
Oligodendroglioma III	2	0
Glioblastoma IV	9	2
Total	119	95 (80%)

TABLE VI

Extracerebral tumors treated with interstitial ^{125}I Curie therapy (Febr. 1979 - March 1981)

	number of cases	
	treated	still living
Ependymoma	3	3
Germinoma	6	4
Pineocytoma	1	1
Craniopharyngioma	5	5
Pituitary Adenoma	2	2
Meningioma	2	2
Teratoma	2	2
Metastasis	6	3
Total	27	22 (81%)

implantation, should, on the basis of our experience with about 1200 cases, be carried out taking into account that, with larger brain tumours of the hemispheres, open surgery with, if possible, total or subtotal removal of tumour should principally be striven for. Should, however, removal or partial resection of a tumour prove impossible because of anticipated severe functional deficit, or should there be, for other reasons, a contra-indication for open surgery, then stereotactic Curie therapy is indicated (Table VII). Further:

5.1. In the case of anaplastic tumours (grade III or IV), the combination of interstitial brachy-Curie therapy and percutaneous radiation therapy allows an effective increase in the local dose, thus improving the palliative result. If the anaplastic tumour is detected at an early stage and if it fulfills the conditions

TABLE VII

Stereotactic Curie therapy – clinical indications

1. Brachy-Curie therapy plus percutaneous irradiation	(a+b) Anaplastic glioma (grade III, IV); malignant ependymoma; melanoma; medulloblastoma; sarcoma; metastasis
2. Curie therapy	(a) Cysts; small-volume glioma (grade I, II); hypophysis-adenoma (b) Glioma (grade I, II); ependymoma; dysontogenetic tumours; extracerebral benign tumours
3. Curie therapy plus percutaneous irradiation	Anaplastic glioma (grade II–III); germinoma; pineoblastoma-cytoma
4. Percutaneous irradiation (including neuro-axis)	Inoperable large-volume anaplastic glioma; multiple metastases; medulloblastoma; germinoma; malignant extracerebral tumours

for small-volume irradiation, brachy-Curie therapy alone is indicated first. Regular CT control is used to decide whether re-implantation or additional percutaneous irradiation is indicated.

5.2. Experience has shown that tumours of grades I and II do not react satisfactorily to the percutaneous irradiation technique. If the boundaries of the tumour can be easily delimited, Curie therapy with additional aids is indicated (cyst puncture, catheter systems, etc.). Long-term, palliative and curative results can be achieved.

5.3. For extended non-delimited hemisphere gliomas of grades I and II, corticoid therapy alone should be applied. A surprisingly long palliative effect can sometimes be achieved with this treatment.

5.4. For the transition to anaplastic tumours (grade II-III), Curie therapy in combination with percutaneous irradiation is preferred.

5.5. Percutaneous radiation therapy, if necessary in combination with radiosensitizers and chemo-cytostatica, is indicated, if at all, whenever the anaplastic tumour (of grade III or IV) is already infiltrating extensively and whenever the general condition and the functional deficits, as well as any spread into the stem and basal ganglia, do not any longer justify brachy-Curie therapy. In such cases, priority must be given to enabling the patient to live a purposeful and worthwhile life rather than to forcing a statistically longest-possible palliative result.

Used with clearly defined indications, computer-aided stereotactic interstitial Curie therapy has proved to constitute real progress in radiotherapy, and it has led to a significant improvement in the survival statistics and even to cures.

ACKNOWLEDGEMENTS

I would like to acknowledge the assistance of J.Schildge, B.Busam and K.Weigel, who co-operated with our group in the evaluation of the results of the therapies presented in Table III, Table IV, and Tables V and VI, respectively.

REFERENCES

[1] MUNDINGER, F.: Stereotaktische Operationen am Gehirn. Grundlagen–Indikationen–Resultate, Hippokrates-Verlag, Stuttgart (1975).

[2] RIECHERT, T., MUNDINGER, F.: Beschreibung und Anwendung eines Zielgerätes für stereotaktische Hirnoperationen (2. Modell), Acta Neurochir., Suppl. 3 (1956) 308-337.

[3] BIRG, W., MUNDINGER, F.: Computer calculations of target parameters for a stereotactic apparatus, Acta Neurochir. 29 (1973) 123-129.

[4] BIRG, W., MUNDINGER, F., KLAR, M.: A computer programme system for stereotactic neurosurgery, Acta Neurochir., Suppl. 24 (1977) 99-108.

[5] MUNDINGER, F., SAUERWEIN, K.: "GammaMed", ein neues Gerät zur interstitiellen, nur einige Minuten dauernden Bestrahlung von Hirngeschwülsten mit Radioisotopen, auch intraoperativ anwendbar, Acta Radiol., Ther., Phys., Biol. 5 (1966) 48-51.

[6] MUNDINGER, F.: "Rationale and methods of interstitial iridium-192-brachy-Curie-therapy and iridium-192 or iodine-125 protracted long term irradiation", in Stereotactic Cerebral Irradiation (INSERM Symp. No.12: SZIKLA, G., Ed.), Elsevier/North-Holland Biomedical Press, Amsterdam (1979) 101-116.

[7] MUNDINGER, F.: "Stereotaktische Therapie nicht resezierbarer intracranieller Tumoren mit Ir-192 und Jod-125", in Kombinierte Chirurgische und Radiologische Behandlung Maligner Tumoren, (WANNENMACHER, M., SCHREIBER, H.W., GAUWERKY, F., Eds), Urban and Schwarzenberg, Munich (1981).

[8] MUNDINGER, F.: "Treatment of brain tumors with radioisotopes", in Progress of Neurological Surgery (KRAYENBÜHL, H., MASPES, M., SWEET, Ch., Eds) Vol. 1, Karger Verlag, Basel, New York (1966) 202-257.

[9] MUNDINGER, F.: "The treatment of brain tumors with interstitially applied radioactive isotopes"., in Radionuclide Applications in Neurology and Neurosurgery (WANG, Yen, PAOLETTI, P., Eds), Charles C. Thomas, Springfield, Illinois (1970) 199-265.

[10] MUNDINGER, F., METZEL, E.: Interstitial radioisotope therapy of intractable diencephalic tumors by the stereotaxic permanent implantation of iridium-192, including bioptic control, Confin. Neurol. 32 (1970) 195-202.

[11] MUNDINGER, F.: "Stereotactic Curie-therapy of pituitary adenomas. A long term follow-up study", in Advances in Stereotactic and Functional Neurosurgery (GILLINGHAM, F.U., HITCHCOCK, E.R., TURNER, U.W., Eds), Acta Neurochir., Suppl. 21 (1974) 169-176.

[12] HILARIS, B.S.: Handbook of Interstitial Brachy-Therapy, Memorial Sloan-Kettering Cancer Center, New York, Publishing Sciences Group, Inc., Acton, Massachussetts (1975).

[13] MUNDINGER, F., BIRG, W.: CT-aided Stereotaxy for Functional Neurosurgery and Deep Brain Implants, Advances in Neurosurgery, Vol.10, Springer Verlag, Berlin, Heidelberg, New York (1981).

[14] MUNDINGER, F., RIECHERT, T.: Hypophysentumoren–Hypophysektomie. Klinik–Therapie–Ergebnisse, Thieme-Verlag, Stuttgart (1967).

[15] MUNDINGER, F., BUSAM, B., BIRG, W., SCHILDGE, J.: "Results of interstitial iridium-192-brachy-Curie-therapy and iridium-192 protracted long term irradiation", in Stereotactic Cerebral Irradiation (INSERM Symp. No. 12: SZIKLA, G., Ed.), Elsevier/North-Holland Biomedical Press, Amsterdam (1979) 303-320.

Progress in Radio-Oncology II, edited by K. H. Kärcher et al., Raven Press, New York © 1982.

Multiple Daily Fractionation Radiotherapy in Association with Hyperthermia and/or Misonidazole

Two years' experience with head and neck cancer

G. ARCANGELI
Istituto Medico e di Ricerca Scientifica,
Rome,
Italy

Abstract

Fifty-one patients with 111 neck lymph node metastases from head and neck cancer have been treated with several combination modalities involving multiple daily fractionation (MDF) radiotherapy alone or combined with hyperthermia (HT) and/or misonidazole chemotherapy (MIS), such that comparable lesions in the same patient underwent at least two different types of combined therapeutic modalities. The local control rate appeared to be much higher in patients treated with the two basic protocols MDF and MDF+MIS than in our historical series of patients with the same lesions treated with a conventional radiotherapy course. When HT was also delivered a complete clearance was obtained in more than 75% of the lesions. This response remained almost constant during the whole follow-up period (18 months). No major general or local toxicity was observed.

1. INTRODUCTION

Multiple daily fractionation radiotherapy had been introduced in our Institute some years ago for irradiating a miscellaneous series of tumours [1]. Several radiobiological criteria can justify the use of this fractionation scheme per se, providing that some treatment parameters (dose per fraction, number of daily fractions, and interval between fractions) are carefully evaluated in order to exploit the biological phenomena determining the differential response [2].

A multiple daily fractionation (MDF) scheme of 3 daily fractions comprising 2, 1.5 and 1.5 Gy on each day, five days a week, with a 4 h interval between the fractions, was employed, giving an average total dose of 60 Gy. Furthermore, the compression of the overall treatment time permitted the concomitant administration of hypoxic cell sensitizers and/or hyperthermia for almost the full length of the radiotherapy course.

Investigators of optimum fractionation and sequence of heat and radiation seem now to have reached some measure of agreement, in favour of employing, at a temperature of 43°C or below, a sequential treatment (that is, hyperthermia delivered 4 h after irradiation) [3,4]. It appears that, with simultaneous administration of both treatment modalities [3] or with temperatures of 45°C or above [5], the optimum therapeutic ratio would only be obtained if the tumour could be heated preferentially with respect to normal tissue.

Some of our patients have been treated with microwave or radiofrequency heating to a temperature of 42.5±0.5 °C for 45 min on every other day, immediately after the second fraction of a thrice-a-day fractionated irradiation therapy. Such a treatment schedule represents a compromise between the two above-mentioned strategies.

About half the patients also received misonidazole, a hypoxic cell sensitizer. This, when administered simultaneously with hyperthermia has, in animal tumours, been shown to give an increase in the enhancement of the radiation response over that observed when radiation is combined with only one of the two modalities [6]. Furthermore, misonidazole shows in humans a serum half-life sufficiently long [7] to be actively present throughout all the daily fractions of an MDF radiotherapy course.

Misonidazole (MIS) was administered 2 h before the first fraction of MDF, at a dose of 1.2 g/m^2 of body surface area for 10 treatment days to a total dose of 12 g/m^2. For this treatment schedule serum levels of about 50 $\mu g/ml$ were constantly observed in our patients. According to experimental observations [8], these levels would give a theoretical enhancement of about 1.5, ignoring any possible effect of reoxygenation.

Our study was conducted on 51 patients with multiple neck node metastases (N_2-N_3) from head and neck cancer. This particular clinical system allowed us to treat comparable lesions in the same patient with at least two different combinations of therapeutic modalities. Patients were assigned to one of the two basic treatment categories, MDF or MDF+MIS; one of the nodes was also heated, so that the nodes were distributed through different therapeutic schemes as shown in Table I.

2. RESULTS

The results obtained with different treatment protocols were compared with a historical series of patients with the same type of tumours treated with conventional fractionation. As expected, the crude survival of these patients seems to be unaffected by the type of treatment (Fig.1).

TABLE I. NODES TREATED BY MULTIMODALITY TREATMENTS

51 patients, multiple neck node metastases, head and neck cancer

Basic therapy	Given alone (No. of nodes)	Given with hyperthermia (No. of nodes)
MDF	35	29
MDF + misonidazole	25	22

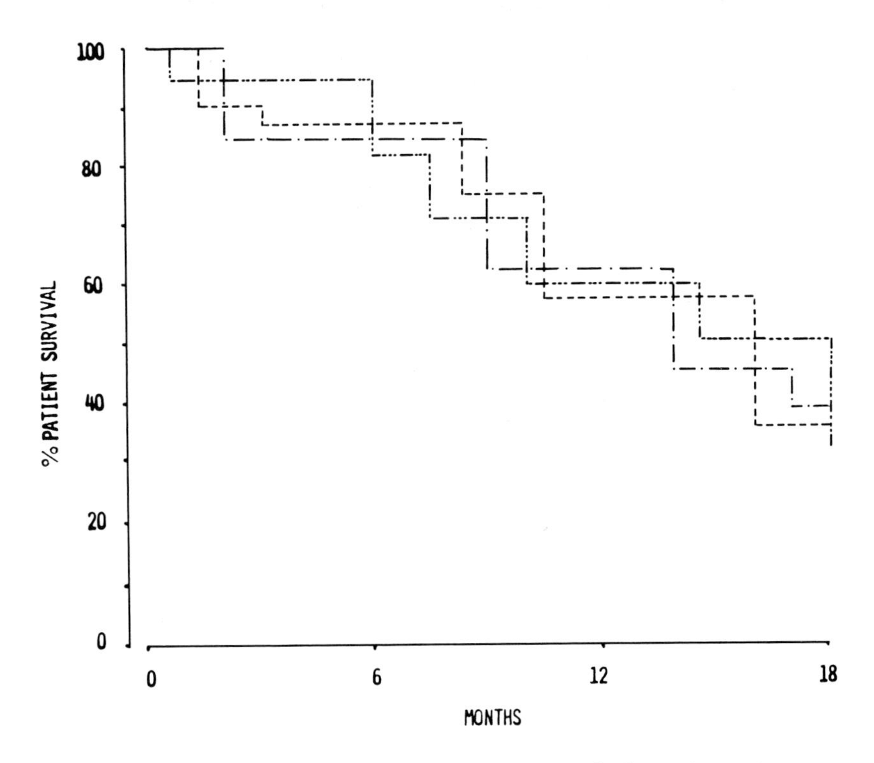

FIG.1. Patient survival (treatments: conventional fractionation —···—··— ; MDF alone -------; MDF+MIS —·—·—).

Nevertheless, the immediate response appears to be much greater in patients treated with multimodality protocols; a complete clearance was obtained in more than 75% of the lesions that were also treated with hyperthermia (Table II). The difference is statistically significant among the several treatment groups. Because of the numerous patient deaths, this difference during the follow-up period is only statistically significant as compared with the historical series for lesions treated also with hyperthermia.

When the per cent success in surviving patients is plotted against the period of observation (Fig.2), local control features among different protocols can be appreciated. In particular, the lesions treated with the two basic protocols (MDF and MDF+MIS) did better than those treated with conventional fractionation. When hyperthermia was also employed, the local control rate was even better and remained at the same level during the period of observation, thus suggesting that recurrence is rarely a cause of death.

3. TOXICITY

Only in 3 out of 22 patients was some mild peripheral neuropathy observed. All these patients recovered completely within 2 months after treatment. A mild immediate nausea, easily controlled by metoclopramide, was present in 80% of the patients. In 4 out of 6 patients treated with misonidazole and irradiated with two cross-firing portals, an early and heavy oropharyngeal

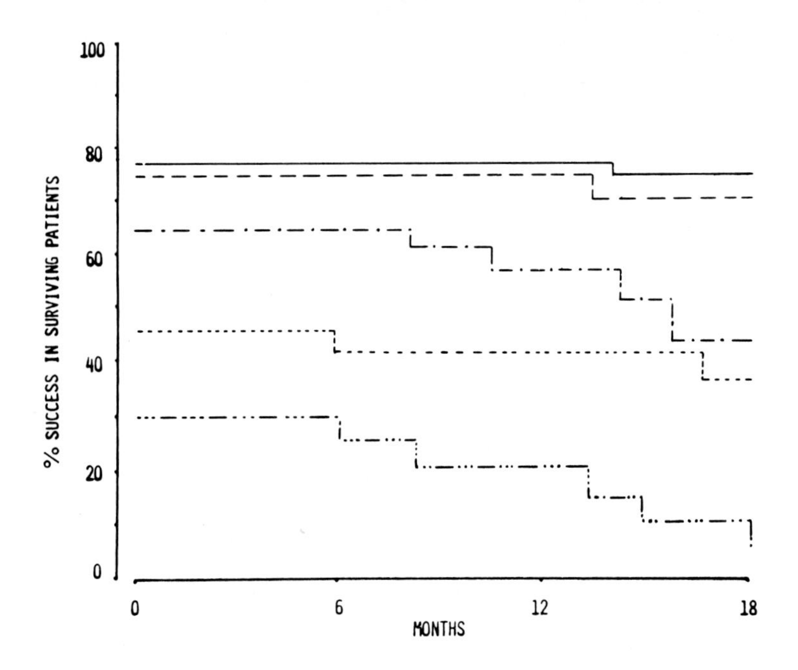

FIG.2. Per cent local successes in surviving patients following different comparative treatment schedules (treatments: conventional fractionation —···—··· ; MDF alone ······· ; MDF+MIS —·—·— ; MDF+HT — — — ; MDF+HT+MIS ———).

TABLE II. RESULTS OF MULTIMODALITY TREATMENTS [a]

Treatment [b]	Complete response at end of X-ray therapy		Complete response at 12 months		Complete response at 18 months		Crude complete response at 18 months	
1. MDF	16/35	0.46	8/19	0.42	4/11	0.36	4/35	0.11
2. MDF+HT	22/29	0.76	12/16	0.75	7/10	0.70	7/29	0.24
3. MDF+MIS	16/25	0.64	8/14	0.57	4/9	0.44	4/25	0.16
4. MDF+HT+MIS	17/22	0.77	10/13	0.77	6/8	0.75	6/22	0.27
5. Conventional	18/60	0.30	6/30	0.20	1/18	0.05	1/60	0.02

[a] Statistical significances:
End of treatment – 1 versus 2, 1 versus 4: $p < 0.05$;
2 versus 5, 3 versus 5, 4 versus 5: $0.05 > p > 0.01$
At 12 months – 2 versus 5, 4 versus 5: $p < 0.05$.
At 18 months – no significant difference ($p = 0.2$) between any pairs of groups.

[b] MDF – multiple daily fractionation of radiation therapy; HT – hyperthermia; MIS – misonidazole.

mucositis was observed. No increased skin radiation-reaction was observed when hyperthermia was also delivered.

REFERENCES

[1] ARCANGELI, G., MAURO, F., MORELLI, D., NERVI, C.: Multiple daily fractionation in radiotherapy: Biological rationale and preliminary clinical experiences, Eur. J. Cancer 15 (1977) 1077.
[2] ARCANGELI, G., MAURO, F.: "Multiple daily fractionation (MDF) radiotherapy: Clinical results", in Biological Bases and Clinical Implications of Tumor Radioresistance (Proc. 2nd Int. Symp., Rome, 1980: FLETCHER, G.H., NERVI, C., WITHERS, H.R., ARCANGELI, G., MAURO, F., TAPLEY, N.DuV., Eds), Masson Publishing USA Inc., New York (1981, in press).
[3] OVERGAARD, J.: "Effect of fractionated hyperthermia and radiation on an experimental tumor and its surrounding skin", in Hyperthermia in Radiation Oncology (Proc. 1st Mtg European Group, Cambridge, UK, 1979: ARCANGELI, G., MAURO, F., Eds) Masson Italia, Milan (1980) 241-245.
[4] STEWART, F.A., DENEKAMP, J.: The therapeutic advantage of combined heat and X-rays on a mouse fibrosarcoma, Br. J. Radiol. 51 (1978) 307.
[5] ARCANGELI, G., CIVIDALLI, A., DE VITA, R., LOVISOLO, G., MAURO, F., PARDINI, M.C.: Temperature and sequence in thermoradiotherapy: Biological evidence in vivo, Bull. Cancer (1981, in press).
[6] OVERGAARD, J.: Effect of misonidazole and hyperthermia on the radiosensibility of a CH3 mouse mammary carcinoma and its surrounding normal tissue, Br. J. Cancer 41 (1980) 10.
[7] DISCHE, S., SAUNDERS, M.I., LEE, M.E., ADAMS, G.E., FLOCKHART, I.R.: Clinical testing of the radiosensitizer Ro-07-0582: Experience with multiple doses, Br. J. Cancer 35 (1977) 567.
[8] ASQUITH, J.C., WATTS, M.E., PATEL, K.B., SMITHEN, C.E., ADAMS, G.E.: Electron-affinic sensitization, V. Radiosensitization of hypoxic bacterial and mammalian cells in vitro by some nitroimidazoles and nitropyrazoles, Radiat. Res. 60 (1974) 108.

Progress in Radio-Oncology II, edited by K. H. Kärcher et al., Raven Press, New York © 1982.

New Types of Fractionation for Optimization of Combinations of Radiotherapy and Chemotherapy

M. TUBIANA, R. ARRIAGADA, J.M. COSSET
Institut Gustave–Roussy,
Villejuif,
France

Abstract

Concomitant administration of cytotoxic drugs and radiation therapy generally enhances the toxic effect on normal tissues without notably improving the control of the tumour. A time interval of one week between the two treatment modalities is, for most drugs, sufficient to reduce the side effects. A delay of chemotherapy until after the completion of radiotherapy allows the occult metastases to increase in size while a similar delay of radiotherapy is also detrimental, as drugs are often not effective on bulky tumours. Taking both these disadvantages into account, a treatment protocol is proposed in which chemotherapy and radiotherapy are given alternately without undue delay of either. Chemotherapy is started with the scheduling of one cycle every month. Radiotherapy is given one week after interrupting chemotherapy and continued until one week before beginning a further cycle of chemotherapy, and so on until completion of radiotherapy. Such a split course of radiotherapy should have an effect on a tumour comparable to that of a conventional fractionation. This protocol has been used on five patients with non-Hodgkin lymphoma and ten patients with oat-cell carcinoma of the lung. The tolerance is satisfactory, but follow-up is too short to allow of assessment of late effects or survival.

Combinations of radiation and drug therapy have been used in human treatment for over 2 decades. Two main conclusions emerge from recent comprehensive reviews of the published data (1, 2, 3).

Firstly, concomitant administration of drug and radiation generally enhances the toxic effects on normal tissues without much improving the control of the tumor. Interaction between the 2 modalities mainly adds to the complications and analysis of the published data show that the therapeutic gain of such concomitant administration appears to be very small if any; often it is more detrimental than beneficial. In particular the hopes of enhancing the effect on the tumor by a so-called synchronisation of the neoplastic cells (4) have not been fulfilled in clinical practice.

Secondly a short gap between the 2 treatment modalities generally reduces the cumulative toxic effects on normal tissues. For most drugs a time interval of one week between their administration and radiation therapy is sufficient (1, 3).

In short, sequential appears more promising than concomitant administration. In sequential regimens the schedule is of great importance. Early administration of both radiotherapy and chemotherapy is advisable for many reasons. Firstly, delaying the start of chemotherapy until after the completion of radiation therapy allows occult metastases to grow and to become less sensitive to chemotherapy. For example if the tumor doubling time is equal to one month, a delay of 2 months allows the metastases to become four fold bigger ; as the size is a very critical parameter, this could significantly reduce the effectiveness of adjuvant therapy. However delaying radiotherapy for a long period is also risky, as cytotoxic drugs generally are not effective on bulky tumors.

Early irradiation is also advantageous in that it prevents the proliferation of drug-resistant cells. Spontaneous mutation to single drug resistance is relatively frequent. In addition to selection of drug resistant cells which were initially present, mutation to drug resistance may occur during treatment. Whatever the mechanism, it is commonly observed, both in man and rodent, that initially drug sensitive tumors become progressively less responsive and ultimately fail to respond during continuous treatment (5).

One of the aims of polychemotherapy was to avoid the consequences of the presence of drug resistant cells. However resistance to as many as 6 separate drugs has been shown to occur in treatment of human or experimental tumors (5). Moreover when many drugs are used in combination, the individual drug doses must be reduced in order to avoid cumulative toxicity, and some of the drugs are only toxic to vital normal cells without contributing to tumor cell kill. This underlines the usefulness of irradiation, an agent without cross resistance with drugs and for which the cross toxicity is minimal.

The mathematical model constructed by Goldie and Coldman (6) shows that when 2 agents are used, scheduling is critical because a delay in the administration of one of them increases the probability of development of a resistant tumor cell line.

METHOD AND PRELIMINARY RESULTS

In order to reconcile the needs for early administration of both agents and for sequential administration we have proposed (3, 7) a new treatment schedule which alternates radiotherapy and chemotherapy.

Chemotherapy is initiated without any delay and the interval between the successive cycles is, as usual, about one month. Radiotherapy starts one week after completion of the first cycle of chemotherapy and is continued for about 2 weeks. It is then interrupted one week prior to the second cycle of chemotherapy and is resumed one week after its completion.

As in 2 weeks it is impossible to deliver more than 25 Gy the radiotherapy course should be carried out as a split course : for example twice 25 Gy in 2 weeks separated by a time interval of about 2 weeks. The TDF of such a treatment is about equal to that of a continuous irradiation delivering 50 Gy in 5 weeks.

Such an alternating protocol avoids concomitant administration of drugs and radiation and does not alter the rhythm of chemotherapy. It has been used at Villejuif in 2 feasibility trials in order to assess both the tolerance of normal tissue and the effects on the tumor.

For anaplastic oat-cell carcinoma of the lung, each chemotherapy cycle includes Adriamycin : 40 mg/m^2, VP 16 : 10 mg/m^2, Methotrexate : 10 mg/kg, Cyclophosphamide : 300 mg/m^2 x 4. The duration of this cycle is 4 days. One week after completion of the first cycle, the irradiation of the brain starts and 15 Gy are delivered in 10 days. The second cycle of chemotherapy begins 10 days after the last session of irradiation. One week after its completion a second series of radiotherapy is given to 2 target volumes : the brain and the tumor mass and mediastinum. 15 Gy in 10 days are given to each volume. Between the second and the third cycle of chemotherapy and between the third and the fourth the irradiation of the mediastinum and of the tumor mass is continued. Thereafter no irradiation is performed between the chemotherapy cycles. Therefore the total dose of radiotherapy is 30 Gy to the brain in 2 series of sessions and 45 Gy to the tumor mass and the mediastinum in 3 series of irradiation. Twelve patients have been treated with this protocol during the past fourteen months. All of them have completed their treatment. The follow-up is too short to allow an evaluation of the therapeutic results; however it can be stated that the immediate tolerance is good and that the early effects on the tumor are promising.

In non-Hodgkin lymphoma of unfavourable histological types, we have undertaken under the auspices of the Radiotherapy-Chemotherapy group of the EORTC, a feasibility trial, comparing a conventional radiotherapy-chemotherapy combination with this new alternated chemotherapy-radiotherapy regimen for stage II patients. Treatment is started with chemotherapy (Adriamycin, Oncovin, Cyclophosphamide, Prednisone). No irradiation is performed between the first and the second cycle in order to allow an assessment of the effect of this polychemotherapy on neoplastic tissues.

Ten days after the end of the second cycle an irradiation of the areas which were initially involved is carried out and 15 Gy are delivered in 12 days. Ten days after completion of this first series of radiotherapy a third cycle is administered. Radiotherapy is performed again between the third and the fourth chemotherapy cycles in order to deliver 30 Gy to all irradiated volumes. A booster irradiation of 15 Gy might be given to bulky masses after the fourth cycle. Chemotherapy is continued up to 8 cycles. Three patients have so far completed this alternating regimen. The immediate tolerance to treatment has been excellent and complete remission was obtained in all of them.

DISCUSSION

Sequential combination of radiotherapy and chemotherapy constitutes one of the few promising avenues for research in current cancer therapy. However despite a large number of studies the optimal scheduling remains to be found. Conventional scheduling in which a full course of radiotherapy, or chemotherapy, is completed before the administration of the other modality has the inconvenience of a long delay for one of the two modalities.

Concomitant administration avoids any delay but enhances the toxic effects on normal tissues. In contrast to what was hoped it was found impossible to take advantage of the synchronizing effect of radiation or drugs resulting from preferential killing and mitotic delay. After exposure to both ionizing radiation and cytotoxic drugs the cell survival percentage depends upon the phase of the cell cycle ; moreover the progress of surviving proliferating cells throughout the cell cycle is also temporarily blocked and the duration of the delay depends upon the phase of the cell cycle in which the cell was at the time of exposure. The change of the age distribution of the surviving cells resulting from these phenomena increases the proportion of cells in one of the phases of the cell cycle, thereby producing a cycling variation of the sensitivity of the surviving cells. It was suggested that this cell reassortment, or synchronization, could be exploited in human cancer therapy. However a critical survey of experimental data showed that it would be extremely difficult to reach such a goal (4) and unfortunately this prediction was confirmed by the discouraging results obtained in clinical trials.

While cell reassortment is the main factor which has to be considered in concomitant administration, repopulation becomes prominent for sequential administration. Acceleration of cell proliferation is observed after chemotherapy and irradiation in both tumors and normal tissues. In normal tissues the lapse of time between drug administration, or irradiation, and the start of regeneration is relatively short ; for example it is of only a few hours for hemopoietic stem cells. In some experimental tumors repopulation is observed during fractionated irradiation. However the proportion of proliferating cells decreases in most human tumors during a course of radiotherapy (8) while it increases in normal tissues due to powerful homeostatic mechanisms. This partly explains the relatively low efficacy of concomitant administration as most drugs are more effective on cycling cells. On the other hand, and this is the important feature for sequential administration, the duration of repopulation is generally shorter in normal tissues than in tumors. For example the number of hemopoietic stem cells and their proliferative rate return to normal in about one week after completion of chemotherapy or radiotherapy whereas in most experimental tumors the proliferative fraction remains increased for about 2 weeks(4).The few data available for human metastases in lung or skin suggest that the phase of accelerated growth rate has a duration of a few weeks or one month and is longer than the regeneration time of bone marrow or of other critical tissues. The differences in timing of repopulation in tumors and normal tissues provide opportunities for a differential effect and with a proper time interval it might be possible to hit preferentially tumor cells while they are still rapidly proliferating (4).

Furthermore the critical tissues generally are different for radiotherapy and chemotherapy. For most polychemotherapy bone marrow failure is the main complication whereas for radiotherapy late effects on connective tissue surrounding the tumor are the limiting factor. Alternating two different treatment modalities with minimal cross toxicity gives more time for recovery of critical normal tissue for each modality.

These theoretical considerations have been recently confirmed by an interesting series of experiments reported by Looney et al. (9). These authors studied a rat hepatoma for which no cure was achieved with either radiotherapy or chemotherapy given alone. They obtained a tumor cure rate of 60 % when 3 series of combined radiation (15 Gy) and Cyclophosphamide (150 mg/kg) were given sequentially and the time between modalities held constant at 7 days. The time interval between the 2 modalities and between successive sequences was relatively critical. For example a time interval of 7 days between radiation and Cyclophosphamide was the most effective in controlling tumor growth with least host toxicity. The cure rate was reduced to 10 % when the time between the first sequence of Cyclophosphamide and radiation and the following one was increased from 7 to 25 days. This reduction in cure rate was probably due to greater tumor repopulation.

It is difficult to extrapolate from rodent tumors to human tumors. This is why critical investigation of the optimal tumor interval should continue. The reactions may vary from tumor to tumor; therefore it is probably preferable to base scheduling on the study of the effects on normal tissues. Further clinical research will show whether the time interval chosen in this first series of patients is the optimum ; it at least has the advantage of not altering the usual chemotherapy schedule and of allowing only a short period for possible tumor repopulation.

REFERENCES

(1) "Combined effects of chemotherapy and radiotherapy on normal tissue tolerance", in Frontiers of Radiation Therapy and Oncology, vol. 13, VAETH J.M., Basel, New York, Karger (1979).
(2) Conference on Combined Modalities Chemotherapy-Chemotherapy, Int.J. Oncol. 5 (1979) 1139.
(3) TUBIANA, M., Les associations radiothérapie-chimiothérapie, J. Eur. de Radioth. 1 (1980) 107.
(4) TUBIANA, M., FRINDEL, E., VASSORT, F., "Critical Survey of Experimental Data on in vivo Synchronization by Hydroxyurea, in Recent Results in Cancer Research, vol 52, GRUDMANN, E., GROSS, R., Berlin, New York, Springer Verlag (1975) 187.
(5) SCHABEL, F.M., SKIPPER, H.E., TRADER, M.W., LASTER, W.R., CORBETT, T.H., GRISWOLD, D.P., "Concept for Controlling Drug Resistant Tumor Cells", in Breast Cancer, Experimental and Clinical Aspects, MOURIDSEN, H.T., PALSHOF, T., Pergamon Press (1980) 199.
(6) GOLDIE, J.H., CODMAN, A.J., a mathematical model for relating the drug sensitivity of tumors to their spontaneous mutation rate. Cancer Treat. Rep. 63 (1979) 1727.
(7) TUBIANA, M., Les associations radiothérapie-chimiothérapie, Bull. Cancer (Paris) 68 (1981) 109.
(8) COURDI, A., TUBIANA, M., CHAVAUDRA, N., MALAISE, E.P., LE FUR, R., Changes in labeling indices of human tumors after irradiation. Int. J. Radiat. Oncology Biol. Phys. 6 (1980) 1639.
(9) LOONEY, W.B., RITENOUR, E.R., HOPKINS, H.A., solid tumor models for the assessment of different treatment modalities : XVI sequential combined modality (cyclophosphamide-radiation) therapy, Cancer 47 (1981) 860.

Hyperthermia

Progress in Radio-Oncology II, edited by K. H. Kärcher et al., Raven Press, New York © 1982.

Thermotolerance

Recent studies on animal tissue of relevance to clinical practice

S.B. FIELD, M.P. LAW, R.G. AHIER, C.C. MORRIS
Medical Research Council Cyclotron Unit,
Hammersmith Hospital,
London,
United Kingdom

Abstract

The induction of thermotolerance by various priming heat treatments has been measured in the skin, intestine and cartilage of rodents. Both the magnitude and time of occurrence of thermotolerance were different in the three tissues. For each tissue, as the priming treatment was reduced, the maximum thermotolerance was also reduced and it occurred earlier. Maximal thermotolerance induced by a single heat treatment was not further increased by subsequent treatment. The time course of thermotolerance after a two priming treatment was identical to that after one. Tolerance to heat used to enhance damage by X-rays was induced in skin providing the combined treatment was heat immediately followed by radiation. The magnitude of this form of thermotolerance was much less than for direct heat damage. If radiation was given first in a combined treatment, no thermal tolerance was observed. These results are consistent with the response of skin to fractionated combined heat and X-ray treatments.

INTRODUCTION

In the clinic hyperthermia is normally given in several fractions, as is the case with radiotherapy. Therefore, in order to plan an effective treatment regime, it is necessary to understand the processes occurring between heat fractions.

There are two important factors to be considered. The first is that repair of heat damage occurs, analogous to repair of sublethal injury after ionizing radiation. The extent of repair is considerable so that no effect of a "sub threshold" heat treatment may be apparent by the time of a subsequent treatment. The second factor is that a heat treatment may induce a transient resistance to additional hyperthermia. This latter phenomenon is known as thermotolerance. Its effect may be very large and must be taken into account in the optimum design of a clinical course of fractionated hyperthermia. There are two basic forms of thermotolerance, known as types I and II [1,2]. Type I results from continuous heating at temperatures below about 43°C. Type II results from a prior heating with subsequent incubation at normal body temperature. Both types have been well documented *in vitro* [3] but few results have been reported for tissues *in situ*.

In addition to causing direct injury to cells and tissues, hyperthermia may also enhance the effects of other anti-cancer agents such as ionizing radiations or chemotherapy. There have been reports of induced resistance to these hyperthermal effects also. The present paper is a report of thermotolerance to heat treatments which either cause direct thermal damage or enhance the effects of X-rays.

DIRECT HEAT DAMAGE

Three organs in rodents have been used at the Cyclotron Unit to investigate induced tolerance to direct thermal injury, i.e. intestine and skin in the mouse, and tail of the rat.

Hyperthermia of the mouse intestine was achieved by immersing an exteriorized loop of gut in a thermostatically controlled bath of Krebs-Ringer solution. At 24 hours after heating crypt survival was assayed in the jejunum. This approach enabled thermal survival curves to be derived following single heat treatments between 42 and 44°C [4]. During continuous heating at 42.5°C or lower, resistance indicated by a reduction in slope of the survival curve developed after 60-80 minutes (Thermotolerance I). At temperatures of 43°C or above this form of tolerance was not apparent. In a second series of experiments a priming treatment was followed at various time intervals by a test treatment (figure 1). A transient resistance to the second treatment was induced, i.e. thermotolerance II . After a priming treatment of 42°C for 1 hour, the peak of resistance occurred between 5 and 10 hours later. At the time of maximum tolerance the slope of the survival curve was increased by a factor of approximately 14 compared to a single treatment [4]. After a priming treatment of 41°C for 1 hour, maximum tolerance had already occurred by the end of the priming treatment but was reduced in magnitude. The time course of development is thus dependent on the priming treatment, as is the rate of decay, as seen in figure 1.

The mouse ear was heated by direct immersion in a water bath. Only thermotolerance II, i.e. to a second (test) treatment, was investigated. The endpoint used was a 50% probability of causing necrosis. The first sign of necrosis normally occurred within a few days, observations being taken for 10 days after hyperthermia. Thermal tolerance was assessed directly by an increase in the length of heating of the test treatment relative to a single treatment to produce the same effect [5]. Thermal tolerance was induced by very small heat treatments, it being significant after only 2 minutes at 43.5°C. The maximum effect obtained was an increase in heating time of the test treatment by more than a factor of 2. After a priming treatment of 43°C for 20 minutes the peak occurred after about 1 day. The peak tolerance was again dependent on the priming treatment, both in magnitude and timing (Figure 1).

Baby rat tails were heated by immersion of the whole tail in a water bath when the animals were 7 days old. Four weeks after treatment the tails were radiographed and the number of remaining vertebrae counted. Reduction in the number of vertebrae was dependent on the duration of heating and the endpoint used was the treatment to reduce the number to 17. In other respects the experimental design was similar to that with the ear studies. It is seen in figure 1 that thermotolerance in the tail is extremely large, reaching a factor of almost 5 after a priming treatment of 43°C for 30 minutes. As with skin and gut, smaller priming treatment leads to an earlier onset of thermotolerance and its maximum value is reduced.

For the three normal tissue systems used the decay of thermotolerance takes several days and may not be complete after 1 week. The curves labelled

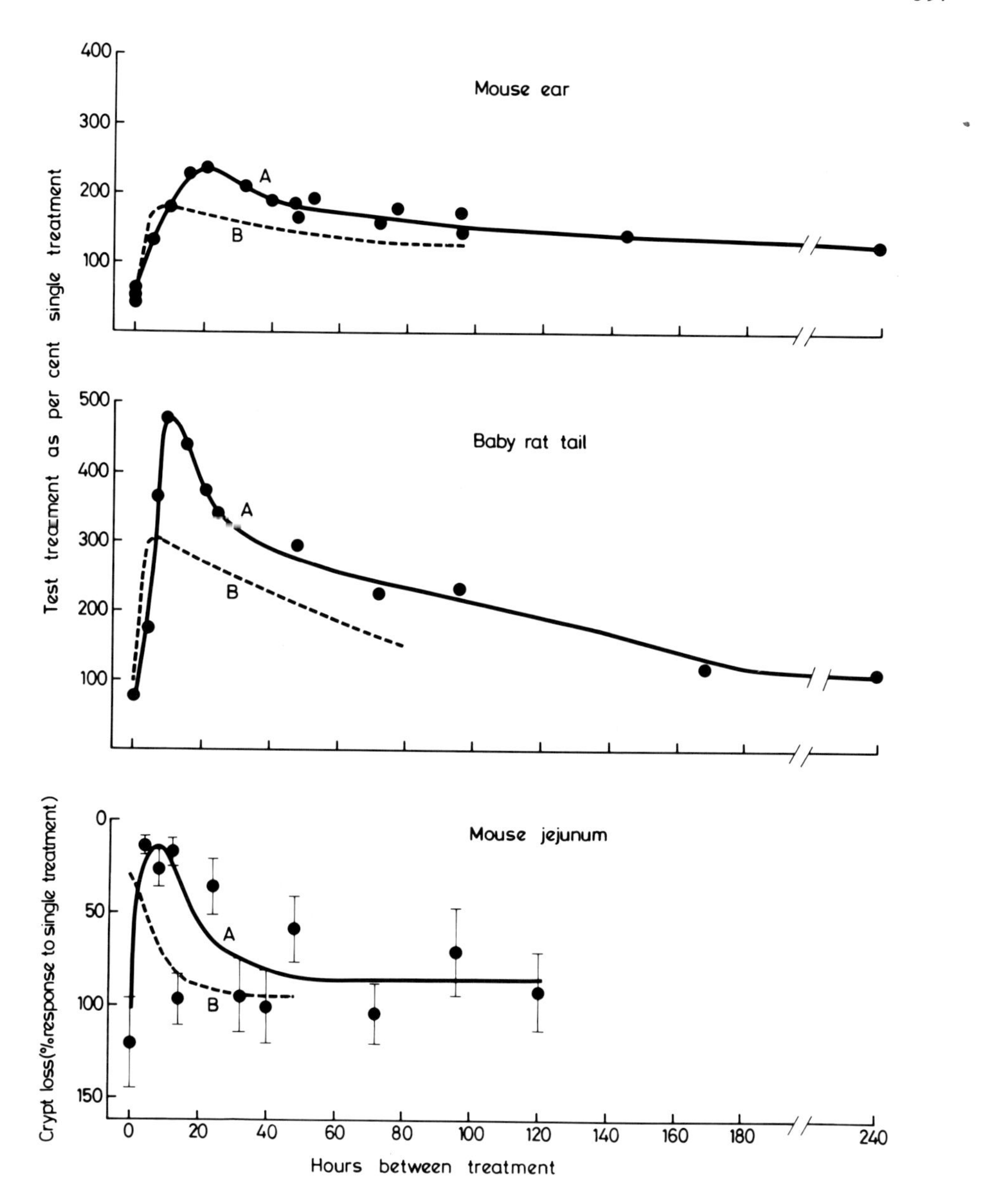

FIG.1. Effect of a priming hyperthermic treatment given at varying times before a test treatment. 100% on the ordinate represents the effect of the test treatment alone. For mouse ear and baby rat tail, thermal tolerance is expressed as the percentage increase in time of heating of the test treatment required to cause a given level of injury. For mouse jejunum, tolerance is expressed as residual damage for a fixed test treatment. The curves labelled A are for priming treatments of 43.5°C for 20 min to mouse ear; 43.0°C for 30 min to rat tail; 42°C for 60 min to mouse jejunum, and are approximately equivalent [6]. The curves labelled B are for priming treatments of 43.5°C for 7 min to mouse ear (interpolated between the 5 and 10 min results); 43.0°C for 10 min to rat tail and 41.0°C for 60 min to mouse jejunum, and are also approximately equivalent priming treatments. Data points are only given for curves A.

A in figure 1 are thought to be after equivalently sized priming treatments. The same is true for curves labelled B [6]. It is seen that the peak value of tolerance varies from tissue to tissue (although it is not possible to compare directly the intestinal results with the other two because of the different experimental design). The time of occurrence of maximum tolerance generally occurs between 6 and 24 hours, varying from tissue to tissue. It normally occurs earlier with cells in vitro.

It is important to question how thermotolerance is affected by repeated hyperthermal treatments. Evidence from in vitro studies indicates that a peak is reached after a few fractions [7]. Fig. 2 shows the results of an experiment using the mouse ear skin, also aimed at answering this question. A priming treatment (43.5°C for 20 minutes) was given to produce maximum tolerance at 22 hours (full line in figure 2). At 22 hours a second treatment (43.5°C for 70 minutes) was given. The second treatment was 40% greater than that which alone would produce a 50% probability of necrosis, but because of the induced tolerance there was, as expected, no visible reaction. A third treatment was given at varying times after the second, and the heating times required to produce necrosis were estimated. The results are indicated by the points in figure 2. Thermal tolerance produced by the second treatment was identical to that from the first. It appears therefore that once the maximum tolerance that can be induced by a single priming treatment is reached, it cannot be increased by successive treatment. Also the time course of thermotolerance was the same for one or two priming treatments. Whether these findings are the same for additional treatments is not yet known.

COMBINED HEAT AND X-RAYS

Fractionation of combined heat and X-rays is already used clinically Little information on thermotolerance for these combined effects exists, although the indications are that less thermotolerance of this type is induced than for heat given alone [2]. However, the phenomenon is important for the combined treatment as illustrated by experiments on the mouse ear. The radiation response of the ear was measured for the combined treatment where heat was given either immediately before X-rays (HX), or after X-rays (XH). Thermotolerance was considerably less for (HX) than for heat alone and its time course may be different [7]. However, no thermal tolerance was observed with the combination (XH) (figure 3). These results are consistent with a fractionation experiment in which a single treatment of heat and X-rays was compared with 2, 5 or 10 fractions of the combined treatment [7]. When the thermal enhancement ratio (TER) for each treatment regime was plotted as a function of dose per fraction, it was seen that for the combination (HX) the TER decreased with decreasing dose per fraction, i.e. with increasing numbers of fractions, presumably due to thermotolerance (figure 4). In contrast, when the combination XH was used there was no significant reduction in TER, consistent with there being no thermotolerance.

DISCUSSION AND CONCLUSION

It is clear from the results presented here and in the literature that thermotolerance may be a very large effect. It may vary significantly in magnitude and time course from one tissue to another.

The mechanism of development of thermotolerance is not clear, but it is known to be influenced by a variety of factors. These include the cell's nutritional status, tonicity, pH, temperature during the development period,

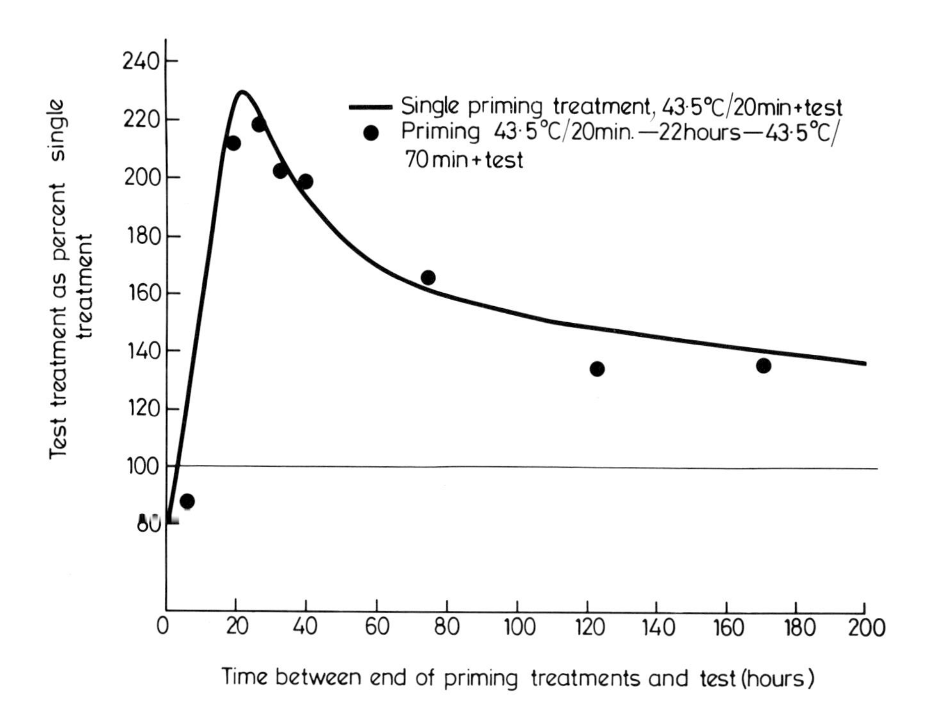

FIG.2. Thermotolerance for mouse ear skin after a single or two priming treatments. The solid line represents the relative increase in the duration of the test treatment at 43.5 °C after a single priming hyperthermia of 43.5 °C for 20 min (curve A in appropriate curve in figure 1). The data points represent the response after two priming treatments, i.e. 43.5 °C for 20 min followed by 43.5 °C for 70 min 22 hours later. The test treatment was given at varying times after the end of the second priming treatment.

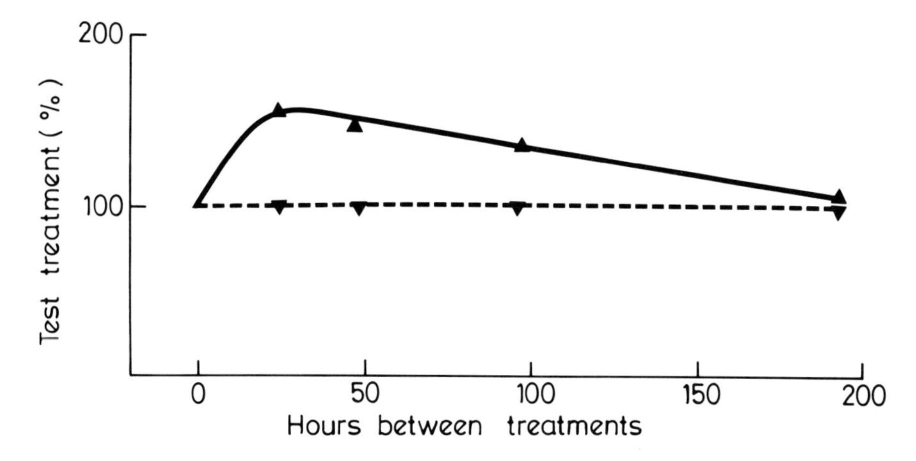

FIG.3. Percentage increase in heating time required to cause a given enhancement of radiation damage to mouse ear skin resulting from a priming heat treatment of 43.5°C for 30 min given at varying times earlier. Δ represents the combination heat followed by X-radiation (HX), ∇ represents the combination XH.

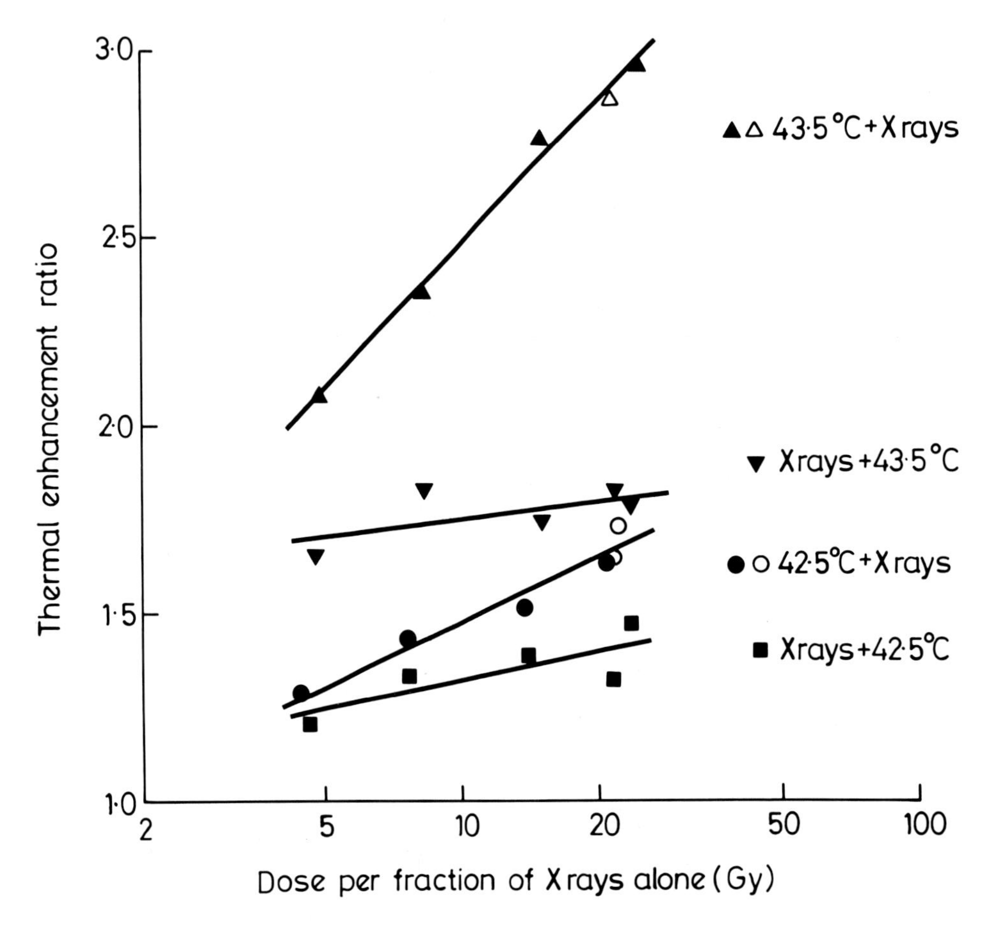

FIG.4. Thermal enhancement ratio as a function of dose per fraction of X-rays alone for mouse ear skin. The largest doses represent single treatments (X-rays alone compared with the combination of heat and X-rays): next largest 2 fractions, 5 fractions and 10 fractions, respectively. For the combination HX there is a marked reduction in TER with increasing number of fractions, presumably due to thermotolerance: for the combination XH there is little change in TER with fractionation (taken from Law [8]).

protein synthesis and presence of polyamines. These factors may all vary between different normal tissues and tumours so that a detailed study of the phenomenon is required if thermotolerance is to become an aid to hyperthermia therapy, rather than a hindrance.

REFERENCES

[1] BAUER, K.D. and HENLE, K.J., Arrhenius analysis of heat survival curves from normal and thermotolerant CHO cells. Radiat. Res. 78 (1979) 251-263.

[2] FIELD, S.B. and ANDERSON, R.L. Thermotolerance: a review of observations and possible mechanisms. J. Nat. Cancer Inst. (1981), in press.

[3] HENLE, K.J. and DETHLEFSEN, L.A. Heat fractionation and thermotolerance: a review. Cancer Research 38 (1978) 1843-1851.

[4] HUME, S.P. and MARIGOLD, J.C.L. Transient, heat-induced thermal resistance in the small intestine of mouse. Radiat. Res. 82 (1980) 526-535.

[5] LAW, M.P., COULTAS, P.C. and FIELD, S.B. Induced thermal tolerance in the mouse ear. Brit. J. Radiol. 52 (1979) 308-314.

[6] LAW, M.P. The induction of thermal resistance in the ear of the mouse by heating at temperatures ranging from 41.5 to 45.5°C. Radiat. Res. 85 (1981) 126-134.

[7] LAW, M.P., AHIER, R.G. and FIELD, S.B. The effect of prior heat treatment on the thermal enhancement of radiation damage in the mouse ear. Brit. J. Radiol. 52 (1979) 315-321.

[8] LAW, M.P. Some effects of fractionation on the response of the mouse ear to combined heat and X rays. Radiat. Res. 80 (1979) 360-368.

Progress in Radio-Oncology II, edited by K. H. Kärcher et al., Raven Press, New York © 1982.

Fractionated Thermo-Radio-Therapy of Solid Experimental Mouse Tumours*

F. DIETZEL, B. GRUNDEI
Radiologisches Institut der Städtischen Krankenanstalten,
Bayreuth

G. LINHART
Abteilung Strahlenbiologie und Strahlenschutz,
Zentrum für Radiologie,
Universität Giessen,
Giessen

Federal Republic of Germany

Abstract

Solid Ehrlich neck tumours of 1 ml size in female NMRI-mice were locally treated with 8 fr X 5 Gy of X-ray irradiation alone or in combination with an immediately following 3 min local 2.4 GHz microwave hyperthermia to 42°C within the tumour. The intervals between the fractions of thermo-radio-therapy (TRT) were 2 days each. Compared with a 1.6 times higher dose of fractionated isolated X-ray irradiation (8 Gy per fraction), which did not result in local tumour control on day 100, but did cause severe local skin and myelon reaction (extreme desquamation, deep ulceration, myelopathia), the combined TRT modality resulted in 13.2% local tumour control without defect of surrounding normal tissue. Thus the therapeutic gain factor is at least 1.6. Furthermore, in experiments with intermittent hyperthermia (only 1 hyperthermia treatment per week), it could be demonstrated that it is not necessary to heat up the tumour with ***each*** fraction. One heat fraction per week combined with three X-ray fractions per week resulted in a clear therapeutic gain with respect to local tumour control.

1. INTRODUCTION

Most of the work concerned with the effect of a combination of hyperthermia and X-irradiation on solid experimental tumours in vivo use high single doses of radiation of about the dose for control of 50% of the tumours (TCD 50) [1,2]. Nearly all experiments administering hyperthermia and X-rays simultaneously, or at least within a short time interval, result in thermal enhancement ratios (TER) of 1.2 to 2.0, depending on the tumour system and hyperthermia and radiation modalities.

Clinically such high single radiation doses of 20-40 Gy are not feasible. Unfortunately, there have been only few experiments [2,3] specifying the TER of fractionated thermo-radio-therapy (TRT).

In particular, statements about the therapeutic gain factor (TGF) of fractionated TRT are contradictory. Whereas Overgaard [2] has found a definite enhancement with fractionated TRT also,

* Supported by grant Di 199/5 from the Deutsche Forschungsgemeinschaft, Bonn, Federal Republic of Germany.

experiments by Stewart and Denekamp [4] lead to the conclusion that any TGF is lost with fractionation. Two questions appear important:

Is it possible to prove a therapeutic gain with fractionated TRT?

Is it necessary to heat with each X-ray fraction to get a full enhancement?

The experiments described below seek to answer them.

2. MATERIAL AND METHODS

2.1. Tumour System

The tumour system used is the Ehrlich neck tumour in NMRI mice, which we have used in experiments with hyperthermia for many years [1,3]. We have already obtained results with high single-dose irradiations with this system. Hence it is possible to compare the results of single-dose TRT with the results of fractionated TRT.

For these experiments, 100 mice served as an untreated control group.

2.2. Hyperthermia Modalities

In our experiments we apply local short-term 2.4 GHz microwave hyperthermia immediately following X-irradiation. Within the exposure time of 3 minutes, the temperature in the centre of the tumour reaches 41.5 to 42 °C. The time/temperature profile is triangular (a):

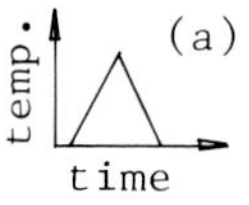

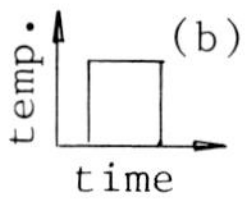

temp. (a) time   temp. (b) time

whereas nearly all other authors try to obtain a rectangular one (b).

A group of 50 mice received fractionated hyperthermia alone to act as a hyperthermia control group.

2.3. Fractionation of Hyperthermia and Radiation

The whole 150 kV X-ray dose of 40 Gy was administered in 8 fractions of 5 Gy each, 3 fractions per week. The overall treatment time is 17 days. Two experimental groups have been compared: in one group, hyperthermia was administered after each X-irradiation, while in the other, hyperthermia was applied only with every third radiation fraction, i.e. an intermittent hyperthermia regimen (Fig.1). When given, hyperthermia was always initiated immediately after X-irradiation.

3. RESULTS

The experimental results are presented in Figs 2-4. Tumour

growth is illustrated in Fig.2, while survival curves are shown in Fig.3. The increase in survival time for 50% of the animals is graphically displayed in Fig.4.

4. DISCUSSION

4.1. Therapeutic Gain

If TRT were to produce qualitatively only the same effects in the tumour and surrounding tissue as higher X-ray doses alone, there would not be any advantage in using TRT.

Commonly, a therapeutic gain is achieved when the damage in the tumour is greater than in surrounding tissue. In nearly all cases, the skin is judged as 'surrounding tissue', using the criterion of moist desquamation dose for 50% of the treatments.

In our experiments, the end-point chosen has not only been the appearance of moist desquamation, but also any degree of radiation damage to skin, spinal cord, and dorsal parts of the lungs, i.e. most of the structures limiting the dose given in clinical radiation therapy. Although tumour is locally under control with 8 fractions of 8 Gy each (Fig.3), the radiation damage appearing afterwards, such as severe ulceration, contractures, as well as radiogenic myelopathia, resulting in paraplegia, finally causes death of the animals (Fig.3).

With fractionated TRT, we have succeeded in obtaining an increase in animal survival rate we could not have obtained with fractionated X-irradiation alone. Fractionated TRT is correlated with diminution of severe side-effects.

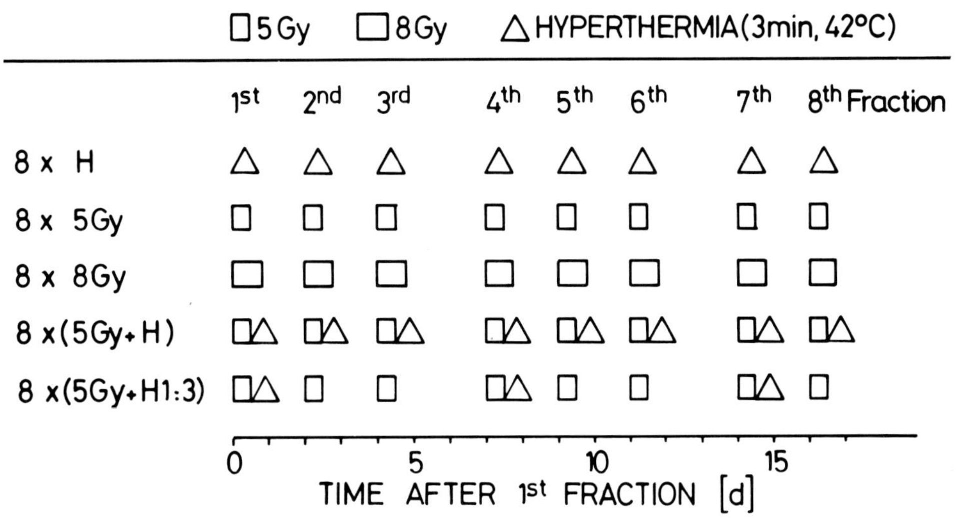

FIG.1. Fractionation modalities. Hyperthermia given alone, X-irradiation given alone at doses of either 5 Gy or 8 Gy, and radiotherapy combined with hyperthermia (5 Gy + H) were administered in 8 fractions over a period of 17 days. With intermittent hyperthermia (5 Gy + H 1:3) one in three X-ray fractions is combined with hyperthermia.

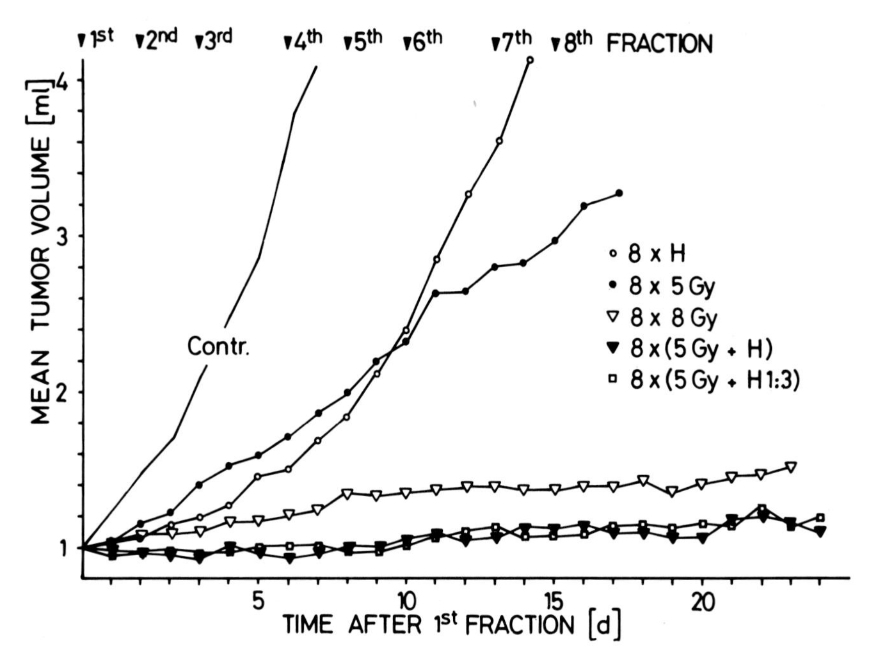

FIG.2. Tumour growth curves. Tumour growth is stopped by 8 fractions of 5 Gy of X-radiation combined with hyperthermia better than with a 1.6 times higher X-ray fractionation alone. With the combined treatment modality, there is no difference between combining each X-ray fraction (5 Gy + H) or one in three X-ray fractions with hyperthermia (5 Gy + H 1:3).

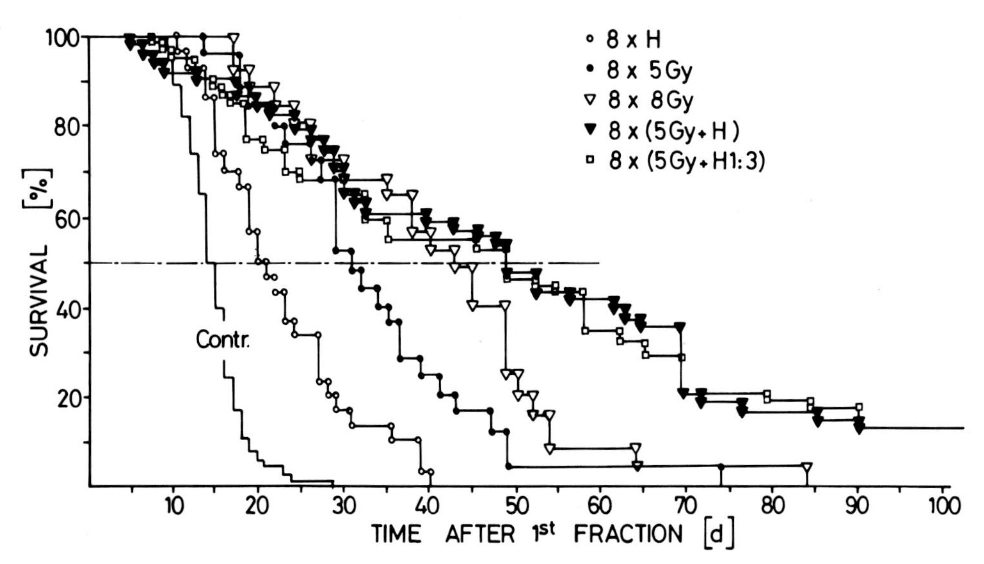

FIG.3. Survival curves. Whereas all animals treated with X-rays or hyperthermia alone died before day 100, 13% of the animals treated with a combined therapy survive. Intermittent hyperthermia (5 Gy + H 1:3), results in the same effects as hyperthermia combined with each X-irradiation.

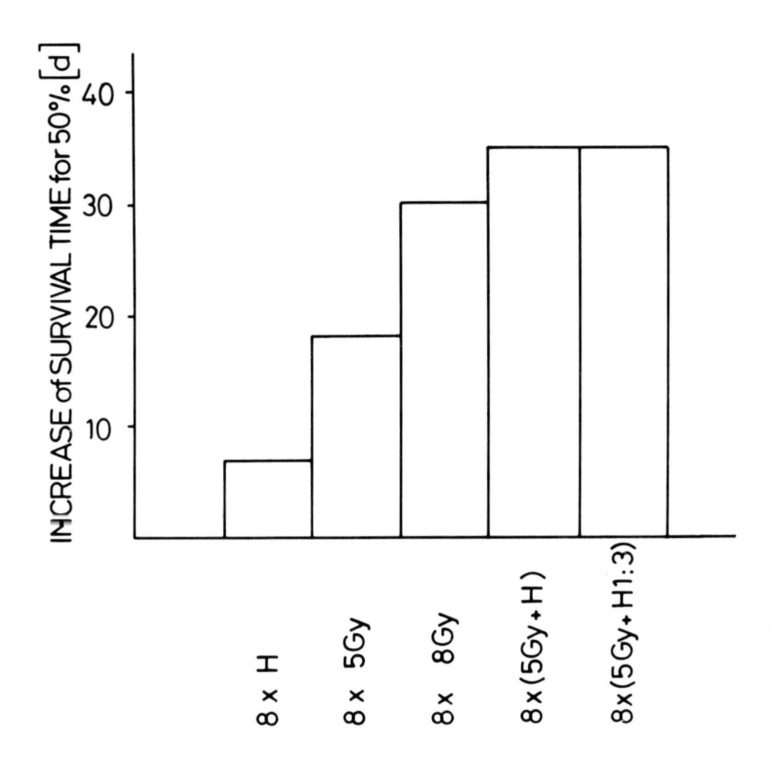

FIG.4. Increase in survival. The increase in the survival time for 50% of the animals (ST 50) with combined therapy exceeds the purely additive effect of hyperthermia (8 X H) and X-irradiation (8 X 5 Gy or 8 X 8 Gy) alone. The effect of a combination of hyperthermia with each X-ray fraction is, however, no greater than the effect of combining one in three X-ray fractions with hyperthermia (intermittent hyperthermia).

For clinical application, the criterion 'cure without side-effects' is of greater value than local tumour control with lethal radiogenic side-effects.

With fractionated TRT the increase of survival time (Fig.4) is greater than 1.6, i.e. the 1.6 times higher X-ray dose is less effective than the lower X-ray dose combined with hyperthermia, both administered with the same fractionation. The enhancement is significantly more than an additive effect of hyperthermia and radiation alone.

When considering cures, the therapeutic gain is even more obvious. Cures without defects only occurred with TRT. Whereas all animals treated with X-rays alone died, 13% of the animals treated with the combined modality survived day 100.

4.2. Intermittent Hyperthermia

The second result of the present experiments is that it is not necessary to combine every X-ray fraction with hyperthermia. Hyperthermia given once a week combined with one in three irradiation fractions produces the same enhancement effect (Figs 2-4).

That intermittent hyperthermia is as effective as hyperthermia combined with every X-ray fraction can be accounted for by thermotolerance. When hyperthermia is administered with every fraction (Fig.2), the thermotolerance reduces the enhancement effect of hyperthermia. Since thermotolerance diminishes with time, the full enhancement effect of hyperthermia is available after a break of one week.

Summarizing, it is not necessary to combine hyperthermia with each radiation fraction; hyperthermia administered once a week is just as effective. This result may be of importance in the practical application of thermo-radio-therapy in the clinic.

REFERENCES

[1] DIETZEL, F.: Thermo-Radio-Therapie, Urban and Schwarzenberg, Munich, Vienna, Baltimore (1978).

[2] DIETZEL, F.: "Fractionated radiotherapy and short-term microwave hyperthermia", Proc. 3rd Int. Symp. Cancer Therapy by Hyperthermia, Drugs and Radiation (Fort Collins, Colorado, 22-26 June 1980), J. Natl. Cancer Inst., Monogr. (1981).

[3] OVERGAARD, J.: "Influence of sequence and interval on the biological response to combined hyperthermia and radiation", Proc. 3rd Int. Symp. Cancer Therapy by Hyperthermia, Drugs and Radiation (Fort Collins, Colorado, 22-26 June 1980), J. Natl. Cancer Inst., Monogr. (1981).

[4] STEWART, F., DENEKAMP, J.: "Why is the therapeutic advantage of combined heat and X-rays lost with fractionation?", Proc. 3rd Int. Symp. Cancer Therapy by Hyperthermia, Drugs and Radiation (Fort Collins, Colorado, 22-26 June 1980), J. Natl. Cancer Inst., Monogr. (1981).

Progress in Radio-Oncology II, edited by K. H.
Kärcher et al., Raven Press, New York © 1982.

Hyperthermia

Clinical results

H.I. BICHER
Henry Ford Hospital,
Detroit, Michigan,
United States of America

Abstract

A large number of patients have now been entered into a phase I/II protocol to examine the effects of fractionated hyperthermia and radiation on tumour response. Included in the study were 11 different histologics with anatomical locations varying between peripheral and superficial metastases to deep-seated, solid tumours. Patients were treated with four fractions of microwave-induced hyperthermia (45.0 ± 0.5°C), each separated by intervals of 72 hours. Treatments were 1.5 hours in duration to a volume encompassing the tumour. Microwaves at frequencies of 915 MHz or 300 MHz were employed. Patients were given a one week rest following the first four treatments, following which a second series of four fractions were administered, again at 72 hour intervals. Each of these fractions consisted of a 400 rad dose of radiation followed within 20 min by hyperthermia (42.5 ± 5°C) for 1.5 hours. To date 121 fields have been treated in 82 patients. New technology allows for the treatment of deep-seated tumours in the neck, chest and pelvis. Follow-up times range from 2 to 19 months. Total regression is seen in 65% of all cases, partial regression in 35% and no response is seen in only 5% of treatments. Six local and four marginal recurrences have been observed. Adverse effects were rare. Based upon our results (above) and those of other investigators, several modifications have been made in the above protocol. Site specific trials are currently in progress to study the feasibility of deep-seated heating with intracavitary antennae as well as to assess tumour response. In addition, a randomized trial to examine the clinical relevance of thermotolerance has been started.

Introduction

The use of hyperthermia as a clinical modality has taken great strides in the past few years as more investigators realized the importance of complete temperature and treatment documentation.

Recent studies [1-7] involving a combination of hyperthermia and x-irradiation have made a serious effort to measure and document the hyperthermia treatments more accurately. In most cases a comparison with radiation alone controls is made. Kim et al.[5] have treated 50 patients with a variety of cutaneous tumors.

Hornback et al.[3] treated 72 patients with advanced cancer using the combined therapy. Of the patients treated with hyperthermia prior to radiation therapy, 53% experienced complete remission of symptoms while in the group of patients treated with heat following radiotherapy, 92% showed complete remission. Again there was no set protocol and the radiation doses varied from 500 to 600 rad per day with total doses from 3000 to 6500 rads.

Manning et al.[6] reported a very limited study combining localized heat and radiation. The response rate for heat+radiation combination was 80-90% compared with 50% response rate for heat-alone and radiation-alone groups. The author suggests a beneficial therapeutic ratio and minimal side effects from the combined treatments.

Another interesting study was reported by Arcangeli et al. [1], in fifteen patients with multiple neck node metastases from head and neck treated with either radiation alone or in combination with hyperthermia. A total of 33 neck nodes were treated, 12 with radiation alone and the rest with the combination.

The radiation schedule resulted in 46% complete responses, which was enhanced to 85% complete responses when combined with hyperthermia.

In a preliminary publication [2] we reported an effective fractionation regime using 45°C regional hyperthermia combined with low dose (1600 rads) x-irradiation, yielding an overall total response rate of 65%. These results are now expanded to include an enlarged series as well as introducing an intracavitary device for the treatment of deep-seated tumors.

The above mentioned clinical studies are both interesting and encouraging. In addition, recent physiological evidence shows a differential "breaking point" in blood flow in tumors as compared to normal tissues which results in dramatic shifts in intratumor pH [8]. These observations may, in part, explain the results of the clinical trial we are reporting here.

Methods

The exact protocol followed has been reported in detail elsewhere [2] (also RTOG protocol #78-06A). Briefly, treatment consisted of 4 fractions of hyperthermia alone followed after a one-week rest by 4 additional fractions of hyperthermia, this time immediately following radiation. All treatments were separated by 72 hours following a Monday-Thursday or Tuesday-Friday pattern. Each hyperthermia treatment was for 1½ hours at the prescribed temperature (45°C alone; 42°C with radiation) and each radiation dose was 400 rad. Therefore, treatment consisted of a total of 8 hyperthermia treatments and 1600 rad over a total period of 5 weeks.

Complete thermometry was performed during every patient treatment employing microthermocouples (100μ). The microthermocouples were implanted in the tumor (whenever possible) and in surrounding or overlying normal tissues. Throughout treatment, temperature readings were taken at 5-minute intervals under "power-off" conditions to eliminate any possible interference artifacts.

Heating was accomplished using either 915 or 300 MHz microwaves delivered with partially dielectric loaded external beam applicators or intracavitary antennae. In all cases air cooling was applied either to the skin (external applicators) or to the jacket of the antenna to minimize normal tissue heating (and hence damage). With the variety of heating equipment available we have been able to heat uniformly externally up to 7cm in depth as well as internally heating the head and neck, mediastinum and pelvis.

Results

At this time 178 patients have been treated at our clinic with a multimodality regime involving hyperthermia administered in multifraction fashion (8 hyperthermia treatments per field). Since many of these patients had multiple tumors, at least 250 tumors have been treated (over 2000 treatment sessions). Not all of them fitted all criteria for inclusion in the specific protocol, but among evaluable results the following can be cited:

> 121 fields (tumors) were treated according to our 8 fraction protocol with 1600 rads in 4 fractions. The final results show almost no toxicity, and a rate of 65% of total responses and 30% partial response.

Further analysis of this series is shown in Tables I-VII.

Table I shows a summary of all the patients treated who completed the entire protocol and were followed up at least two months. Table II provides a breakdown of the summarized data by histology. From this table it is clear that every histological type treated does respond to this therapy.

TABLE I. SUMMARY OF RESULTS ON 82 PATIENTS

Fields treated	121	
Total response	79	65.5%
Partial response	36	29.7%
No response	6	5.0%
Recurrences:		
Local	5	
Marginal	3	
Complications:		
Skin burns	2	(completely healed)
Tongue and pharynx burns	2	(completely healed)
Grand seizure	1	(neck treatment, epileptic patient)

TABLE II. RESULTS BY HISTOLOGY

Histology	No. of fields	Response	Follow-up (months)
Malignant melanoma	19	9 total 7 partial 3 no response	2-14
Malignant lymphoma	8	8 total	2-9
Squamous cell carcinoma	25	9 total 15 partial 1 no response	2-8
Adenocarcinoma	60	48 total 10 partial 2 no response	2-9
Other (transitional cell, basal cell, glioma, sarcoma)	9	5 total	2-11
Summary	121	79 total 36 partial 6 no response	2-14

[a] Total response: No tumor at 2 months follow-up and thereafter.
Partial response: Tumor decreased in size to half or less at 2 months follow-up.

Table III reports the results of our toxicity study employing the intracavitary microwave antenna system. Following 212 treatment sessions of 1½ hours each, the only observed toxicity was one central pneumonitis. Since response is only evaluated after 2 months at this time, only 14 patients are evaluable (Table IV). Even in these patients with deep seated tumors (mediastinum, pelvis) only 14% failed to respond.

Tables V-VII evaluate response to the combined modality in different anatomical locations. In head and neck recurrences, breast and chest wall, and skin tumors only a small percentage (9%, 10% and 3% respectively) failed to respond to combined hyperthermia and radiation while total responses varied from 46% to 76%, yielding our reported average of 65.5% (Table I).

Discussion

As seen in the detailed response breakdown shown in Table II, the hyperthermia-radiation fractionation regime chosen seems to be at least partially successful in a wide variety of tumors. Detailed examination of the data shows essentially no treatment toxicity with the antenna applicators (Table III) since 212 sessions (318 hours) of treatment resulted in only one case of minimal toxicity. During these treatments (Table IV) tumor response was seen in all but 2 cases. Site-specific analysis (Tables V-VII) also shows the relative effectiveness of this therapy regardless of anatomical location.

TABLE III. RESULTS WITH INTRACAVITARY ANTENNA: TOXICITY

10 treatments completed, < 2-month follow-up	
10 hyperthermia treatments of 1½ hours each	= 100 sessions
14 treatments completed, ≥ 2-month follow-up (see Table IV)	
8 hyperthermia treatments of 1½ hours each	= 112 sessions
Total:	212 sessions
Toxicity: only one, central pneumonitis	

TABLE IV. RESULTS WITH INTRACAVITARY ANTENNA: RESPONSE

Treatments: 1600 rads in four fractions given over 2 weeks; 8 hyperthermia sessions of 1½ hours each.

No. of patients	Total response	Partial response	No response	Recurrences
14 (see Table III)	6	6	2	1 (local)

TABLE V. HEAD AND NECK PATIENTS

No. of patients	Total response	Partial response	No response	Recurrences
22	10	10	2	1 (neck)

TABLE VI. BREAST AND CHEST WALL

No. of fields	Total response	Partial response	No response	Incomplete response
29	19	7	3	3

TABLE VII. SKIN TUMORS

No. of fields	Total response	Partial response	No response
33	25	7	1

The fractionation regime employed in this study (regional hyperthermia plus low dose radiation) should be compared with those employed in other reported clinical trials.

The study reported here as well as the results of other investigators tend to indicate the relative effectiveness and lack of overall adverse effects from combined hyperthermia and radiation. Further prospective, site-specific trials are now planned or in progress to further evaluate both the safety and effectiveness of fractionated hyperthermia and radiation. In addition, the patients already treated will continue to be followed at 2-month intervals.

REFERENCES

[1] ARCANGELI,G.,BARNI,E.,DIVIDALLI,A, et al. "Effectiveness of microwave hyperthermia combined with ionizing radiation: Clinical results on neck node metastases", in Int.J.Radiat. Oncol.Biol.Phys. 6 (1980) 143-148.
[2] BICHER,H.I.,SANDHU,T.S.,HETZEL,F.W. "Hyperthermia and radiation in combination: A clinical fractionation regime", in Int.J.Radiat.Oncol.Biol.Phys. 6 (1980) 867-870.
[3] HORNBACK,N.B.,SHUPE,R.E.,HOMAYON,S.,et al. "Preliminary clinical results of combined 433 MHz microwave therapy and radiation therapy on patients with advanced cancer", in Cancer 40 (1977) 2354-2863.
[4] JOHNSON,R.J.R.,SANDHU,T.S.,HETZEL,F.W.,et al. "A pilot study to investigate the therapeutic ratio of 41.5-42.0°C hyperthermia radiation", in Int.J.Radiat.Oncol.Biol.Phys. 5 (1979) 947-953.
[5] KIM,J.H.,HAHN,E.W.,BENJAMIN,F.J. "Treatment of superficial cancers by combination hyperthermia and radiation therapy", in Clin.Bul. 9 (1979) 13-16.
[6] MANNING,M.R.,CETAS,T.,BOONE,M.L.M.,MILLER,R.C. "Clinical hyperthermia: Results of the phase I clinical trial combining localized hyperthermia with or without radiation", (Abstr.) in Int.J.Radiat.Oncol.Biol.Phys. 5 (1979) S2:173.
[7] U,R.,NOELL,K.T.,WOODWARD,K.T.,et al. "Microwave-induced local hyperthermia in combination with radiotherapy of human malignant tumors", in Cancer 45 (1980) 638-646.
[8] BICHER,H.I.,HETZEL,F.W.,SANDHU,T.S. et al. "Effects of hyperthermia on normal and tumor microenvironment in Radiology 137 (1980) 523-530.

Progress in Radio-Oncology II, edited by K. H. Kärcher et al., Raven Press, New York © 1982.

The Biological Basis for Clinical Treatment with Combined Hyperthermia and Radiation*

A proposal for an EORTC study

J. OVERGAARD
Radiumstationen,
The Institute of Cancer Research
and Department of Oncology and Radiotherapy,
Aarhus,
Denmark

Abstract

The biological rationale for the use of hyperthermia as an adjuvant to radiotherapy depends on the ability to achieve selective tumour heating. A simultaneous heat and radiation treatment utilizing hyperthermic radiosensitization gives the highest thermal enhancement, but to the same degree in tumour and normal tissue. Such a treatment is therefore only advantageous if selective tumour heating can be achieved. If both tumour and surrounding normal tissue are heated, a sequential treatment schedule (i.e. heat 4 hours after radiation) should be used. This is based on the almost selective hyperthermic cytotoxicity against acidic and chronically hypoxic radioresistant tumour cells. Such treatment yields a smaller thermal enhancement in the tumour, but does not increase the radiation response in normal tissue. A clinical study designed to explore this biological rationale is described.

1. INTRODUCTION

Abundant experimental and early clinical experience indicates that hyperthermia may be useful as an adjuvant to radiotherapy in the treatment of solid tumours [1-5]. The biological rationale for this combined therapy has been extensively studied in experimental models [2,6,7]. This has created a basis for clinical application which is briefly reviewed in the following. In addition, a proposal for an EORTC study is outlined.

2. INTERACTION BETWEEN HEAT AND RADIATION

To understand the biological interaction between heat and radiation, it is important to realize that at least two different types of interaction exist.

---

* Supported by the Danish Cancer Society (Grant No. 24/79) and "Ingeborg and Leo Danin's Foundation for Scientific Research".

2.1. Radiosensitizing Effect

Firstly hyperthermia has a radiosensitizing effect. This includes direct radiosensitization, a decreased repair of sublethal and potentially lethal radiation damage and an enhanced killing of cells in relatively radioresistant phases of the cell cycle [1-6]. A decrease in OER which has been suggested may, however, be an artefact [2,5,6].

The radiosensitizing effect of hyperthermia is related to the time of application between the two modalities [2-4,6,7]. In general, maximum sensitization is obtained by true simultaneous treatment, and any time elapsing between the modalities will tend to reduce this effect. This reduction is most prominent if hyperthermia is given after radiation, and the hyperthermic radiosensitization disappears completely if the two modalities are given separated by an interval longer than 3-4 hours (in the given sequence). With the opposite sequence the decrease is more protracted and variable. Hyperthermic radiosensitization seems to be responsible for the high TER values seen with simultaneous heat and radiation treatment. Since almost identical TER values are obtained in tumours and normal tissue [4,6,7] (Fig.1), it is likely that direct radiosensitization and reduced repair of sublethal damage are the main mechanisms behind this effect.

2.2. Cytotoxic Effect

Secondly, heat has a cytotoxic effect in tumour cells. This cytotoxicity is enhanced by certain environmental conditions, and cells in areas with chronic hypoxia, acidity and nutritional deprivation, typical of the poorly vascularized parts of solid tumours, are considerably more sensitive to hyperthermic destruction than cells in a normal environment [2-8]. Furthermore, heat itself may enhance such environmental conditions by changing the tumour metabolism into a more anaerobic type - especially by reducing the blood flow in heated tumours [5,8,9]. That such heat-sensitive chronically hypoxic cells are also the most radioresistant may indirectly influence the response to combined heat and radiation treatment, since a smaller radiation dose may be adequate to control the remaining oxygenated tumour cells. Indeed, recent studies have shown that the fraction of clonogenic hypoxic cells in a solid tumour was markedly reduced after a moderate heat treatment that itself did not induce any severe damage in normal tissue [10]. The increased TER values observed in tumours given sequential treatment with intervals greater than 4 hours between heat followed by radiation are probably a consequence of this cytotoxicity (Fig.2), which has no well-defined time-relationship to the radiation treatment.

3. IMPLICATIONS FOR CLINICAL TREATMENT

The experimental studies indicate that two different treatment schedules may be useful for clinical treatment, depending on the ability to heat the tumour selectively.

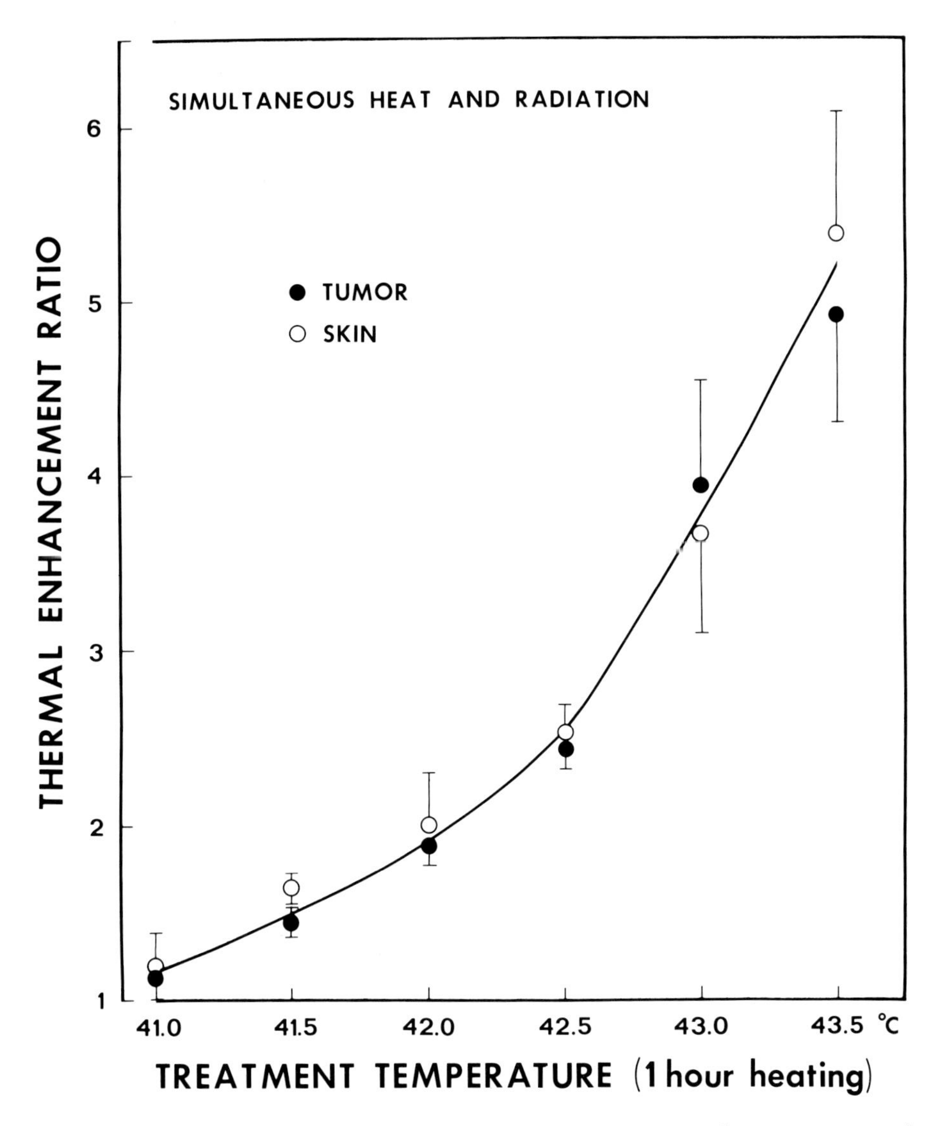

FIG.1. Thermal enhancement ratio (TER) as a function of temperature in a mouse mammary carcinoma and its surrounding skin. The TER values increase markedly with increasing temperature. No difference was observed between the TER for skin and tumour response when hyperthermia and radiation were given simultaneously (vertical bars are 95% confidence limits).

3.1. Simultaneous Application of Heat and Radiation

Assuming that selective tumour heating can be obtained, the optimal therapeutic effect will be achieved through simultaneous application of heat and radiation, utilizing the hyperthermic radiosensitization. It is important that hyperthermia is given in connection with all radiation fractions in order to achieve maximum sensitization [4,6,11]. Whether such a fractionated treatment is influenced by the development of thermal tolerance has not been settled [12]. However, experimental studies indicate that thermal enhancement ratios in fractionated regimens may be of approximately the same magnitude as those found in comparable single treatments [11] (Fig.3).

A simultaneous treatment results in the highest thermal enhancement ratio, but to the same degree in tumour and normal tissue, i.e. without improving the therapeutic ratio [2,7,11] (Fig.1). Nevertheless, such treatment may in principle also be applicable even if both tumour and surrounding tissue are exposed to hyperthermia, or if a higher temperature can be obtained within the tumour. This situation may occur especially during heating of large tumours [13], and it is probably due to a reduced blood flow and/or a heat-dependent increase in the blood flow in the surrounding normal tissue [9,14]. In such cases, an appreciable therapeutic gain may be obtained owing to the steep temperature gradient for thermal enhancement [4,6,7] (Fig.1). Unfortunately, it is currently not possible to predict in which tissues or tumours such differential heating patterns may occur. Furthermore, whenever radiation is given up to the tolerance level, it is important to know the exact TER values for a given heat treatment in order to avoid unacceptable radiation damage.

3.2. Sequential Application of Radiation and Heat

If selective or semi-selective tumour heating cannot be secured, the treatment rationale must be based on the sequential application of radiation and hyperthermia, utilizing the hyperthermic cytotoxic destruction of acidic, nutritionally deprived and chronically hypoxic radioresistant tumour cells. Such treatment should be given in a sequence where hyperthermia is given about four hours <u>after</u> radiation. With this schedule no enhancement of the radiation response in normal tissue is expected, and an improved therapeutic gain may be achieved due to the selective destruction of radioresistant tumour cells [2,6,7,11]. The degree of hyperthermia used with this schedule is limited by the heat tolerance in the critical normal tissue involved. In most well-vascularized tissues local heating to about 43°C for 1 hour is normally acceptable. With this schedule the optimum effect may be achieved without giving hyperthermia in connection with all radiation fractions. In fact, it is possible that several fractions of hyperthermia may not result in any better response than a single large fraction, due to the development of thermal

FIG.2. Thermal enhancement ratio as a function of time interval and sequence between hyperthermia (42.5°C for 1 h) and radiation treatments of a mammary carcinoma and its surrounding skin. Maximum TER values are observed after simultaneous treatment and any interval reduces the thermal effect. In the tumour there is a persistent thermal enhancement with intervals of $\geq$ 4 h with a TER value of about 1.5, independent of the sequence. In the skin, if heat is given before radiation, a prolonged thermal enhancement slightly below the tumour response is observed. If radiation is given before heat, the TER for skin is markedly reduced within one hour, and after 4 hours no thermal enhancement of the skin is retained. This results in the maximum therapeutic effect if both tumour and skin are heated (vertical bars are 95% confidence limits).

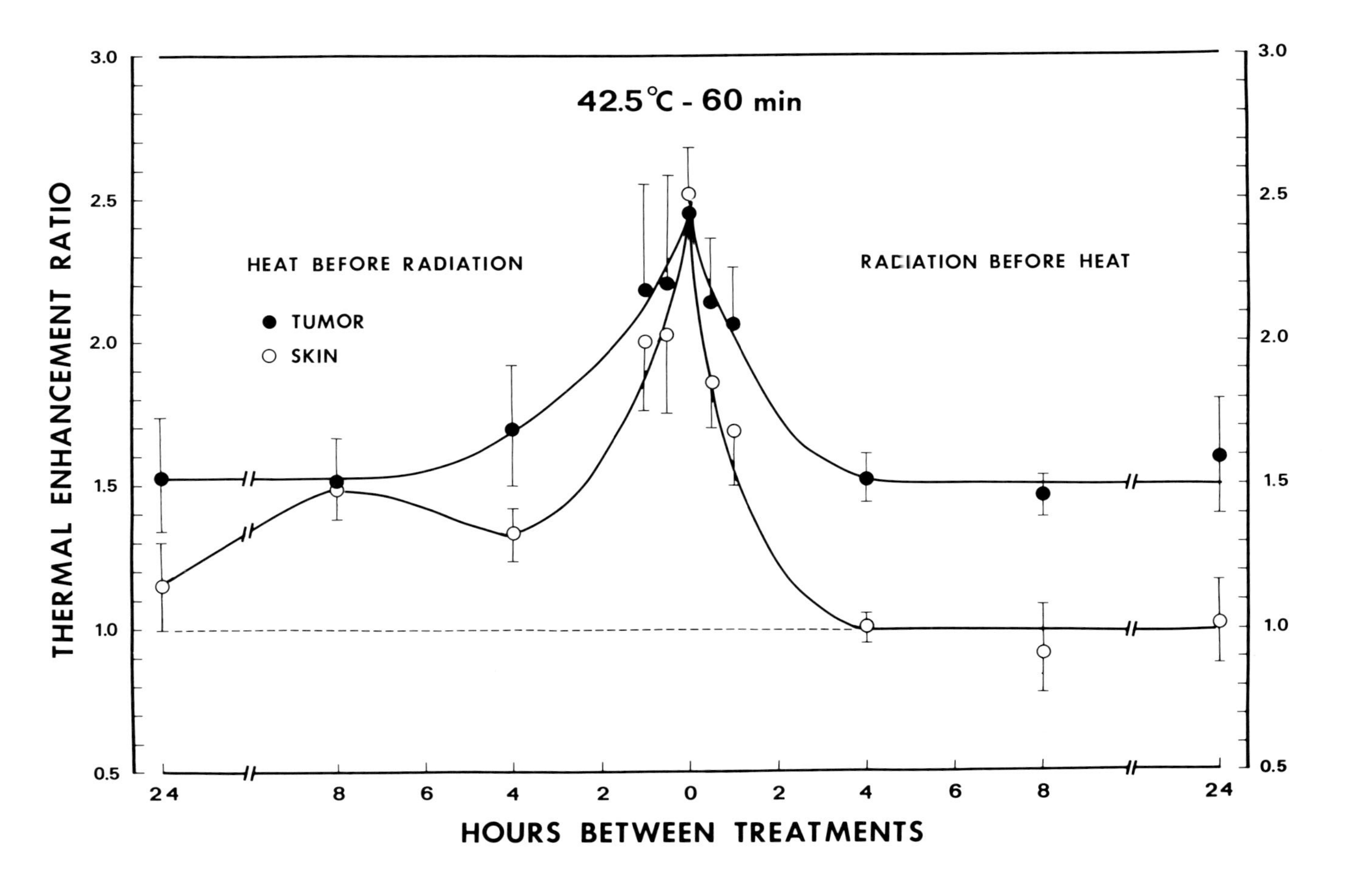
42.5°C - 60 min
HEAT BEFORE RADIATION
RADIATION BEFORE HEAT
TUMOR
SKIN
THERMAL ENHANCEMENT RATIO
HOURS BETWEEN TREATMENTS
3.0
2.5
2.0
1.5
1.0
0.5
24
8
6
4
2
0

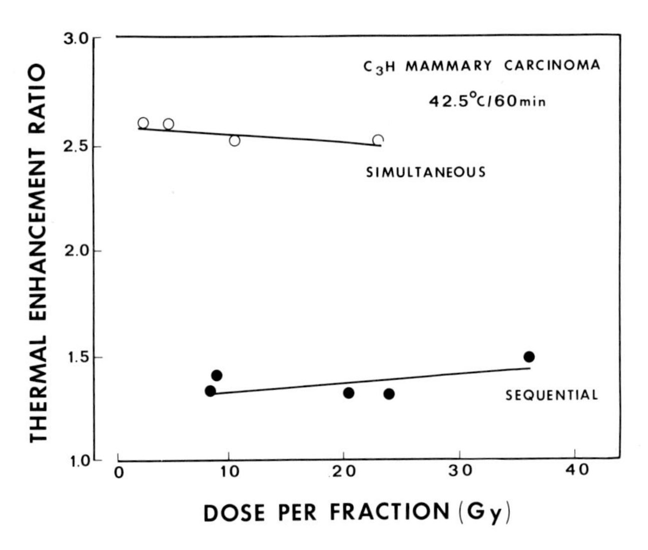

FIG.3. Thermal enhancement ratio as a function of dose per fraction in a murine mammary carcinoma. Neither simultaneous nor sequential treatment (heat 4 h after radiation) seems to have any significant fraction effect in this tumour. Data from experiments with treatment given in 1, 2, 5 or 10 fractions.

tolerance [6,11,15]. However, this has not been clarified in detail and is an important area for future research. To avoid any hyperthermic radiosensitization of the normal tissue, an interval of several days should be allowed between hyperthermia and a subsequent radiation treatment [11].

4. CLINICAL EXPERIENCE

Whereas there appear to be well-defined biological rationales for the application of heat and radiation, these rationales depend on whether or not a tumour can be heated selectively. Unfortunately, such selective heating does not seem to be possible with most of the presently available techniques, although active skin cooling together with the previously mentioned vascular changes may tend to accumulate relatively more heat in the tumour [16-18].

Despite the increasing number of early clinical trials, clinical data on direct comparisons of the effect of radiation and of the effect of a combination of heat and radiotherapy is sparse [4,11,16-22]. However, these studies all show a remarkable heat-induced improvement in the radiation response (Table I). All but one of these investigations have applied heat and radiation with short time intervals between them and, with the exception of our preliminary study [11], the question addressed regarding the clinical application of radiation and heat either simultaneously or sequentially has not been studied.

5. EORTC HYPERTHERMIA PROTOCOL

The EORTC hyperthermia group, which was founded in 1978, has been a European forum for the discussion of hyperthermia in cancer

TABLE I. EFFECT OF ADJUVANT HYPERTHERMIA ON RADIATION RESPONSE IN HUMAN TUMOURS

Study	No. of patients/tumours	Frequency of complete response	
		Radiation alone [a]	Radiation + heat [b]
Arcangeli et al. [16]	39	47%	85%
U et al. [21]	7	14%	85%
Overgaard [11]	22	37%	77%
Kim et al. [22]	86	33%	80%
Bede et al. [19]	24	0%	9%
Corry et al. [20]	6	0%	40%
All	184	31%	74%

[a] Identical radiation dose in both arms.

[b] The specific heat treatment and fractionation schedule differ among the studies.

treatment. The group has now initiated its first protocol. Based on the above-mentioned biological and clinical experience, the objective of the protocol was chosen to evaluate a biological rationale for a combined heat and radiation therapy by comparing, in tumour and normal tissue, the effect of radiation alone with the effect of combined radiation and hyperthermia treatment, given either simultaneously or sequentially.

There are no restrictions on the heating techniques used since the study also plans to analyse the feasibility of obtaining local heating in superficial tumours using various heating techniques.

Patients with multiple recurrent or metastatic lesions accessible for treatment will be included. The individual tumours in each patient will be randomized to treatment in one of the following schedules:

(1) Radiation alone (4 fractions of 6 Gy each, 2 fractions per week);

(2) Simultaneous radiation and hyperthermia (4 fractions of 6 Gy, each fraction followed immediately by hyperthermia at 43.5°C for 30 min);

(3) Sequential radiation and hyperthermia (4 fractions of 6 Gy, each fraction followed after 4 hours by hyperthermia at 43.5°C for 30 min).

Tumour and normal tissue response will be observed with emphasis on an improved tumour response for the combined treatment and with respect to differences in the radiation reaction in the surrounding and overlying skin.

It is realized that this initial protocol is simple and may yield only limited information, especially in view of the use of different heating techniques. However, this reflects the current clinical situation. Moreover, more stringent criteria for the treatment technique might seriously limit the number of centres and patients joining the protocol.

It is planned to use the experience gained from this study subsequently as the basis for a more detailed study in which thermal enhancement ratios in tumours and normal tissues will be studied by means of dose/response data. An extension of the study to include misonidazole in the two types of treatment schedules seems also to be a valuable area of clinical research in view of some promising experimental results already reported [11,23].

REFERENCES

[1] DEWEY, W.C., FREEMAN, M.L., RAAPHORST, G.P., CLARK, E.P., WONG, R.S.L., HIGHFIELD, D.P., SPIRO, I.J., TOMASOVIC, S.P., DENMAN, D.L., COSS, R.A.: "Cell biology of hyperthermia and radiation", in Radiation Biology in Cancer Research (MEYN, R.E., WITHERS, H.R., Eds), Raven Press, New York (1980) 589-621.

[2] FIELD, S.B., BLEEHEN, N.M.: Hyperthermia in the treatment of cancer, Cancer Treat. Rev. 6 (1979) 63.

[3] OVERGAARD, J.: "The effect of local hyperthermia alone and in combination with radiation, on solid tumors", in Cancer Therapy by Hyperthermia and Radiation (STREFFER, C., et al., Eds), Urban and Schwarzenberg, Baltimore, Munich (1978) 49-61.

[4] OVERGAARD, J.: "Hyperthermic modification of the radiation response in solid tumors", in Biological Basis and Clinical Implications of Tumor Radioresistance (FLETCHER, G.H., NERVI, C., WITHERS, H.R., ARCANGELI, G., MAURO, F., TAPLEY, N. duV., Eds), Masson Publishing USA, New York (1981).

[5] SUIT, H., GERWECK, L.E.: Potential for hyperthermia and radiation therapy, Cancer Res. 39 (1979) 2290.

[6] OVERGAARD, J.: Influence of sequence and interval on the biological response to combined hyperthermia and radiation, Natl. Cancer Inst. Monogr. (1981, in press).

[7] OVERGAARD, J.: Simultaneous and sequential hyperthermia and radiation treatment of an experimental tumor and its surrounding normal tissue in vivo, Int. J. Radiat. Oncol., Biol. Phys. 6 (1980) 1507.

[8] OVERGAARD, J.: Effect of hyperthermia on malignant cells in vivo. A review and hypothesis, Cancer 39 (1977) 2637.

[9] BICHER, H.I., HETZEL, F.W., SANDHU, T.S., FRINKA, S., VAUPEL, P., O'HARA, M.D., O'BRIEN, T.: Effects of hyperthermia on normal and tumor microenvironment, Radiology 137 (1980) 523.

[10] OVERGAARD, J.: Effect of hyperthermia on the hypoxic fraction in an experimental mammary carcinoma in vivo, Br. J. Radiol. 54 (1981) 245.

[11] OVERGAARD, J.: Fractionated radiation and hyperthermia. Experimental and clinical studies, Cancer, 48 (1981) 1116-1123.

[12] FIELD, S.B., ANDERSON, R.L.: Thermotolerance: a review of observations and possible mechanisms, Natl. Cancer Inst. Monogr. (1981, in press).

[13] STORM, F.K., HARRISON, W.H., ELLIOTT, R.S., MORTON, D.L.: Normal tissue and solid tumor effects of hyperthermia in animal models and clinical trials, Cancer Res. 39 (1979) 2245.

[14] SONG, C.W., KANG, M.S., RHEE, J.G., LEVITT, S.H.: Effect of hyperthermia on vascular function in normal and neoplastic tissues, Ann. N.Y. Acad. Sci. 335 (1980) 35.

[15] NIELSEN, O.S., OVERGAARD, J., KAMURA, T.: Influence of thermotolerance on the interaction between hyperthermia and radiation in a solid tumour in vivo, Br. J. Radiol. submitted 1982.

[16] ARCANGELI, G., BAROCAS, A., MAURO, F., NERVI, C., SPANO, M., TABOCCHINI, A.: Multiple daily fractionation (MDF) radiotherapy in association with hyperthermia and/or misonidazole: experimental and clinical results, Cancer 45 (1980) 2707.

[17] JOHNSON, R.J.R., SANDHU, T.S., HETZEL, F.W., SONG, S.Y., BICHER, H.I., SUBJECK, J.R., KOWAL, H.S.: A pilot study to investigate skin and tumor thermal enhancement ratios of 41.5-42.0°C hyperthermia with radiation, Int. J. Radiat. Oncol., Biol. Phys. 5 (1979) 947.

[18] MARMOR, J.B., HAHN, G.M.: Combined radiation and hyperthermia in superficial human tumors, Cancer 46 (1980) 1986.

[19] BEDE, Z., DIVEN, Z., SHOEXION, W.: "The clinical effects of radiofrequency diathermy and radiation in bladder cancer: a preliminary report", in Cancer Therapy by Hyperthermia, Drugs and Radiation (Proc. 3rd Int. Symp. Fort Collins, Colorado, 1980).

[20] CORRY, P., BARLOGIE, B., SPANOS, W., ARMOUR, E., BARKLEY, H., GONZALES, M.: "Approaches to clinical application of combinations of nonionizing and ionizing radiations", in Radiation Biology in Cancer Research (MEYN, R.E., WITHERS, H.R., Eds), Raven Press, New York (1980) 637-644.

[21] U, R., NOELL, T.K., WOODWARD, K.T., WORDE, B.T., FISHBURN, R.I., MILLER, L.S.: Microwave-induced local hyperthermia in combination with radiotherapy of human malignant tumors, Cancer 45 (1980) 638.

[22] KIM, J.H., HAHN, E.W., ANTICH, P.P.: Radiofrequency hyperthermia for clinical cancer therapy, Natl. Cancer Inst. Monogr. (1981, in press).

[23] OVERGAARD, J.: Effect of misonidazole and hyperthermia on the radiosensitivity of a C3H mouse mammary carcinoma and its surrounding normal tissue, Br. J. Cancer 41 (1980) 10.

Progress in Radio-Oncology II, edited by K. H. Kärcher et al., Raven Press, New York © 1982.

434 MHz as an Adjuvant in Cancer Therapy

A survey of results obtained and the biochemical knowledge derived from the use of this therapy

J.A.G. HOLT
Radiotherapy and Oncology Centre,
Wembley, Western Australia,
Australia

Abstract

Combined treatment using X-ray therapy and 434 MHz electromagnetic radiation (VHF) has improved the palliative response, the local tumour control and the survival rate in several types of cancer. Comparison between the series of Hodgkin's disease, non-Hodgkin's lymphoma, rectal, breast, head and neck, bladder and prostatic cancer has been by historical or simultaneous serial evaluation. A series of breast cancers treated after biopsy reveals only one failure to eliminate the local cancer and involved regional nodes. Two thirds survive without any clinical evidence of active cancer and have no radiation sequelae in the treated breasts. Hodgkin's and non-Hodgkin's lymphoma usually respond in all stages of the disease even if resistant to all conventional agents. Patients with recurrent cancer of rectum survive 50% longer when VHF is used; 2 patients without operation survive, apparently cancer free, 4 and 1 years, respectively. Bladder cancer confined to the pelvis can always have excellent palliation: the bladder can be cleared in approximately 80% of the cases. Head and neck cancer has three times the survival rate 3 years after VHF therapy. Prostatic cancer responds well and palliative remission in patients with widespread metastases for 1 to 3 years is possible.

1. INTRODUCTION

The adverse effects of electrical currents on cancer have been recorded by Kottke [1] and Schwan and Piersol [2]. The beneficial effects of electromagnetic radiations were used by Denier [3], who combined 375 MHz and X-radiation to treat cancer successfully. Zimmer et al. [4] showed differences in the electrical characteristics of normal and cancer cells and that the latter selectively absorbed more energy than normal cells do from uniform fields of non-ionizing electromagnetic radiation. Schwan and Piersol [5] provided data for normal cells whilst Joines [6] established the parameters which applied to cancer cells. Cancer cells selectively absorb almost five times more energy than normal cells do when exposed to 300 MHz radiation. The chance observation by Schwartz [7] that 434 MHz could cause some deposits of porcine cancer to become liquefied was the stimulus for the construction of a device to irradiate humans in an attempt to treat their cancer. One of these machines was installed in our practice in June 1974 and put into clinical operation.

Normal people irradiated with 434 MHz (called VHF hereafter) exhibit a common pattern of absorption and reflection.

This pattern is altered by the presence of cancer. VHF radiation is moderately and selectively absorbed in cancer, thus creating differential hyperthermia in the living subject.

Calculations of the radiosensitivity values for various cancers under conditions of in-air, under three atmospheres absolute oxygen pressure and hyperthermia induced by whole-body wax bath, and 27 MHz and VHF irradiation, have been made [8]. Since VHF causes a much greater increase in sensitivity at a temperature of approximately 39°C than wax bath heating to 41.8°C, it must have thermal and non-thermal effects [9].

2. EQUIPMENT AND METHODS

The original equipment used 12 conventional Erbotherm VHF 69 generators, while a later unit was specially constructed to our design. In the former, each generator produced a nominal power output of 200 W at 434 MHz which fed a dipole in a plastic case via a coaxial cable. The 12 dipoles were arranged in three concentric rings of 4 dipoles each. They thus surrounded an octagonal-sided cylindrical space some 60 cm long and 54 cm in diameter. This dipole assembly was mounted on a vertical column and could be used to scan a patient standing therein [10].

Because the patient frequently found a need for support, the assembly was rebuilt in horizontal orientation. The patient was accommodated on a plywood couch top supported on nylon ropes.

The second piece of equipment was designed as a result of the experience gained during the first four years of operation of the Erbotherm generators. Four dipoles are used to surround the patient, who is supported on a polypropylene couch top. With this system the dipole array remains stationary whilst the patient can be moved with the couch through the array, at a constant rate of approximately 4 cm per second. The dose can be altered by varying the power supplied to one or more dipoles. Four generators each supply a dipole with a nominal power between 0.1 and 2 kW.

Measurements have been carried out which compare the output of the generators when each in turn is feeding into 'free space', into various phantoms and into patients with and without cancer. Absolute measurements of stray fields were made using a Narda broad-band isotropic monitor Model 8305 with probe Model 8323 covering 300 MHz to 18 GHz. Relative levels of stray-field intensity were monitored on a Polarad STU 1B wave analyser. Power meters were inserted at appropriate points in the aerial cables to monitor the forward and reflected VHF intensities.

3. VHF RADIATION

434 MHz generators have been widely used in Europe for 20 years or more as conventional equipment for physiotherapists without reports of adverse effects on patients or staff. VHF before, during and after X-ray therapy has been tried. Simultaneous VHF and X-ray therapy produces the most rapid objective regressions of cancer. In the first series of head and neck cancer, VHF was given before X-ray therapy because it was hoped to measure the temperatures created while the X-ray therapy was

being delivered, while the specific non-thermal effects of VHF on cancer were also utilized. This specific non-thermal effect has recently been confirmed in animal experiments by Johnson et al. [11].

This combined therapy has been in use by Hornback et al., who report similar examples of VHF-induced radiosensitivity increases [12,13]. Hornback and his team have demonstrated that their best clinical results have come from X-ray therapy followed by VHF. The optimum timing and dose of VHF have yet to be determined, but a regime of 120 to 160 rads applied uniformly throughout the cancer volume followed by 4 applications of 1 to 2 minutes each in the four-antenna equipment appears the best one tried so far. One of the applications immediately prior to the X-ray therapy is used when response appears delayed. Each application of VHF is with 1 kW nominal power applied to each dipole, and the treatments are spaced at half-hour intervals.

4. RESULTS

4.1. Hodgkin's Disease

Twelve patients with Hodgkin's disease [14] who had active recurrences resistant to conventional treatment all obtained complete remission of their disease. Eight are alive without evidence of disease between 6 and 17 months later. One died at 7 months from trauma, one died at 8 months of aplastic anaemia (he had had 8 years of cytotoxics prior to VHF therapy), one is lost to follow-up and one had further metastatic disease which is being successfully treated.

4.2. Non-Hodgkin's Disease

Eighteen patients who were undergoing conventional therapy prior to the introduction of VHF were all suffering from unresponsive widespread active lymphoma and were treated between March 1974 and December 1975 [15-17]. At December 1980, six had died, two in 1976 and 1978 of acute lymphoblastic crises, one in 1976 from marrow failure, two in 1978 and 1979 after withdrawing from treatment due to severe radiation effects (nausea, enteritis, pneumonitis) and one in 1976 from multiple skin cancers after treatment for and remission of an acute lymphoblastic crisis 18 months before. The remaining 12 are all symptom-free, 10 normal to all investigation and two have had recurrences treated during 1980. Of the 10 'normal' patients, one was treated with glucose analogues, hypoglycaemia and streptokinase only, the remaining nine with large field X-ray therapy and VHF. The average course was 8 daily treatments of 150 rads each to a volume not exceeding half the body.

Since December 1975, seven more widespread lymphomata have been treated by similar combined courses. One patient died of septicaemia following an attempt in 1976 to treat the whole body in a single 10-day course of treatment. The other six are all in complete remission, one treated in 1976, one in 1977, one in 1978, two in 1979 and one in 1980.

4.3. Rectal Cancer

A trial was made comparing proven recurrent cancer in 24 patients treated by X-ray therapy with 24 patients treated by combined methods [18]. In the first group 19 had had abdomino-perineal resections, 2 colostomy only and 3 biopsy only, compared with 18, 5 and 1 respectively in the combined therapy group. The median ages were 59 and 57 in the two groups, and the crude survivals were 16.3 and 26.9 months at $p = 0.005$. VHF had thus produced a 50% increase in survival time and two patients remain alive without evidence of actual cancer at 60 months. All patients treated by X-ray therapy alone were dead at 42 months from the date of first treatment. In addition, only two patients in the VHF group failed to obtain complete relief of pain, while 13 failed to obtain even reasonable palliation of their symptoms from X-ray therapy alone.

Two patients refused surgery twice and were treated solely with combined methods. Both obtained complete regression of their primary and regional nodes and survive 4 years and 1 year. The 4-year survivor is being investigated for liver metastases.

4.4. Breast Cancer

(A) Twenty-six patients suffering from Stage 1 to 4 breast cancer have had regional combined therapy after biopsy. This has controlled the primary disease and regional nodes in all but one person. Ten have developed multiple metastases and nine have since died: average survival of these Stage 3 and 4 cancers was 29 months from first treatment. The average radiation dose used was 3200 rads over approximately 25 treatments, 12 of which were combined with VHF. No complications or sequelae have been revealed. Combined treatment usually restores the breast to a normal appearance and consistency without fibrosis. Nipple retraction and peau d'orange disappear and visible evidence of radiation therapy disappears within 3 to 6 months. Of the 16 survivors with apparently healed, intact breasts, four were treated in 1974, three in 1976, four in 1977, four in 1978 and one in 1979. Several others have been treated with equally good results in the last 15 months. Six were Stage 1, five Stage 2, three Stage 3 and two Stage 4.

(B) Between July 1974 and July 1979, 44 Stage 1 and 2 patients who have been operated upon by mastectomy and axillary sampling or clearance were treated by giving 3000 rads to the chest wall, internal mammary, axillary and supraclavicular regions over 15 treatments interspersed with 6 to 9 treatments to the whole area with combined therapy to a total of 1200 rads. They were compared with a historical series in which 44 similar post-operative patients were treated with 25 treatments of X-ray therapy to 5000 rads ± 500 rads.

Within similar observation times, three of the combined therapy group and nine of the non-VHF group developed recurrences in the scars and/or supraclavicular areas. Seventeen of the combined group and 25 of the non-VHF group developed distant metastases.

(C) As in the cases with rectal cancer, the survival time is approximately 50% better in the combined therapy group when treatment is given for widespread metastatic breast cancer. This is achieved by using larger fields or half- or whole-body therapy combined with VHF. Patients with widespread bone, liver, soft-tissue and/or intrathoracic spread survived an average of 17 months from their first treatment with these techniques compared to a historical group of 27 patients who had an average survival of 11 months when treated by X-ray therapy alone: neither group had additional hormone therapy.

4.5. Head and Neck Cancer

Fifty-two patients with head and neck cancer treated sequentially by combined therapy were compared with 52 consecutive patients treated on hyperbaric oxygen and 52 treated in air, respectively [19]. At 3 years, the crude cancer-free survival of the hyperbaric oxygen group was nearly twice (29%) and that of the VHF group was three times (54%) that of the in-air, conventionally-treated group (19%). Since these results were published, a further 96 patients have been treated at all stages, resulting in clearance of the primary disease in 71 of them. Nasopharyngeal cancers respond excellently; of 11 treated between January 1977 and January 1980, all have had control and healing of the primary site and neck nodes, two have had clearance of lung and liver metastases, respectively, and survive more than 1 year later, two more are alive after treatment for other metastases and seven appear to be completely controlled at 1, 1, 2, 2, 2, 3 and 3 years, respectively.

4.6. Bladder Cancer

(A) Eight patients with proven muscular invasion (T2) have all achieved complete clearance of the bladder and remain clear between 1 and 5 years later.

(B) Seventeen patients with perivesical growths (T3) obtained clearance and restoration of bladder function, in 13 with excellent palliation, and cessation of bleeding in the remainder.

(C) Twenty-two patients with fixed pelvic growths (T4) or post-surgical recurrences or multiple metastases all obtained palliation with 10 doses of 150 rads plus VHF delivered on alternate days. All cases of haematuria were improved, most controlled. Pain relief was only absent in one patient. Three patients, bed-ridden with multiple pelvic and spine metastases, were restored to mobility and survived an average of 9 months in comfort. One patient who had gross pelvic disease and multiple lung metastases had control of all her disease for 1 year with the therapy described above. Two patients with bowel obstruction following recurrences after cystectomy had relief of pain and obstruction and survived 1½ and 2 years, respectively, after even lower doses of X-ray therapy. No T1 patients were referred for treatment. All patients seen with bladder cancer since 1975 were treated with VHF because survivals for Stages T2, T3 and T4 are at least three times as long after combined therapy as after any other method. We believe that all early bladder cancer should be treated by combined techniques before any other method is attempted - in the expectation of 90% to 100% control of this disease.

4.7. Prostate Cancer

(A) Five patients with disease localized to the gland and regional nodes have all obtained control and are alive and well without surgery or hormone therapy at 6, 5, 5, 4 and 3 years, respectively, without complications. X-ray doses averaged 3600 rads in 20 to 25 treatments, 12 of them being combined with VHF.

(B) Eleven patients with disease confined to the lumbar spine, pelvic bones and contents have all had remissions exceeding 2 years without hormones, three surviving disease-free over 4 years.

(C) Nineteen patients treated when widespread metastases were present (i.e. liver, lungs, bones above L1 vertebra, brain) have all obtained relief of symptoms and have an average survival to date of 15 months, without using hormones.

4.8. Primary Brain Cancer

In early studies it was established that primary gliomata would nearly all improve with very small exposures to VHF alone [10]. With 4 antennae, each powered by an Erbe VHF 69 generator surrounding the head, a daily exposure of 15 seconds with a nominal radiated power of 180 W per antenna is adequate to cause clinical improvements. After 5 to 10 such doses, combined therapy is used to try and eliminate the residue. In 16 patients suffering from malignant gliomata of grades 2, 3 or 4, astrocytoma and glioblastoma, 14 had relief of symptoms and neurological improvement and eight had a normal CT scan within 6 months of treatment. Total X-ray doses between 1200 and 4800 rads were used to obtain these remissions. Four patients remain alive and well at 4, 3, 2 and 2 years, respectively. The others developed recurrences; retreatment in all was of good palliative value and the average survival was 7 months.

4.9. Other Cancers

At least 90% of all types of cancer obtain palliative relief from simple low-dose regimes of combined therapy, without morbidity. Malignant schwannoma, myxoliposarcoma, fibrosarcoma, chondro and osteosarcoma, and mesothelioma can all be made to respond with X-ray doses not exceeding a total of 2000 rads when combined with VHF [16,17].

5. GLUCOSE AND OXYGEN EFFECTS ON CANCER AND NORMAL CELLS

A series of patients suffering from cutaneous primary and metastatic cancers and the scars from biopsies therefrom on the limbs were exposed to VHF at free-in-air intensities of between 10 and 20 mW/cm^2. Full details are recorded [8] and the conclusions appear important. It was discovered that severe limb hypoglycaemia below 0.1 µl/1 created by systemic insulin injections, or anoxia created by tourniquet and exercise, or both, delayed or prevented normal wound healing. In the presence of normal glucose and oxygen levels, VHF will increase the speed of normal cell healing and appears to have no further effects once the skin is fully healed. These effects are already well known and utilized in surgical practice [20,21]. However, only hypoglycaemia prevented VHF stimulating the growth of the cancer deposits. Anoxia alone was ineffective in this regard.

Some possible conclusions from these findings appear to be:

(a) At least part of the energy for cancer cell growth is derived from the anaerobic utilization of glucose;

(b) Normal cell repair can only occur in the presence of glucose and oxygen. The sophisticated control mechanisms can only operate in the presence of oxygen;

(c) Under VHF stimulation cancer cells lose all their characters of differentiation and function, i.e. they become 'primitive', yet without the potential of embryo cells to form more adult structures.

It has been shown [22] that the nucleus is always anoxic, but when the nuclear membrane breaks during mitosis the whole cell assumes anoxic metabolism until the membrane is reformed [23]. It is reasonable to suggest that the energy for mitosis is probably derived from anaerobic glucose metabolism. However, the control of all functions demands the energy derived from aerobic glucose metabolism [8].

It is proposed, therefore, that cancer is a defect in the linkage between the aerobic and the anaerobic glucose metabolic systems. Energy derived from aerobic glucose metabolism is used to control the energy which, derived from anaerobic glucose metabolism, is essential for mitosis. Any sublethal cellular injury which releases the anaerobic glucose metabolic energy functions from control by the aerobic glucose metabolic pathways may cause cancer.

6. CONCLUSIONS

It is the specific electrical differences in conductivity and the dielectric constant between cancer and normal cells which permit 434 MHz electromagnetic radiation (VHF) to be used to:

(1) Stimulate normal human growth and repair processes;

(2) Distinguish between apparently identical cancer and primitive cells;

(3) Produce specific changes in reflection/absorption of VHF which identify cancer; the interaction between 434 MHz and living cancer is a resonance phenomenon;

(4) Create non-specific thermal and specific non-thermal effects in cancer;

(5) Increase the radiosensitivity of cancer by 1 or 2 decades;

(6) Design effective therapy regimes for all types and stages of human cancer using combined X-ray therapy and VHF.

In addition, it is a tool which can investigate theories of cellular organization and control and be able to pinpoint sites of action of conventional cytotoxic drugs.

REFERENCES

[1] KOTTKE, F.J.: in Therapeutic Heat and Cold, 2nd ed. (LICHT, S., Ed.), Licht, New Haven, Connecticut (1975) 488.

[2] SCHWAN, H.P., PIERSOL, G.M.: Absorption of electromagnetic energy in body tissues; review and critical analysis, physiologic and clinical aspects, Am. J. Phys. Med. 34 (1955) 425.

[3] DENIER, A.: Essai de traitement de tumeurs inopérables. Leur radio-sensibilisation par les micro-ondes de 80 cm associées au surdosage roentgenotherapique rendu possible par les ondes infrarouges, Arch. d'Electric Med. 44 (1936) 403.

[4] ZIMMER, R.P., ECKER, H.A., POPOVIC, V.P.: Selective electromagnetic heating of tumours in animals in deep hypothermia, Trans. MTT IEEE 19 2 (1971) 238.

[5] SCHWAN, H.P., PIERSOL, G.M.: The absorption of electromagnetic energy in body tissues, Am. J. Phys. Med. 33 (1954) 371.

[6] JOINES, W.T.: "Optimizing the heat treatment of tumours by microwaves", in Clinical Prospects for Hypoxic Cell Sensitisers and Hyperthermia, Caldwell, W.L., Durand, R.E. (Publ.) Univ. of Wisconsin, 1300 University Ave., Madison, Wisconsin 53706, USA (1977) 229.

[7] SCHWARTZ, W.: personal communication, 1974.

[8] HOLT, J.A.G.: The cause of cancer. Biochemical defects in the cancer cell demonstrated by the effects of electromagnetic radiation and glucose and oxygen, Med. Hypoth. 5 (1979) 109.

[9] HOLT, J.A.G.: Increase of x-ray sensitivity of cancer after exposure to 434 MHz electromagnetic radiation, J. Bioeng. 1 (1977) 479.

[10] HOLT, J.A.G.: The use of V.H.F. radiowaves in cancer therapy, Australas. Radiol. 19 3 (1975) 223.

[11] JOHNSON, R.J.R., SUBJECK, J.R., KOWAL, H., YAKAR, D., MOREAU, D.: "Hyperthermia in cancer therapy", in Treatment of Radio-resistant Cancers (ABE, M., SAKAMOTO, K., PHILLIPS, T.L., Eds), Elsevier/North Holland Biomedical Press, Amsterdam (1979) 77-80.

[12] HORNBACK, N.B., SHUPE, R.E., SHIDNIA, H., JOE, B.T., SAYOC, E., MARSHALL, C.: Preliminary clinical results of combined 433 megahertz microwave therapy and radiation therapy on patients with advanced cancer, Cancer 40 (1977) 2854.

[13] HORNBACK, N.B., SHUPE, R.E., SHIDNIA, H., JOE, B.T., SAYOC, E., GEORGE, R., MARSHALL, C.: Radiation and microwave therapy in the treatment of advanced cancer, Radiology 130 (1979) 459.

[14] HOLT, J.A.G.: Alternative therapy for recurrence Hodgkin's disease, Br. J. Radiol. 53 (1980) 1061-1067.

[15] NELSON, A.J.M., HOLT, J.A.G.: Microwave adjuvant to radiotherapy and chemotherapy for advanced lymphoma, Med. J. Aust. 1 (1980) 311.

[16] NELSON, A.J.M., HOLT, J.A.G.: The problem of clinical hyperthermia, Australas. Radiol. 21 (1977) 21.

[17] HOLT, J.A.G., NELSON, A.J.M.: Four years of microwave in cancer therapy, J. Belge Radiol. 62 (1979) 467.

[18] CASSIDY, B.: "Survival in cancer of the rectum 1975–1979. A comparison between the results of x-ray therapy alone and in combination with V.H.F.", read at Ann. Gen. Mtg Coll. Radiol. Australasia, Perth, Western Australia, 1980.
[19] NELSON, A.J.M., HOLT, J.A.G.: Combined microwave therapy, Med. J. Aust. 2 (1978) 88.
[20] DOYLE, J.R., SMART, B.V.: Stimulation of bone growth by short wave diathermy, J. Bone Joint Surg. 45A (1963) 46.
[21] BURNEY, F., Ed.: Electrical Stimulation of Bone Growth and Repair, Springer-Verlag, Berlin (1977).
[22] STERN, H.: On the intranuclear environment, Science 121 (1955) 144.
[23] STERN, H., TIMOREN, S.: The position of the cell nucleus in pathways of hydrogen transfer: Cytochrome C, flavoproteins, glutathione and ascorbic acid, J. Gen. Physiol. 38 1 (1954) 41–52.

Progress in Radio-Oncology II, edited by K. H. Kärcher et al., Raven Press, New York © 1982.

Experience in Clinical Hyperthermia and Radiation Therapy in Patients with Advanced Cancer

N.B. HORNBACK, R.E. SHUPE, Carol MARSHALL, Reba BAKER
Department of Radiation Oncology,
Indiana University School of Medicine,
Indianapolis, Indiana,
United States of America

Abstract

During the past six years the Department of Radiation Oncology at Indiana University School of Medicine has been investigating the use of hyperthermia in combination with radiation in the treatment of advanced cancer patients. 356 patients received a total of 4282 hyperthermia treatments given with various types of standard hospital diathermy equipment. Temperature measurements from 39–45°C were produced locally for periods of one to three hours. Radiation doses of 3000–3500 rads in previously irradiated tumours and 6000–6500 rads in tumours not previously irradiated were given either prior to or following the hyperthermia treatments. 95% of the patients had either partial or complete relief of symptoms and 88% of the patients experienced either a partial or complete objective response. The basic physics of heat loss and the dynamic principles of heat production and heat loss in physiological tissue prevented consistent temperature elevation throughout the tumour. While still in its infancy, the use of hyperthermia in combination with radiation in the clinic has offered relief of symptoms in advanced cancer patients with a minimum of morbidity and offers the possibility of a valuable adjunct to radiation therapy.

1. INTRODUCTION

It has been well established in the laboratory, utilizing animal tumour models and tumour tissue cultures, that heat potentiates the effects of ionizing radiation against malignant tumour cells [1-13]. The degree of potentiation is dependent not only upon the height of the temperature reached, but the time for which the temperature was applied [3,9,10,14,15]. Timing of the radiation/heat sequence appeared to affect this enhanced sensitivity, but it is currently unclear as to exactly what the proper sequence should be. There is considerable information on the subject [7,16-23], but much of the laboratory benchwork, while of interest from a scientific standpoint, provides little useful information to the practising clinician.

The most effective temperature to produce this enhanced tumouricidal effect remains an enigma; however, temperatures of 41°C are usually used as the lower limits of 'hyperthermia' and thus the majority of benchwork performed in the area of hyperthermia begins at this temperature. In clinical practice, a whole-body core temperature of 41°C (106°F) approaches tolerance level and temperatures of 42°C (108°F), even for short periods of time, are poorly tolerated by even healthy adults, let alone acute or chronically ill advanced cancer patients. Temperatures of 45°C can be tolerated locally for a few minutes before excessive normal tissue damage occurs; however, this unfortunately has little bearing on the overall long-term control of the cancer patient, as the cause of failure in the majority of cancer patients is eventually secondary to widespread metastasis. When utilizing a potentially new and useful therapeutic method such as hyperthermia, it is important that toxicity and response studies first be conducted in carefully controlled clinical settings prior to any widespread use of the new modality. Extreme care must be taken when subjecting the advanced cancer patient to experimental modes of therapy, as the patient's resistance against the malignancy has already been compromised and to induce additional pain and suffering through a new treatment method in these individuals without a promise of increased useful life is to be vigorously condemned. With the above in mind, the Department of Radiation Oncology at the Indiana University School of Medicine has, since 1974, treated a wide variety of tumours in patients with advanced cancer with a combination of RF frequencies. The following is a summary of our experience to date.

2. PATIENTS

Since 1974, 356 patients with advanced cancer have received a total of 4282 hyperthermia treatments in the Department of Radiation Oncology at the Indiana University School of Medicine. The patient population were of a mixed socio-economic status with a wide variety of histological types of tumour. In addition, most patients had recurrences following surgery, radiation, multi-drug therapy or a combination thereof. Eligibility for entrance into the study for combined radiation and hyperthermia consisted of the following criteria:

(1) Proven histological evidence of recurrent malignant, persistent or radioresistant disease;

(2) Measurable tumour size;

(3) Symptoms from tumour;

(4) Previous determination by an independent oncologist that the tumour could be expected to be refractory to conventional treatment methods.

Approval of the Human Experimental Research Committee at Indiana University was obtained and each patient was apprised of the experimental nature of the treatment programme.

Of the 356 patients treated, 196 were males, 160 were females, and ranged from 4 to 87 years of age. Major histological tumour types are given in Table I. Table II lists the primary site of tumours.

3. METHODS

3.1. Radiation

Various combinations of hyperthermia and radiation have been used during the past six years. Patients who were treated for

TABLE I. TREATMENT BY HISTOLOGY

Histological type	No. of patients
Squamous cell carcinoma	131
Adenocarcinoma	113
Sarcoma	20
Miscellaneous malignant tumours	92

TABLE II. TREATMENT BY PRIMARY SITE

Primary site	No. of patients
Gastro-intestinal (other)	73
Gynaecological	69
Head and neck	58
Pancreas	26
Oesophagus	24
Breast	13
Sarcomas (soft-tissue)	12
Genito-urinary	11
Lung	9
Bone	7
Unknown	10
Miscellaneous	44

recurrences following high-dose radiation therapy received a maximum of 3000-4000 rads at standard fractionation (180-200 rad/d). In a few cases of historically radioresistant tumours, i.e. osteogenic sarcomas, chondrosarcomas or malignant melanomas, the patients received up to a maximum of 6000 rads at standard fractionation doses of 200 rad/d. The radiation therapy was given utilizing either a 4 MV photon beam, a 25 MV photon beam, or electron beam therapy from 7-25 MeV, depending upon the location of the tumour.

3.2. Hyperthermia

The treatment equipment has been previously described in detail in other publications [24,25], but will be briefly reported. The hyperthermia generator equipment utilized in patient therapy consisted of either a single or multiple 433.92 MHz diathermy units, a standard single 2450 MHz diathermy unit, or a 27.12 MHz RF generator either using condenser antennae or inductor antennae to produce heat at the desired level. Initially, the heat therapy was given for 20 minutes immediately prior to the radiation therapy. After conducting several animal tumour laboratory studies utilizing various X-ray and heat sequences, it became evident that, to produce the most destructive effects on the tumour, heat therapy should be given following the radiation therapy.

3.3. Temperature Monitoring and Measurement During Therapy

At the beginning temperature monitoring was undertaken using standard mercury thermometers as well as thermistor and thermocouple temperature probes. Because all metal-containing probes are influenced by the RF field (even if placed at right angles to the electric field), temperature measurement was accomplished by inserting the probes through thin-walled polyethylene catheters after the RF power had been turned off. Utilizing this technique and placing several probes in the same tumour, temperatures within the same tumour were found to vary as much as 2°C despite the application of heat-induction methods that should have given better temperature uniformity. This variation in the temperatures within the same tumour could be accounted for by several factors. Variation in heat production by RF fields (which produce both standing waves and reinforcement waves within the tumour), the basic physical properties of heat loss, i.e. convective, conductive, thermal radiation and evaporation occurring within and from the tumour and, most importantly, the physiological changes present within the tumour itself - cooling by blood flow, metabolic activity of the tumour, etc. Despite the difficulties and questionable reliability of the accuracy of temperature measurements, all patients with accessible tumours had multiple temperature measurements, and careful recordings were made during each hyperthermia treatment session. The temperature produced varied from 39 to 45°C depending upon location of the tumour and type of heat-generating equipment used. This temperature was usually maintained in the study for a minimum of one hour and in some cases for up to three hours.

4. EVALUATION OF RESPONSE TO THERAPY

Since the patient population consisted of advanced recurrent cancer patients, the most reliable parameters for evaluating tumour treatment, i.e. tumour-free survival rate, could not be utilized, and tumour response (partial or complete) and relief of symptoms (partial or complete) would have to be utilized. Partial tumour response was defined as greater than 50% reduction of the visible or palpable tumour mass and complete tumour response was defined as complete regression of all visible or palpable tumour within the treatment area. Partial symptomatic response was defined as a major reduction in pain medication required by the patient, and complete symptomatic response was defined as complete cessation of all medication for pain.

5. RESULTS

5.1. Tumour Response

The response rates are seen in Table III. The response rates were not related to histological type, site of primary or previous therapy received. A few patients had dramatic relief of symptoms following only one or two heat treatments; other patients experienced no relief of symptoms. Exophytic tumours, regardless of histological type of tumour, appeared to respond more rapidly than infiltrative types of tumours, similar to the response seen with radiation alone. The type of previous therapy did not appear to influence response to combined therapy. Failure to control the lesion with previous radiation therapy alone was not a determinant for response, as tumour response was often noted even when lower doses of radiation were combined with heat. From a clinical standpoint, the temperature reached within the tumour did not correlate well with response rates. Some patients with elevation of tumour temperature to 45°C locally for 30 to 40 minutes had no response to therapy while other patients' tumours treated at lower temperatures (39-40°C for 20-30 minutes) had a dramatic response to therapy.

One group of patients, Stage III-B carcinoma of the cervix, was compared with a similar group of patients who had received radiation alone. While not a randomized clinical study, since all carcinoma of the cervix patients were treated under similar conditions by the same group of radiation therapists, utilizing the same equipment and identical treatment planning techniques, the study is suggestive that local control rates can be improved by the application of local heat. The results of this subgroup of patients are listed in Table IV.

TABLE III. RESPONSE RATES

	None	Partial	Complete	Total response
Subjective response	8%	35%	67%	92%
Objective response	12%	33%	56%	88%

TABLE IV. LOCAL CONTROL RATES WITH AND WITHOUT HEAT: CARCINOMA OF THE CERVIX, STAGE III-B [a]

Cobalt-60 (46 patients)	35%
25 MeV (15 patients)	53%
25 MeV + heat (17 patients)	71%

[a] FIGO staging system.

6. TOXICITY

6.1. Acute

Symptoms during hyperthermia treatment sessions varied with patients and area treated. No pain was associated with the treatments except early in the study, when two patients experienced pain in an area where the connecting wire from the temperature probe crossed the normal skin to enter the tumour. An additional three patients developed second-degree burns in normal skin under two conditions. One when the antennae was in direct contact with the skin, and two when liquid (sweat) was present on the skin. The above acute reactions during treatment are easily prevented by ensuring that the metal-containing wire from the probes is not in direct contact with the skin and that a thin absorbing material such as cloth is used whenever antennae must be in contact with the patient. All patients experience some degree of tiredness following the treatments and, where large volumes of tissue were subjected to hyperthermia, the patient became weak and tired. All patients recovered within 24 hours from these symptoms. There did not appear to be enhanced nausea with the addition of hyperthermia to X-ray therapy.

6.2. Late Effects of Radiation on Normal Tissue

Several patients have now been followed for a minimum of four years and there does not appear to be any increased damaging effect on normal tissues in the irradiated area. It was the opinion of some of the clinic physicians that the hyperthermia offered a protective mechanism for normal skin and that the expected irradiation changes were less than what was to be expected when irradiation was given alone.

7. DISCUSSION

There is no question that hyperthermia alone can produce necrosis in tumours and that hyperthermia enhances the effectiveness of radiation in the treatment of laboratory animals. How to utilize this information to its best advantage in treating human cancers remains an enigma. Dramatic tumour responses can be seen in some patients and no response is seen in others despite careful attempts at duplicating the treatments from patient to patient.

Since our major goal was to evaluate tolerance of advanced cancer patients to various hyperthermia applications, the study was successful in that it has been clearly established that patients can tolerate local hyperthermia up to 45°C in tumours without anaesthesia and that the hyperthermia does not appear to increase normal tissue damage from radiation alone. Long-term effects could only be evaluated in a few patients, and there did not appear to be any long-term damage of normal structures that was not to be expected from radiation alone. No suggestion of increased tendency for metastasis to occur was evident although the material to study this aspect was limited as most patients already had widespread disease when first seen.

Eighty-five per cent of all patients treated either had partial or complete tumour regression. Since all patients treated received ionizing radiation as well as varying degrees of hyperthermia, it was impossible to determine from this study to what degree hyperthermia enhanced the radiation response to malignant tumours. However, several patients who had previously been refractory to high-dose radiation were re-treated using low-dose radiation (3000-3500 rads) accompanied by hyperthermia and had complete disappearance of the tumour, suggesting a combined effect.

It was rewarding to note the degree of symptomatic relief that could be obtained in some patients after they had been considered unsuitable for additional conventional therapy. A lessening of symptoms in 90% of the patients treated was encouraging.

Inter-institutional Phase III Hyperthermia Protocol studies required to produce the necessary information to determine the effectiveness of hyperthermia are slow to develop for several reasons:

(1) The wide variety of Tyrolean hyperthermia equipment available and the multitude of techniques and schedules used in currently applied clinical hyperthermia;

(2) The lack of information concerning the most effective temperatures to be reached and the period of time the temperature should be held;

(3) Failure to have reliable, consistent, non-perturbing temperature measuring devices;

(4) Failure to understand the scientific basis for the destruction of tumour cells by hyperthermia.

Despite the above difficulties in developing the required scientific information, there is no question that symptomatic patients with advanced cancer have been helped with combined therapy and have experienced a minimum of untoward reactions. The continued use of hyperthermia with radiation would seem to be indicated for a selected group of patients who are suitable for this type of experimental form of treatment.

REFERENCES

[1] ALLEN, F.M.: Biological modification of effects of roentgen rays. II High temperature and related factors, Am. J. Roentgenol. 73 (1955) 836.

[2] ARON, I., SOKOLOFF, B.: Combined roentgenotherapy and ultra short waves, Am. J. Surg. 36 (1937) 533-543.

[3] BEN-HUR, E., ELKIND, M.D., BRONK, B.V.: Thermally enhanced radioresponse of cultured Chinese hamster cells: Inhibition of repair of sublethal damage and enhancement of lethal damage, Radiat. Res. 58 (1974) 38-51.

[4] CATER, D.B., SILVER, I.A., WATKINSON, D.A.: Combined x-ray and microwave therapy, Annu. Rep. Br. Emp. Cancer Campaign 40 (1962) 375.

[5] CATER, D.B., SILVER, I.A., WATKINSON, D.A.: Combined therapy with 220 kV Roentgen and 10 cm microwave heating in rat hepatoma, Acta Radiol. 2 (1964) 321-336.

[6] GERNER, E.W., CONNOR, W.G., BOONE, M.L.M., DOSS, J.D., MAYER, E.G., MILLER, R.C.: The potential of localized heating as an adjunct to radiation therapy, Radiology 116 (1975) 433-439.
[7] HAHN, E.W., ALFIERI, A.A., KIM, J.H.: Increased cures using fractionated exposures of x irradiation and hyperthermia, in the local treatment of the Ridgeway osteogenic sarcoma in mice, Radiology 113 (1974) 199-202.
[8] MUCKLE, D.S.P.: Response of a musculoskeletal tumor to combined radiation therapy, Acta Radiol., Ther., Phys., Biol. 13 (1974) 297-306.
[9] OVERGAARD, K., OVERGAARD, J.: Investigations on the possibility of a thermic tumor therapy-II. Action of combined heat-roentgen treatment on a transplanted mouse mammary carcinoma, Eur. J. Cancer 8 (1972) 573-575.
[10] OVERGAARD, K., OVERGAARD, J.: Radiation sensitizing effect of heat, Acta Radiol., Ther., Phys., Biol. 13 (1974) 501-511.
[11] PORTIS, B.: Effect of short wave length roentgen rays and diathermy on transplantable rat carcinoma, Trans. Chicago Pathol. Soc. 12 (1927) 15-22.
[12] ROBINSON, J.E., WIZENBERG, M.J., McCREADY, W.A.: Combined hyperthermia and radiation suggest an alternative to heavy particle therapy for reduced oxygen enhancement ratios, Nature (London) 251 (1974) 521-522.
[13] SELAWRY, O.S., CARLSON, J.C., MOORE, G.E.: Tumor response to ionizing rays at elevated temperature; a review and discussion, Am. J. Roentgenol. 80 (1958) 833.
[14] ALFIERI, A.A., HAHN, E.W., KIM, J.H.: The relationship between the time of fractionated and single doses of radiation and hyperthermia on the sensitization of an in vivo mouse tumor, Cancer 36 (1975) 893-903.
[15] HOFER, K.G., CHAPPIN, D.A., HOFER, M.G.: Effect of hyperthermia on the radiosensitivity of normal and malignant cells in mice, Cancer 38 (1976) 279-287.
[16] CORRY, P.M., ROBINSON, S., GETZ, S.: Hyperthermia effects on DNA repair mechanisms, Radiology 123 (1977) 475-482.
[17] DURAND, R.E.: Potentiation of radiation lethality by hyperthermia in a tumor model: Effects of sequence, degree, and duration of heating, Int. J. Radiat. Oncol., Biol. Phys. 4 (1978) 401-403.
[18] GILLETTE, E.L., ENSLEY, B.A.: Effect of heating order on radiation response of mouse tumor and skin, Int. J. Radiat. Oncol., Biol. Phys. 5 (1979) 209-213.
[19] HARISIADIS, L., SUNG, D., KESSARIS, N., HALL, E.J.: Hyperthermia and low dose-rate irradiation, Radiology 129 (1978) 195-198.
[20] HARRIS, J., MURTHY, A.K., BELLI, J.A.: The effect of delay between heat and x-irradiation on the survival response of plateau phase V-79 cells, Int. J. Radiat. Oncol., Biol. Phys. 2 (1977) 515-519.
[21] SAPARETO, S.A., RAAPHORST, G.P., DEWEY, W.C.: Cell killing and the sequencing of hyperthermia and radiation, Int. J. Radiat. Oncol., Biol. Phys. 5 (1979) 343-347.
[22] STEWART, B.S., DENEKAMP, J.: The therapeutic advantage of combined heat and x-rays on a mouse fibrosarcoma, Br. J. Radiol. 51 (1978) 307-316.
[23] THRALL, D.W., GILLETTE, E.L.: Combination of heat and ionizing radiation on C3H mouse skin and the C3H mouse mammary adeno CA in vivo; significance of the order of application and quantitation of the heat (abstr.), Radiat. Res. 59 (1974) 186.
[24] HORNBACK, N.B., SHUPE, R., SHIDNIA, H., JOE, B.T., SAYOC, E., GEORGE, R., MARSHALL, C.: Radiation and microwave therapy in the treatment of advanced cancer, Radiology 130 (1979) 459-464.
[25] HORNBACK, N.B., SHUPE, R.E., SHIDNIA, H., JOE, B.T., SAYOC, E., MARSHALL, C.: Preliminary clinical results of combined 433 Megahertz microwave therapy and radiation therapy on patients with advanced cancer, Cancer 40 6 (1977) 2854-2863.

Progress in Radio-Oncology II, edited by K. H. Kärcher et al., Raven Press, New York © 1982.

Clinical Applications of Hyperthermia with Radiation Therapy*

R. JOHNSON, R. SCOTT, T. BURKE,
R. KRISHNAMSETTY, J. SUBJECK
Roswell Park Memorial Institute,
Buffalo, New York,
United States of America

Abstract

Since 1976, 126 patients have been given 1008 heat applications. Fifty-three patients received 915 MHz microwave hyperthermia following radiotherapy. Forty of the 53 patients had sufficient follow up for evaluation. These patients were treated on two protocols. In the first protocol, patients with tumours within 4 cm of the skin surface, with a poor prognosis using conventional radiation alone, were given 400 rad fractions twice a week followed by local hyperthermia to 43.5°C for 30 minutes. In the second protocol, post-radiation hyperthermia was given twice per week to potentially curable patients wtih tumours within 4 cm of the skin surface receiving conventional 200 rad/d five times a week. These patients were treated to doses of 5500 to 6500 rads. Thirty-seven of the 53 patients had a complete tumour response, with follow up varying from three to fourteen months. Twenty-four of these patients had two lesions, one of which was heated while the other served as a radiated control. Nineteen of the 24 had a complete response to hyperthermia, while 14 of the 24 with radiation alone had a complete response. A longer follow up is required to determine if the radiation efficacy has been improved. More rapid tumour regression was usually noted in the hyperthermia-treated patients. Morbidity as measured by skin reaction was not usually increased in the heated field since air surface cooling was utilized.

INTRODUCTION

Hyperthermia may be used as an adjunctive treatment with radiation either during or immediately adjacent to radiation therapy or at a time separate from the radiation therapy.

When differential tumor heating can be obtained, heating during or immediately after radiation therapy will achieve the maximum therapeutic efficacy. If differential tumor heating is not obtained, such as may occur with regional heating, adjusting the timing of heating following radiation and the intervals between either hyperthermia and radiation or hyperthermia may have the potential to improve the therapeutic ratio.

The clinical objectives of this work are:

1. To determine the toxicity and efficacy of hyperthermia after a 400 rad fraction of radiation.
2. To determine whether the addition of hyperthermia will improve the local control rate of curable lesions treated with conventional fractionation.

* This work has been supported in part by National Cancer Institute Grant No. CA-14058.

CLINICAL METHODS

RTOG 78-06 - RPMI Protocol

Schedule I - This protocol determines the effect of the addition of hyperthermia on the control rate of superficial advanced tumors and also the effect of timing of hyperthermia after radiation on both tumor and normal tissue response. Patients with two or more lesions may be included in this protocol so that in some instances controls will be present in the same patient.

Schedule II - This protocol was designed to assess whether the addition of heat twice a week to conventional fractionation for potentially curable primary or secondary superficial tumors of the head and neck or breast will improve the long term tumor control. Whenever available, a second tumor in the same patient is used as an unheated radiated control. The heat treatments are started after 1000 rads of radiation to preclude the possibility that minimally radiated cells might be disseminated by the heat treatments.

Radiation Therapy

Radiation was given using either electrons or photons of appropriate energy. For the Schedule I patients, 400 rad fractions were administered at the 90% level to doses ranging from 3200 rads to 6000 rads. When paired lesions were present in the same patient, the radiated alone control has either received the same dose as the hyperthermia or has been carried to a higher dose when the control lesion has not responded equally to the hyperthermia.

The Schedule II patients with potentially curable lesions have been planned using conventional fractionation of 200 rads per day to a tumor dose in the range of 6000 to 6600 rads. Identical doses have been given to both the heated lesion and unheated control.

Heating Method

Heat applied using 915 MHz microwaves with RPMI dielectric filled waveguides was administered. An RPMI designed gantry system facilitates positioning of the patient. Skin surface cooling is achieved using surface air jets mounted with velco to the applicator head. Power is controlled manually and adjusted to maintain a tumor temperature, whenever possible, of 43.5°C for 30 minutes. Air cooling stops when power is switched off.

Thermometry

Thermometry has presented a major problem when using microwaves or RF heating systems. Our past experience has been in the use of microthermocouples in 20 gauge needles, thermistors, liquid crystal probes and, most recently, the Bowen probe. We have been unable to record the temperatures during treatment due to the interference of the conducting probes. In order to obtain an accurate temperature reading, we obtained cooling curves from the thermocouples immediately on turning the power off and extrapolate that to the time the power was turned off. A standard calculator has been programmed to allow immediate temperature determinations to be made from the cooling curves. Although the Bowen probe, as used by B.S.D., allows measurements to be made during hyperthermia, its size makes its insertion difficult and presents potential errors to real tissue measurement due to vascular damage which such a large probe can cause.

Blood Flow and pO_2 Assessment

An investigation is in progress to determine any relationships between tumor and normal tissue cooling curves, pO_2 and tumor regression. One minute cooling curves are obtained during the first and last treatments on all patients in the middle and at the end of the hyperthermia exposures using the same power level.

Oxygen probes are implanted into tumors of selected patients so that pressure measurements may be taken before, during and after x-ray treatments, and 43°C hyperthermia. Implantation over a several day period is performed since comparisons from one day to the next are valid only if the same microcirculatory zone is measured. Critical to meeting this requirement is the use of a Teflon-coated oxygen probe newly developed [1]. This device is biologically non-reactive, withstands an implantation period beyond one week and has a small enough tip size to measure microcirculatory zones undergoing changes produced by hyperthermia, x-rays or 100% oxygen breathing. The probe is implanted together with a 200 micron thermocouple through a 22 gauge plastic cannula

RESULTS

In this report 53 patients received microwave hyperthermia following radiotherapy. Forty of the 53 patients had sufficient follow up for evaluation. Thirty-seven of the 53 patients had a complete tumor response with a follow up varying from 3 to 14 months.

Table I shows the results of the addition of hyperthermia to radiotherapy for curable lesions of the breast and head and neck regions. For those patients available for a one year follow up, ten patients out of eleven had a complete response for breast lesions and all head and neck lesions had a complete response.

Table II shows the results of treating those breast cases which had two lesions. The tumor response at the end of treatment was generally more complete when hyperthermia was added. The one year follow up shows all

TABLE I. LONG TERM FOLLOW-UP RESULTS OF RPMI PILOT STUDY: HYPERTHERMIA PLUS RADIOTHERAPY

	BREAST	HEAD AND NECK
	Total	Total
End of Treatment	1 - No response 14 - Partial 3 - Complete	0 - No response 6 - Partial 3 - Complete
Four Month Follow Up	0 - No response 2 - Partial 16 - Complete	0 - No response 1 - Partial 8 - Complete
One Year Follow Up	0 - No response 1 - Partial 10 - Complete	0 - No response 0 - Partial 6 - Complete

heated lesions to respond completely whereas two of the unheated controls showed either partial or no response.

Table III shows those head and neck tumors which were multiple with one lesion as an unheated control. The response at the end of treatment was similar to the breast lesions, with a greater degree of response in the heated lesions than the radiotherapy control. At one year follow up, all hyperthermia treated lesions had a complete response with three radiotherapy controls having partial response only.

The skin reaction in the microwave field has not generally exceeded that of the unheated skin providing skin cooling has been used. On three occasions, small superficial heat blisters had been caused due to local hot

TABLE II. RPMI PILOT STUDY:
BREAST CARCINOMA - PAIRED LESIONS

	Hyperthermia Plus Radiotherapy	Radiotherapy Alone
	Total	Controls
End of Treatment	0 - No response 7 - Partial 2 - Complete	4 - No response 4 - Partial 1 - Complete
Four Month Follow Up	0 - No response 1 - Partial 8 - Complete	1 - No response 4 - Partial 4 - Complete
One Year Follow Up	0 - No response 0 - Partial 5 - Complete	1 - No response 1 - Partial 3 - Complete

TABLE III. RPMI PILOT STUDY:
HEAD AND NECK CARCINOMA - PAIRED LESIONS

	Hyperthermia Plus Radiotherapy	Radiotherapy Alone
	Total	Controls
End of Treatment	0 - No response 4 - Partial 1 - Complete	2 - No response 3 - Partial 0 - Complete
Four Month Follow Up	0 - No response 1 - Partial 4 - Complete	0 - No response 4 - Partial 1 - Complete
One Year Follow Up	0 - No response 0 - Partial 4 - Complete	0 - No response 3 - Partial 1 - Complete

spots. Late skin changes show no apparent increase in the microwave fields. Some patients failed to complete their treatment due to their inability to remain still for the duration of the treatment.

DISCUSSION

The pilot studies show that in no instances was the result of heat and radiation worse than radiation alone. The follow up of the patients with paired lesions, although limited, suggest that there will be an improvement in long term control of some tumors when heat is added. It is important that patients now be selected for hyperthermia who have long term survival since early regression may have no relation to long term control. We have noted, however, that the tumors which have not responded completely with heat and radiation have also shown minimal regression by the completion of treatment.

Repeated heat doses in the 43.5°C range may increase the vascularity of normal tissue and decrease that of some tumor tissue. The tumor pH would be expected to fall differentially during heating periods in many tumors with a poor vascular supply. A differential blood flow between tumor and normal tissue would be required for regional heating to improve the therapeutic efficacy unless the tumor cell pH is selectively lowered.

Methods of deep heating require development if regional heating does improve the therapeutic efficacy. Various claims have been made for different methods of deep heating; however, no method has yet completed clinical thermal dosimetry.

The major problem confronting the radiotherapists using hyperthermia, with the exception of ultrasound, is thermometry. Perturbing probes may be extremely misleading. The current non-perturbing probes are large and may in turn give false readings due to vascular damage or are such a size that they cannot be repeatedly used clinically. The new invasive non-perturbing fiber optic probes should alleviate this problem to a degree, but the ultimate aim should be to use non-invasive thermometry.

The protocol utilizing 400 rad plus heat twice a week was designed in order to minimize possible effects of thermal tolerance and utilize the maximum effect of heat radiation sensitization with the large fractions. The fewer large fractions will also have the disadvantage of being less effective for reoxygenation unless a radiation sensitizer was administered. It would appear that the therapeutic ratio might be improved by utilizing a second heat treatment at an optimum time, which might be shorter than 72 hours if thermal tolerance is inhibited _in-vivo_ by a low pH.

There is no current clinical data to suggest that a therapeutic gain can be achieved if cellular conditions are similar for tumor and normal tissue cells. Whether the magnitude of any differential thermal tolerance effect obtained by varying the intervals between either heat and radiation or between treatments if of a sufficient magnitude to be discernible clinically is unknown. The clinical therapeutic efficacy would be expected to increase if tumor cells are heated more than the normal tissue cells and/or maintained at a lower pH.

Thermal tolerance, the split dose recovery phenomena in hyperthermia, is of significant interest because the recovery of thermal resistance in the heated cells greatly surpasses the resistance of unheated cells. _In-vitro_ the development of thermotolerance can lead to a protection ratio of orders of magnitude [2]. Thermotolerance has also been observed in-vivo [3], although the levels of protection induced and relative normal and tumor tissue inducibility is unclear. However, this phenomenon might be of clinical importance if modest levels of thermotolerance are induced in-vivo. Relative inducibilities of normal and tumor cells in-vitro and in-vivo as well as induction and decay kinetics may yield valuable clues to assist the therapist in scheduling thermal doses, in determining length of treatments, treatment temperature, etc., in order to optimize the therapeutic ratio.

Because of the potential magnitude of the thermotolerance phenomenon, it is of importance to understand the biological basis for its induction. A thermotolerance phenomenon has also been observed in yeast [4]. In this study, the induction of thermal resistance was found to correlate strongly with the synthesis of heat shock proteins. Using a temperature sensitive mutant yeast cell, these authors further demonstrated that the development of thermoresistance in this system requires the transport of RNA from the nucleus to the cytoplasm. In a similar study in mammalian cells, Subjeck et al. [5,6] has shown a similar strong correlation between thermotolerance induction and heat shock protein induction. The fact that heat shock proteins may be responsible for thermotolerance gives no information as to the role which these proteins play in the development of thermotolerance. If this proposal proves to be correct, then work correlating thermotolerance with cellular changes may provide clues for the elucidation of heat shock protein function. Since these proteins are present in significantly reduced quantities in non-heated cells, it is also conceivable that the control levels of these proteins may predict inherent cell thermosensitivities to primary heat doses. Since heat shock proteins are inducible under conditions other than heat shock [7], it is possible that thermotolerance may be inducible under other circumstances.

SUMMARY

The radiation therapeutic efficacy is increased when differential heating can be achieved without impairing reoxygenation. A knowledge of basic mechanisms of thermal tolerance and radiation tolerance is required if these factors are to be utilized in improving the therapeutic ratio. Thermometry and heat production systems require development before the results of regional heating can be assessed.

REFERENCES

[1] BURKE, T.R., JOHNSON, R.J.R., MAROSAN, Cs.B., KRISHNAMSETTY, R., SAKO, K., KARAKOUSIS, C., WOJTAS, F.: Implantable oxygen microelectrode suitable for medium-term investigations of post-surgical tissue hypoxia and changes in tumor tissue oxygenation produced by radiotherapy, J. Med. (1981, in press).

[2] LI, G.C., HAHN, G.M.: A proposed operational model of thermotolerance based on effects of nutrients on the initial treatment temperature, Cancer Res. 40 (1980) 4501-4508.

[3] HENLE, K.J., DETHLEFSEN, L.A.: Heat fractionation and thermotolerance: a review, Cancer Res. 38 (1978) 1843-1851.

[4] McALISTER, L., FINKELSTEIN, D.B.: Heat shock proteins and thermal resistance in yeast, Biochem. Biophys. Res. Comm. 93 3(1980).

[5] SUBJECK, J.R., SCIANDRA, J., REPASKY, E., JOHNSON, R.J., :"Induction of new protein synthesis following hyperthermia at 45 °C and 41 °C in CHO cells", Radiation Research (Proc. 29th Annu. Mtg., Minneapolis, Minnesota 31 May - 4 June, 1981).
[6] SUBJECK, J.R., SCIANDRA, J., JOHNSON, R.J.: Thermotolerance and heat shock proteins, Br. J. Radiol., submitted 1981.
[7] LEVINSON, W., OPPERMANN, H., JACKSON, J.: Transition series metals and sulphydryl reagents induce the synthesis of four proteins in eukaryotic cells, Biochim. Biophys. Acta 606 (1980) 170-180.

Progress in Radio-Oncology II, edited by K. H. Kärcher et al., Raven Press, New York © 1982.

Whole-Body Hyperthermia at Moderate Temperatures in the Treatment of Malignant Disease

H. MADOC-JONES, J.J. SANTORO,
E.S. STERNICK, D.F.H. WALLACH
Department of Therapeutic Radiology,
Tufts-New England Medical Center,
Boston, Massachusetts,
United States of America

Abstract

Moderate-temperature (40–41°C) hyperthermia in combination with localized radiotherapy is undergoing a Phase I trial at Tufts–New England Medical Center. Hyperthermia is achieved by depositing 27 MHz RF energy primarily into the great vessels of the trunk and abdomen, using a computer-assisted system. Circulation of warm air reduces heat loss from the skin. Patients treated have either disseminated or locally advanced and uncontrollable malignant disease. The treatment plan calls for 9 courses of hyperthermia over 3 weeks, i.e. 3 treatments/week, with local radiotherapy (according to departmental policy for the particular tumour) 5 times/week. Unsedated patients are brought to 40 ± $\frac{1}{2}$°C in 30–40 min, maintained at this temperature for 45 min and irradiated immediately thereafter. Of 9 patients, 6 completed 8 or more courses of hyperthermia without serious side-effects. Three patients did not complete the treatment plan because of complications of their disease. The rationale for 40–41°C hyperthermia is presented and fundamental research in this area is described. Clinical laboratory investigations on patients subjected to whole-body hyperthermia are summarized.

1. INTRODUCTION

At Tufts-New England Medical Center, a Phase I trial of whole-body hyperthermia at moderate temperatures is under way in combination with localized radiotherapy. The choice of temperatures between 40 and 41°C was based on both basic and clinical considerations. Thus, whole-body heating to 40-41°C can be achieved simply and safely [1-3] as an adjunct to fractionated chemotherapy or radiation therapy, whereas higher temperatures, e.g. 42°C, have resulted in unacceptable rates of serious complications including cardiovascular and neurologic sequelae, and even death [4-6].

Available information shows that many tumour cell types, both in vitro and in vivo, are suppressed or damaged by heat more readily than are normal cells [7-11]. Growth inhibition of most heat-sensitive, neoplastic-cell mutants occurs at 41°C [12-18]. Furthermore, significant modulations of microtubule stability occur at 41°C [19] and known membrane protein state changes [20] and heat enhancement of drug permeation [21,22] occur at 41°C. In addition, the membrane potential, which is known to affect permeation of many chemical agents, has been shown to be thermosensitive in the moderate heat range [23-26] and differences in thermal responses have been observed between normal and neoplastic cells [24]. Finally, significant enhancement of cancer chemotherapeutic effects has been observed at target temperatures of 40°C by limb perfusion [27] and, more recently, by whole-body hyperthermia [3]. In the light of the above, temperatures at which cells are killed by heat alone (above 42°C) may not be necessary.

2. CLINICAL DEVELOPMENT OF MODERATE-TEMPERATURE WHOLE-BODY HYPERTHERMIA

Whole-body heating in the 40-41°C range was pioneered at Freiburg, Federal Republic of Germany, and has been shown to be relatively simple and to be well tolerated as an adjunct to fractionated chemotherapy and/or radiation therapy [1-3].

The whole-body hyperthermia unit used in Freiburg is a prototype manufactured by Siemens. A copy of this unit has been constructed at the Tufts-New England Medical Center (Fig.1). The patient is placed in a plastic box and core heating to 40-41°C is achieved by depositing 27 MHz RF energy primarily into the great vessels of the trunk (and abdomen). Circulation of warm air reduces heat loss from the skin. The desired core temperature is reached within 30-40 min and individuals are readily maintained at the target temperature for 60 min or more. When the patient is brought back to ambient conditions the temperature returns rapidly to normal.

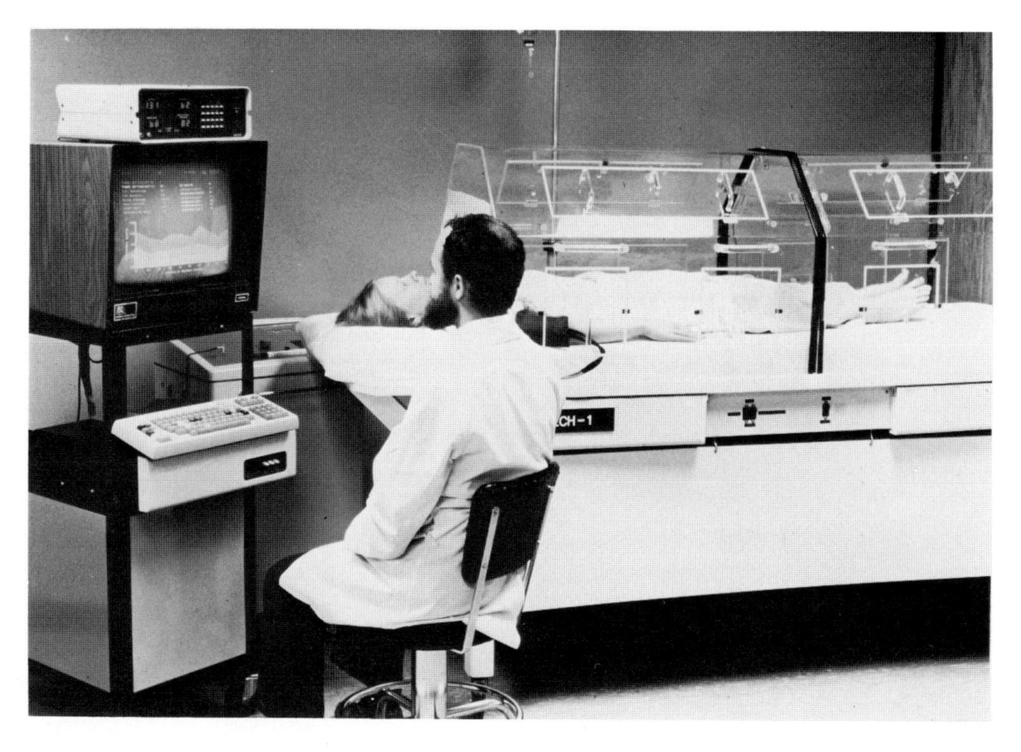

FIG.1. Whole-body hyperthermia apparatus at Tufts-New England Medical Center.

Computerized techniques are being used to measure and record the temperature of the patient and to monitor other physiological parameters such as blood pressure, heart rate etc. We are in the process of developing our computer system to control patient temperature using optical temperature probes that are not affected by the RF current.

Because heating is concentrated in the body 'core', many physiological compensatory mechanisms are minimized. Heating is therefore well-tolerated and cardiovascular or other ill-effects are avoided. Since the heating is concentrated in the body core, the procedure has obvious potential in the therapy of deep-seated tumours of the thorax, abdomen or pelvis. At Freiburg over 70 patients have been treated by this method without ill-effects (over 500 hours total at target temperature). We now report the initial results of our Phase I trial.

TABLE I. PATIENTS TREATED TO MAY 1981

Diagnosis	WBH/(weeks) [a]	Total tumour dose (rads)/(weeks)	Survival [b]
1. Malignant mesothelioma	12/4	6000/6	12 months (expired)
2. Metastatic thymoma	9/3	5000/6	6 months (expired)
3. Metastatic un-differentiated large-cell carcinoma	9/3	(a) 5800/6 (chest wall) (b) 3000/2 (mediastinum)	8 months
4. Metastatic rectal carcinoma	3/1	500/5	8 months
5. Primary lung carci-noma or metastasis from breast	2/1	1000/5	1 month (expired)
6. Recurrent colon carcinoma	8/3	4500/5	7 months
7. Lung adenocarcinoma	8/3	4000/4	2 months
8. Lung adenocarcinoma	2/1	6600/7	3 months
9. Lung adenocarcinoma	9/4	6600/7	

[a] WBH/(weeks): Courses of whole-body hyperthermia/number of weeks.

[b] Survival: Interval since last radiation treatment. No.1 died of progressive disease; No.2 died of myocardial infarction; No.5 died of progressive disease.

3. PHASE I TRIAL OF WHOLE-BODY HYPERTHERMIA (40-41°C) AND FRACTIONATED RADIOTHERAPY

By May 1981, we had treated nine patients with either disseminated or locally advanced and uncontrollable malignant disease (Table I). Patients are housed in our Clinical Study Unit during their hyperthermia schedule. They are heated to 40.0 ± 0.5°C in 30-40 min and are maintained at the target temperature for 45 min. Thereafter the patients are covered with blankets and wheeled from the hyperthermia tank to the irradiation room. Radiation therapy is administered, according to departmental policy appropriate for the particular tumour, while the patients are still hyperthermic. Our programme is to give 9 courses of hyperthermia over a period of three weeks, i.e. 3 treatments per week, with conventional radiotherapy on 5 days per week (hyperthermia and radiotherapy on Monday, Wednesday and Friday; radiation alone on Tuesday and Thursday).

Of the nine patients, six completed eight or more courses of hyperthermia. Three patients did not complete their scheduled hyperthermia because of complications of their disease rather than difficulties with hyperthermia. There were no serious side-effects, and the minor side-effects are given in Table II. Relaxation counselling has proven very useful in alleviating anxiety and claustrophobia.

4. CONCLUSION

Nine patients have received up to nine courses each of moderate-temperature hyperthermia while receiving conventional radiotherapy. The patients have tolerated this well without any serious side-effects or complications. It is therefore reasonable to pursue this trial and to develop a Phase II trial for moderate-temperature whole-body hyperthermia as an adjunct to radiation therapy.

TABLE II. SIDE-EFFECTS OF WHOLE-BODY HYPERTHERMIA
Number of patients - 9; total courses of hyperthermia - 62

Side-effect	Episodes	Patients
Anxiety/claustrophobia [a]	all	all
Dyspnoea	2	2
Hyperventilation	3	2
Indigestion	9	1
Reflux oesophagitis	1	1
Skin burn [b]	1	1

[a] Anxiety alleviated by 'relaxation techniques'.

[b] Single episode due to unsatisfactory RF coil.

5. FUTURE WORK

We are initiating a Phase I trial of moderate-temperature hyperthermia with concurrent chemotherapy. In addition, our clinical hyperthermia programme has a laboratory research arm looking at several biochemical and physiological responses to 27 MHz RF hyperthermia and trying to separate these from the effects of pyrogens. In this we are examining acute-phase proteins, complement activation, endogenous pyrogens, neutrophil phagocytic properties and natural killer cell mechanisms in samples of patients made hyperthermic. In addition, we are looking at the effect of hyperthermia on the blood flow of tumours treated with high-energy photons on the betatron. This can be measured by monitoring the wash-out of ^{15}O generated during betatron therapy [28,29].

REFERENCES

[1] FABRICIUS, H.-A., NEUMANN, H., STAHN, R., ENGELHARDT, R., LOHR, G.W.: Klinisch-chemische und immunologische Veränderungen bei Gesunden nach einer einstündigen 40°C-Ganzkörperhyperthermie, Klin. Wochenschr. 56 (1978) 1049-1056.

[2] NEUMANN, H., ENGELHARDT, R., FABRICIUS, H.-A., STAHN, R., LOHR, G.W.: Klinisch-chemische Untersuchungen an Tumorpatienten unter Zytostatika- und Ganzkörperhyperthermiebehandlung, Klin. Wochenschr. 57 (1979) 1311-1315.

[3] NEUMANN, H., FABRICIUS, H.-A., ENGELHARDT, R.: Moderate whole-body hyperthermia in treatment of small cell carcinoma of the lung, J.Natl. Cancer Inst. (1981, in press).

[4] LARKIN, J.M., EDWARDS, W.S., SMITH, D.E.: Systemic thermotherapy: Description of a method of physiological tolerance in clinical subjects, Cancer 40 (1977) 3155-3181.

[5] BARLOGIE, B., CORRY, P.M., YIP, E., LIPPMAN, L., JOHNSTON, D.A., KALIL, K., TENCZYNSKI, T.R., REILLY, E., LAWSON, R., DOSIK, G., RIGOR, B., HANKENSON, R., FREIREICH, E.J.: Total body hyperthermia with and without chemotherapy for advanced human neoplasms, Cancer Res. 39 (1979) 1481-1489.

[6] BULL, J.M., LEES, D., SCHUETTE, W., WHANG-PENG, J., SMITH, R., BYNUM, G., ATKINSON, E.R., GOTTDIENER, J.S., GRALNICK, H.R., SHAWKER, T.H., DEVITA, V.T., Jr.: Whole-body hyperthermia: a phase-I trial of a potential adjuvant to chemotherapy, Ann. Intern. Med. 90 (1979) 317-323.

[7] WALLACH, D.F.H.: Basic mechanisms in tumor thermotherapy, J. Mol. Med. 2 (1977) 381-403.

[8] MORICCA, G., CAVALIERE, R., CAPUTO, A., MONDOVI, B.: Hyperthermia treatment of tumours: experimental and clinical applications, Recent Results Cancer Res. 59 (1977) 112-152.

[9] OVERGAARD, J.: Effect of hyperthermia on malignant cells in vivo. A review and a hypothesis, Cancer 39 (1977) 2637-2646.

[10] STREFFER, C.A., VANBUEINNGEN, D., DIETZEL, F., ROTTINGER, E., ROBINSON, L.E., SCHERER, E., SEEBER, S., TROTT, K.R., Eds: Cancer Therapy by Hyperthermia and Radiation, Urban and Schwarzenberg, Baltimore, Munich (1978).

[11] FIELD, S.B., BLEEHEN, N.M.: Hyperthermia in the treatment of cancer, Cancer Treat. Rev. 1979 6 (1980) 63-94.

[12] RENGER, H.C., BASILICO, C.: Mutation causing temperature-sensitive expression of cell transformation by a tumor virus (SV40-3T3 mouse cells - growth control), Proc. Natl. Acad. Sci. USA 69 (1972) 109-114.

[13] RENGER, H.C., BASILICO, C.: Temperature-sensitive simian virus 40 transformed cells: phenomena accompanying transitions from the transformed to the "normal" state, J. Virol. 11 (1973) 702-708.
[14] DI MAYORCA, G., GREENBLATT, M., TRAUTHEN, T., SOLLER, A., GIORDANO, R.: Malignant transformation of BHK $_{21}$ clone 13 cells in vitro by nitrosamines - A conditional state, Proc. Natl. Acad. Sci USA 70 (1973) 46-49.
[15] YAMAGUCHI, N., WEINSTEIN, I.B.: Temperature-sensitive mutants of chemically transformed epithelial cells, Proc. Natl. Acad. Sci. USA 72 (1975) 214-218.
[16] BOUCK, N., DI MAYORCA, G.: Somatic mutation as the mechanism of malignant transformation by chemical carcinogens, Nature (London) 264 (1976) 23-30.
[17] MIYASHITA, K., KAKUNAGA, T.: Isolation of heat- and cold-sensitive mutants of Chinese hamster lung cells affected in their ability to express the transformed state, Cell 5 (1975) 131-138.
[18] BASILICO, C.: Temperature-sensitive mutations in animal cells, Adv. Cancer Res. 24 (1977) 223-266.
[19] LIN, P.-S., TURI, A., KWOCK, L., LU, R.C.: Hyperthermic effect on microtubule organization, J. Natl. Cancer Inst. (1981, in press).
[20] WALLACH, D.F.H., MIKKELSEN, R.B., KWOCK, L.: "Plasma membranes as targets and mediators in tumor chemotherapy", in Molecular Actions and Targets for Cancer Chemotherapeutic Agents (2nd Ann. Bristol-Myers Symp. Cancer Research, New Haven, Connecticut, 1979: SARTORELLI, A., Ed.), (1981) 433-451.
[21] GOSS, P., PARSONS, P.G.: The effect of hyperthermia and melphalan on survival of human fibroblast strains and melanoma cell lines, Cancer Res. 37 (1977) 152-156.
[22] THUNING, C.A., BAKER, N.A., WARREN, J.: Synergistic effect of combined hyperthermia and a nitrosourea in treatment of a murine ependymoblastoma, Cancer Res. 40 (1980) 2726-2790.
[23] MIKKELSEN, R.B., WALLACH, D.F.H.: Periodate induced cross-linking of concanavalin A - reactive membrane proteins of rabbit thymocytes, Cell Biol. Int. Rep. 1 (1977) 51-55.
[24] YI, P.N.: Cellular ion content changes during and after hyperthermia, Biochem. Biophys. Res. Commun. 91 (1979) 177-182.
[25] MIKKELSEN, R., KOCH, B.: Thermosensitivity of the membrane potential of normal and transformed hamster lymphocytes, Cancer Res. 1 (1980) 239-260.
[26] MIKKELSEN, R.B., KOCH, B.: Membrane potential thermosensitivity of normal and simian virus 40 transformed lymphocytes, J. Natl. Cancer Inst. (1981, in press).
[27] STEHLIN, J.S., GIOVANELLA, B.C., DE IPOLYI, P.D., MUNEZ, R.L., ANDERSON, B.A.: Results of hyperthermia perfusion for melanoma of the extremities, Surg. Gynecol. Obstet. 140 (1975) 339-348.
[28] HUGHES, W.L., NUSSBAUM, G.H., EMAMI, B., CONNOLLY, R., REILLY, P.: Photon activation ^{15}O decay studies of tissue perfusion, Science 204 (1979) 1215-1217.
[29] TEN HAKEN, R., NUSSBAUM, G., EMAMI, B., HUGHES, W.: Photon activation ^{15}O decay studies of tumor blood flow, Med. Phys. (1981, in press).

Progress in Radio-Oncology II, edited by K. H.
Kärcher et al., Raven Press, New York © 1982.

Interactions Between Misonidazole, Hyperthermia, and Some Cytotoxic Drugs on Multicellular Tumour Spheroids

N.M. BLEEHEN, Jane E. MORGAN
University Department and
Medical Research Council Unit of Clinical Oncology and Radiotherapeutics,
The Medical School,
Cambridge,
United Kingdom

Abstract

Multicellular tumour spheroids of the EMT6 cell line have been used to investigate the interaction of hyperthermia with bleomycin and adriamycin, by clonogenic assay and measurement of growth delay. Enhancement of bleomycin damage was seen, but mainly additive effects with adriamycin. No interactions were seen with the EMT6 tumour in vivo under the conditions selected. Misonidazole pretreatment under hypoxia produced marked enhancement of the effect of subsequent oxic hyperthermia. This sensitization was removed when pretreated spheroids were subsequently incubated in oxygen at 37°C before the heat. Incubation at 0°C produced variable effects, depending on spheroid size. Addition of cysteamine during the pretreatment increased the hyperthermic cell kill, whilst its addition only during the hyperthermia protected. Possible explanations for these observations are discussed.

1. INTRODUCTION

The multicellular spheroid grown in suspension culture was originally proposed by Sutherland and his colleagues (1) as a model which is intermediate in organisation between monolayer cell cultures and solid tumours in vivo. Response to treatment may be measured by trypsin disaggregation and clonogenic assay of the surviving fraction, or, as suggested by Yuhas et al. (2), by regrowth delay.

With increase in size, structural changes occur. Three separate zones become apparent. Central necrosis is separated from a rim of viable cells on the outside by an annulus of pyknotic cells, and there is a relationship between the size of these zones with oxygen deprivation and their responses to radiation. Sutherland and Durand have presented a comprehensive review of the data available up to 1976 (3) which should be consulted for further details.

Hyperthermia in the treatment of cancer is currently of considerable experimental and clinical interest (4). There are several reasons why it might be of value, some of which make the tumour spheroid model particularly appropriate for in vitro studies. Thus, hypoxic cells are at least as sensitive as well oxygenated cells to the effect of heat; nutritionally deprived cells and cells at low pH are especially sensitive; cells in S-phase which tend to be resistant to X-rays are sensitive to heat and actions therefore may be complimentary in effect; and finally, enhanced responses may be seen from the interaction of heat with radiation and some cytotoxic drugs.

Sutherland (5) reported differential cell killing by heat across the spheroid, with the central cells being more sensitive, whilst Durand (6) showed increasing thermal resistance with increase in spheroid size and considerably less sensitivity than that seen in single cells. Enhanced radiation damage by hyperthermia has been reported (6,7).

Sutherland and his co-workers (8) reported marked resistance to the cytotoxicity of adriamycin by large (600-1000 μm diameter) EMT6/Ro spheroids when compared with similar cells growing in exponential monolayer. They explained this, at least in part, by demonstrating a gradient in adriamycin concentration from the peripheral to central regions of the spheroid. However, differences in the kinetic state, hypoxic status and the phenomena resulting from cell association must also play their part.

Selective cytotoxicity of misonidazole for hypoxic cells has been demonstrated by several groups of workers (9,10) and subsequently investigated by many others, including our own. Hyperthermic potentiation of the hypoxic cytotoxicity of misonidazole has been seen in vitro (11,12) and in vivo (12). Sridhar and Sutherland (13) investigated this phenomenon in spheroids and demonstrated enhanced cytotoxicity, not only in the radiobiologically hypoxic cells but also in aerobic cells, an effect also reported by us in vivo (12).

It has also been shown that pretreatment of cells in vitro with misonidazole may enhance their sensitivity to heat and to some chemotherapeutic agents (11). Tumours in vivo may show a similar enhanced response to some anti-cancer drugs following pretreatment (14,15).

We have recently been investigating the effect on EMT6 spheroids of the interactions of hyperthermia with bleomycin (16,17) and adriamycin (17). We have also more recently studied the result of misonidazole pretreatment on the heat sensitivity of these spheroids at different sizes (18) and the influence of cysteamine on the interaction. This paper summarises some of these results, together with new data.

2. MATERIALS AND METHODS

2.1 Cells

Separate sublines of the EMT6 tumour were used for monolayer cultures (EMT6/Ca/M/CC) and for growth as multicellular spheroids in vitro or tumours in vivo in Balb/C mice (EMT6/Ca/VJAC). Further details are described elsewhere (12,16).

2.2 Multicellular spheroids

2.2.1 *Preparation*

Spheroids were grown from the VJAC subline maintained by alternating growth as a solid tumour in vivo with four passages in vitro.

Cells from one of the in vitro passages were used to grow up spheroids according to the method of Yuhas et al. (2) and as described for our laboratory by Twentyman (19). Briefly, single cells are grown in Eagle's medium with 10% foetal calf serum in tissue culture flasks base coated with agar to prevent adhesion to the plastic surface. After incubation at 37^{0}C they reach a size of 200-300 μm by day 6. To produce larger spheroids, single spheroids are then placed into agar coated wells on plastic multi-dishes and incubated for a further 6 days, at which time they reach a diameter of 650-800 μm.

2.2.2 *Pretreatment*

Pretreatment was carried out following transfer of spheroids into continuously stirred glass spinner vessels containing 100 ml complete medium with or without 5 mM misonidazole at 37^{0}C. Continuous gassing with 5% CO_2/95% N_2 (O_2 < 10 ppm) at a rate in excess of 500 ml/min was used to produce hypoxia. At the end of the pretreatment period, spheroids were washed twice with fresh medium before further treatment.

2.2.3 *Heat treatment*

Treatments at 37^{0}C were carried out in an incubator. Temperatures above this were achieved by immersion of containers, with the spheroids in appropriately gassed complete medium together with drugs as required.

2.2.4 *Drugs*

Misonidazole (Roche Laboratories), Bleomycin (Lundbeck) and Adriamycin (Farmitalia) were all dissolved in Hanks balanced salt solution and then appropriately diluted before use.

2.2.5 *Assay methods*

After treatment, spheroids were washed in fresh medium. They were then disaggregated by addition of 0.075% trypsin at 37^{0}C for 15 minutes and assayed for clonogenic fraction by in vitro colony formation. Alternatively, individual spheroids were placed in wells of plastic multi-dishes and growth delay measured microscopically with an eye-piece graticule. Full details of the assay methods are reported elsewhere (19).

3. RESULTS

3.1 Bleomycin (BLM)

Exponential monolayer cultures of EMT6 cells in the presence of 2-20 μg/ml BLM show an enhanced cytotoxicity following incubation for 1h at 43^{0}C. This is much greater than that seen for hyperthermia or drug alone. A similar effect was seen with 200-300 μm spheroids when the effect was assayed either by measurement of the surviving fraction immediately after treatment or by regrowth delay (Fig.1). In contrast, no increase in cytotoxicity was seen when EMT6, as intramuscular leg tumours (120-170 mm^3 volume), was treated with 0.06 mg/g BLM injected i.p. and treated for 1 h at an intra-tumour temperature of around 43^{0}C.

Pretreatment of spheroids at 40^{0}C for 3-6 h induced both thermal and drug tolerance as measured by both clonogenic assay and growth delay.

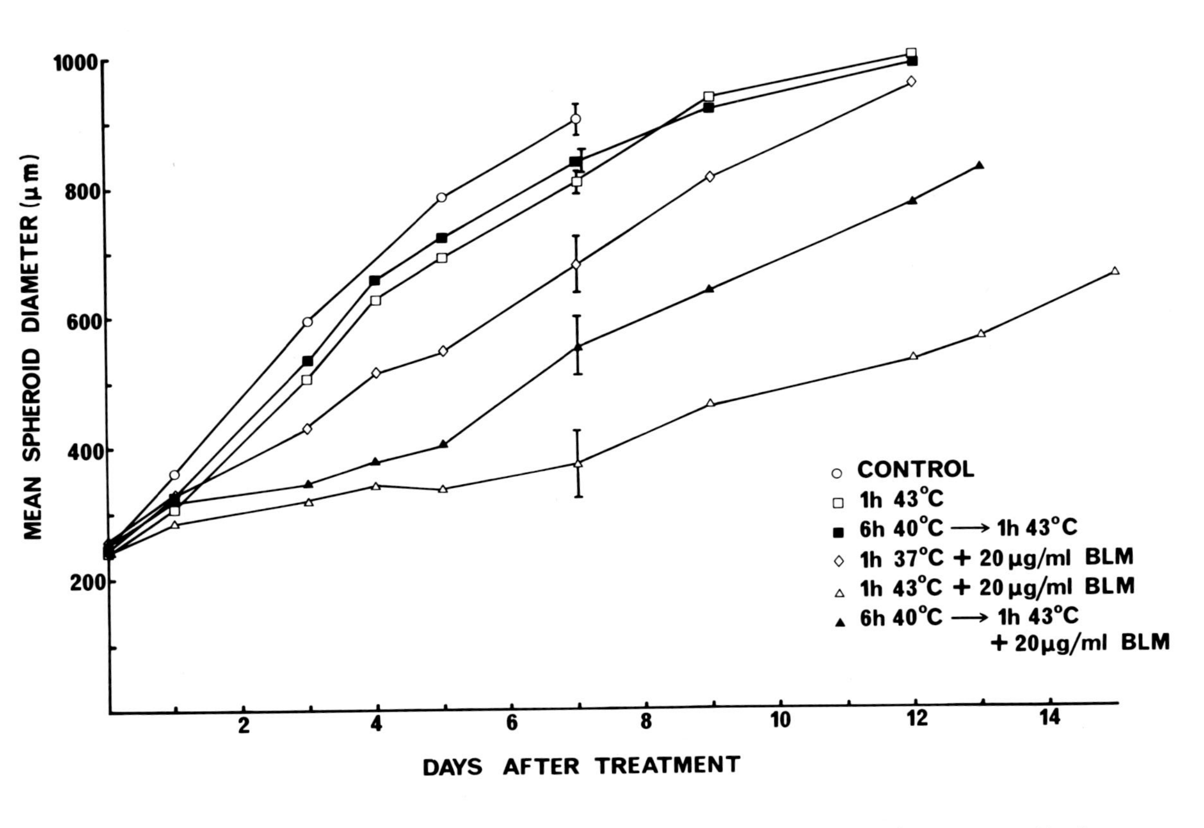

FIG.1. Growth curves for EMT6 speroids treated with BLM on day zero. Each point represents the mean spheroid diameter from groups of 8-12 spheroids. Limits shown are 2•SE.

Preliminary experiments have failed to demonstrate such tolerance with the solid tumours.

3.2 Adriamycin (ADM)

In contrast to the above results we were unable to demonstrate any effect greater than an independent action of both ADM and heat (1 h at 43°C) on 200-300 μm spheroids when measured either by clonogenic assay or growth delay. However, after prolonged exposure for 6 h at 43°C to 1 μg/ml ADM an enhanced effect was seen by both methods.

3.3 Misonidazole (MISO)

The effect of hypoxic MISO pretreatment on the subsequent heat sensitivity of spheroids has been investigated under a variety of conditions. Hypoxic incubation with 5 mM MISO significantly enhanced the response of both 200 μm and 650 μm spheroids to subsequent heat at 43°C as measured by both assay methods, and is dependent on the duration of pretreatment (2-6 h). Oxic incubation at 37°C of washed spheroids between pretreatment with MISO and subsequent hyperthermia results in a progressive loss of heat sensitisation. Thus, in large spheroids, there is a 20 fold recovery after 6 h at 37°C before 1½ h at 43°C. Small spheroids showed recovery of a lesser magnitude.

In an attempt to inhibit this recovery of sublethal damage, spheroids were held at 0°C instead of 37°C during the oxic period before heat. This showed differences between small and large spheroids. In the former, recovery was inhibited and indeed there was even a trend to increased subsequent heat sensitivity. However, with large spheroids, an unexpected finding was the progressive fall in surviving fraction when pretreated spheroids were held at 0°C without any subsequent treatment. Regrowth assays confirmed both sets of observations.

In contrast to pretreatment under hypoxia, similar work under oxic conditions, at comparable levels of MISO cytotoxicity, failed to demonstrate any increased thermal sensitivity.

3.4 Cysteamine (CYS)

We have studied the effect of CYS to see if it can interact with putative toxic products from MISO pretreatment. Simultaneous hypoxic pretreatment of large spheroids with 5 mM CYS and MISO increased the heat sensitization previously seen with MISO alone. However, if CYS was present only during the heating period, protection was observed.

4. DISCUSSION

The multicellular tumour spheroid is a useful model with which to investigate complex schedules. Precision of the treatment conditions is better than with in vivo tumours, whilst structural and metabolic changes may mimic those in vivo. However, we have demonstrated, both with BLM and ADM, that spheroid responses following combined heat and drug treatment appear to be closer to those seen from in vitro monolayer cultures than the in vivo tumour response. We have no explanation for this observation other than that the in vivo heating by waterbath immersion is not as effective in producing a uniform temperature throughout the tumour (12). Also, penetration of drugs into tumours, and particularly ADM, may also not be uniform and be very dependent on exposure time. This is discussed further elsewhere (17).

Our studies on hypoxic MISO pretreatment of spheroids demonstrate enhancement of the cytotoxicity of subsequent exposure to heat. More than 1 h pretreatment is required and this may relate to several factors. We believe that full hypoxia is achieved within 0.5 h of gassing but have no direct measurements of oxygen tension to confirm this. Our data would suggest that build up of a cytotoxic reduction product is required. Such cytotoxic intermediates have been proposed (20). Incubation at 37^{0}C before hyperthermia results in progressive loss of heat sensitization suggesting either repair of sublethal damage or removal of the reduction products from the spheroid. Our studies with small spheroids held at 0^{0}C instead of 37^{0}C suggest that an active repair process may be involved, although the other mechanism cannot be excluded. We have no adequate explanation for the increased cytotoxicity seen when large spheroids are held at 0^{0}C after MISO pretreatment. It may be that active repair processes are stopped in a situation where diffusion of cytotoxic products from the greater volume of the large spheroid is also reduced. We have measured the overall concentration of MISO remaining in spheroids after hypoxic pretreatment and find it to be low ($\sim$ 5% of the original MISO concentration). It does not seem likely therefore that retained MISO can explain the observed effects.

Cysteamine may protect against cell killing due to the nitro-reduction products of MISO (9). This protection may be pH dependent (21). Our observations of a protective effect when CYS is present during the heating, but not pre-heating, is consistent with its action as a radical scavenger. However, one might then have expected this also to happen when it is present during the pretreatment period. This difference may relate to changes in pH in the spheroids under the different conditions.

5. CONCLUSIONS

Our studies with hyperthermia and drugs on spheroids have demonstrated that it can produce useful data on interactions. However, the spheroid model is not free of its own artefacts and may still not be fully predictive of in vivo effects. This difference may be due to real differences in qualitative responses, or might just be quantitative differences due to the selection of inappropriate test conditions. The possibility of using both clonogenic and regrowth assays may be a real advantage where recovery from potentially lethal damage occurs, as with both heat and some drugs.

REFERENCES

(1) SUTHERLAND, R.M., McCREDIE, J.A., INCH, W.R., Growth of multicellular spheroids in tissue culture as a model of nodular carcinomas, J. Natl. Cancer Inst., 46 (1971) 113.

(2) YUHAS, J.M., LI, A.P., MARTINEZ, A.O., LADMAN, A.J., A simplified method for production and growth of multicellular tumour spheroids, Cancer Res., 37 (1977) 3639.

(3) SUTHERLAND, R.M., DURAND, R.E., Radiation response of multicell spheroids - an in vitro tumour model, Current Topics Radiation Res., 11 (1976) 87.

(4) FIELD, S.B., BLEEHEN, N.M., Hyperthermia in the treatment of cancer, Cancer Treat. Rev., 6 (1979) 63.

(5) SUTHERLAND, R.M., "Effect of hyperthermia on reoxygenation in multicell spheroid systems", in Cancer Therapy by Hyperthermia (Proc. Int. Symp.) Amer. Coll. Radiol., Washington D.C. (1975) 99.

(6) DURAND, R.E., "Effect of hyperthermia and/or irradiation in multicell spheroid systems", in Cancer Therapy by Hyperthermia and Radiation (Proc. Int. Symp.) Amer. Coll. Radiol., Washington D.C. (1975) 101.

(7) LUCKE-HUHLE, J., DERTINGER, H., Kinetic response of an in vitro 'tumour model' (V79 spheroids) to 42^{0}C hyperthermia, Europ. J. Cancer, 13 (1977) 23.

(8) SUTHERLAND, R.M., EDDY, H.A., BAREHAM, B., REICH, K., VANANTWERP, D., Resistance to adriamycin in multicellular spheroids, Int. J. Radiat. Oncol. Biol. Phys., 5 (1979) 1225.

(9) HALL, E.J., ASTOR, M., GEARD, C., BIAGLOW, J., Cytotoxicity of Ro-07-0582; enhancement by hyperthermia and protection by cysteamine, Br. J. Cancer, 35 (1977) 809.

(10) SUTHERLAND, R.M., BAREHAM, B.J., REICH, K.A. Cytotoxicity of hypoxic cell sensitizers in multicell spheroids, Cancer Clin. Trials, 3 (1980) 73.

(11) STRATFORD, I.J., ADAMS, G.E., HORSMAN, M.R. and 4 others, The interaction of misonidazole with radiation, chemotherapeutic agents or heat, Cancer Clin. Trials, 3 (1980) 231.

(12) BLEEHEN, N.M., HONESS, D.J., MORGAN, J.E., Interaction of hyperthermia and the hypoxic cell sensitizer Ro-07-0582 on the EMT6 mouse tumour, Br. J. Cancer, 35 (1977) 299.

(13) SRIDHAR, R., SUTHERLAND, R.M., Hyperthermic potentiation of cytotoxicity of Ro 07-0582 in multicell spheroids, Int. J. Radiat. Oncol. Biol. Phys., 2 (1977) 531.

(14) ROSE, C.M., MILLAR, J.L., PEACOCK, J.H., PHELPS, T.A., STEPHENS, T.C., "Differential enhancement of melphalan cytotoxicity in tumour and normal tissue by misonidazole", in Radiation Sensitizers: Their Use in the Clinical Management of Cancer, BRADY, L.W., Ed., Masson, U.S.A. (1980) 250.

(15) TANNOCK, I.F., In vivo interaction of anti-cancer drugs with misonidazole or metronidazole: cyclophosphamide and BCNU, Br. J. Cancer, 42 (1980) 871.

(16) MORGAN, J.E., BLEEHEN, N.M., "A comparison of the interaction between hyperthermia and bleomycin on the EMT6 tumour as a monolayer or spheroids in vitro and in vivo", in Hyperthermia in Radiation Oncology (Proc. 1st Meeting of European Group, Cambridge, 1979: ARCANGELI, G., MAURO, F., Ed.), Masson, Milano (1980) 165.

(17) MORGAN, J.E., BLEEHEN, N.M., Response of EMT6 multicellular tumour spheroids to hyperthermia and cytotoxic drugs, Br. J. Cancer, 43 (1981) 384.

(18) MORGAN, J.E., BLEEHEN, N.M. Heat sensitivity of EMT6 multicellular tumour spheroids following misonidazole pretreatment, J. Natl. Cancer Inst. (1981) in press.

(19) TWENTYMAN, P.R., The response to chemotherapy of EMT6 spheroids as measured by growth delay and by cell survival, Br. J. Cancer, 42 (1980) 297.

(20) WONG, T.W., WHITMORE, G.F., GULYAS, S., Studies on the toxicity and radiosensitizing ability of misonidazole under conditions of prolonged incubation, Radiat. Res., 75 (1978) 541.

(21) PALCIC, B., SKOV, K.A., SKARSGAARD, L.D., "Effect of reducing agents on misonidazole cytotoxicity", in Radiation Sensitizers: Their Use in the Clinical Management of Cancer, BRADY, L.W., Ed., Masson, U.S.A., (1980) 438.

Progress in Radio-Oncology II, edited by K. H. Kärcher et al., Raven Press, New York © 1982.

Summary Paper

Innovations in Clinical Radiation Therapy Research

L.W. BRADY
Department of Radiation Therapy and Nuclear Medicine,
Hahnemann Medical College,
Philadelphia, Pennsylvania,
United States of America

Abstract

After a brief historical introduction to radiation therapy of cancer, the four subject headings considered at the Second International Meeting on Progress in Radio-Oncology, held in Baden, Austria, in May 1981, are examined, namely high-LET radiations, radiation sensitizers (and protectors), hyperthermia, and 'altered' fractionation regimens. The information presented during the meeting — in the papers, in the ensuing discussions and in the round-table discussions of the main topics — is summarized.

Roentgen described X-rays in his publication of 8 November 1895, and the Curies' their discovery of radium in 1896. Almost immediately the biological effects were recognized and, in January 1896, external beam radiation therapy was used to treat a patient with carcinoma of the breast, with a positive objective response from the radiation therapy. The first patient to have been cured by radiation therapy was reported in 1899; the first patient with cancer of the cervix to be cured by radiation therapy was in 1913, reported by Margaret Cleaves.

Clinical radiation therapy had a long and painful gestation period from 1895 to the early 1920s. Many significant and important advances were made during this period. Becquerel, in 1901, noted an ulceration on his abdominal skin adjacent to a pocket in which he carried a glass tube of radium salt. Voigt in Germany utilized radiation therapy for the treatment of cancer pain with excellent relief. Despeignes, in France, treated a patient with gastric cancer in 1896, and many other areas were treated with favourable response. However, the techniques were inconstant and often non-reproducible.

Technological advances were much more rapid than basic biological knowledge. By 1913, Coolidge had developed an X-ray tube with a peak energy of 140 kV. By 1922, 200 kV X-rays were available for deep therapy. It was at this point that clinical radiation therapy was born, when Regaud, Coutard and Hautant presented evidence at the International Congress of Oncology in Paris that advanced laryngeal cancer could be cured without disastrous treatment-produced sequelae. By 1934, Coutard had developed a protracted fractionated scheme which remains the basis for current radiation therapy.

Following these developments, ionizing radiation was more precisely defined. Treatment planning and delivery became accurate and reproducible. X-ray generators operating at 800 to 1000 kV were installed for medical use as early as 1932 and these were followed by cyclotrons, synchro-cyclotrons, betatrons, bevatrons, linear accelerators, and nuclear reactors. Cobalt-60, caesium-137, iridium-192, iodine-125, supplemented brachytherapy techniques using radium-226.

Today, clinical radiation therapy stands on a firm foundation of basic understanding of ionizing radiation and its effect on tissue, and the biology of that effect in normal tissues and tumour. This explosive growth in knowledge of the physics of radiation therapy, clinical treatment planning, the utilization of computers in radiation therapy, as well as basic information in radiation biology and how it might be implemented in clinical situations, are well known.

The Second International Meeting on Progress in Radio-Oncology has presented major and important applications of basic physics and biologic data and their application in clinical practice. These new techniques, now being implemented in general clinical practice, offer major opportunities toward improving the potential for cure of many cancers not cured in the past.

CLINICAL APPLICATIONS OF FAST NEUTRONS

Within six years of the discovery of the neutron by Chadwick in 1932, fast neutrons were applied by Stone et al. in the treatment of patients with cancer. This programme was carried out with the cyclotron built by Lawrence at the University of California, Berkeley. Radiobiological experiments carried out by Lawrence et al. in 1936, Aebersold in 1939 and Axelrod et al. in 1941 suggested that neutrons might be more effective in destroying tumour than are photons.

The initial results of the treatment programme at the University of California, Berkeley, seemed to permit some optimism, since regression of advanced tumours was observed in all patients. Stone, however, observed that the late effects on normal tissues were such that further work with the neutron beam was not warranted, and therefore its use clinically was abandoned.

Severe late effects on normal tissues were shown later to be due to lack of knowledge. It was not initially realized that dependence of the RBE of neutrons was related to the size of the individual dose fractions. Brennan and Phillips, and subsequently Sheline et al., reviewed the original patient material, confirming that late normal-tissue damage was caused by unacceptably high doses. These reviews, together with data accrued by Fowler and Morgan on reactions of pig skin, and by Barendsen in 1968 on experimental tumour and normal-tissue reactions renewed the clinical interest in fast-neutron therapy.

The results accrued by Catterall working at Hammersmith Hospital with the Medical Research Council Cyclotron, starting in 1968, focused on tumours that were localized, superficial and advanced. These studies, reported in 1975 and 1977, demonstrated an increase in local tumour control from 19% to 76% in favour of neutrons. Even though the arms of the trial were not completely comparable with respect to total dose, morbidity and complications, the results were good enough to suggest that other investigations be carried out by other centres world-wide. At this time (1981), clinical investigations are in progress in some 20 centres in Europe, Japan and the United States of America.

Even though the modern evaluation of fast-neutron therapy dates from the early 1960s and includes several hundred patients, most with advanced malignant tumours, its place in medicine has not been clearly defined. But then optimal uses of conventional photon therapy have not yet been determined despite the treatment of several hundred thousand patients since the discovery of X-rays and natural radioactive sources in 1895-1896.

Mammalian cells in both normal tissues and in tumours respond differently to fast neutrons and to high-energy photons. With fast neutrons, hypoxia exerts a much less protective influence on induced cellular lethality. There is also less variation in post-radiation recovery from sublethal damage and the age of proliferative cells has less influence on radiation-induced damage.

The evidence accumulated to date suggests that local or regional control of epidermoid carcinomas arising in several head and neck structures and of metastatic cervical adenopathy may be greater after neutron therapy than after conventional photon therapy. The data accumulated from the patients with astrocytomas Grade III and IV treated by neutrons demonstrated that there was no improvement in survivals but that the tumour was destroyed in most patients.

These basic radiobiological differences between photon beam therapy and neutron beam therapy suggest more frequent control of some tumours but uncertainty about therapeutic results. The current technology is adequate for the initial biological and clinical investigations.

It was on the basis of these data that the National Cancer Institute in the United States of America placed investigations for neutron beam therapy in a high-priority category. It is the hope that therapeutic advantages will be documented in the research centres and that this will lead to the establishment of regional neutron-therapy centres.

There are compelling reasons to demonstrate the efficacy of neutron beam therapy among the treatment programmes for cancer. The limited data accumulated to date suggest, but do not document, therapeutic advantages in selected situations. There is substantial risk that currently planned studies will be inconclusive because of limited accession of patients in a short period to the few funded research centres in the United States.

Emerging from evaluation of the data to date is the fact that: (i) every neutron beam currently used in the world is a mixed beam; (ii) practically all machines are different with respect to RBE, energy, penumbral characteristics, outputs, etc.; (iii) there is a crying need for better machines of greater flexibility and reliability, for better treatment planning and for better treatment schedules; (iv) there is a need to evaluate the concept of 'shrinking field' during treatment; and (v) there must be a longer follow-up and a proper designation of tumour sites to be treated.

Therefore, this programme of research in the evaluation of neutron beam therapy needs input from a co-ordinated effort on a world-wide basis.

CLINICAL STUDIES OF PION RADIATION THERAPY

Investigations utilizing the pion beam at the University of New Mexico/Los Alamos National Laboratory have been under way now for seven years. The primary objective of the programme relates to the evaluation of negative pi-mesons (pions) in the management of various solid tumours. The various tumour sites selected have been those not managed well by current conventional treatment programmes and where there is a high incidence of local regional recurrence.

Phase II studies have been continued with particular emphasis on treatment of patients with astrocytomas, Grade III-IV, unresectable carcinomas of the pancreas, and Stage III and IVA squamous cell carcinomas of the uterine cervix. Phase III trials are currently under way for squamous cell carcinoma of the oral cavity, nasopharynx, oropharynx and hypopharynx, as well as for inoperable or locally recurrent adenocarcinoma of the rectum and T_3/T_4 transitional carcinomas of the bladder.

Bush et al. have reported the results for 96 patients treated with curative intent between November 1976 and April 1980. In this group, crude survival rates are 14 out of 15 patients (93%) for carcinomas of the prostate, 14/36 (39%) for head and neck cancers, 6/23 (26%) for brain, 0/11 for carcinomas of the pancreas, and 4/12 (33%) for other sites including oesophagus, lung, stomach, urinary bladder, rectum, uterine cervix and skin. Local control rates were 100% for carcinomas of the prostate, 50% for head and neck cancers, 13% for brain primary tumours, 0% for pancreas and 58% for other sites. Local control rates for patients treated with pions alone parallel those for patients treated with pions plus additional conventional radiation therapy and/or surgery. Thirty-three patients have been treated since July 1980, 6 of whom have expired and 27 survived. Five of these patients had carcinomas of the pancreas, with none being disease-free by computed tomography criteria.

The continued emphasis of the University of New Mexico/Los Alamos Laboratory programme is to develop sophisticated techniques for treatment planning, continued accession of patients to the various protocols, continued analysis of acute reactions and their relationship to total dose, fractionation, protraction, hyperfractionation as well as other associated factors, and the continued evaluation of severe chronic effects related to pion radiation therapy alone.

The continued exploration of this technique in cancer management will necessitate close and careful co-operation among all operators of pion facilities in the world in order to be able to include appropriate patients in Phase I/II/III studies. By pooling the data available from all institutions involved in such programmes, answers to the multitude of questions relative to its efficacy will be answered in an appropriate clinical time-frame.

RADIATION SENSITIZERS AND PROTECTORS COMBINED WITH RADIATION THERAPY IN CANCER MANAGEMENT

Radiation sensitizers are agents which can increase the lethal properties of ionizing radiation when administered in conjunction with radiation therapy. They increase radiation sensitivity without being innately toxic and can give rise to significant increases in the radiation sensitivity of neoplasms over normal tissues. Much laboratory work, mostly in vitro, has shown that many chemical compounds can act as radiation sensitizers. The practicality depends upon the exploitation of the differences between normal and malignant cells when radiation sensitizers are used. Many tumours contain hypoxic cells, which are relatively resistant to the lethal effects of ionizing radiation. In human tumours, these hypoxic cells present a potential barrier to successful treatment. The application of chemical radiosensitizers active against hypoxic cells offers great potential for improvement in cancer management using radiation therapy.

Although many tumours are inoperable even in an early localized stage, radiation therapy may have a major impact upon the potential for cure. The use of hypoxic cell radiation sensitizers in conjunction with radiation therapy in these as well as in other more difficult-to-treat tumours may yield a significant improvement in survival. Moreover, if the use of radiosensitizers could reduce the radiation dose required to eradicate a tumour, a large body-area could be safely irradiated. The good results obtained with extended-field radiotherapy in diseases like seminoma or lymphoma could be applicable to other more resistant solid tumours when combined with radiation sensitizers.

Future studies of misonidazole will need to explore a whole range of schedules and use of intravenous as well as oral routes. It has already been shown that the drug retains some of its radiosensitizing capacity when administered with fractionated radiation dosages, thus permitting comparison with conventional, fractionated radiation therapy.

A combination of high-LET radiation and misonidazole has also shown promising results in cultured cells and in transplantable tumours in mice. Such results may afford another way of using radiosensitizers in the clinic. Other possibilities that need to be explored are the combination of these radiation sensitizing agents with hyperthermia or with chemotherapy. Finally, the possibility remains of combining radiosensitizers with radioprotection agents, all of which need to be explored clinically.

In summary, hypoxic cell radiosensitizers offer a promising new treatment modality. Extensive trials with misonidazole are under way, while a search for more effective and less toxic agents has been initiated. Whether hypoxic cell radiosensitizers will yield improvements in the results of radiation therapy of many malignancies will be determined in the future years.

The utilization of radioprotectors offers a unique and exciting prospect of using differential effects in normal tissues versus tumour tissue, since the former tissues show a greater uptake of these sulphydryl-containing compounds than the latter, with a concomitantly lower protective efficiency of these compounds in hypoxic cells. Studies will need to elaborate the dose schedules, the radiation programmes, the routes of administration, as well as their efficacy in various tumour sites, in order to identify the particular area in which they are most useful. It has also been demonstrated that sulphydryl-containing radioprotective compounds are active in protecting cells against the toxic effects of alkylating-agent chemotherapy, and this area needs to be explored in depth.

The utilization of all these agents - radiation therapy, cytotoxic chemotherapy, hypoxic cell sensitizers and radiation protectors - needs to be explored as to the appropriate circumstances for using any combination of them that will result in significant improvements in the therapeutic ratio and in the long-term control of tumours.

HYPERTHERMIA IN CANCER MANAGEMENT

Reports of hyperthermia in medical application have appeared anecdotally throughout the medical literature for hundreds of years. In the late 19th century, numerous reports appeared regarding regression or permanent cure of malignant disease by concurrent erysipelas infection with febrile episodes. Coley developed a bacterial toxin utilized to induce fever, reporting numerous instances of tumour regression or cure as a consequence of the febrile episode induced by Coley's toxin. However, as the toxin became more purified, the magnitude of the febrile episode diminished and less striking results were obtained.

Sporadic interest was maintained in hyperthermia into the early 20th century, but at that point it became discredited. Within the last decade a renewed interest in hyperthermia has led to a proliferation of laboratory and clinical studies. Serious clinical testing has been under way for several years at various medical facilities, utilizing a variety of methods for induction of hyperthermia. Thus, while hyperthermia as a potential modality for cancer therapy has existed for some time, the exact science of hyperthermia is presently in its infancy.

Major problems exist as to the standardization of equipment for achieving the elevations in temperature, problems relative to: (i) thermometry, (ii) homogeneity of the distribution of the heat within the tumour itself, (iii) the appropriate temperature to which the tumour must be taken, (iv) preservation of normal tissue, (v) the selection of tumours to be treated by this technique, and (vi) which tumours of the various histologic types would be more responsive to such a treatment programme.

Most of the data thus far accumulated have been collected by treatment of patients who were preterminal or terminal, and in whom all forms of conventional modality treatment had failed.

The National Cancer Institute in the United States of America has initiated a programme to evaluate the various equipments presently available in an effort to standardize the techniques, to standardize the methods for thermometry and to begin accumulating data in a pilot venture that would be utilized for phase I, phase II and phase III clinical trials.

Basic problems exist as to the technique for measuring the temperature within the tumour. Many presently-used instruments for thermometry interfere with appropriate and precise measurement of temperature.

Various tumours offer particularly unique kinds of situations for consideration of programmes combining hyperthermia and radiation therapy. These are tumours not readily controlled by other treatment modalities, such as oesophageal carcinomas, pancreatic carcinomas, advanced stage gynaecologic malignancies, etc.

Other tumours offer unique opportunities to evaluate the impact of hyperthermia versus hyperthermia plus radiation therapy versus no treatment. These would be patients with mycosis fungoides, patients who have diffuse superficial disease with long anticipated life spans. Not treating a portion of their disease for a period of up to six to eight weeks would not significantly endanger the patient nor would it withdraw the modalities from that patient in the future.

The utilization of hyperthermia in clinical cancer management stands at the threshold of great potential. However, more precise technical and biological data are necessary before it can be widely implemented in general clinical management.

HYPERFRACTIONATION IN RADIATION THERAPY FOR CANCER

The basic aim in clinical experimental radiation therapy is to improve the therapeutic ratio. Multiple attempts have been made to do this utilizing hyperbaric oxygen therapy (at two to three atmospheres pressure) with radiation therapy, split-course techniques, combined chemotherapy and radiation therapy, high-LET radiation programmes, as well as hypoxic cell sensitizers given together with radiation therapy. To date, there has been no clear-cut difference in results or any significant improvement over the traditional methods of conventional and daily-fractionated photon beam radiation therapy.

The foundation on which the clinical practice of radiation therapy in the 1980s is based is the proposal put forth by Coutard and Baclesse in the 1920s concerning standard fractionated daily radiation therapy treatment given over finite periods of time. The empirical observations indicate that curative courses of radiation therapy are possible and that they can be carried out with a minimum toxicity arising from the treatment. The advances that have been made in the last ten years indicate major improvements in the management of many tumours at different sites, including carcinoma of the cervix, Hodgkin's disease, carcinoma of the larynx, carcinoma of the bladder, etc. These accomplishments have been made possible by better understanding of radiation-therapy physics, radiation biology and the availability of high-energy equipment for external beam radiation therapy.

Experimental evidence exists that hyperfractionation can improve the therapeutic ratio and give better normal-tissue tolerance of the treatment programme. Hyperfractionation, however, has never been given an adequate clinical trial.

In the last two years, the Radiation Therapy Oncology Group has conducted a pilot study utilizing hyperfractionation in patients with head and neck cancer. In January 1978, data were presented on 60 patients with malignant tumours of the head and neck and oesophagus who had been accessioned into the study from three institutions. The protocol required that each patient receive 6000 rads in 20 treatment days, with two fractions of 150 rads administered twice daily, separated by four to six hours. An analysis of the results was performed in March 1978. At this time, it became apparent that four different treatment methods had evolved within the several institutions and among the 60 patients. The four treatment plans were:

1. 6000 rads given in 20 days, 40 fractions delivered at 150 rads, two fractions daily;
2. 6000 rads given over six to seven weeks, 40 fractions, with 150 rads given twice daily, but with a planned two-week rest during the course of treatment;
3. 6000 rads given in 25 treatment days, 250 rads daily, with 125 rads for each treatment fraction;
4. 5000 rads given over four to five weeks, 250 rads daily, with 125 rads for each fraction. This was followed after one month by surgery.

The treatment programmes carried out in groups 3 and 4 were based on the potential utilization of surgery following the radiotherapy and the physician's objection to a higher daily dose.

It became apparent, after reviewing the results of normal-tissue tolerance and patient acceptance, that the plan of 125 rads twice daily to either a preoperative level of 5000 rads or a definitive level of 6000 rads in five weeks were suitable arms for a randomized study. On the other hand, the plan of 6000 rads delivered with twice-daily fractions in a shorter period of time seemed to produce severe, intolerable mucositis unless a two-week break intervened.

From this study carried out by the Radiation Therapy Oncology Group emerged the following conclusions:

(a) 6000 rads can be administered over five weeks at a rate of 125 rads twice daily without severe reaction and is acceptable to most patients;

(b) Late sequelae such as necrosis, fibrosis and loss of salivary function are reduced;

(c) The local tumour clearance and control is equal to the conventional photon treatment and may in fact be superior;

(d) After 5000 rads of twice-daily fractions, surgery can be carried out with no increase in complications apparent, and there may be less difficulty with fibrosis and bleeding.

The data from these pilot studies require an intense programme of randomization for clinical trial in order to substantiate the points that have been made based on the pilot study. Data from the University of Texas M.D. Anderson Hospital for patients having advanced head and neck cancer substantiate essentially the same conclusions. Several other published studies have been reviewed, involving higher daily doses delivered in three fractions per day separated by four to five hours.

It is clear that further investigation is needed with these techniques. Major co-operation among various clinical centres would allow for the numbers of patients necessary in order to evaluate the efficacy of each of the techniques and to substantiate the conclusions indicated by the pilot study. This is a programme that can be carried out in almost any institution with the usual array of facilities, resources and manpower. Results from this study may prove to be as effective as many of the other more complicated manipulations being investigated in radiation therapy.

Subject Index